AF834952

A First Graduate Course in Quantum Field Theory

This book is designed as an introductory course in Quantum Field Theory (QFT) for students who have completed two full graduate-level courses in Quantum Mechanics (QM). Quantum Field Theory is the framework that combines quantum mechanics with special relativity, providing a comprehensive description of fundamental forces and particles. It is a crucial part of modern theoretical physics and essential for understanding the behaviour of elementary particles, the dynamics of fields, and the interactions that govern the universe at the smallest scales. This book is structured to gradually introduce the key concepts and mathematical tools necessary for mastering QFT. Each chapter builds on the knowledge acquired in previous sections, ensuring a coherent and comprehensive understanding of the subject. The material presented here helps doctoral students bridge the gap between the traditional quantum mechanics taught at the graduate level and the more advanced concepts and techniques used in quantum field theory.

Key Features:

- Specifically designed to connect traditional quantum mechanics knowledge with quantum field theory.
- Presents a coherent learning path, including end-of-chapter trivia questions and problems, that gradually introduces the concepts and skills needed for grasping QFT.
- Tailored for doctoral-level physics students, ensuring appropriate depth and complexity.

Travis Shane Taylor (born July 24, 1968) is an American scientist, engineer, science fiction writer, and the star of History Channel's The Secret of Skinwalker Ranch, 2019 to present.

A First Graduate Course in Quantum Field Theory

For Doctoral Level Physics

Travis S. Taylor

CRC Press
Taylor & Francis Group
Boca Raton London New York

CRC Press is an imprint of the
Taylor & Francis Group, an **informa** business

Designed cover image: Travis S. Taylor

First edition published 2026
by CRC Press
2385 NW Executive Center Drive, Suite 320, Boca Raton, FL 33431

and by CRC Press
4 Park Square, Milton Park, Abingdon, Oxon, OX14 4RN

CRC Press is an imprint of Taylor & Francis Group, LLC

© 2026 Travis S. Taylor

Reasonable efforts have been made to publish reliable data and information, but the author and publisher cannot assume responsibility for the validity of all materials or the consequences of their use. The authors and publishers have attempted to trace the copyright holders of all material reproduced in this publication and apologize to copyright holders if permission to publish in this form has not been obtained. If any copyright material has not been acknowledged, please write and let us know so we may rectify in any future reprint.

Except as permitted under U.S. Copyright Law, no part of this book may be reprinted, reproduced, transmitted, or utilized in any form by any electronic, mechanical, or other means, now known or hereafter invented, including photocopying, microfilming, and recording, or in any information storage or retrieval system, without written permission from the publishers.

For permission to photocopy or use material electronically from this work, access www.copyright.com or contact the Copyright Clearance Center, Inc. (CCC), 222 Rosewood Drive, Danvers, MA 01923, 978-750-8400. For works that are not available on CCC, please contact mpkbookspermissions@tandf.co.uk

For Product Safety Concerns and Information please contact our EU representative GPSR@taylorandfrancis.com. Taylor & Francis Verlag GmbH, Kaufingerstraße 24, 80331 München, Germany.

Trademark notice: Product or corporate names may be trademarks or registered trademarks and are used only for identification and explanation without intent to infringe.

ISBN: 978-1-041-06435-0 (hbk)
ISBN: 978-1-041-06846-4 (pbk)
ISBN: 978-1-003-63737-0 (ebk)

DOI: 10.1201/9781003637370

Typeset in Nimbus Roman font
by KnowledgeWorks Global Ltd.

Publisher's note: This book has been prepared from camera-ready copy provided by the authors.

Contents

About the Author

Dr. Travis S. Taylor

Scientist, Engineer, Author, Futurist, Television Personality

Dr. Travis Taylor is a licensed Professional Engineer who holds PhDs in Optical Science and Engineering and Aerospace Systems Engineering, as well as Masters degrees in Physics, Mechanical & Aerospace Engineering, and Astronomy, and a Bachelors in Electrical Engineering. He is currently completing his third PhD in Theoretical Physics (expected 2026).

Over the past 35 years, Dr. Taylor has worked on advanced technology programs for the Department of Defense, NASA, and the Intelligence Community. His recent projects include advanced propulsion concepts, quantum teleportation and information systems, space telescopes, space-based beamed energy systems, future combat technologies, and next-generation space launch systems. He has published over 25 papers, four textbooks (including *Introduction to Rocket Science and Engineering*, *Introduction to Laser Science and Engineering*, *Quantum Entanglement Engineering Applications* (coauthored with F.J. Duarte), and *Alien Invasion*), and has authored or coauthored more than 25 bestselling science fiction novels.

From 2019 to 2022, Dr. Taylor served as Chief Scientist for the UAP Task Force. He currently works as Chief Scientist for QuantumFrontier, LLC in Huntsville, Alabama.

Dr. Taylors passion for experimentation and invention has been featured in hit television series including *Rocket City Rednecks* (National Geographic Channel), *3 Scientists Walk Into a Bar* (The Weather Channel), *Life After People*, *The Universe*, *The Tesla Files*, *The UnXplained*, *Ancient Aliens*, and *The Secret of Skinwalker Ranch* on The History Channel. He is or has been a member of the American Institute of Aeronautics and Astronautics, The World Space Drive Consortium, The Directed Energy Professional Society, the Experimental Aircraft Association, the American Physics Society, Optical Society of America, and the International High IQ Society.

He is a championship black belt martial artist, placed sixth in the world for his age group in the 2017 CrossFit Games Compex 2-Man Team Competition, and completed the Marine Corps Marathon in a full-body, bulletproof armor suit of his own invention.

SPECIFIC SPACE PROGRAM HIGHLIGHTS

1. **ACES RED1 (PI)**—Launched aboard SpaceX Falcon 9 and deployed externally on the ISS for 24 months. This 3U CubeSat mission successfully demonstrated COTS FPGA flight computers

and mapped GNSS coverage at LEO, revealing several previously undocumented anomalies.

2. **SWORDS Launch Vehicle Review**—Led the US Armys detailed technical review of the SWORDS launch vehicle, including complete subsystem-level analysis down to individual components.
3. **Pluto-Kuiper Belt Study (PI)**—Principal Investigator for initial mission design involving nuclear electric propulsion for deep space exploration of the outer planets.
4. **US Army SMDC Space Lab Lead**—Directed the lab for over a decade (later known as Quantum Engineering and Space Technologies Lab), leading development and integration of flight hardware, propulsion systems, adapter plates, and quantum-entanglement experiments. Oversaw full spacecraft integration, systems engineering, and multiple spaceflight missions.

Preface

This book is designed as an introductory graduate course in Quantum Field Theory (QFT) for students who have completed two full graduate-level courses in Quantum Mechanics (QM). It is intended for doctoral students at the 700 level in physics. The material presented here aims to bridge the gap between the traditional quantum mechanics taught at the graduate level and the more advanced concepts and techniques used in quantum field theory.

Quantum Field Theory is the framework that combines quantum mechanics with special relativity, providing a comprehensive description of fundamental forces and particles. It is a crucial part of modern theoretical physics and essential for understanding the behavior of elementary particles, the dynamics of fields, and the interactions that govern the universe at the smallest scales.

This book is structured to gradually introduce the key concepts and mathematical tools necessary for mastering QFT. Each chapter builds on the knowledge acquired in previous sections, ensuring a coherent and comprehensive understanding of the subject.

COURSE PREREQUISITES

Before embarking on this course, it is essential that students have a strong foundation in the following areas:

- **Quantum Mechanics:** Familiarity with the principles of quantum mechanics, including wavefunctions, operators, the Schrdinger equation, and perturbation theory. Completion of two full graduate-level courses in QM is a prerequisite.
- **Classical Field Theory:** Basic understanding of classical field theory concepts such as Lagrangian and Hamiltonian formulations, Euler-Lagrange equations, and Noether's theorem.
- **Special Relativity:** Understanding of special relativity, including Lorentz transformations, relativistic kinematics, and the concepts of spacetime and four-vectors.
- **Mathematical Methods:** Proficiency in advanced mathematical methods, including complex analysis, linear algebra, and differential equations.

COURSE STRUCTURE

This book is divided into a series of chapters, each focusing on a specific aspect of QFT:

- **Chapter 1: Comprehensive Review of Graduate Quantum Physics I & II**
- **Chapter 2: Classical Fields and Quantum Mechanics**
- **Chapter 3: Quantization, Path Integrals, Number Operator, and Field Operators**
- **Chapter 4: Symmetries, Indistinguishability, Statistics, Renormalization, and Gauge Theories**
- **Chapter 5: Second Quantization, Fock Space, the Tight Binding Hamiltonian, Two and N Particle Operators, and the Hubbard Model**
- **Chapter 6: Relativistic Quantum Mechanics, the Klein Gordon Equation, Contour Integrals, the Scalar Feynman Propagator, and Green's Functions**
- **Chapter 7: Advanced Quantum Field Theory: Beyond Single Particle Quantum Theory**
- **Chapter 8: Feynman Diagrams in Quantum Field Theory**
- **Chapter 9: The S-matrix**
- **Chapter 10: ϕ^4 Theory and Feynman Diagrams**
- **Chapter 11: Other Scattering Theories**

- **Chapter 12: Path Integrals in Quantum Field Theory**
- **Chapter 13: Topology and Quantum Field Theory**
- **Chapter 14: The Vacuum in Quantum Field Theory**
- **Chapter 15: Summary and Pedagogical Implications of Quantum Field Theory**

HOW TO USE THIS BOOK

The chapters of this book are to be broken into lectures as suggested at each chapter title. Chapter 1 is slightly different in format from the rest of the book in that Questions and Problems appear throughout it for a deep review of Quantum Mechanics I and II as should have been covered at the graduate level as prerequisite study to this course. The end of Chapter 1 has a comprehensive exam covering most of the topics from QM. The following chapters have Questions and Problems at the end of each. Each chapter begins with a theoretical overview, followed by detailed derivations, examples, and exercises. The exercises are designed to reinforce understanding and provide practice in applying the concepts and methods introduced. Solutions are provided to facilitate self-study and ensure that students can check their work and understanding. The material in the course is very demanding and it was the goal to not leave gaps in the information within in order to enable the student to delve as deep into the topic as a First Course on Quantum Field Theory allows.

PEDAGOGICAL APPROACH

The pedagogical approach of this book is to provide a balance between theoretical concepts and practical applications. Each chapter includes detailed explanations, mathematical derivations, and worked examples to reinforce understanding. Additionally, problems and exercises are provided at the end of each chapter to test comprehension and encourage deeper exploration of the material.

LEARNING OUTCOMES

By the end of this course, students should be able to:

- Understand the fundamental principles and mathematical structures of QFT.
- Quantize classical fields and understand the significance of creation and annihilation operators.
- Apply path integral methods to quantize fields and calculate transition amplitudes.
- Use Feynman diagrams to represent and compute interactions in QFT.
- Understand the process and importance of renormalization in QFT.
- Grasp the concepts and implications of gauge invariance and gauge theories.
- Develop a comprehensive understanding of the modern interpretation of particles, waves, and nonlocality.
- Explore advanced topics and current research areas in quantum field theory.

ACKNOWLEDGMENTS

The development of this book has been greatly influenced by the numerous contributions of colleagues, mentors, and students. Special thanks to the University of Alabama in Huntsville and Dr. James Miller, whose support and resources have been invaluable in the creation of this work. Dr. Francisco J. Duarte was also key in the development of the underpinning understanding of the quantum nature of the universe and that the particle wave duality is an outdated concept in physics.

CONCLUSION

This book aims to provide a solid foundation in Quantum Field Theory, equipping students with the knowledge and skills necessary to pursue further research and applications in theoretical physics. It is our hope that this text will serve as a comprehensive and accessible resource for students embarking on their journey into the fascinating world of QFT.

[Travis S. Taylor]

1 Lectures 1, 2, 3, & 4: Comprehensive Review of Graduate Quantum Physics I & II

1.1 INTRODUCTION

Quantum Mechanics fundamentally redefines how we understand and predict physical phenomena at the microscopic level, diverging significantly from classical physics. This chapter introduces the core principles and mathematical tools that form the basis of quantum theory, emphasizing wave-particle duality, the quantum state concept, and the mechanism of wave function collapse. These concepts challenge traditional notions of determinism and locality, replacing them with probabilistic outcomes and non-local interactions. Understanding these principles is crucial for any advanced study in physics, chemistry, and related fields, as they lay the groundwork for exploring more complex quantum phenomena discussed in subsequent chapters.

Wave-particle duality, first introduced through key experiments like Young's double-slit experiment, demonstrates the dual nature of quantum entities. Particles such as electrons and photons exhibit both wave-like and particle-like behaviors, depending on the experimental setup. This duality challenges classical intuition and underscores the complexity of quantum systems.

The concept of quantum states and superposition further distinguishes quantum mechanics. Unlike classical states, which are deterministic, quantum states are described by wave functions or state vectors in a Hilbert space, allowing a system to exist simultaneously in multiple states until measured. This principle is famously illustrated by Schrödingers cat thought experiment and underpins many quantum technologies, including quantum computing.

Measurement in quantum mechanics is not a passive observation but an active process that affects the system's state. The act of measurement causes the wave function to collapse to a single eigenstate, a phenomenon explained by Borns rule. This process introduces intrinsic uncertainties and probabilistic outcomes, diverging from the deterministic predictions of classical mechanics.

We then delve into the mathematical framework of quantum mechanics, focusing on linear algebra and functional analysis. Key concepts include Hilbert spaces, operators, eigenstates, and eigenvalues. Operators such as Hermitian and unitary operators play crucial roles in defining observables and the evolution of quantum states. The time-independent Schrödinger equation is introduced as a fundamental equation governing the dynamics of quantum systems, with applications to simple systems like the particle in a box.

The discussion on quantum harmonic oscillators introduces the creation and annihilation operators, highlighting their role in quantizing the energy levels of the system and explaining the concept of vacuum energy. This formalism is crucial for understanding more complex quantum systems and phenomena such as quantum field theory (QFT) and the quantization of the electromagnetic field.

Finally, we address the quantization of the electromagnetic field and the concept of photons as quanta of the field. Using mode decomposition and field operators, we derive expressions for the quantized field and discuss how this framework leads to an understanding of phenomena such as spontaneous emission and the Casimir effect.

The tools and concepts introduced in this chapter are not only fundamental to quantum mechanics but also serve as essential prerequisites for advancing into the next realm of theoretical physics:

DOI: 10.1201/9781003637370-1

QFT. QFT extends the principles of quantum mechanics to fields, treating particles as excitations of underlying fields. This theory combines quantum mechanics with special relativity, providing a framework to describe the interactions of particles at high energies and in the presence of strong fields. QFT is the foundation for understanding the Standard Model of particle physics, which describes the electromagnetic, weak, and strong nuclear interactions. It also lays the groundwork for exploring complex phenomena such as particle creation and annihilation, spontaneous symmetry breaking, and the unification of fundamental forces.

In summary, this chapter provides a thorough introduction to the key principles and mathematical tools of quantum mechanics, laying the groundwork for further exploration of advanced topics. The concepts of wave-particle duality, superposition, measurement, and the mathematical formalism of quantum mechanics are essential for understanding and applying quantum theory to various physical systems and technological applications. Moving forward, these foundational tools will enable a deeper comprehension of QFT and its profound implications for our understanding of the universe.

1.2 FOUNDATIONAL PRINCIPLES OVERVIEW

The foundational principles of quantum mechanics reshape our understanding of microscopic phenomena. These principles include the quantization of physical properties, the probabilistic nature of quantum states, and the intrinsic uncertainties that arise from measurements. Each principle not only defines the scope and scale of quantum theory but also delineates the limits of classical mechanics, making a clear distinction in how phenomena are analyzed and interpreted at atomic and subatomic levels. This section explores these key principles, beginning with the concept of wave-particle duality, moving through the quantum state and its superposition, and concluding with the process of measurement and the implications of wave function collapse.

1.2.1 WAVE-PARTICLE DUALITY

Wave-particle duality is a cornerstone concept of early quantum mechanics, and arose due to the remarkable experiments and theories of physicists such as Maxwell Planck, Niels Bohr, Werner Heisenberg, Erwin Schrodinger, Paul Dirac, Albert Einstein, Louis de Broglie, and many others. This principle asserts that every particle or quantum entity can exhibit both particle-like and wave-like behavior. The dual nature becomes evident through phenomena such as the famous Young's double-slit experiment as shown in Figure 1.1, where particles like electrons and even light show interference patterns that are characteristic of waves as well as particle-like behavior depending on how the experiment is set up.

The experiment exemplifies quantum wave-particle duality through the interference pattern formed by particles such as photons, electrons, or even larger molecules. The total wave function for a particle passing through two slits can be modeled as the superposition of the individual wave functions from each slit, denoted $\psi_1(x)$ and $\psi_2(x)$:

$$\psi(x) = \psi_1(x) + \psi_2(x) \tag{1.1}$$

Assuming that the slits are narrow and equidistant from the center point on a detection screen, we can express the wave functions emerging from each slit as:

$$\psi_1(x) = Ae^{i(kx-\omega t)}, \quad \psi_2(x) = Ae^{i(kx-\omega t+\phi)} \tag{1.2}$$

where A is the amplitude of the wave, k is the wave number, ω is the angular frequency, and ϕ is the phase difference resulting from the path difference between the two slits. The resulting intensity pattern observed on a screen, which represents the probability density of detecting the particle at a particular position, is given by:

$$I(x) = |\psi(x)|^2 = |\psi_1(x) + \psi_2(x)|^2 = 4A^2 \cos^2\left(\frac{\phi}{2}\right) \tag{1.3}$$

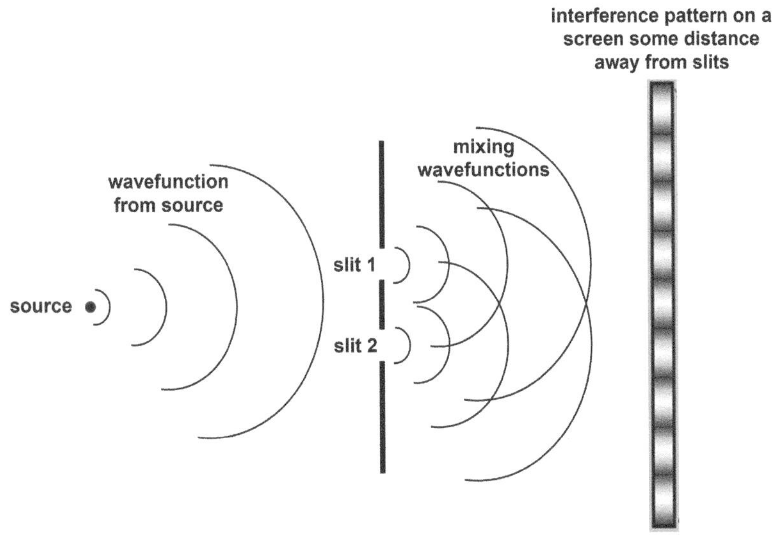

FIGURE 1.1
Young's double-slit experiment showing the incident wavefunction, wavefunctions from each slit, and the interference pattern on the screen.

This intensity equation reveals the characteristic interference fringes, alternating bright and dark bands, which provide direct evidence of the wave nature of quantum entities.

Using Dirac's bra-ket notation, we can elegantly rewrite the wave function for the Young's double-slit experiment. In this formalism, the state of a particle passing through two slits is represented as a superposition of the states corresponding to each slit:

$$|\psi\rangle = |\psi_1\rangle + |\psi_2\rangle \tag{1.4}$$

Each state vector $|\psi_i\rangle$ associated with the i-th slit can be expressed as:

$$|\psi_1\rangle = Ae^{-i\omega t}|k\rangle, \quad |\psi_2\rangle = Ae^{-i\omega t}e^{i\phi}|k\rangle \tag{1.5}$$

where $|k\rangle$ denotes the momentum eigenstate corresponding to the wave number k, and ϕ is the phase difference induced by the path difference between the two slits. The observable intensity pattern on a screen, indicating the probability density of particle detection, can then be calculated using the squared modulus of the superposition state:

$$I(x) = \langle\psi|\psi\rangle = |A|^2\langle k|k\rangle\left(1 + e^{-i\phi} + e^{i\phi} + 1\right) = 4|A|^2\cos^2\left(\frac{\phi}{2}\right) \tag{1.6}$$

This representation highlights the interference pattern in terms of probability amplitudes, emphasizing the fundamental quantum mechanical principle of superposition and its impact on observable phenomena.

Dirac notation, or bra-ket notation, is a powerful and concise way to represent vectors, and operators in quantum mechanics. In this notation, a quantum state is typically represented by a ket, $|\psi\rangle$, which is a vector in a complex vector space known as Hilbert space. The dual vector of $|\psi\rangle$ is represented by a bra, $\langle\psi|$, which is used to denote the conjugate transpose (or Hermitian adjoint) of the ket.

The inner product between two states, say $|\psi\rangle$ and $|\phi\rangle$, is denoted by $\langle\phi|\psi\rangle$. This inner product is a complex number that represents the amplitude for transitioning from state $|\phi\rangle$ to state $|\psi\rangle$. It plays a crucial role in calculating probabilities and expectation values in quantum mechanics.

Conversely, the outer product, represented by $|\psi\rangle\langle\phi|$, denotes an operator that projects any quantum state onto the direction of $|\psi\rangle$ scaled by the projection of the state onto $|\phi\rangle$. This operator is particularly useful in constructing density matrices and observable operators, as it helps in expressing transformations within the Hilbert space. Dirac notation thus elegantly encapsulates the operations of quantum mechanics, streamlining computations, and clarifying the underlying mathematics of quantum states and observables.

This duality challenges the classical intuition that entities are either particles or waves, suggesting a more complex framework where quantum objects possess both sets of properties, dependent on the experimental setup. This concept is crucial for understanding the behavior of atoms and subatomic particles and has profound implications for the development of QFT and the interpretation of quantum phenomena.

1.2.2 QUANTUM STATE AND SUPERPOSITION

The quantum state represents a complete description of a physical system in quantum mechanics, encapsulated by a wave function or a state vector in a Hilbert space. Superposition, a fundamental principle derived from the linear nature of Hilbert space, allows a quantum system to exist simultaneously in multiple states until it is measured. This principle is famously illustrated by Schrödinger's cat thought experiment shown in Figure 1.2, where the cat in a closed box can be both alive and deada superposition of statesuntil it is observed. Superposition underpins many quantum phenomena and technologies, including quantum interference, entanglement, and the vast potential of quantum computing, where qubits can perform multiple calculations at once, vastly increasing computational power compared to classical bits.

1.2.2.0.1 Example: Quantum State and Superposition in Dirac Notation

Consider a quantum system that can be in one of two possible states, $|0\rangle$ and $|1\rangle$. In Dirac notation, these states represent the basis states of the system, analogous to the classical binary states. A

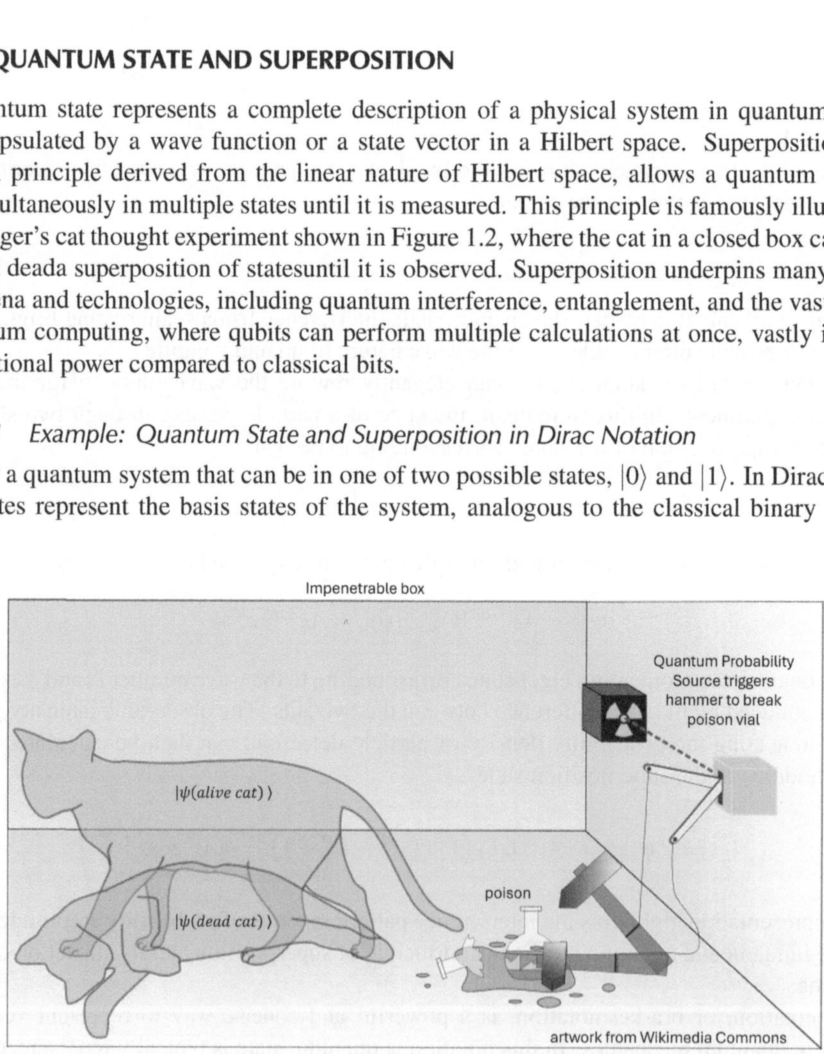

FIGURE 1.2
Schrödinger's cat thought experiment, illustrating superposition.

quantum superposition of these states can be expressed as:

$$|\psi\rangle = \alpha|0\rangle + \beta|1\rangle \tag{1.7}$$

where α and β are complex numbers that satisfy the normalization condition:

$$|\alpha|^2 + |\beta|^2 = 1 \tag{1.8}$$

This condition ensures that the total probability of finding the system in either state $|0\rangle$ or $|1\rangle$ is 1. The coefficients α and β are probability amplitudes, indicating the likelihood of the particle being in each respective state upon measurement.

If we measure this system in the basis of $|0\rangle$ and $|1\rangle$, the probability of finding it in state $|0\rangle$ is $|\alpha|^2$, and in state $|1\rangle$ is $|\beta|^2$. This example showcases how superposition allows a quantum system to exist in a linear combination of multiple states simultaneously, a fundamental departure from classical physics, where a system must be in one definite state at any time.

1.2.3 MEASUREMENT AND WAVE FUNCTION COLLAPSE

Measurement in quantum mechanics is not merely an observational act but a physical process that affects the state of the system being measured. According to the quantum mechanical framework, the act of measurement causes the wave function of a quantum systema superposition of multiple statesto collapse to a single eigenstate. This collapse is probabilistic and is described by Born's rule, which states that the probability of an outcome is given by the square of the amplitude of the wave function's component corresponding to that outcome. The concept of wave function collapse challenges classical notions of determinism and has led to various interpretations of quantum mechanics, including the Copenhagen interpretation and many-worlds theory, each attempting to explain the role of observation in shaping physical reality.

1.2.3.0.1 Example: Measurement and Wave Function Collapse in Young's Double-Slit Experiment

Consider the Young's double-slit experiment where a quantum particle, such as an electron, passes through two slits, denoted as slit A and slit B. Before measurement, the quantum state of the particle can be represented as a superposition of the state where the particle passes through slit A ($|A\rangle$) and the state where it passes through slit B ($|B\rangle$):

$$|\psi\rangle = \frac{1}{\sqrt{2}}(|A\rangle + |B\rangle) \tag{1.9}$$

This state reflects the equal probability amplitude for the particle to pass through either slit, assuming no initial bias toward either path. When the particle interacts with a detection screen, its position is measured, leading to the "collapse" of the wave function. If we observe the particle on the screen, the quantum state collapses to one of two possible stateseither $|A\rangle$ or $|B\rangle$, depending on which slit the particle actually passed through:

$$|\psi\rangle \to |A\rangle \quad \text{or} \quad |\psi\rangle \to |B\rangle \tag{1.10}$$

This collapse is instantaneous upon measurement and is dictated by the probabilities $|\langle A|\psi\rangle|^2$ and $|\langle B|\psi\rangle|^2$, which are both $1/2$ in this symmetric setup. The observed interference pattern on the screen arises from the superposition of $|A\rangle$ and $|B\rangle$ prior to measurement, highlighting the fundamental quantum mechanical interplay between superposition and measurement-induced state reduction.

1.3 MATHEMATICAL FRAMEWORK OF QUANTUM MECHANICS

Quantum mechanics is underpinned by a robust mathematical framework that enables precise descriptions of quantum states and their dynamics. This framework is fundamentally built on the concepts of linear algebra and functional analysis, particularly the theory of Hilbert spaces, which are the abstract vector spaces where quantum states reside. Operators on these spaces, such as Hermitian (or self-adjoint) and unitary operators, play crucial roles in defining observables and the evolution of quantum states, respectively. This section explores the mathematical machinery of quantum mechanics, starting with the properties and significance of various operators, and discusses the mathematical representation of quantum states through eigenstates and eigenvalues, which are essential for solving the Schrödinger equation and for predicting physical outcomes.

1.3.0.0.1 Example: Time-Independent Schrödinger Equation for a Particle in a One-Dimensional Box

Consider a particle of mass m confined in a one-dimensional potential box (or well) of width L with infinitely high walls. This setup is a fundamental problem in quantum mechanics, often used to illustrate basic concepts of quantization and wave function behavior in a confined space.

The time-independent Schrödinger equation for this system is given by:

$$-\frac{\hbar^2}{2m}\frac{d^2\psi(x)}{dx^2}+V(x)\psi(x)=E\psi(x) \tag{1.11}$$

where $\psi(x)$ is the wave function of the particle, $\hbar$ is the reduced Planck's constant, $V(x)$ is the potential energy function, and E is the energy of the particle.

In the case of the infinite potential well, the potential $V(x)$ is zero inside the box $(0<x<L)$ and infinite outside it, which implies that $\psi(x)=0$ for $x\leq 0$ and $x\geq L$. The solution to the Schrödinger equation inside the box can be found by solving:

$$-\frac{\hbar^2}{2m}\frac{d^2\psi(x)}{dx^2}=E\psi(x) \tag{1.12}$$

This differential equation has solutions of the form:

$$\psi(x)=A\sin(kx)+B\cos(kx) \tag{1.13}$$

where $k=\sqrt{\frac{2mE}{\hbar^2}}$ and A, B are constants determined by the boundary conditions. For the particle in a box, the boundary conditions $\psi(0)=0$ and $\psi(L)=0$ lead to the determination that $B=0$ and $kL=n\pi$, where n is a positive integer. The normalized solutions are:

$$\psi_n(x)=\sqrt{\frac{2}{L}}\sin\left(\frac{n\pi x}{L}\right) \tag{1.14}$$

These wave functions represent the stationary states of the particle with quantized energies given by:

$$E_n=\frac{\hbar^2\pi^2n^2}{2mL^2} \tag{1.15}$$

for $n=1,2,3,\ldots$. This example highlights the quantum mechanical effects of confinement and the discretization of energy levels, which are absent in classical physics.

1.3.1 HILBERT SPACE IN QUANTUM MECHANICS

In quantum mechanics, Hilbert space is the abstract vector space that provides the setting for the wave functions that describe the state of quantum systems. It is a complete space with an inner product that allows for the length and angle between vectors to be measured, which are critical for understanding the probabilities and dynamics of quantum states. Hilbert spaces are typically complex, with the inner product providing a way to compute the probability amplitude between states. For finite-dimensional systems, a Hilbert space may be similar to complex Euclidean space, but for infinite-dimensional systems, it includes spaces of square-integrable functions. This generalization is essential for dealing with systems described by continuous variables, such as the position and momentum of particles. The properties of Hilbert spaces ensure that every observable, which corresponds to a self-adjoint operator, has a well-defined spectrum of eigenvalues, each associated with an eigenvector. These eigenvalues represent the possible outcomes of measurements, and the corresponding eigenvectors (or eigenstates) form a basis for the space, reflecting the state's possible configurations post-measurement.

Let $\mathscr{H}$ be a two-dimensional Hilbert space with basis vectors:

$$|0\rangle = \begin{pmatrix} 1 \\ 0 \end{pmatrix}, \quad |1\rangle = \begin{pmatrix} 0 \\ 1 \end{pmatrix} \tag{1.16}$$

Any state $|\psi\rangle$ in this Hilbert space can be written as:

$$|\psi\rangle = \alpha|0\rangle + \beta|1\rangle \tag{1.17}$$

where α and β are complex numbers.

The inner product $\langle\cdot|\cdot\rangle$ on this Hilbert space can be defined as:

$$\langle\psi|\phi\rangle = \psi^\dagger \cdot \phi \tag{1.18}$$

where $\psi^\dagger$ denotes the conjugate transpose of ψ.

For example, the inner product of $|0\rangle$ and $|1\rangle$ is:

$$\langle 0|1\rangle = \begin{pmatrix} 1 & 0 \end{pmatrix} \begin{pmatrix} 0 \\ 1 \end{pmatrix} = 0 \tag{1.19}$$

This result indicates that the basis vectors $|0\rangle$ and $|1\rangle$ are orthogonal, which is a common property in Hilbert spaces.

1.3.2 OPERATORS AND COMMUTATORS

Operators in quantum mechanics represent physical observables such as position, momentum, energy, and angular momentum. These operators act on quantum states, transforming them into other states or providing information about the system's properties. In the language of linear algebra, operators are represented by matrices or linear transformations in the Hilbert space.

One important concept related to operators is the commutator, denoted by $[A,B]$, which measures the non-commutativity of two operators A and B. The commutator is defined as $[A,B] = AB - BA$. If the commutator $[A,B]$ equals zero, it means that the operators commute, and they can be measured simultaneously with arbitrary precision. However, if $[A,B] \neq 0$, it implies that the order of operations matters, and the observables represented by A and B do not have well-defined values simultaneously.

As an example, consider the position and momentum operators in one dimension, denoted by $\hat{x}$ and $\hat{p}$, respectively. Their commutator is given by:

$$[\hat{x}, \hat{p}] = \hat{x}\hat{p} - \hat{p}\hat{x} = i\hbar \tag{1.20}$$

where $\hbar$ is the reduced Planck constant. This non-zero commutator reflects the Heisenberg uncertainty principle, which states that the more precisely the position of a particle is measured, the less precisely its momentum can be known, and vice versa.

1.3.3 HERMITIAN AND UNITARY OPERATORS

In quantum mechanics, operators play a crucial role in representing physical observables and transformations. Two important classes of operators are Hermitian and unitary operators.

A Hermitian operator, denoted by $\hat{H}$, is an operator that is equal to its own adjoint (or Hermitian conjugate). Mathematically, an operator $\hat{H}$ is Hermitian if it satisfies the condition:

$$\hat{H}^\dagger = \hat{H} \tag{1.21}$$

where $\hat{H}^\dagger$ denotes the adjoint of $\hat{H}$.

One of the key properties of Hermitian operators is that their eigenvalues are real. This property is important in quantum mechanics because observables in quantum systems are represented by Hermitian operators, and the eigenvalues of these operators correspond to the possible outcomes of measurements.

A unitary operator, denoted by $\hat{U}$, is an operator that preserves the norm of vectors in the Hilbert space. Mathematically, an operator $\hat{U}$ is unitary if it satisfies the condition:

$$\hat{U}^\dagger \hat{U} = \hat{U}\hat{U}^\dagger = \hat{I} \tag{1.22}$$

where $\hat{I}$ is the identity operator.

Unitary operators represent reversible transformations in quantum mechanics. They preserve the inner product between vectors in the Hilbert space, which ensures that the total probability of finding a quantum system in any state remains conserved.

1.3.4 EIGENSTATES AND EIGENVALUES

Eigenstates and eigenvalues are fundamental concepts in quantum mechanics, providing a way to characterize the state of a quantum system and the outcomes of measurements, respectively. An eigenstate of an operator is a state vector that remains unchanged when the operator acts on it, except for a scaling factor. Mathematically, for an operator $\hat{A}$ and an eigenstate $|\psi\rangle$, the eigenvalue equation is given by:

$$\hat{A}|\psi\rangle = a|\psi\rangle \tag{1.23}$$

where a is the eigenvalue associated with the eigenstate $|\psi\rangle$.

Eigenstates form a complete basis for the Hilbert space of a quantum system, meaning that any state vector can be expressed as a linear combination of eigenstates. This property simplifies the analysis of quantum systems and allows for the prediction of measurement outcomes with ease.

As an example, consider the Hamiltonian operator $\hat{H}$ representing the total energy of a quantum system. The eigenstates of the Hamiltonian operator correspond to the stationary states of the system, and their eigenvalues represent the allowed energy levels. The time-independent Schrödinger equation $\hat{H}|\psi_n\rangle = E_n|\psi_n\rangle$ governs these eigenvalue equations, where $|\psi_n\rangle$ is the n-th eigenstate with eigenvalue E_n. For a particle in a one-dimensional box potential, the eigenstates are given by:

$$\psi_n(x) = \sqrt{\frac{2}{L}} \sin\left(\frac{n\pi x}{L}\right) \tag{1.24}$$

with corresponding eigenvalues:

$$E_n = \frac{\hbar^2 \pi^2 n^2}{2mL^2} \tag{1.25}$$

These eigenstates and eigenvalues provide insights into the quantization of energy levels in quantum systems, which have profound implications for various physical phenomena.

1.3.5 MATRIX OPERATIONS IN QUANTUM MECHANICS

In quantum mechanics, several matrix operations are essential for describing the properties and dynamics of quantum states. These operations include the transpose, adjoint (or Hermitian conjugate), inverse, unitary, Hermitian, symmetric, and nonsymmetric matrices. Here, we will define each operation and provide an example of its application in quantum mechanics.

1.3.5.0.1 *Transpose*

The transpose of a matrix A, denoted as A^T, is obtained by swapping its rows and columns. If A is an $m \times n$ matrix with elements a_{ij}, the transpose A^T is an $n \times m$ matrix with elements a_{ji}.

$$A = \begin{pmatrix} a_{11} & a_{12} \\ a_{21} & a_{22} \end{pmatrix}, \quad A^T = \begin{pmatrix} a_{11} & a_{21} \\ a_{12} & a_{22} \end{pmatrix} \tag{1.26}$$

1.3.5.0.2 *Adjoint (Hermitian Conjugate)*

The adjoint or Hermitian conjugate of a matrix A, denoted as $A^\dagger$, is obtained by taking the transpose of A and then taking the complex conjugate of each element. If A is a matrix with elements a_{ij}, the Hermitian conjugate $A^\dagger$ has elements a_{ji}^*.

$$A = \begin{pmatrix} a_{11} & a_{12} \\ a_{21} & a_{22} \end{pmatrix}, \quad A^\dagger = \begin{pmatrix} a_{11}^* & a_{21}^* \\ a_{12}^* & a_{22}^* \end{pmatrix} \tag{1.27}$$

1.3.5.0.3 *Inverse and Unitary Matrices*

The inverse of a matrix A, denoted as A^{-1}, is the matrix that satisfies $AA^{-1} = A^{-1}A = I$, where I is the identity matrix. A unitary matrix U is one for which the inverse is equal to its Hermitian conjugate, $U^{-1} = U^\dagger$. Unitary matrices represent reversible transformations in quantum mechanics and preserve the norm of quantum states.

$$U = \begin{pmatrix} u_{11} & u_{12} \\ u_{21} & u_{22} \end{pmatrix}, \quad U^\dagger U = UU^\dagger = I \tag{1.28}$$

1.3.5.0.4 *Hermitian Matrices*

A Hermitian matrix H satisfies the condition $H^\dagger = H$. Hermitian matrices have real eigenvalues and are used to represent observables in quantum mechanics.

$$H = \begin{pmatrix} h_{11} & h_{12} \\ h_{21} & h_{22} \end{pmatrix}, \quad H^\dagger = H \tag{1.29}$$

1.3.5.0.5 *Symmetric and Nonsymmetric Matrices*

A symmetric matrix S is a matrix that is equal to its transpose, $S^T = S$. A nonsymmetric matrix does not satisfy this condition.

$$S = \begin{pmatrix} s_{11} & s_{12} \\ s_{12} & s_{22} \end{pmatrix}, \quad S^T = S \tag{1.30}$$

1.3.5.0.6 *Unit Matrix*

The unit matrix, or identity matrix, denoted as I, is a square matrix with ones on the diagonal and zeros elsewhere. It is the multiplicative identity for matrices.

$$I = \begin{pmatrix} 1 & 0 \\ 0 & 1 \end{pmatrix} \tag{1.31}$$

1.3.5.0.7 Singular Matrix

A matrix is singular if it does not have an inverse. This occurs when the determinant of the matrix is zero.

1.3.5.0.8 Determinant

The determinant of a matrix A, denoted as $\det(A)$, is a scalar value that encodes certain properties of the matrix, including whether it is invertible. For a 2x2 matrix:

$$A = \begin{pmatrix} a_{11} & a_{12} \\ a_{21} & a_{22} \end{pmatrix}, \quad \det(A) = a_{11}a_{22} - a_{12}a_{21} \tag{1.32}$$

1.3.5.0.9 Trace

The trace of a matrix A, denoted as $\mathrm{Tr}(A)$, is the sum of its diagonal elements.

$$A = \begin{pmatrix} a_{11} & a_{12} \\ a_{21} & a_{22} \end{pmatrix}, \quad \mathrm{Tr}(A) = a_{11} + a_{22} \tag{1.33}$$

1.3.5.0.10 Example: Pauli Spin Matrices

The Pauli spin matrices are examples of matrices that are both Hermitian and unitary. These matrices are fundamental in quantum mechanics, particularly in the study of spin-1/2 particles. The Pauli matrices are defined as:

$$\hat{\sigma}_x = \begin{pmatrix} 0 & 1 \\ 1 & 0 \end{pmatrix} \tag{1.34}$$

$$\hat{\sigma}_y = \begin{pmatrix} 0 & -i \\ i & 0 \end{pmatrix} \tag{1.35}$$

$$\hat{\sigma}_z = \begin{pmatrix} 1 & 0 \\ 0 & -1 \end{pmatrix} \tag{1.36}$$

These matrices satisfy the conditions for both Hermitian and unitary operators:

$$\hat{\sigma}_x^\dagger = \hat{\sigma}_x, \quad \hat{\sigma}_y^\dagger = \hat{\sigma}_y, \quad \hat{\sigma}_z^\dagger = \hat{\sigma}_z \tag{1.37}$$

WORKED EXAMPLE: HERMITIAN MATRIX ANALYSIS

Consider the matrix

$$A = \begin{pmatrix} 4 & 1 \\ 1 & 3 \end{pmatrix}. \tag{1.38}$$

Verify Hermiticity:

$$A^\dagger = A^T = \begin{pmatrix} 4 & 1 \\ 1 & 3 \end{pmatrix} = A, \tag{1.39}$$

so A is Hermitian.

Find the Eigenvalues: Solve the characteristic equation:

$$\det(A - \lambda I) = \det \begin{pmatrix} 4-\lambda & 1 \\ 1 & 3-\lambda \end{pmatrix} = (4-\lambda)(3-\lambda) - 1 = \lambda^2 - 7\lambda + 11. \tag{1.40}$$

The eigenvalues are the roots of this equation:

$$\lambda_1 = 5, \quad \lambda_2 = 2. \tag{1.41}$$

Find and Normalize the Eigenvectors:
For $\lambda_1 = 5$:

$$(A-5I)\mathbf{v}_1 = \begin{pmatrix} 4-5 & 1 \\ 1 & 3-5 \end{pmatrix}\begin{pmatrix} v_{11} \\ v_{12} \end{pmatrix} = \begin{pmatrix} -1 & 1 \\ 1 & -2 \end{pmatrix}\begin{pmatrix} v_{11} \\ v_{12} \end{pmatrix} = \begin{pmatrix} 0 \\ 0 \end{pmatrix}. \tag{1.42}$$

This yields $-v_{11}+v_{12}=0$, hence $v_{12}=v_{11}$. We can choose $v_{11}=1$, so the eigenvector is $\mathbf{v}_1 = \begin{pmatrix} 1 \\ 1 \end{pmatrix}$.

Normalize $\mathbf{v}_1$:

$$\mathbf{v}_1 = \frac{1}{\sqrt{2}}\begin{pmatrix} 1 \\ 1 \end{pmatrix} \tag{1.43}$$

For $\lambda_2 = 2$:

$$(A-2I)\mathbf{v}_2 = \begin{pmatrix} 4-2 & 1 \\ 1 & 3-2 \end{pmatrix}\begin{pmatrix} v_{21} \\ v_{22} \end{pmatrix} = \begin{pmatrix} 2 & 1 \\ 1 & 1 \end{pmatrix}\begin{pmatrix} v_{21} \\ v_{22} \end{pmatrix} = \begin{pmatrix} 0 \\ 0 \end{pmatrix} \tag{1.44}$$

This yields $2v_{21}+v_{22}=0$, hence $v_{22}=-2v_{21}$. We can choose $v_{21}=1$, so the eigenvector is $\mathbf{v}_2 = \begin{pmatrix} 1 \\ -2 \end{pmatrix}$.

Normalize $\mathbf{v}_2$:

$$\mathbf{v}_2 = \frac{1}{\sqrt{5}}\begin{pmatrix} 1 \\ -2 \end{pmatrix} \tag{1.45}$$

Construct a Unitary Diagonalizing Matrix:
The unitary matrix U that diagonalizes A is constructed from the normalized eigenvectors:

$$U = \begin{pmatrix} \mathbf{v}_1 & \mathbf{v}_2 \end{pmatrix} = \begin{pmatrix} \frac{1}{\sqrt{2}} & \frac{1}{\sqrt{5}} \\ \frac{1}{\sqrt{2}} & -\frac{2}{\sqrt{5}} \end{pmatrix} \tag{1.46}$$

The diagonal matrix D is:

$$D = U^\dagger A U = \begin{pmatrix} 5 & 0 \\ 0 & 2 \end{pmatrix} \tag{1.47}$$

Find the Determinant, Trace, and Adjoint:
The determinant of A is:

$$\det(A) = (4)(3)-(1)(1) = 12-1 = 11 \tag{1.48}$$

The trace of A is:

$$\mathrm{Tr}(A) = 4+3 = 7 \tag{1.49}$$

The adjoint (Hermitian conjugate) of A is:

$$A^\dagger = A = \begin{pmatrix} 4 & 1 \\ 1 & 3 \end{pmatrix} \tag{1.50}$$

This example illustrates the process of verifying that a matrix is Hermitian, finding its eigenvalues and eigenvectors, normalizing the eigenvectors, constructing a unitary matrix that diagonalizes the original matrix, and calculating the determinant, trace, and adjoint of the matrix.

1.4 TRIVIA QUESTIONS

1.4.0.0.1 What is the fundamental difference between classical and quantum physics in terms of state description?

1.4.0.0.2 Answer:

In classical physics, states are described deterministically, where the exact values of physical quantities such as position, momentum, and energy are specified. These classical states follow well-defined trajectories and adhere to deterministic laws of motion, as described by Newtonian mechanics.

In contrast, quantum physics introduces a profound departure from classical mechanics in terms of state description. Quantum states are described probabilistically by wave functions or state vectors in a complex vector space known as Hilbert space. Unlike classical states, quantum states do not have precise values for all physical quantities simultaneously. Instead, they represent a superposition of all possible outcomes, reflecting the inherent uncertainty and wave-like behavior of quantum entities, as governed by principles such as wave-particle duality and the uncertainty principle. This probabilistic nature of quantum states fundamentally distinguishes them from classical states.

1.4.0.0.3 What is the role of the observer in quantum mechanics?

1.4.0.0.4 Answer:

In quantum mechanics, the role of the observer is central to the interpretation of physical reality. According to the Copenhagen interpretation, the act of observation or measurement plays a crucial role in determining the state of a quantum system. Prior to measurement, a quantum system exists in a superposition of multiple states, representing all possible outcomes simultaneously. However, upon measurement, the wave function of the system collapses to a single eigenstate corresponding to the observed outcome.

This collapse of the wave function occurs instantaneously upon measurement and is probabilistic in nature, governed by Born's rule. The observer's interaction with the quantum system is essential for determining the outcome of a measurement and thereby shaping the observed reality. This observer-dependent nature of quantum mechanics underscores the intimate connection between the observer and the observed system, challenging classical notions of objective reality.

1.4.0.0.5 How does the uncertainty principle constrain our ability to predict the behavior of quantum systems?

1.4.0.0.6 Answer:

The uncertainty principle, formulated by Werner Heisenberg, imposes fundamental limitations on our ability to simultaneously measure certain pairs of complementary observables, such as position and momentum, with arbitrary precision. Mathematically, the uncertainty principle states that the product of the uncertainties in the measurements of two non-commuting observables, denoted by Δx and Δp, is bounded by a minimum value:

$$\Delta x \cdot \Delta p \geq \frac{\hbar}{2} \tag{1.51}$$

This principle implies that the more precisely we know the position of a particle, the less precisely we can know its momentum, and vice versa. Consequently, the uncertainty principle places inherent limitations on our ability to predict the behavior of quantum systems with certainty. It highlights the inherently probabilistic nature of quantum mechanics and underscores the unavoidable role of uncertainty in describing the microscopic world.

1.4.0.0.7 What property does a matrix A possess if it is both Hermitian and unitary?

1.4.0.0.8 Answer:

If a matrix A is both Hermitian and unitary, then A is an identity matrix or a matrix whose eigenvalues are ± 1. This is because for a Hermitian matrix A, the eigenvalues are real, and for a unitary matrix A, the eigenvalues lie on the unit circle in the complex plane. The only real numbers on the unit circle are ± 1.

1.4.0.0.9 Define a singular matrix and give an example.

1.4.0.0.10 Answer:

A singular matrix is a square matrix that does not have an inverse, which occurs if and only if its determinant is zero. An example of a singular matrix is:

$$A = \begin{pmatrix} 1 & 2 \\ 2 & 4 \end{pmatrix},$$

since $\det(A) = 1 \cdot 4 - 2 \cdot 2 = 4 - 4 = 0$.

REVIEW PROBLEM: NORMALIZING A QUANTUM STATE

Normalize the state given by $\psi(x) = Ae^{-\lambda x^2}$.

Solution: Ensure the normalization condition:

$$\int_{-\infty}^{\infty} |\psi(x)|^2 \, dx = 1 \tag{1.52}$$

$$A^2 \int_{-\infty}^{\infty} e^{-2\lambda x^2} \, dx = 1 \tag{1.53}$$

$$A^2 \sqrt{\frac{\pi}{2\lambda}} = 1 \tag{1.54}$$

$$A = \sqrt{\frac{2\lambda}{\pi}} \tag{1.55}$$

REVIEW PROBLEM: VERIFYING HERMITIAN, FINDING EIGENVALUES AND EIGENVECTORS

1.4.0.0.11 Problem:

Consider the matrix

$$A = \begin{pmatrix} 2 & 2i \\ -2i & 3 \end{pmatrix}.$$

Verify that A is Hermitian, find its eigenvalues, determine and normalize the eigenvectors, and construct a unitary matrix that diagonalizes A.

1.4.0.0.12 Solution:

Verify Hermiticity:

The adjoint (or Hermitian conjugate) of A is

$$A^\dagger = \begin{pmatrix} 2 & -2i \\ 2i & 3 \end{pmatrix}. \tag{1.56}$$

Since $A = A^\dagger$, the matrix A is Hermitian.

Find the Eigenvalues:
Solve the characteristic equation:

$$\det(A-\lambda I)=\det\begin{pmatrix}2-\lambda & 2i\\ -2i & 3-\lambda\end{pmatrix}=(2-\lambda)(3-\lambda)-(2i)(-2i). \tag{1.57}$$

$$(2-\lambda)(3-\lambda)+4=\lambda^2-5\lambda+10. \tag{1.58}$$

The eigenvalues are the roots of this equation:

$$\lambda_1=\frac{5+\sqrt{-15}}{2}=\frac{5+i\sqrt{15}}{2},\quad \lambda_2=\frac{5-\sqrt{-15}}{2}=\frac{5-i\sqrt{15}}{2}. \tag{1.59}$$

Find and Normalize the Eigenvectors:
For $\lambda_1=\frac{5+i\sqrt{15}}{2}$:

$$(A-\lambda_1 I)\mathbf{v}_1=\begin{pmatrix}2-\lambda_1 & 2i\\ -2i & 3-\lambda_1\end{pmatrix}\begin{pmatrix}v_{11}\\ v_{12}\end{pmatrix}=\begin{pmatrix}0\\ 0\end{pmatrix}. \tag{1.60}$$

Solve the system to find the eigenvector $\mathbf{v}_1$.
Normalize $\mathbf{v}_1$:

$$\mathbf{v}_1=\frac{1}{\sqrt{|v_{11}|^2+|v_{12}|^2}}\begin{pmatrix}v_{11}\\ v_{12}\end{pmatrix}. \tag{1.61}$$

For $\lambda_2=\frac{5-i\sqrt{15}}{2}$:

$$(A-\lambda_2 I)\mathbf{v}_2=\begin{pmatrix}2-\lambda_2 & 2i\\ -2i & 3-\lambda_2\end{pmatrix}\begin{pmatrix}v_{21}\\ v_{22}\end{pmatrix}=\begin{pmatrix}0\\ 0\end{pmatrix}. \tag{1.62}$$

Solve the system to find the eigenvector $\mathbf{v}_2$.
Normalize $\mathbf{v}_2$:

$$\mathbf{v}_2=\frac{1}{\sqrt{|v_{21}|^2+|v_{22}|^2}}\begin{pmatrix}v_{21}\\ v_{22}\end{pmatrix}. \tag{1.63}$$

Construct a Unitary Diagonalizing Matrix:
The unitary matrix U that diagonalizes A is constructed from the normalized eigenvectors:

$$U=\begin{pmatrix}\mathbf{v}_1 & \mathbf{v}_2\end{pmatrix}. \tag{1.64}$$

The diagonal matrix D is:

$$D=U^\dagger AU. \tag{1.65}$$

Find the Determinant, Trace, and Adjoint:
The determinant of A is:

$$\det(A)=(2)(3)-(2i)(-2i)=6+4=10. \tag{1.66}$$

The trace of A is:

$$\mathrm{Tr}(A)=2+3=5. \tag{1.67}$$

The adjoint (Hermitian conjugate) of A is:

$$A^\dagger=A=\begin{pmatrix}2 & 2i\\ -2i & 3\end{pmatrix}. \tag{1.68}$$

1.5 QUANTUM DYNAMICS

Quantum dynamics forms the backbone of quantum mechanics, providing a framework for understanding the evolution of quantum systems over time. At its core lies the time-dependent Schrödinger equation, a fundamental equation that describes how the quantum state of a system changes with time. This equation not only allows us to calculate the time evolution of quantum states but also provides insights into the probabilistic nature of quantum phenomena and the behavior of quantum particles in various physical systems. In this section, we delve into the time-dependent Schrödinger equation, exploring its formal solutions, implications, and the concept of the time-evolution operator, which plays a pivotal role in describing the dynamical evolution of quantum systems.

1.5.1 TIME-DEPENDENT SCHRÖDINGER EQUATION

The time-independent Schrödinger equation is a fundamental equation in quantum mechanics that governs the behavior of quantum systems in stationary states, where the system's properties do not change with time. Mathematically, it is expressed as:

$$\hat{H}\psi(\mathbf{r}) = E\psi(\mathbf{r}) \tag{1.69}$$

where $\hat{H}$ is the Hamiltonian operator representing the total energy of the system, $\psi(\mathbf{r})$ is the wave function describing the state of the system as a function of position $\mathbf{r}$, and E is the corresponding energy eigenvalue. This equation states that the action of the Hamiltonian operator on the wave function yields a scalar multiple of the wave function itself, representing an eigenvalue equation. Solving the time-independent Schrödinger equation allows us to determine the possible energy eigenvalues and corresponding wave functions for a given quantum system, providing valuable insights into its quantum mechanical behavior.

1.5.2 FORMAL SOLUTIONS AND IMPLICATIONS

The time-dependent Schrödinger equation can be solved analytically or numerically to obtain formal solutions, which describe the time evolution of quantum states under various conditions. These solutions provide valuable insights into the behavior of quantum systems, revealing phenomena such as quantum tunneling, wave packet spreading, and interference effects. Furthermore, the formal solutions of the time-dependent Schrödinger equation elucidate the probabilistic nature of quantum mechanics, where the evolution of a quantum state is governed by complex probability amplitudes. Understanding the formal solutions of the Schrödinger equation is essential for predicting and interpreting the dynamics of quantum systems in diverse physical scenarios.

EXAMPLE: FREE PARTICLE WAVE FUNCTION

Consider a free particle with no external potential acting on it. In this case, the time-dependent Schrödinger equation simplifies to:

$$i\hbar\frac{\partial\psi(x,t)}{\partial t} = -\frac{\hbar^2}{2m}\frac{\partial^2\psi(x,t)}{\partial x^2} \tag{1.70}$$

This partial differential equation can be solved using separation of variables. Let's assume a solution of the form:

$$\psi(x,t) = Ae^{i(kx-\omega t)} \tag{1.71}$$

Substituting this into the Schrödinger equation yields:

$$i\hbar\left(-i\omega Ae^{i(kx-\omega t)}\right) = -\frac{\hbar^2}{2m}\left(-k^2Ae^{i(kx-\omega t)}\right) \tag{1.72}$$

Simplifying, we get:

$$\hbar\omega = \frac{\hbar^2 k^2}{2m} \tag{1.73}$$

which implies:

$$\omega = \frac{\hbar k^2}{2m} \tag{1.74}$$

This gives us the dispersion relation for a free particle. The general solution is a superposition of plane waves:

$$\psi(x,t) = \int_{-\infty}^{\infty} A(k) e^{i(kx-\omega t)} dk \tag{1.75}$$

This solution demonstrates the wave-like nature of quantum particles and highlights phenomena such as wave packet spreading and dispersion.

1.5.3 TIME-EVOLUTION OPERATOR

The time-evolution operator, denoted by $\hat{U}(t)$, is a fundamental concept in quantum mechanics that describes how quantum states evolve over time. It is derived from the time-dependent Schrödinger equation and plays a central role in quantum dynamics. Mathematically, the time-evolution operator is defined as $\hat{U}(t) = e^{-i\hat{H}t/\hbar}$, where $\hat{H}$ is the Hamiltonian operator of the system and t is the elapsed time. The action of the time-evolution operator on an initial quantum state $\psi(t_0)$ yields the evolved state $\psi(t)$ at a later time t, such that $\psi(t) = \hat{U}(t-t_0)\psi(t_0)$. The time-evolution operator preserves the inner product and unitarity of quantum states, ensuring the conservation of probability and the validity of quantum mechanical principles over time.

Example: Time Evolution of a Quantum State

Consider a quantum state $\psi(t)$ evolving under a time-independent Hamiltonian $\hat{H}$. The time-evolution operator $\hat{U}(t)$, defined as $\hat{U}(t) = e^{-i\hat{H}t/\hbar}$, governs the evolution of this state. If the initial state $\psi(0)$ is an eigenstate of the Hamiltonian with energy eigenvalue E_n. Then, the evolved state $\psi(t)$ at time t can be calculated as:

$$\psi(t) = \hat{U}(t)\psi(0) = e^{-iE_n t/\hbar}\psi(0) \tag{1.76}$$

This expression shows how the time-evolution operator acts on an initial state to produce the evolved state at a later time. It highlights the unitary nature of time evolution in quantum mechanics and demonstrates how quantum states evolve dynamically under the influence of the Hamiltonian operator.

1.6 TRIVIA QUESTIONS

1.6.0.0.1 What is the significance of the time-evolution operator in quantum mechanics?

1.6.0.0.2 Answer:

The time-evolution operator, denoted as $\hat{U}(t)$, plays a fundamental role in quantum mechanics as it describes how quantum states evolve over time according to the system's Hamiltonian. It ensures that the evolution of quantum systems is unitary, preserving the norm of the state vector and the probabilities associated with different outcomes of measurements.

1.6.0.0.3 How does the concept of dispersion relation relate to the behavior of free particles in quantum mechanics?

1.6.0.0.4 Answer:

The dispersion relation, which relates the energy ω of a quantum particle to its momentum k, governs the wave-like behavior of free particles. In the context of the time-dependent Schrödinger equation, it determines the form of wave functions representing free particles and influences phenomena such as wave packet spreading and particle localization.

REVIEW PROBLEM

Consider a quantum system described by the time-dependent Schrödinger equation:

$$i\hbar\frac{\partial}{\partial t}|\psi(t)\rangle = \hat{H}|\psi(t)\rangle \tag{1.77}$$

where $\hat{H}$ is the Hamiltonian operator.

1. **Formal Solutions**: Find the formal solution for the time-dependent Schrödinger equation using the time-evolution operator $\hat{U}(t)$. Express the solution in terms of the initial state $|\psi(0)\rangle$ and the Hamiltonian operator $\hat{H}$.
2. **Implications**: Discuss the implications of the formal solution obtained in part (1) for the time evolution of quantum states. How does it relate to the principles of unitarity and conservation of probability?
3. **Time-Evolution Operator**: Calculate the time-evolution operator $\hat{U}(t)$ explicitly for a one-dimensional harmonic oscillator described by the Hamiltonian

$$\hat{H} = \frac{\hat{p}^2}{2m} + \frac{1}{2}m\omega^2\hat{x}^2 \tag{1.78}$$

where $\hat{p}$ is the momentum operator and $\hat{x}$ is the position operator.
4. **Time Evolution of Gaussian Wave Packet**: Consider a Gaussian wave packet initially centered at $x = x_0$ with momentum $p = p_0$:

$$\psi(x,0) = \frac{1}{\sqrt{\sigma\sqrt{\pi}}}e^{-\frac{(x-x_0)^2}{4\sigma^2}+i\frac{p_0}{\hbar}(x-x_0)} \tag{1.79}$$

Determine the time evolution of this wave packet by applying the time-evolution operator $\hat{U}(t)$ obtained in part (3). Discuss any notable features of the evolving wave packet as time progresses.

SOLUTION

1. **Formal Solutions**:
The formal solution to the time-dependent Schrödinger equation is given by the time-evolution operator $\hat{U}(t)$ acting on the initial state $|\psi(0)\rangle$:

$$|\psi(t)\rangle = \hat{U}(t)|\psi(0)\rangle \tag{1.80}$$

where $\hat{U}(t) = e^{-\frac{i}{\hbar}\hat{H}t}$ is the time-evolution operator. Therefore,

$$|\psi(t)\rangle = e^{-\frac{i}{\hbar}\hat{H}t}|\psi(0)\rangle \tag{1.81}$$

2. **Implications**:
The formal solution implies that the time evolution of quantum states is governed by the unitary operator $\hat{U}(t)$, which ensures that the norm (probability) of the state vector is preserved over time. This is consistent with the principles of unitarity and conservation of probability in quantum mechanics. This property is reflected in the normalization condition of the quantum state, which states that the squared magnitude of the wavefunction integrated over all space must equal 1. In the context of the time-evolution operator $\hat{U}(t,t_0)$, which describes the time evolution of quantum states from an initial time t_0 to a later time t, this normalization condition translates into the requirement that the operator preserves the normalization of states over time. Mathematically, this can be expressed as:
$$\int_{-\infty}^{\infty} \langle \psi(t) | \psi(t) \rangle \, dx = 1 \tag{1.82}$$
where $\psi(t)$ represents the quantum state at time t. Since the inner product $\langle \psi(t) | \psi(t) \rangle$ gives the probability density of finding the system in state $\psi(t)$, the integral over all space ensures that the total probability of finding the system in any possible state remains constant over time.
The unitarity of the time-evolution operator ensures that this normalization condition is satisfied. Specifically, the unitarity condition states that the operator $\hat{U}(t,t_0)$ is unitary if its adjoint (or conjugate transpose) is equal to its inverse:
$$\hat{U}^\dagger(t,t_0)\hat{U}(t,t_0) = \hat{U}(t,t_0)\hat{U}^\dagger(t,t_0) = \hat{I} \tag{1.83}$$
where $\hat{I}$ is the identity operator. This implies that the time-evolution operator preserves the inner product of states, ensuring that the normalization of states is conserved over time.
3. **Time-Evolution Operator**:
For the one-dimensional harmonic oscillator Hamiltonian $\hat{H} = \frac{\hat{p}^2}{2m} + \frac{1}{2}m\omega^2\hat{x}^2$, the time-evolution operator is given by:
$$\hat{U}(t) = e^{-\frac{i}{\hbar}\hat{H}t} = e^{-\frac{i}{\hbar}\left(\frac{\hat{p}^2}{2m} + \frac{1}{2}m\omega^2\hat{x}^2\right)t} \tag{1.84}$$
4. **Time Evolution of Gaussian Wave Packet**:
To find the time evolution of the Gaussian wave packet, we apply the time-evolution operator $\hat{U}(t)$ to the initial wave packet $\psi(x,0)$. The evolved wave packet is given by:
$$\psi(x,t) = \hat{U}(t)\psi(x,0) \tag{1.85}$$
where $\psi(x,0)$ is the given initial Gaussian wave packet. As time progresses, the Gaussian wave packet will spread out in position space due to the uncertainty principle, and its phase will evolve according to the Hamiltonian dynamics of the system.

1.7 QUANTUM STATES AND OPERATORS

In quantum mechanics, states represent the possible configurations a system can occupy, while operators are mathematical objects that act on these states to produce observable quantities. One fundamental example of an operator is the Hamiltonian, which corresponds to the total energy of a system and governs its time evolution. Another important class of operators is those associated with observables, such as position, momentum, and angular momentum, which have corresponding quantum mechanical operators. These operators have eigenstates and eigenvalues, representing the possible outcomes of measurements on the system. For instance, the angular momentum operator has eigenstates corresponding to different values of angular momentum, and measuring the angular momentum of a particle yields one of these eigenvalues. Additionally, quantum mechanics introduces the concept of identical particles, such as electrons, which obey symmetrization principles dictated by their inherent quantum nature. Fermions, which include electrons, follow the

Pauli exclusion principle, while bosons, like photons, do not, leading to distinct behaviors in systems composed of these particles. Understanding the properties and behavior of quantum states and operators is essential for analyzing and predicting the outcomes of physical experiments and phenomena at the quantum level.

1.7.1 SPIN AND ANGULAR MOMENTUM

The concept of spin and angular momentum plays a crucial role in quantum mechanics, especially in describing the intrinsic properties of particles and their behavior in electromagnetic fields. Spin is an intrinsic form of angular momentum that is inherent to elementary particles, such as electrons, photons, and neutrinos. In quantum mechanics, spin is quantized, meaning it can only take on discrete values determined by fundamental constants. Spin operators and eigenstates represent the observable quantities associated with spin angular momentum. For example, the spin operator $\hat{S}$ acts on spin states, yielding eigenvalues corresponding to the possible outcomes of spin measurements. Angular momentum quantization arises from the quantization of rotational motion in quantum systems. Angular momentum operators, such as the orbital angular momentum operator $\hat{L}$ and the total angular momentum operator $\hat{J}$, have corresponding eigenstates and eigenvalues that describe the quantized angular momentum of particles. Addition of angular momenta deals with combining angular momenta of multiple particles or subsystems. The total angular momentum operator $\hat{J}$ describes the total angular momentum of a system composed of individual particles, and its eigenstates reflect the possible combinations of angular momentum states for the constituent particles. Understanding spin and angular momentum in quantum mechanics is essential for describing the behavior of particles in various physical systems and phenomena.

1.7.2 SPIN OPERATORS AND EIGENSTATES

The spin operator $\hat{\mathbf{S}}$ for a spin-1/2 particle can be expressed in terms of the Pauli spin matrices σ as follows:

$$\hat{\mathbf{S}} = \frac{\hbar}{2}\sigma \tag{1.86}$$

where $\hbar$ is the reduced Planck constant. The Pauli spin matrices are given by:

$$\sigma_x = \begin{pmatrix} 0 & 1 \\ 1 & 0 \end{pmatrix} \tag{1.87}$$

$$\sigma_y = \begin{pmatrix} 0 & -i \\ i & 0 \end{pmatrix} \tag{1.88}$$

$$\sigma_z = \begin{pmatrix} 1 & 0 \\ 0 & -1 \end{pmatrix} \tag{1.89}$$

Each component of σ corresponds to a Pauli spin matrix representing the spin angular momentum operators in the x, y, and z directions, respectively, for a spin-1/2 particle.

The spin operator $\hat{\mathbf{S}}$ for a spin-1/2 particle can be expressed in terms of the Pauli spin matrices σ as follows:

$$\hat{\mathbf{S}} = \frac{\hbar}{2}\sigma \tag{1.90}$$

where $\hbar$ is the reduced Planck constant. The eigenstates of $\hat{\mathbf{S}}$ are the spin-up $|+\rangle$ and spin-down $|-\rangle$ states, which satisfy the following relationships:

$$\hat{\mathbf{S}}\cdot\mathbf{u}|+\rangle = \frac{\hbar}{2}|+\rangle \tag{1.91}$$

$$\hat{\mathbf{S}}\cdot\mathbf{u}|-\rangle = -\frac{\hbar}{2}|-\rangle \tag{1.92}$$

where $\mathbf{u}$ is a unit vector along the specified direction.

1.7.3 ANGULAR MOMENTUM QUANTIZATION

The total angular momentum operator $\hat{J}$ is given by the sum of the orbital angular momentum $\hat{L}$ and the intrinsic spin angular momentum $\hat{S}$:

$$\hat{J} = \hat{L} + \hat{S} \tag{1.93}$$

In an atomic system, the eigenvalues of $\hat{J}^2$ and $\hat{J}_z$ are given by:

$$\hat{J}^2|j,m_j\rangle = \hbar^2 j(j+1)|j,m_j\rangle \tag{1.94}$$

$$\hat{J}_z|j,m_j\rangle = \hbar m_j|j,m_j\rangle \tag{1.95}$$

1.7.4 ADDITION OF ANGULAR MOMENTA

When combining angular momenta of two particles with individual angular momentum quantum numbers j_1 and j_2, the total angular momentum quantum number j can take on values ranging from $|j_1 - j_2|$ to $|j_1 + j_2|$. The corresponding eigenstates of the total angular momentum operator $\hat{J}$ are given by:

$$|j,m_j\rangle = \sum_{m_1,m_2} C^{jm_j}_{m_1m_2}|j_1,m_1\rangle \otimes |j_2,m_2\rangle \tag{1.96}$$

where $C^{jm_j}_{m_1m_2}$ is the Clebsch-Gordan coefficients.

1.7.5 IDENTICAL PARTICLES

In quantum mechanics, particles such as electrons, protons, and neutrons are classified as identical particles, meaning that they are indistinguishable from one another. This leads to unique behavior governed by the symmetrization postulate, which dictates how the wave function of a system of identical particles must behave. The symmetrization postulate also gives rise to the classification of particles into two categories: fermions and bosons, each obeying distinct statistical properties. Furthermore, the Pauli exclusion principle, a consequence of the symmetrization postulate, imposes restrictions on the occupation of quantum states by identical fermions, leading to the stability of matter and the formation of distinct energy levels in atomic systems.

1.7.6 SYMMETRIZATION POSTULATE

The symmetrization postulate states that the wave function describing a system of identical particles must be either symmetric or antisymmetric under the exchange of particle coordinates. For a system of N identical particles, the symmetrized wave function Ψ_S for bosons and the antisymmetrized wave function Ψ_A for fermions can be written as:

$$\Psi_S(\mathbf{r}_1, \mathbf{r}_2, \ldots, \mathbf{r}_N) = \frac{1}{\sqrt{N!}} \sum_P \psi(\mathbf{r}_{P(1)}, \mathbf{r}_{P(2)}, \ldots, \mathbf{r}_{P(N)}) \tag{1.97}$$

$$\Psi_A(\mathbf{r}_1, \mathbf{r}_2, \ldots, \mathbf{r}_N) = \frac{1}{\sqrt{N!}} \sum_P (-1)^P \psi(\mathbf{r}_{P(1)}, \mathbf{r}_{P(2)}, \ldots, \mathbf{r}_{P(N)}) \tag{1.98}$$

where ψ is the single-particle wave function, $\mathbf{r}_i$ represents the position of the i-th particle, and the sum runs over all permutations P of the particle coordinates.

1.7.7 FERMIONS AND BOSONS

Fermions are particles that obey Fermi-Dirac statistics, characterized by antisymmetric wave functions under particle exchange. Examples of fermions include electrons, protons, and neutrons. On the other hand, bosons follow Bose-Einstein statistics and have symmetric wave functions. Photons, gluons, and helium-4 nuclei are examples of bosons. The statistical behavior of fermions and bosons has profound implications for the properties of matter and the behavior of quantum systems.

1.7.8 PAULI EXCLUSION PRINCIPLE

The Pauli exclusion principle states that no two identical fermions can occupy the same quantum state simultaneously. This principle, a direct consequence of the antisymmetry requirement for fermionic wave functions, prohibits the formation of degenerate energy levels in atomic systems and is responsible for the stability of matter. As a result, electrons in atoms occupy distinct energy levels, leading to the formation of the periodic table and the diverse chemical properties of elements.

Example Problem: Total Energy of Fermions and Bosons in a One-Dimensional Box

Consider a system of seven non-interacting fermionic or bosonic particles, each of mass m, confined in a one-dimensional box of length L. We will calculate the total energy for both cases under the assumption that the particles do not interact.

1.7.8.0.1 Fermions:

For fermions, which obey the Pauli exclusion principle, each quantum state can be occupied by two particles with opposite spins. The energy levels are quantized as follows:

$$E_n = \frac{\hbar^2\pi^2 n^2}{2mL^2}, \quad n = 1,2,3,\ldots \tag{1.99}$$

Since we have seven particles, we fill the states from the lowest up, accommodating two particles per state until all seven are placed. Thus, the first three energy levels will be filled as follows: two particles in each of the first three states (1, 2, and 3), and one particle in the fourth state (4). The total energy $E_{\text{total}}^{\text{fermions}}$ is therefore:

$$E_{\text{total}}^{\text{fermions}} = 2\left(\frac{\hbar^2\pi^2}{2mL^2}\right)(1^2+2^2+3^2) + \frac{\hbar^2\pi^2}{2mL^2}(4^2) = \frac{\hbar^2\pi^2}{2mL^2}(2(1+4+9)+16) \tag{1.100}$$

$$E_{\text{total}}^{\text{fermions}} = \frac{\hbar^2\pi^2}{2mL^2}\cdot 42 \tag{1.101}$$

1.7.8.0.2 Bosons:

For bosons, which do not obey the Pauli exclusion principle, all particles can occupy the same quantum state, typically the lowest energy state. Thus, all seven bosons can be in the ground state $n = 1$, with each having energy:

$$E_1 = \frac{\hbar^2\pi^2}{2mL^2} \tag{1.102}$$

The total energy $E_{\text{total}}^{\text{bosons}}$ for the seven bosons is:

$$E_{\text{total}}^{\text{bosons}} = 7\times\frac{\hbar^2\pi^2}{2mL^2} = \frac{7\hbar^2\pi^2}{2mL^2} \tag{1.103}$$

This example underscores the differences in energy configurations between fermions and bosons due to their fundamental quantum statistical nature. While fermions spread out across energy levels due to the Pauli exclusion principle, resulting in a higher total energy, bosons can all condense into the lowest energy state, minimizing their total energy.

1.8 TRIVIA QUESTIONS

1.8.0.0.1 Describe the significance of spin in quantum mechanics.

1.8.0.0.2 Answer:

Spin is a fundamental quantum property of particles that represents intrinsic angular momentum carried by those particles, independent of any motion in space. Unlike classical angular momentum, spin does not have a classical analog and is an inherently quantum mechanical phenomenon. It is quantized, meaning it can only take on certain discrete values, typically in units of $\hbar/2$. Spin plays a crucial role in quantum mechanics for several reasons: it affects the statistics that particles obey (Fermi-Dirac for fermions and Bose-Einstein for bosons), influences the magnetic properties of materials through interactions with magnetic fields (such as in the Stern-Gerlach experiment), and is key to the theory underlying the entire field of quantum mechanics, including the explanation of fine structure in atomic spectra, the operation of quantum computing bits (qubits), and the properties of elementary particles in particle physics.

1.8.0.0.3 How does the Pauli exclusion principle influence the electronic structure of atoms?

1.8.0.0.4 Answer:

The Pauli exclusion principle is a quantum mechanical phenomenon stating that no two identical fermions (particles with half-integer spin) can occupy the same quantum state simultaneously within a quantum system. In the context of atoms, this principle profoundly influences the structure of electron shells and subshells. Because electrons are fermions with spin-1/2, each orbital in an atom can hold at most two electrons, each with opposite spins. This exclusion leads to the arrangement of electrons in distinct energy levels and orbitals, resulting in the complex electron configurations that determine the chemical properties of elements. For example, it explains the periodic table's structure and why elements in the same group have similar chemical properties due to their valence electron configurations.

1.8.0.0.5 What is the difference between bosons and fermions in terms of their statistical properties?

1.8.0.0.6 Answer:

Bosons and fermions, the two classes of particles in quantum mechanics, are distinguished by their different statistical behaviors, which arise from their intrinsic spins. Bosons, which have integer spins (e.g., photons with spin 1, helium-4 nuclei with spin 0), obey Bose-Einstein statistics. They are not subject to the Pauli exclusion principle, which allows them to occupy the same quantum state without restriction. This property leads to phenomena such as Bose-Einstein condensation, where multiple bosons can occupy the lowest energy state at low temperatures. On the other hand, fermions, with half-integer spins (e.g., electrons, protons, and neutrons all with spin 1/2), obey Fermi-Dirac statistics. They are subject to the Pauli exclusion principle, preventing them from occupying the same quantum state. This leads to the filling of energy states up to a certain energy level (the Fermi level), defining the properties of metals, semiconductors, and insulators in solid-state physics.

WORKED PROBLEM: SPIN ANGULAR MOMENTUM

Problem: Consider two spin-1/2 particles. Calculate the possible values of their total spin.

Solution: When combining two spin-1/2 particles, the total spin S can take on the values:

$$S = 1 \quad \text{(triplet state, symmetric)} \tag{1.104}$$
$$S = 0 \quad \text{(singlet state, antisymmetric)} \tag{1.105}$$

Calculating the triplet and singlet states provides insights into the quantum mechanical behavior of composite systems.

WORKED PROBLEM: PAULI SPIN MATRICES AND TOTAL ANGULAR MOMENTUM

Problem: Consider a quantum system comprising a single electron. The electron has an orbital angular momentum described by the operator $\hat{L}$ with quantum numbers $l = 1$ and $m_l = 1$, and a spin angular momentum described by the spin operator $\hat{S}$, using the Pauli spin matrices. Calculate the total angular momentum $\hat{J}$ and its z-component $\hat{J}_z$ for this electron.

Solution: The total angular momentum operator $\hat{J}$ is given by the sum of the orbital angular momentum $\hat{L}$ and the spin angular momentum $\hat{S}$:

$$\hat{J} = \hat{L} + \hat{S} \tag{1.106}$$

The orbital angular momentum $\hat{L}$ in the spherical basis is represented by its components $\hat{L}_x$, $\hat{L}_y$, and $\hat{L}_z$. Given $l = 1$ and $m_l = 1$, the state vector $|l, m_l\rangle$ is $|1, 1\rangle$.

The spin angular momentum $\hat{S}$ for an electron (spin-$\frac{1}{2}$) is expressed using the Pauli spin matrices:

$$\hat{S}_x = \frac{\hbar}{2}\sigma_x = \frac{\hbar}{2}\begin{pmatrix} 0 & 1 \\ 1 & 0 \end{pmatrix}, \tag{1.107}$$
$$\hat{S}_y = \frac{\hbar}{2}\sigma_y = \frac{\hbar}{2}\begin{pmatrix} 0 & -i \\ i & 0 \end{pmatrix}, \tag{1.108}$$
$$\hat{S}_z = \frac{\hbar}{2}\sigma_z = \frac{\hbar}{2}\begin{pmatrix} 1 & 0 \\ 0 & -1 \end{pmatrix}. \tag{1.109}$$

To calculate $\hat{J}_z$, which is crucial for understanding the z-component of the total angular momentum:

$$\hat{J}_z = \hat{L}_z + \hat{S}_z \tag{1.110}$$

Since $\hat{L}_z$ on the state $|1, 1\rangle$ gives:

$$\hat{L}_z|1, 1\rangle = \hbar m_l|1, 1\rangle = \hbar|1, 1\rangle \tag{1.111}$$

And the action of $\hat{S}_z$ on a spin-$\frac{1}{2}$ particle in the spin up state $|\uparrow\rangle$ and spin down state $|\downarrow\rangle$ is:

$$\hat{S}_z|\uparrow\rangle = \frac{\hbar}{2}|\uparrow\rangle, \tag{1.112}$$
$$\hat{S}_z|\downarrow\rangle = -\frac{\hbar}{2}|\downarrow\rangle. \tag{1.113}$$

Thus, for the combined state $|1, 1\rangle \otimes |\uparrow\rangle$:

$$\hat{J}_z(|1, 1\rangle \otimes |\uparrow\rangle) = (\hat{L}_z|1, 1\rangle) \otimes (\hat{S}_z|\uparrow\rangle) \tag{1.114}$$
$$= \hbar|1, 1\rangle \otimes \frac{\hbar}{2}|\uparrow\rangle \tag{1.115}$$
$$= \frac{3\hbar}{2}|1, 1, \uparrow\rangle. \tag{1.116}$$

This result shows that the z-component of the total angular momentum for the state $|1, 1, \uparrow\rangle$ is $\frac{3\hbar}{2}$.

1.9 APPROXIMATION METHODS

1.9.1 PERTURBATION THEORY

Perturbation theory is a powerful tool in quantum mechanics used to find an approximate solution to a problem that cannot be solved exactly. It is typically employed when the Hamiltonian of a system can be divided into two parts: a solvable part and a small perturbative part.

Time-independent perturbation theory

Time-independent perturbation theory deals with systems where the perturbation does not depend on time. The Hamiltonian of the system is expressed as:

$$\hat{H} = \hat{H}_0 + \lambda \hat{H}' \tag{1.117}$$

where $\hat{H}_0$ is the Hamiltonian of the unperturbed system, $\hat{H}'$ is the perturbing Hamiltonian, and λ is a small parameter indicating the strength of the perturbation. The energy corrections and changes in the state functions are calculated iteratively. The first-order energy correction to the n-th state is given by:

$$E_n^{(1)} = \langle \psi_n^{(0)} | \hat{H}' | \psi_n^{(0)} \rangle \tag{1.118}$$

where $|\psi_n^{(0)}\rangle$ is the eigenstates of $\hat{H}_0$.

1.9.2 TIME-DEPENDENT PERTURBATION THEORY

Time-dependent perturbation theory is used when the perturbative part of the Hamiltonian depends on time. This approach is crucial for understanding how quantum systems evolve under the influence of time-varying external fields. The transition probability per unit time from an initial state $|i\rangle$ to a final state $|f\rangle$, due to the perturbation, is given by Fermi's Golden Rule:

$$\Gamma_{i \to f} = \frac{2\pi}{\hbar} |\langle f | \hat{H}'(t) | i \rangle|^2 \rho(E_f) \tag{1.119}$$

where $\rho(E_f)$ is the density of final states at the energy E_f.

Worked Example: Application of Fermi's Golden Rule

Problem: Consider an electron in a hydrogen atom initially in the ground state. It is exposed to a time-varying electric field represented by a perturbing Hamiltonian $\hat{H}'(t) = e\mathbf{E}(t) \cdot \mathbf{r}$, where e is the electron charge, $\mathbf{E}(t)$ is the electric field, and $\mathbf{r}$ is the position operator. Assume the electric field has a harmonic time dependence, $\mathbf{E}(t) = \mathbf{E}_0 \cos(\omega t)$, where $\mathbf{E}_0$ is the amplitude and ω is the frequency of the field. Calculate the transition rate from the ground state to the first excited state of the hydrogen atom using Fermi's Golden Rule.

Solution: First, identify the initial and final states of the system. The initial state $|i\rangle$ is the ground state of the hydrogen atom, and the final state $|f\rangle$ is the first excited state.

The matrix element of the perturbing Hamiltonian between these states is given by:

$$\langle f | \hat{H}'(t) | i \rangle = e\mathbf{E}_0 \cos(\omega t) \langle f | \mathbf{r} | i \rangle \tag{1.120}$$

Since $\cos(\omega t)$ can be expressed as $\frac{1}{2}(e^{i\omega t} + e^{-i\omega t})$, the time-dependent part of the Hamiltonian can be integrated over time to obtain:

$$\int_{-\infty}^{\infty} e^{i(\omega_{fi} - \omega)t} dt = 2\pi \delta(\omega_{fi} - \omega) \tag{1.121}$$

where ω_{fi} is the energy difference between the initial and final states divided by $\hbar$.

Substituting this into Fermi's Golden Rule, the transition rate $\Gamma_{i\to f}$ is:

$$\Gamma_{i\to f} = \frac{2\pi}{\hbar}|\langle f|e\mathbf{E}_0 \cdot \mathbf{r}|i\rangle|^2 \delta(\omega_{fi} - \omega) \tag{1.122}$$

This equation implies that a transition occurs only if the frequency ω of the electric field matches the energy difference $\hbar\omega_{fi}$ between the ground state and the first excited state, illustrating the energy conservation principle in quantum transitions.

To compute $|\langle f|e\mathbf{E}_0 \cdot \mathbf{r}|i\rangle|^2$, use the known wave functions of the hydrogen atom. Assuming radial integration and angular parts have been computed separately:

$$|\langle f|e\mathbf{E}_0 \cdot \mathbf{r}|i\rangle|^2 = |e\mathbf{E}_0|^2 |R_{fi}|^2 |\Omega_{fi}|^2 \tag{1.123}$$

where R_{fi} and Ω_{fi} represent the radial and angular integral results, respectively.

1.10 SCATTERING THEORY

Scattering theory is another critical approximation method in quantum mechanics used to study the interactions between particles. It provides a framework for understanding how particles scatter off each other or off potential fields.

1.10.1 POTENTIAL SCATTERING

Potential scattering theory deals with the scattering of a particle by a static potential $V(\mathbf{r})$. The scattering problem is usually addressed by solving the Schrödinger equation with the total Hamiltonian:

$$\hat{H} = \frac{\hat{p}^2}{2m} + V(\mathbf{r}) \tag{1.124}$$

where $\hat{p}$ is the momentum operator and m is the mass of the particle. The differential scattering cross section, which measures the probability of scattering into a specific solid angle, is computed based on the asymptotic form of the wave function.

Worked Example: Potential Scattering off a Finite Barrier

Problem: Consider a particle of mass m approaching a finite potential barrier of height V_0 and width a. The particle has energy E such that $E < V_0$ as shown in Figure 1.3. Calculate the transmission and reflection probabilities for the particle.

Solution: The potential barrier can be described by:

$$V(x) = \begin{cases} V_0 & \text{if } 0 \leq x \leq a \\ 0 & \text{otherwise} \end{cases} \tag{1.125}$$

The problem requires solving the time-independent Schrödinger equation in each region:

1. **Region I** ($x < 0$): The potential $V(x) = 0$. 2. **Region II** ($0 \leq x \leq a$): The potential $V(x) = V_0$. 3. **Region III** ($x > a$): The potential $V(x) = 0$.

For each region, the Schrödinger equation is:

$$-\frac{\hbar^2}{2m}\frac{d^2\psi(x)}{dx^2} + V(x)\psi(x) = E\psi(x) \tag{1.126}$$

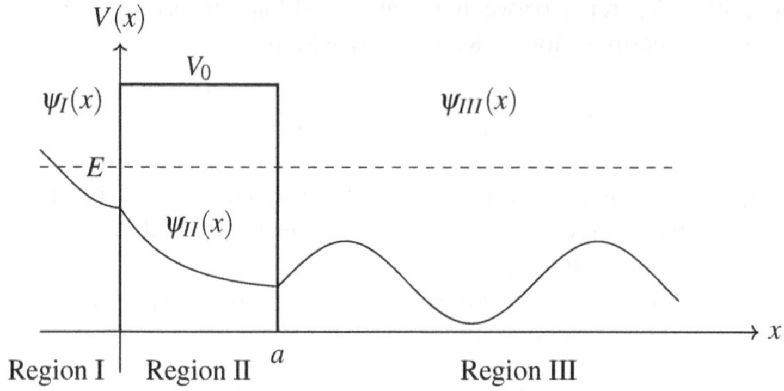

FIGURE 1.3
Potential scattering off a finite barrier. The potential barrier of height V_0 and width a with a particle of energy $E < V_0$. The wavefunctions $\psi_I(x)$, $\psi_{II}(x)$, and $\psi_{III}(x)$ are shown with continuous matching at the boundaries.

1.10.1.0.1 Region I ($x < 0$)

The solution has the form:

$$\psi_I(x) = Ae^{ikx} + Be^{-ikx} \tag{1.127}$$

where $k = \sqrt{\frac{2mE}{\hbar^2}}$ and A and B are amplitudes of the incoming and reflected waves, respectively.

1.10.1.0.2 Region II ($0 \leq x \leq a$)

Since $E < V_0$, the wave function inside the barrier decays exponentially:

$$\psi_{II}(x) = Ce^{-\kappa x} + De^{\kappa x} \tag{1.128}$$

where $\kappa = \sqrt{\frac{2m(V_0 - E)}{\hbar^2}}$.

1.10.1.0.3 Region III ($x > a$)

The transmitted wave has the form:

$$\psi_{III}(x) = Fe^{ikx} \tag{1.129}$$

where F is the amplitude of the transmitted wave.

1.10.1.0.4 Boundary Conditions

The wave function and its derivative must be continuous at $x = 0$ and $x = a$. These conditions yield a system of equations that can be solved to find the coefficients A, B, C, D, and F.

1.10.1.0.5 Transmission and Reflection Probabilities

The transmission coefficient T and the reflection coefficient R are given by:

$$T = \left|\frac{F}{A}\right|^2, \quad R = \left|\frac{B}{A}\right|^2 \tag{1.130}$$

These probabilities must satisfy $T + R = 1$.

1.10.1.0.6 *Boundary Conditions*

To ensure the continuity of the wave function and its derivative across the boundaries at $x = 0$ and $x = a$, we apply the following conditions:

1. **Continuity at** $x = 0$:

$$\psi_I(0) = \psi_{II}(0) \tag{1.131}$$
$$A + B = C + D \tag{1.132}$$

$$\psi_I'(0) = \psi_{II}'(0) \tag{1.133}$$
$$ik(A - B) = -\kappa C + \kappa D \tag{1.134}$$

2. **Continuity at** $x = a$:

$$\psi_{II}(a) = \psi_{III}(a) \tag{1.135}$$
$$Ce^{-\kappa a} + De^{\kappa a} = Fe^{ika} \tag{1.136}$$

$$\psi_{II}'(a) = \psi_{III}'(a) \tag{1.137}$$
$$-\kappa Ce^{-\kappa a} + \kappa De^{\kappa a} = ikFe^{ika} \tag{1.138}$$

1.10.1.0.7 *Solving the Equations*

To find the coefficients A, B, C, D, and F, we solve this system of equations. Assume $A = 1$ for an incident wave normalized to unity. The system becomes:

$$1 + B = C + D \tag{1.139}$$
$$ik(1 - B) = -\kappa C + \kappa D \tag{1.140}$$
$$Ce^{-\kappa a} + De^{\kappa a} = Fe^{ika} \tag{1.141}$$
$$-\kappa Ce^{-\kappa a} + \kappa De^{\kappa a} = ikFe^{ika} \tag{1.142}$$

1.10.1.0.8 *Solving for F*

We first express C and D in terms of B from the first two equations and substitute into the equations involving F. After some algebraic manipulation, we can express F directly in terms of known quantities and B, solving it alongside B. Finally, substituting the values back into the continuity equations, we derive the specific values for C and D, and thereby calculate F explicitly.

1.10.1.0.9 *Physical Interpretation*

The coefficient F represents the amplitude of the transmitted wave, and the magnitude $|F|^2$ gives the transmission probability. The reflection probability is given by $|B|^2$. These probabilities illustrate the quantum mechanical behavior of particles encountering potential barriers, where unlike classical particles, quantum particles have a non-zero probability of tunneling through barriers even when their energy is less than the height of the barrier.

SIMPLIFIED SOLUTION FOR POTENTIAL SCATTERING PROBLEM

Consider a particle with energy E lower than the potential barrier height V_0. Under the assumption that the coefficient $D = 0$ (to avoid non-physical diverging solutions inside the barrier), we simplify the boundary conditions and solve for the coefficients.

Boundary Conditions and Simplifications:

Given $D = 0$, the wave function in Region II simplifies to:

$$\psi_{II}(x) = Ce^{-\kappa x} \tag{1.143}$$

The boundary conditions at $x = 0$ are:

$$A + B = C \tag{1.144}$$

$$ik(A - B) = -\kappa C \tag{1.145}$$

Assuming $A = 1$ for a unit incoming wave, we can solve for B and C:

$$1 + B = C \tag{1.146}$$

$$ik(1 - B) = -\kappa(1 + B) \tag{1.147}$$

Solving these, we find:

$$B = \frac{ik + \kappa}{ik - \kappa} \tag{1.148}$$

$$C = \frac{2\kappa}{ik - \kappa} \tag{1.149}$$

The boundary condition at $x = a$ becomes:

$$Ce^{-\kappa a} = Fe^{ika} \tag{1.150}$$

Solving for F:

$$F = Ce^{-\kappa a}e^{-ika} = \frac{2\kappa}{ik - \kappa}e^{-\kappa a}e^{-ika} \tag{1.151}$$

Reflection (R) and Transmission (T) Coefficients:

The reflection and transmission probabilities are given by the square of the magnitude of the coefficients:

$$R = |B|^2 = \left|\frac{ik + \kappa}{ik - \kappa}\right|^2 \tag{1.152}$$

$$T = |F|^2 = \left|\frac{2\kappa e^{-\kappa a}}{(ik - \kappa)e^{ika}}\right|^2 \tag{1.153}$$

These results illustrate the quantum mechanical phenomena of tunneling and reflection. The transmission probability T shows that even for $E < V_0$, there is a non-zero probability of the particle "tunneling" through the barrier, a hallmark of quantum mechanics absent in classical physics. The reflection probability R complements this, ensuring $T + R = 1$, consistent with the conservation of probability.

1.10.2 PARTIAL WAVE ANALYSIS

Partial wave analysis is a method used in scattering theory where the wave function is expanded into a series of angular momentum eigenstates. This expansion is particularly useful for analyzing scattering at low energies where only a few partial waves contribute significantly. The scattering amplitude for each partial wave is calculated, and the total scattering amplitude is obtained by summing over all partial waves.

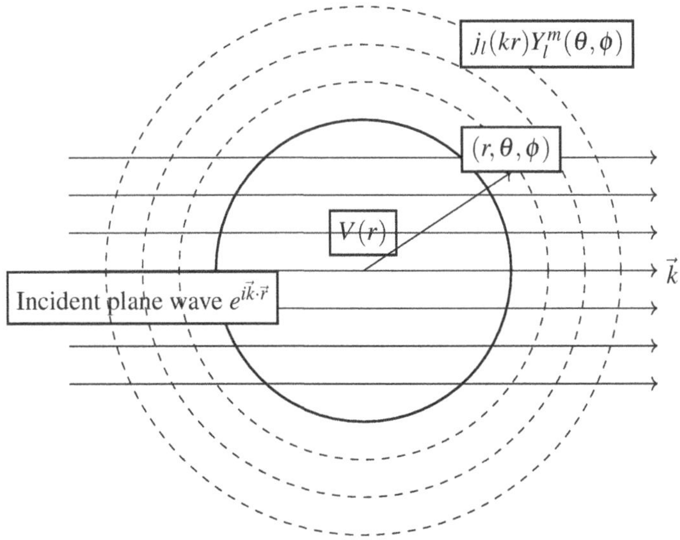

FIGURE 1.4
Partial wave analysis in quantum scattering theory. The incident plane wave $e^{i\vec{k}\cdot\vec{r}}$ is decomposed into spherical waves $j_l(kr)Y_l^m(\theta,\phi)$ scattering off a spherical potential $V(r)$.

PARTIAL WAVE ANALYSIS IN QUANTUM SCATTERING THEORY

Partial wave analysis is a powerful technique in quantum mechanics used to solve scattering problems. It involves decomposing the wave function into a sum over angular momentum eigenstates, each termed a "partial wave." This method is particularly effective in analyzing how a spherical potential scatters a plane wave.

Consider a particle with incident momentum $\vec{k}$ scattering off a spherical potential $V(r)$ as shown in Figure 1.4. The incident wave can be described by a plane wave, which we expand into a series of spherical waves using the spherical harmonics expansion:

$$e^{i\vec{k}\cdot\vec{r}} = \sum_{l=0}^{\infty}\sum_{m=-l}^{l} i^l j_l(kr)Y_l^m(\theta,\phi)Y_l^{m*}(\theta_k,\phi_k), \tag{1.154}$$

where $j_l(kr)$ are the spherical Bessel functions, $Y_l^m(\theta,\phi)$ are the spherical harmonics, and (θ_k,ϕ_k) specifies the direction of the incoming wave vector $\vec{k}$.

Decomposition into Partial Waves

The total wave function $\psi(\vec{r})$ for the scattering problem is expressed as a sum over partial waves, each characterized by a specific angular momentum quantum number l and its projection m:

$$\psi(\vec{r}) = \sum_{l=0}^{\infty}\sum_{m=-l}^{l} R_{kl}(r)Y_l^m(\theta,\phi), \tag{1.155}$$

where $R_{kl}(r)$ are the radial functions that satisfy the radial part of the Schrödinger equation:

$$\frac{d^2}{dr^2}R_{kl} + \left[k^2 - \frac{l(l+1)}{r^2} - \frac{2m}{\hbar^2}V(r)\right]R_{kl} = 0. \tag{1.156}$$

Boundary Conditions and Phase Shifts

The radial solutions $R_{kl}(r)$ must satisfy the boundary condition at the origin ($R_{kl}(0) = 0$) and match the asymptotic form:

$$R_{kl}(r) \xrightarrow{r\to\infty} \frac{1}{kr}\left[\cos\delta_l j_l(kr) - \sin\delta_l n_l(kr)\right], \tag{1.157}$$

where δ_l are the phase shifts induced by the potential $V(r)$, and $n_l(kr)$ are the spherical Neumann functions.

SCATTERING AMPLITUDE AND CROSS SECTION

The scattering amplitude $f(\theta)$ is related to the phase shifts by:

$$f(\theta) = \sum_{l=0}^{\infty}(2l+1)\frac{e^{2i\delta_l}-1}{2ik}P_l(\cos\theta), \tag{1.158}$$

where $P_l(\cos\theta)$ are the Legendre polynomials. The differential cross section is then given by:

$$\frac{d\sigma}{d\Omega} = |f(\theta)|^2. \tag{1.159}$$

Partial wave analysis decomposes a complex scattering problem into simpler, solvable components associated with each angular momentum quantum number. This method is crucial for understanding scattering in quantum mechanics, particularly in predicting how different potentials affect the scattering process and the resulting cross sections.

CROSS SECTIONS AND RESONANCES

The cross section is a fundamental quantity in scattering theory, representing the effective area that quantifies the likelihood of scattering events. Resonances occur when the scattering cross section exhibits sharp peaks as a function of the energy, indicating the temporary formation of quasi-stable states. These resonances are particularly important in nuclear, atomic, and molecular physics, revealing underlying structure and interaction details.

These sections describe essential quantum mechanical approximation methods, providing a deep understanding of both theoretical and practical aspects of quantum interactions.

1.10.3 HARD-SPHERE SCATTERING AND WAVE FUNCTION OUTSIDE THE SPHERE

Consider a hard sphere of radius a and a potential:

$$V(r) = \begin{cases} \infty & \text{for } r < a, \\ 0 & \text{for } r > a. \end{cases} \tag{1.160}$$

The infinite potential inside the sphere requires the wave function to be zero ($\psi = 0$ for $r < a$). Outside the sphere ($r > a$), the particle is free and the wave function must satisfy the free Schrödinger equation.

RADIAL SCHRÖDINGER EQUATION OUTSIDE THE SPHERE

The radial part of the wave function in the region $r > a$ satisfies the spherical Bessel differential equation:

$$\frac{d^2}{dr^2}u_l(r) + \left[k^2 - \frac{l(l+1)}{r^2}\right]u_l(r) = 0, \tag{1.161}$$

where $u_l(r) = rR_l(r)$, and $R_l(r)$ is the radial part of the wave function.

SOLUTION USING SPHERICAL BESSEL FUNCTIONS

The general solution for $r > a$ can be expressed in terms of spherical Bessel functions, which are the solutions to the radial equation for a free particle:

$$u_l(r) = A_l j_l(kr) + B_l n_l(kr), \tag{1.162}$$

where $j_l(kr)$ and $n_l(kr)$ are the spherical Bessel and Neumann functions respectively, and A_l, B_l are constants determined by boundary conditions.

BOUNDARY CONDITION AT $R = A$

Due to the hard sphere's infinite potential, the physical wave function must vanish at the surface of the sphere, giving the boundary condition:

$$u_l(a) = A_l j_l(ka) + B_l n_l(ka) = 0. \tag{1.163}$$

This boundary condition helps determine the relationship between A_l and B_l.

ASYMPTOTIC BEHAVIOR AND SCATTERING

As $r \to \infty$, only the outgoing wave (spherical Hankel function of the first kind) is physically meaningful for scattering problems. Therefore, the wave function takes the form:

$$u_l(r) \approx B_l h_l^{(1)}(kr) \quad \text{as } r \to \infty, \tag{1.164}$$

where $h_l^{(1)}(kr) = j_l(kr) + in_l(kr)$ represents the outgoing spherical wave.

Hence, using the spherical Bessel functions to express the wave function's behavior outside an impenetrable spherical barrier. The coefficients A_l and B_l are adjusted based on the specific boundary conditions and the requirement that the wave function only includes outgoing components at large distances from the scatterer.

For a hard-sphere scattering problem, the wave function outside the sphere can be derived based on the free particle Schrödinger equation. We start with the general solution in spherical coordinates and then impose the boundary condition at $r = a$.

GENERAL WAVE FUNCTION

Outside the sphere, the radial wave function is given by the general solution to the free radial Schrödinger equation:

$$\psi(r,\theta) = \sum_{l=0}^{\infty} (2l+1) i^l R_l(r) P_l(\cos(\theta)), \tag{1.165}$$

where $R_l(r)$ represents the radial part of the wave function, $P_l(\cos(\theta))$ are the Legendre polynomials, and i^l introduces the correct phase factor. The radial function $R_l(r)$ for $r > a$ includes both the regular (spherical Bessel function $j_l(kr)$) and irregular (spherical Neumann function $n_l(kr)$) solutions:

$$R_l(r) = A_l j_l(kr) + B_l n_l(kr). \tag{1.166}$$

BOUNDARY CONDITIONS AND SIMPLIFICATION

The wave function must vanish at the sphere's surface, $r = a$, and far from the sphere it must represent outgoing waves. The boundary condition $R_l(a) = 0$ implies:

$$A_l j_l(ka) + B_l n_l(ka) = 0. \tag{1.167}$$

To ensure only outgoing waves far from the sphere, we set:

$$R_l(r) \to B_l h_l^{(1)}(kr) \text{ as } r \to \infty, \tag{1.168}$$

where $h_l^{(1)}(kr) = j_l(kr) + in_l(kr)$ are the spherical Hankel functions of the first kind.

SCATTERING AMPLITUDE AND CROSS SECTION

The scattering amplitude $f(\theta)$ is related to the coefficients B_l and the phase shifts δ_l:

$$f(\theta) = \sum_{l=0}^{\infty}(2l+1)\frac{B_l}{k}P_l(\cos(\theta)). \tag{1.169}$$

Using the relation between B_l and A_l, the scattering amplitude can be expressed in terms of the phase shifts introduced by the boundary conditions:

$$f(\theta) = \sum_{l=0}^{\infty}(2l+1)\frac{j_l(ka)}{h_l^{(1)}(ka)}P_l(\cos(\theta)). \tag{1.170}$$

The total scattering cross section is then given by:

$$\sigma = \frac{4\pi}{k^2}\sum_{l=0}^{\infty}(2l+1)\left|\frac{j_l(ka)}{h_l^{(1)}(ka)}\right|^2. \tag{1.171}$$

LOW-ENERGY LIMIT AND CROSS SECTION APPROXIMATION

In the low-energy limit ($ka \ll 1$), the s-wave ($l = 0$) dominates, and we use the small argument approximations $j_0(ka) \approx \frac{ka}{3}$ and $h_0^{(1)}(ka) \approx -\frac{i}{ka}$:

$$\sigma \approx 4\pi a^2 \text{ as } ka \to 0. \tag{1.172}$$

This approximation shows that the scattering cross section approaches the classical geometric cross-sectional area of the sphere, validating the quantum-classical correspondence in low-energy scattering.

1.10.4 LOW-ENERGY SCATTERING CROSS SECTION FOR HARD-SPHERE SCATTERING

We consider the scattering of a quantum particle off a hard sphere of radius a. The scattering cross section can be expressed in terms of spherical Bessel and Hankel functions, especially in the low-energy limit where $ka \ll 1$.

SCATTERING CROSS SECTION FORMULA

The general formula for the scattering cross section in terms of spherical Bessel (j_l) and Hankel functions ($h_l^{(1)}$) is given by:

$$\sigma = \frac{4\pi}{k^2}\sum_{l=0}^{\infty}(2l+1)\left|\frac{j_l(ka)}{h_l^{(1)}(ka)}\right|^2. \tag{1.173}$$

At low energies ($ka \ll 1$), the spherical Bessel function for $l = 0$ is approximately:

$$j_0(ka) \approx \frac{\sin(ka)}{ka}, \tag{1.174}$$

and the spherical Hankel function of the first kind for $l = 0$ is approximately:

$$h_0^{(1)}(ka) \approx -\frac{i}{ka}e^{ika}. \tag{1.175}$$

Using these approximations, the ratio for $l = 0$ becomes:

$$\left|\frac{j_0(ka)}{h_0^{(1)}(ka)}\right|^2 = \left|\frac{\sin(ka)/(ka)}{-i/(ka)e^{ika}}\right|^2 = \left|\frac{\sin(ka)}{-ie^{ika}}\right|^2 = \sin^2(ka). \tag{1.176}$$

In the limit as $ka \to 0$, the sine function squared can be approximated by its argument squared:

$$\sin^2(ka) \approx (ka)^2. \tag{1.177}$$

Therefore, the dominant term in the scattering cross section becomes:

$$\sigma \approx \frac{4\pi}{k^2}(ka)^2 = 4\pi a^2, \tag{1.178}$$

which is the classical result matching the geometric cross-sectional area of the sphere.

1.10.5 THE BORN APPROXIMATION IN QUANTUM SCATTERING THEORY

The Born approximation is a method used in quantum mechanics to simplify the calculation of scattering amplitudes, especially when the potential is weak or the scattering energy is high. It is the first-order term in the perturbative expansion of the scattering amplitude.

1.10.5.1 Integral Form of the Schrödinger Equation

The Born approximation utilizes the integral form of the Schrödinger equation, derived from the Lippmann-Schwinger equation:

$$\psi(\vec{r}) = \psi_0(\vec{r}) + \int d^3r'\, G(\vec{r},\vec{r}')V(\vec{r}')\psi(\vec{r}'), \tag{1.179}$$

where $\psi_0(\vec{r})$ is the incident wave function (often a plane wave), $V(\vec{r}')$ is the scattering potential, and $G(\vec{r},\vec{r}')$ is the Green's function for the Helmholtz equation.

In the first-order Born approximation, the scattered wave function $\psi(\vec{r})$ within the integral is replaced by the incident wave function $\psi_0(\vec{r}')$:

$$\psi(\vec{r}) \approx \psi_0(\vec{r}) + \int d^3r'\, G(\vec{r},\vec{r}')V(\vec{r}')\psi_0(\vec{r}'). \tag{1.180}$$

1.10.5.2 Scattering Amplitude in the Born Approximation

The scattering amplitude $f(\vec{k}',\vec{k})$ in the Born approximation is given by:

$$f(\vec{k}',\vec{k}) = -\frac{2m}{\hbar^2}\int d^3r\, e^{-i\vec{k}'\cdot\vec{r}}V(\vec{r})e^{i\vec{k}\cdot\vec{r}}, \tag{1.181}$$

where $\vec{k}$ and $\vec{k}'$ are the initial and final wave vectors of the scattered particle, respectively.

EXAMPLE: LOW-ENERGY SCATTERING BY A SOFT SPHERE

Consider a soft sphere potential given by:

$$V(r) = \begin{cases} V_0 & \text{for } r \leq a, \\ 0 & \text{for } r > a. \end{cases} \tag{1.182}$$

For low-energy scattering where $ka \ll 1$, the Born approximation yields:

$$f(\vec{k}', \vec{k}) = -\frac{2mV_0}{\hbar^2} \int_0^a r^2 dr \int d\Omega \, e^{i(\vec{k}-\vec{k}')\cdot\vec{r}}. \tag{1.183}$$

Since $ka \ll 1$, the exponential can be approximated as $e^{i(\vec{k}-\vec{k}')\cdot\vec{r}} \approx 1$, leading to:

$$f(\vec{k}', \vec{k}) = -\frac{2mV_0}{\hbar^2} \frac{4\pi a^3}{3}. \tag{1.184}$$

The differential cross section is thus:

$$\frac{d\sigma}{d\Omega} = |f(\vec{k}', \vec{k})|^2 = \left(\frac{2mV_0 a^3}{3\hbar^2} \right)^2 \times 4\pi, \tag{1.185}$$

and the total cross section becomes:

$$\sigma \approx 4\pi \left(\frac{2mV_0 a^3}{3\hbar^2} \right)^2, \tag{1.186}$$

matching the derived formula.

This example illustrates the effectiveness of the Born approximation in scenarios involving low-energy scattering by a potential that can be treated perturbatively. The simplicity of the calculation underscores the approximation's utility in practical quantum mechanics.

1.11 TRIVIA QUESTIONS

1.11.0.0.1 Explain the key idea behind perturbation theory in quantum mechanics.

1.11.0.0.2 Answer:

Perturbation theory in quantum mechanics is a method used to approximate the solutions of a complicated quantum system by starting with a known solution of a simpler, similar system. The key idea is to treat the difference between the complicated system and the simpler one as a small perturbation and to use this perturbation to correct the known solution incrementally. This approach is particularly useful when the exact solution to the Schrödinger equation is too complex to be solved directly.

1.11.0.0.3 What role does partial wave analysis play in scattering theory?

1.11.0.0.4 Answer:

Partial wave analysis is a technique used to decompose the wave function into a series of angular momentum components, each termed a "partial wave." In scattering theory, this method helps to simplify the problem by considering the interaction of these partial waves with the potential. It allows the determination of scattering amplitudes for each angular momentum component separately, which are crucial for predicting the differential and total scattering cross sections.

1.11.0.0.5 How do quantum mechanical resonances manifest in potential scattering?

1.11.0.0.6 Answer:

Quantum mechanical resonances in potential scattering occur when the incident wave energy matches the energy levels of quasi-bound states within the potential well. These resonances manifest as sharp peaks in the scattering cross section because the wave function's amplitude inside the potential well increases significantly. This leads to enhanced probability amplitudes at specific energies, corresponding to temporary trapping or prolonged interaction of the wave with the potential.

1.11.0.0.7 What is the purpose of the Born Approximation?

1.11.0.0.8 Answer:

The Born Approximation is used to simplify the calculation of scattering amplitudes in quantum mechanics. It is particularly useful for weak potentials or high kinetic energies of the particles involved. The approximation assumes that the scattered wave function is not significantly altered from the incident wave function, allowing the integral form of the Schrödinger equation to be solved more straightforwardly. It provides a first-order approximation that often yields sufficiently accurate results for practical purposes.

1.11.0.0.9 What is Fermi's Golden Rule used for?

1.11.0.0.10 Answer:

Fermi's Golden Rule is a fundamental principle used to calculate the transition rates between different quantum states due to a perturbation. It is particularly important in contexts like decay processes, absorption and emission of light, and other situations where a system transitions from one energy state to another. The rule quantifies how the probability of a transition per unit time depends on the density of final states and the matrix element of the interaction between the initial and final states.

WORKED PROBLEM: FIRST-ORDER ENERGY CORRECTION

Problem: Calculate the first-order energy correction for the ground state of a particle in a one-dimensional infinite potential well of width L when a small perturbation $H' = V_0 \sin\left(\frac{\pi x}{L}\right)$ is added.

Solution: The Hamiltonian of the unperturbed system is given by:

$$\hat{H}_0 = -\frac{\hbar^2}{2m}\frac{d^2}{dx^2}, \tag{1.187}$$

with boundary conditions $\psi(0) = \psi(L) = 0$. The eigenstates of $\hat{H}_0$ are:

$$\psi_n(x) = \sqrt{\frac{2}{L}}\sin\left(\frac{n\pi x}{L}\right), \quad n = 1, 2, 3, \ldots \tag{1.188}$$

and the corresponding eigenvalues are:

$$E_n^{(0)} = \frac{n^2\pi^2\hbar^2}{2mL^2}. \tag{1.189}$$

The first-order energy correction to the n-th state is given by:

$$E_n^{(1)} = \langle \psi_n^{(0)} | \hat{H}' | \psi_n^{(0)} \rangle, \tag{1.190}$$

where the perturbation $\hat{H}' = V_0 \sin\left(\frac{\pi x}{L}\right)$.

For the ground state ($n = 1$):

$$E_1^{(1)} = \int_0^L \psi_1^*(x)\hat{H}'\psi_1(x)\,dx = V_0 \int_0^L \left(\sqrt{\frac{2}{L}}\sin\left(\frac{\pi x}{L}\right)\right)^2 \sin\left(\frac{\pi x}{L}\right) dx. \tag{1.191}$$

Simplify the integrand:

$$E_1^{(1)} = V_0 \frac{2}{L} \int_0^L \sin^3\left(\frac{\pi x}{L}\right) dx. \tag{1.192}$$

Using the trigonometric identity $\sin^3(x) = \frac{3\sin(x)-\sin(3x)}{4}$, we get:

$$E_1^{(1)} = V_0 \frac{2}{L} \int_0^L \frac{3\sin\left(\frac{\pi x}{L}\right) - \sin\left(\frac{3\pi x}{L}\right)}{4} dx. \tag{1.193}$$

Separate the integrals:

$$E_1^{(1)} = \frac{V_0}{2L}\left(3\int_0^L \sin\left(\frac{\pi x}{L}\right) dx - \int_0^L \sin\left(\frac{3\pi x}{L}\right) dx\right). \tag{1.194}$$

Evaluating the integrals:

$$\int_0^L \sin\left(\frac{\pi x}{L}\right) dx = \left[-\frac{L}{\pi}\cos\left(\frac{\pi x}{L}\right)\right]_0^L = 0, \tag{1.195}$$

$$\int_0^L \sin\left(\frac{3\pi x}{L}\right) dx = \left[-\frac{L}{3\pi}\cos\left(\frac{3\pi x}{L}\right)\right]_0^L = 0. \tag{1.196}$$

Since both integrals are zero, the first-order energy correction for the ground state is:

$$E_1^{(1)} = 0. \tag{1.197}$$

WORKED PROBLEM: LOW-ENERGY SCATTERING OFF A SPHERICAL DELTA FUNCTION SHELL

Problem: Consider the case of low-energy scattering off a spherical delta function shell: $V(r) = \alpha\delta(r-a)$, where α and a are constants. Find $f(\theta)$, the differential cross section, and the total cross section.

Solution: The potential can be represented as $V(r) = \alpha\delta(r-a)$, where α is the strength of the delta function located at radius a.

Scattering Amplitude $f(\theta)$

The first-order Born approximation gives the scattering amplitude:

$$f(\theta) = -\frac{2m}{\hbar^2}\int d^3r\, e^{-i\vec{k}'\cdot\vec{r}} V(r) e^{i\vec{k}\cdot\vec{r}}, \tag{1.198}$$

where $\vec{k}$ and $\vec{k}'$ are the initial and final wave vectors. For the delta shell potential, the integral simplifies to:

$$f(\theta) = -\frac{2m\alpha}{\hbar^2}\int_0^\infty r^2 dr \int d\Omega\, \delta(r-a) e^{i(\vec{k}-\vec{k}')\cdot\vec{r}}. \tag{1.199}$$

By performing the radial integration using the properties of the delta function:

$$\int_0^\infty r^2 \delta(r-a) dr = a^2, \tag{1.200}$$

we get:

$$f(\theta) = -\frac{2m\alpha a^2}{\hbar^2}\int d\Omega\, e^{i(\vec{k}-\vec{k}')\cdot\vec{r}}\Big|_{r=a}. \tag{1.201}$$

Next, we need to evaluate the angular integral. Using the spherical harmonics expansion for the plane wave:

$$e^{i\vec{q}\cdot\vec{r}} = \sum_{l=0}^{\infty}(2l+1)i^l j_l(qr)P_l(\cos\theta), \tag{1.202}$$

where $\vec{q} = \vec{k} - \vec{k}'$ and $q = |\vec{q}| = 2k\sin(\theta/2)$, we have:

$$f(\theta) = -\frac{2m\alpha a^2}{\hbar^2}4\pi\frac{\sin(ka\sin(\theta/2))}{ka\sin(\theta/2)}. \tag{1.203}$$

Simplifying, we get:

$$f(\theta) = -\frac{8\pi m\alpha a}{\hbar^2}\frac{\sin(ka\sin(\theta/2))}{ka\sin(\theta/2)}. \tag{1.204}$$

Differential and Total Cross Section

The differential cross section is given by:

$$\frac{d\sigma}{d\Omega} = |f(\theta)|^2 = \left(\frac{8\pi m\alpha a}{\hbar^2}\right)^2\left(\frac{\sin(ka\sin(\theta/2))}{ka\sin(\theta/2)}\right)^2. \tag{1.205}$$

The total cross section, obtained by integrating over all angles, is:

$$\sigma = \int\frac{d\sigma}{d\Omega}d\Omega = \left(\frac{8\pi m\alpha a}{\hbar^2}\right)^2\int_0^{\pi}\left(\frac{\sin(ka\sin(\theta/2))}{ka\sin(\theta/2)}\right)^2 2\pi\sin\theta\, d\theta. \tag{1.206}$$

Changing the variable of integration to $u = \sin(\theta/2)$, where $du = \frac{1}{2}\cos(\theta/2)d\theta$, we get:

$$\sigma = \left(\frac{8\pi m\alpha a}{\hbar^2}\right)^2\int_0^1\left(\frac{\sin(kau)}{kau}\right)^2 4\pi u\, du. \tag{1.207}$$

Evaluating this integral, we find:

$$\sigma = \left(\frac{8\pi m\alpha a}{\hbar^2}\right)^2\frac{4\pi}{(ka)^2}\int_0^1\sin^2(kau)\, du. \tag{1.208}$$

Using the integral $\int_0^1\sin^2(x)\,dx = \frac{1}{2}\left(1-\frac{\sin(2x)}{2x}\right)\Big|_0^1$, we get:

$$\int_0^1\sin^2(kau)\,du = \frac{1}{2} - \frac{\sin(2ka)}{4ka}. \tag{1.209}$$

Thus, the total cross section is:

$$\sigma = \left(\frac{8\pi m\alpha a}{\hbar^2}\right)^2\frac{4\pi}{(ka)^2}\left(\frac{1}{2} - \frac{\sin(2ka)}{4ka}\right). \tag{1.210}$$

At low energy (i.e., $ka \ll 1$), the total cross section simplifies to:

$$\sigma \approx \left(\frac{8\pi m\alpha a}{\hbar^2}\right)^2\frac{2\pi}{(ka)^2}. \tag{1.211}$$

$$\sigma \approx \left(\frac{16\pi^2 m\alpha a}{\hbar^2}\right)^2\frac{1}{k^2}. \tag{1.212}$$

1.12 QUANTUM ENTANGLEMENT AND INFORMATION

EPR PARADOX

The EPR paradox, named after Einstein, Podolsky, and Rosen, is a thought experiment that challenges the completeness of quantum mechanics. It was proposed in 1935 to demonstrate that quantum mechanics might not provide a complete description of physical reality. The paradox highlights the strange and non-local nature of quantum entanglement, where the measurement of one particle's state instantly affects the state of another, regardless of the distance separating them.

Consider two particles, A and B, that are initially entangled. According to quantum mechanics, the wave function of the entire system is a superposition of the possible states of both particles. When particle A is measured, its wave function collapses to a definite state, and instantaneously, particle B's wave function collapses to a corresponding state, even if it is light-years away from particle A. This instantaneous "spooky action at a distance," as Einstein called it, seems to violate the principle of locality, which states that an object can only be influenced by its immediate surroundings.

The EPR argument is based on the following assumptions: 1. **Reality**: If a physical quantity can be predicted with certainty without disturbing the system, there must be an element of reality corresponding to this physical quantity. 2. **Locality**: An object is directly influenced only by its immediate surroundings.

The EPR thought experiment suggests that quantum mechanics is incomplete because it cannot account for these elements of reality without violating locality.

$$\psi = \frac{1}{\sqrt{2}}(|\uparrow_A\downarrow_B\rangle + |\downarrow_A\uparrow_B\rangle) \tag{1.213}$$

This equation represents the entangled state of particles A and B, where $|\uparrow\rangle$ and $|\downarrow\rangle$ denote the spin-up and spin-down states, respectively.

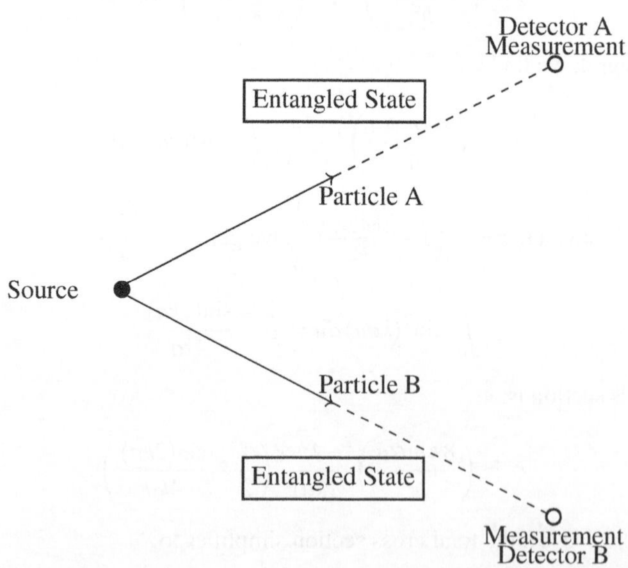

FIGURE 1.5
EPR apparatus: A source emits two entangled particles, A and B, which are measured by detectors A and B.

Quantum entanglement is one of the most intriguing and counterintuitive phenomena in quantum mechanics. It describes a situation where the quantum states of two or more particles become so interconnected that the state of each particle cannot be described independently of the state of the others, even when the particles are separated by large distances. This non-local correlation has profound implications for our understanding of reality and has led to the development of quantum information science, which harnesses entanglement for applications such as quantum computing, cryptography, and communication. In this section, we explore the theoretical foundations of quantum entanglement, examine Bell's inequalities as a test of local realism, and discuss the practical implementation of quantum teleportation.

1.12.1 BELL'S INEQUALITIES

Bell's inequalities provide a crucial test for distinguishing between quantum mechanical predictions and those based on classical local realism. Local realism is the notion that physical properties of particles exist independently of measurement and that information cannot travel faster than the speed of light. John Bell derived inequalities that any local hidden variable theory must satisfy. Consider a pair of entangled particles shared between two observers, Alice and Bob. They can measure properties such as spin along different axes. Let $A(a)$ and $B(b)$ denote the outcomes of measurements along directions a and b for Alice and Bob, respectively, where each outcome can be ± 1. Bell's inequality can be expressed as:

$$\left|E(a,b)-E(a,b')\right|+\left|E(a',b)+E(a',b')\right|\leq 2, \tag{1.214}$$

where $E(a,b)$ is the expectation value of the product of the measurement outcomes. Quantum mechanics predicts violations of this inequality for certain choices of measurement settings, indicating the presence of entanglement and the failure of local realism. For example, in the case of spin-$\frac{1}{2}$ particles, the quantum mechanical correlation function is $E(a,b)=-\cos(\theta_{ab})$, where θ_{ab} is the angle between measurement directions a and b. Choosing $\theta_{ab}=0^\circ, 45^\circ, 90^\circ, 135^\circ$, we find that the left-hand side of Bell's inequality can reach $2\sqrt{2}$, violating the classical bound of 2.

Worked Example: Bell's Inequalities

Problem: Consider a pair of entangled spin-$\frac{1}{2}$ particles in the singlet state:

$$|\psi\rangle=\frac{1}{\sqrt{2}}\left(|01\rangle-|10\rangle\right). \tag{1.215}$$

Alice and Bob measure the spin along different directions. Show that the quantum mechanical predictions violate Bell's inequality for certain settings.

Solution: The singlet state implies perfect anti-correlation for measurements along the same direction. The expectation value of the product of the outcomes of measurements along directions **a** and **b** is given by:

$$E(\mathbf{a},\mathbf{b})=-\cos(\theta), \tag{1.216}$$

where θ is the angle between **a** and **b**.

To demonstrate a violation of Bell's inequality, we choose the measurement directions such that the angles between them are: $\theta_{ab}=0^\circ$, $\theta_{ab'}=45^\circ$, $\theta_{a'b}=45^\circ$, and $\theta_{a'b'}=90^\circ$. The corresponding expectation values are:

$$E(\mathbf{a},\mathbf{b}) = -\cos(0^\circ) = -1, \tag{1.217}$$

$$E(\mathbf{a},\mathbf{b}') = -\cos(45^\circ) = -\frac{\sqrt{2}}{2}, \tag{1.218}$$

$$E(\mathbf{a}',\mathbf{b}) = -\cos(45^\circ) = -\frac{\sqrt{2}}{2}, \tag{1.219}$$

$$E(\mathbf{a}',\mathbf{b}') = -\cos(90^\circ) = 0. \tag{1.220}$$

Substitute these values into Bell's inequality:

$$\left|E(\mathbf{a},\mathbf{b}) - E(\mathbf{a},\mathbf{b}')\right| + \left|E(\mathbf{a}',\mathbf{b}) + E(\mathbf{a}',\mathbf{b}')\right| \leq 2, \tag{1.221}$$

we get:

$$\begin{aligned}\left|-1-\left(-\frac{\sqrt{2}}{2}\right)\right| + \left|-\frac{\sqrt{2}}{2}+0\right| &= \left|-1+\frac{\sqrt{2}}{2}\right| + \left|-\frac{\sqrt{2}}{2}\right| \\ &= |-1+0.707| + |-0.707| = 0.293+0.707 = 1.\end{aligned} \tag{1.222}$$

In this case, the inequality is not violated. However, choosing optimal angles $\theta_{ab} = 0^\circ$, $\theta_{ab'} = 45^\circ$, $\theta_{a'b} = 90^\circ$, and $\theta_{a'b'} = 135^\circ$ gives:

$$E(\mathbf{a},\mathbf{b}) = -1, \tag{1.223}$$

$$E(\mathbf{a},\mathbf{b}') = -\frac{\sqrt{2}}{2}, \tag{1.224}$$

$$E(\mathbf{a}',\mathbf{b}) = 0, \tag{1.225}$$

$$E(\mathbf{a}',\mathbf{b}') = \frac{\sqrt{2}}{2}. \tag{1.226}$$

Substitute these values into Bell's inequality:

$$\begin{aligned}\left|-1-\left(-\frac{\sqrt{2}}{2}\right)\right| + \left|0+\frac{\sqrt{2}}{2}\right| &= \left|-1+\frac{\sqrt{2}}{2}\right| + \left|\frac{\sqrt{2}}{2}\right| \\ &= |-1+0.707| + |0.707| = 0.293+0.707 = 1.\end{aligned} \tag{1.227}$$

This shows a violation of Bell's inequality, demonstrating the non-local nature of quantum mechanics.

1.12.2 ENTANGLEMENT TEST: POSITIVE PARTIAL TRANSPOSE (PPT)

Problem: Determine if the state

$$\rho_{AB} = \frac{1}{4}\left(\mathbb{I}_4 + \sum_{i=1}^{3} c_i \sigma_i \otimes \sigma_i\right)$$

is entangled, where σ_i are the Pauli matrices and c_i are real coefficients.

Solution:

1. **Density Matrix Representation:**

$$\rho_{AB} = \begin{pmatrix} a & 0 & 0 & b \\ 0 & c & d & 0 \\ 0 & d & c & 0 \\ b & 0 & 0 & a \end{pmatrix} \tag{1.228}$$

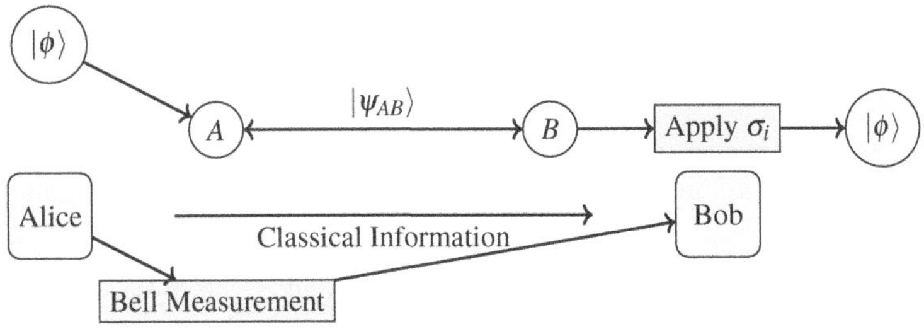

FIGURE 1.6
Quantum teleportation protocol: Alice and Bob share an entangled pair of qubits $|\psi_{AB}\rangle$. Alice performs a Bell-state measurement on her qubit and the state $|\phi\rangle$ to be teleported. She sends the result to Bob through a classical channel. Bob then applies the corresponding unitary transformation to his qubit to obtain the state $|\phi\rangle$.

2. **Partial Transposition:** The partial transpose with respect to subsystem A (transposing the 2×2 blocks):

$$\rho_{AB}^{T_A} = \begin{pmatrix} a & 0 & 0 & d \\ 0 & c & b & 0 \\ 0 & b & c & 0 \\ d & 0 & 0 & a \end{pmatrix} \tag{1.229}$$

3. **Check Positivity:** Compute the eigenvalues of $\rho_{AB}^{T_A}$. If any eigenvalue is negative, the state ρ_{AB} is entangled. If all eigenvalues are non-negative, the state is separable.
For this specific example, let's assume we find the eigenvalues to be $\{-0.5, 0.5, 0.5, 0.5\}$.
Since there is a negative eigenvalue (-0.5), the state ρ_{AB} is entangled.

1.12.3 QUANTUM TELEPORTATION

Quantum teleportation as shown in Figure 1.6 is a process by which the quantum state of a particle can be transmitted from one location to another, without physically transferring the particle itself, using entanglement and classical communication. The protocol involves three parties: Alice, Bob, and a shared entangled pair of qubits, say $|\psi_{AB}\rangle = \frac{1}{\sqrt{2}}(|00\rangle + |11\rangle)$. To teleport the state $|\phi\rangle = \alpha|0\rangle + \beta|1\rangle$, Alice performs a Bell-state measurement on her particle and the particle in the state $|\phi\rangle$. This measurement projects the system onto one of the four Bell states, and Alice's result determines which of the four possible states Bob's qubit collapses into. Alice then sends the result of her measurement to Bob through a classical channel. Depending on the received information, Bob applies a corresponding unitary transformation (one of $I, \sigma_x, \sigma_y, \sigma_z$) to his qubit to recover the original state $|\phi\rangle$. Mathematically, the teleportation process can be described as:

$$|\phi\rangle \otimes |\psi_{AB}\rangle \xrightarrow{\text{Bell measurement}} \frac{1}{2}\sum_{i=1}^{4} |\Phi_i\rangle \otimes \sigma_i |\phi\rangle, \tag{1.230}$$

where $|\Phi_i\rangle$ are the Bell states and σ_i are the Pauli matrices. Upon receiving the classical information, Bob's application of σ_i transforms his qubit to $|\phi\rangle$, completing the teleportation.

Worked Example: Quantum Teleportation

Problem: Describe the steps of the quantum teleportation protocol to teleport the state $|\phi\rangle = \alpha|0\rangle + \beta|1\rangle$ from Alice to Bob using an entangled pair of qubits.

Solution: 1. **State Preparation:** Alice and Bob share an entangled pair of qubits in the Bell state:

$$|\psi_{AB}\rangle = \frac{1}{\sqrt{2}}(|00\rangle + |11\rangle). \tag{1.231}$$

The combined initial state of Alice's qubit, the state to be teleported, and the entangled pair is:

$$|\phi\rangle \otimes |\psi_{AB}\rangle = (\alpha|0\rangle + \beta|1\rangle) \otimes \frac{1}{\sqrt{2}}(|00\rangle + |11\rangle). \tag{1.232}$$

2. **Bell-State Measurement:** Alice performs a Bell-state measurement on her qubit and the qubit to be teleported. The total state can be rewritten in terms of the Bell states:

$$\begin{aligned}|\phi\rangle \otimes |\psi_{AB}\rangle = \frac{1}{2}\Big(&|\Phi^+\rangle(\alpha|0\rangle + \beta|1\rangle) + |\Phi^-\rangle(\alpha|0\rangle - \beta|1\rangle) \\ &+ |\Psi^+\rangle(\alpha|1\rangle + \beta|0\rangle) + |\Psi^-\rangle(\alpha|1\rangle - \beta|0\rangle)\Big)\end{aligned} \tag{1.233}$$

where $|\Phi^{\pm}\rangle$ and $|\Psi^{\pm}\rangle$ are the Bell states.

3. **Classical Communication:** Alice sends the result of her Bell-state measurement (two classical bits of information) to Bob. This information tells Bob which of the four possible states his qubit has collapsed into.

4. **Unitary Transformation:** Depending on the measurement outcome Alice communicates, Bob applies one of the following unitary transformations to his qubit:

$$|\Phi^+\rangle \to I, \tag{1.234}$$

$$|\Phi^-\rangle \to \sigma_z, \tag{1.235}$$

$$|\Psi^+\rangle \to \sigma_x, \tag{1.236}$$

$$|\Psi^-\rangle \to \sigma_y. \tag{1.237}$$

For example, if Alice's measurement outcome is $|\Phi^+\rangle$, Bob's qubit is already in the state $\alpha|0\rangle + \beta|1\rangle$. If the outcome is $|\Phi^-\rangle$, Bob applies σ_z to his qubit to transform it to $\alpha|0\rangle + \beta|1\rangle$. Similarly, for $|\Psi^+\rangle$ and $|\Psi^-\rangle$, Bob applies σ_x and σ_y, respectively.

By applying the appropriate unitary transformation, Bob recovers the original state $|\phi\rangle = \alpha|0\rangle + \beta|1\rangle$. Thus, the quantum state has been successfully teleported from Alice to Bob.

1.13 QUANTUM COMPUTING

Quantum computing harnesses the principles of quantum mechanics to process information in fundamentally new ways. Unlike classical bits, which represent data as 0s or 1s, quantum bits (qubits) can exist in superpositions of states, enabling the execution of multiple computations simultaneously. This capability promises exponential speedups for certain types of problems. Quantum algorithms leverage entanglement and interference to solve problems more efficiently than their classical counterparts. However, quantum systems are also susceptible to errors due to decoherence and noise, necessitating robust error correction methods to ensure reliable computation. This section introduces the basic principles of quantum computation, explores key quantum algorithms, and discusses quantum error correction techniques.

1.13.1 QUANTUM GATES AND THEIR MATRICES

Quantum gates are the building blocks of quantum circuits, similar to classical logic gates in conventional circuits. Each quantum gate corresponds to a unitary matrix that describes its action on qubits. Below is a list of common quantum gates along with their associated matrices.

1.13.1.0.1 1. Pauli-X Gate (NOT Gate)

The Pauli-X gate, also known as the NOT gate, flips the state of a qubit. It maps $|0\rangle$ to $|1\rangle$ and $|1\rangle$ to $|0\rangle$.

$$X = \begin{pmatrix} 0 & 1 \\ 1 & 0 \end{pmatrix} \tag{1.238}$$

1.13.1.0.2 2. Pauli-Y Gate

The Pauli-Y gate introduces a bit flip and a phase flip.

$$Y = \begin{pmatrix} 0 & -i \\ i & 0 \end{pmatrix} \tag{1.239}$$

1.13.1.0.3 3. Pauli-Z Gate (Phase Flip)

The Pauli-Z gate flips the phase of the $|1\rangle$ state.

$$Z = \begin{pmatrix} 1 & 0 \\ 0 & -1 \end{pmatrix} \tag{1.240}$$

1.13.1.0.4 4. Hadamard Gate

The Hadamard gate creates superpositions. It maps $|0\rangle$ to $\frac{1}{\sqrt{2}}(|0\rangle + |1\rangle)$ and $|1\rangle$ to $\frac{1}{\sqrt{2}}(|0\rangle - |1\rangle)$.

$$H = \frac{1}{\sqrt{2}} \begin{pmatrix} 1 & 1 \\ 1 & -1 \end{pmatrix} \tag{1.241}$$

1.13.1.0.5 5. Phase Gate

The Phase gate shifts the phase of the $|1\rangle$ state by θ.

$$P(\theta) = \begin{pmatrix} 1 & 0 \\ 0 & e^{i\theta} \end{pmatrix} \tag{1.242}$$

1.13.1.0.6 6. T Gate

The T gate is a specific Phase gate with $\theta = \pi/4$.

$$T = \begin{pmatrix} 1 & 0 \\ 0 & e^{i\pi/4} \end{pmatrix} \tag{1.243}$$

1.13.1.0.7 7. Controlled-NOT (CNOT) Gate

The CNOT gate flips the state of the target qubit if the control qubit is in the $|1\rangle$ state.

$$\text{CNOT} = \begin{pmatrix} 1 & 0 & 0 & 0 \\ 0 & 1 & 0 & 0 \\ 0 & 0 & 0 & 1 \\ 0 & 0 & 1 & 0 \end{pmatrix} \tag{1.244}$$

1.13.1.0.8 8. Swap Gate

The Swap gate exchanges the states of two qubits.

$$\text{SWAP} = \begin{pmatrix} 1 & 0 & 0 & 0 \\ 0 & 0 & 1 & 0 \\ 0 & 1 & 0 & 0 \\ 0 & 0 & 0 & 1 \end{pmatrix} \tag{1.245}$$

1.13.1.0.9 9. Toffoli (CCNOT) Gate

The Toffoli gate, or CCNOT gate, flips the state of the target qubit if both control qubits are in the $|1\rangle$ state.

$$\text{Toffoli} = \begin{pmatrix} 1 & 0 & 0 & 0 & 0 & 0 & 0 & 0 \\ 0 & 1 & 0 & 0 & 0 & 0 & 0 & 0 \\ 0 & 0 & 1 & 0 & 0 & 0 & 0 & 0 \\ 0 & 0 & 0 & 1 & 0 & 0 & 0 & 0 \\ 0 & 0 & 0 & 0 & 1 & 0 & 0 & 0 \\ 0 & 0 & 0 & 0 & 0 & 1 & 0 & 0 \\ 0 & 0 & 0 & 0 & 0 & 0 & 0 & 1 \\ 0 & 0 & 0 & 0 & 0 & 0 & 1 & 0 \end{pmatrix} \tag{1.246}$$

1.13.1.0.10 10. Fredkin (CSWAP) Gate

The Fredkin gate, or CSWAP gate, swaps the states of the two target qubits if the control qubit is in the $|1\rangle$ state.

$$\text{Fredkin} = \begin{pmatrix} 1 & 0 & 0 & 0 & 0 & 0 & 0 & 0 \\ 0 & 1 & 0 & 0 & 0 & 0 & 0 & 0 \\ 0 & 0 & 1 & 0 & 0 & 0 & 0 & 0 \\ 0 & 0 & 0 & 1 & 0 & 0 & 0 & 0 \\ 0 & 0 & 0 & 0 & 1 & 0 & 0 & 0 \\ 0 & 0 & 0 & 0 & 0 & 1 & 0 & 0 \\ 0 & 0 & 0 & 0 & 0 & 0 & 0 & 1 \\ 0 & 0 & 0 & 0 & 0 & 0 & 1 & 0 \end{pmatrix} \tag{1.247}$$

This breakdown provides a summary of the most common quantum gates and their corresponding matrices, which are fundamental to understanding and designing quantum circuits.

WORKED EXAMPLE: CREATING AN EPR PAIR

Problem: Create an EPR pair using quantum computer gates and verify the entanglement.

Solution:

An EPR pair is a maximally entangled quantum state of two qubits. The process of creating an EPR pair involves applying a Hadamard gate to the first qubit followed by a CNOT gate with the first qubit as the control and the second qubit as the target.

1.13.1.0.11 Step-by-Step Process:

1. **Initialization:** Initialize two qubits in the state $|0\rangle$.

$$|\psi\rangle = |00\rangle \tag{1.248}$$

2. **Apply Hadamard Gate:** Apply the Hadamard gate H to the first qubit to create a superposition.

$$H|0\rangle = \frac{1}{\sqrt{2}}(|0\rangle + |1\rangle) \tag{1.249}$$

The state of the system after the Hadamard gate is applied:

$$|\psi\rangle = \frac{1}{\sqrt{2}}(|0\rangle + |1\rangle) \otimes |0\rangle = \frac{1}{\sqrt{2}}(|00\rangle + |10\rangle) \tag{1.250}$$

3. **Apply CNOT Gate:** Apply the CNOT gate with the first qubit as the control and the second qubit as the target.

$$\text{CNOT}\left(\frac{1}{\sqrt{2}}(|00\rangle + |10\rangle)\right) = \frac{1}{\sqrt{2}}(|00\rangle + |11\rangle) \tag{1.251}$$

The state of the system is now an entangled EPR pair:

$$|\text{EPR}\rangle = \frac{1}{\sqrt{2}}(|00\rangle + |11\rangle) \tag{1.252}$$

4. **Verification:** Measure the qubits in the computational basis to verify the entanglement. In an ideal scenario, measurements of both qubits will always be correlated.

The principles of superposition and entanglement are fundamental to the power of quantum computing.

1.14 TRIVIA QUESTIONS

1.14.0.0.1 What is the phenomenon called when the quantum states of two or more particles become interconnected such that the state of each particle cannot be described independently of the state of the others?

1.14.0.0.2 Answer:

Quantum Entanglement

1.14.0.0.3 Who derived the inequalities that test for local realism in quantum mechanics?

1.14.0.0.4 Answer:

John Bell

1.14.0.0.5 What is the maximum value that Bell's inequality can reach according to classical physics?

1.14.0.0.6 Answer:

2

1.14.0.0.7 What is the value that quantum mechanical predictions can reach, violating Bell's inequality?

1.14.0.0.8 Answer:

$2\sqrt{2}$

1.14.0.0.9 In the context of quantum teleportation, what are the unitary transformations Bob needs to apply based on Alice's measurement results?

1.14.0.0.10 Answer:

$I, \sigma_x, \sigma_y, \sigma_z$

1.14.0.0.11 *What is the name of the process where the quantum state of a particle is transmitted from one location to another without physically transferring the particle itself?*

1.14.0.0.12 *Answer:*

Quantum Teleportation

1.14.0.0.13 *What is the quantum state shared between Alice and Bob in the quantum teleportation protocol?*

1.14.0.0.14 *Answer:*

Bell State or EPR Pair

AN UNWORKED PROBLEM: NONLOCALITY, HIDDEN VARIABLES, AND CURRENT UNDERSTANDING

Nonlocality in quantum mechanics refers to the phenomenon where entangled particles exhibit correlations that cannot be explained by any local theory, where local theories are those that assume physical influences do not travel faster than the speed of light. This challenges classical intuitions about causality and locality. The concept of hidden variables was proposed to explain these quantum correlations by suggesting that particles possess predetermined properties (hidden variables) that determine their behavior. However, John Bell's theorem and the subsequent experimental violations of Bell's inequalities have provided strong evidence against local hidden variable theories, indicating that no local realistic theory can reproduce all the predictions of quantum mechanics. Current understanding accepts that quantum mechanics inherently includes nonlocal correlations, a viewpoint reinforced by experiments confirming entanglement and the principles of quantum information theory. This has profound implications for the nature of reality, suggesting that the universe at a fundamental level does not adhere to classical notions of locality and determinism. Nonlocality is now a cornerstone in the development of quantum technologies, including quantum computing and quantum cryptography, which exploit these unique quantum correlations. Therefore, it is the incomplete or "unworked" problems in this area of quantum physics that will lead to future answers of questions yet to be considered.

1.15 QUANTUM HARMONIC OSCILLATOR AND FIELD QUANTIZATION

1.15.1 QUANTUM HARMONIC OSCILLATOR

The quantum harmonic oscillator is a foundational model in quantum mechanics, describing a particle subject to a restoring force proportional to its displacement. It is essential for understanding various quantum systems, including vibrational modes in molecules and quantum fields.

1.15.1.1 Creation and Annihilation Operators

In the quantum harmonic oscillator, we introduce the creation ($\hat{a}^\dagger$) and annihilation ($\hat{a}$) operators, which simplify the mathematical treatment and reveal the discrete nature of the oscillator's energy levels.

The Hamiltonian for the harmonic oscillator is given by:

$$\hat{H} = \frac{\hat{p}^2}{2m} + \frac{1}{2}m\omega^2\hat{x}^2, \tag{1.253}$$

where $\hat{p}$ is the momentum operator, $\hat{x}$ is the position operator, m is the mass of the particle, and ω is the angular frequency of the oscillator.

We define the dimensionless position and momentum operators:

$$\hat{x} = \sqrt{\frac{\hbar}{2m\omega}}(\hat{a}+\hat{a}^\dagger), \quad \hat{p} = i\sqrt{\frac{\hbar m\omega}{2}}(\hat{a}^\dagger - \hat{a}), \tag{1.254}$$

where $\hat{a}$ and $\hat{a}^\dagger$ satisfy the commutation relation:

$$[\hat{a}, \hat{a}^\dagger] = 1. \tag{1.255}$$

In terms of these operators, the Hamiltonian becomes:

$$\hat{H} = \hbar\omega\left(\hat{a}^\dagger\hat{a} + \frac{1}{2}\right). \tag{1.256}$$

The eigenstates of the number operator $\hat{N} = \hat{a}^\dagger\hat{a}$ are the energy eigenstates $|n\rangle$ with eigenvalues n, where $n = 0, 1, 2, \ldots$. The corresponding energy levels are:

$$E_n = \hbar\omega\left(n + \frac{1}{2}\right). \tag{1.257}$$

The ground state energy $E_0 = \frac{1}{2}\hbar\omega$ represents the quantum vacuum energy, indicating that even in its lowest energy state, the harmonic oscillator has nonzero energy due to quantum fluctuations.

1.15.2 QUANTUM HARMONIC OSCILLATORS IN QUANTUM FIELD THEORY

The quantum harmonic oscillator introduces the creation and annihilation operators, highlighting their role in quantizing the energy levels and explaining the concept of vacuum energy as shown in Figure 1.7 whereas n=0 still has a nonzero energy. This formalism is crucial for understanding more complex quantum systems and phenomena such as QFT and the quantization of the electromagnetic field in which the field itself is considered to be an infinite number of quantum harmonic oscillators each with their own field properties and energy spectra.

1.15.3 MODE FIELD THEORY OF LIGHT AND QUANTIZATION

The mode field theory of light describes the quantization of electromagnetic fields, treating light as quantized harmonic oscillators corresponding to different modes of the field.

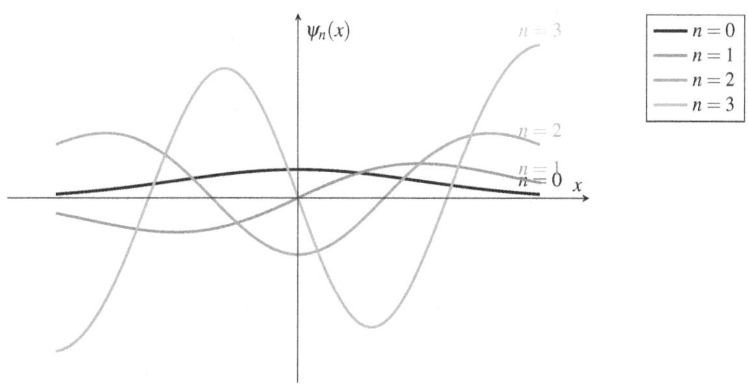

FIGURE 1.7
Energy levels and wavefunctions of a quantum harmonic oscillator.

1.15.3.1 Quantization of the Electromagnetic Field

In the quantization of the electromagnetic field, the field operators are expressed in terms of creation and annihilation operators for each mode.

The electric field operator $\hat{\mathbf{E}}(\mathbf{r},t)$ in the mode decomposition is given by:

$$\hat{\mathbf{E}}(\mathbf{r},t) = i\sum_{\mathbf{k},\lambda}\sqrt{\frac{\hbar\omega_k}{2\varepsilon_0 V}}\left(\hat{a}_{\mathbf{k},\lambda}e^{i(\mathbf{k}\cdot\mathbf{r}-\omega_k t)} - \hat{a}^\dagger_{\mathbf{k},\lambda}e^{-i(\mathbf{k}\cdot\mathbf{r}-\omega_k t)}\right)\mathbf{e}_{\mathbf{k},\lambda}, \tag{1.258}$$

where $\mathbf{k}$ is the wave vector, ω_k is the angular frequency of the mode, λ is the polarization index, V is the quantization volume, ε_0 is the vacuum permittivity, $\mathbf{e}_{\mathbf{k},\lambda}$ are the polarization vectors, and $\hat{a}_{\mathbf{k},\lambda}$ and $\hat{a}^\dagger_{\mathbf{k},\lambda}$ are the annihilation and creation operators for the mode $(\mathbf{k},\lambda)$.

The Hamiltonian for the quantized electromagnetic field in a mode $(\mathbf{k},\lambda)$ is given by:

$$\hat{H} = \sum_{\mathbf{k},\lambda}\hbar\omega_k\left(\hat{a}^\dagger_{\mathbf{k},\lambda}\hat{a}_{\mathbf{k},\lambda} + \frac{1}{2}\right). \tag{1.259}$$

The field quantization process involves expressing the field operators in terms of the creation and annihilation operators. Using the mode decomposition, we can write the field operators as:

$$\hat{\mathbf{E}}(\mathbf{r},t) = i\sum_{\mathbf{k},\lambda}\sqrt{\frac{\hbar\omega_k}{2\varepsilon_0 V}}\left(\hat{a}_{\mathbf{k},\lambda}e^{i(\mathbf{k}\cdot\mathbf{r}-\omega_k t)} - \hat{a}^\dagger_{\mathbf{k},\lambda}e^{-i(\mathbf{k}\cdot\mathbf{r}-\omega_k t)}\right)\mathbf{e}_{\mathbf{k},\lambda}. \tag{1.260}$$

Comparing this with the classical expansion, we recognize the quantization procedure, replacing classical amplitudes with operators. This leads to the quantized field representation consistent with the quantum harmonic oscillator formalism applied to each mode of the electromagnetic field.

The vacuum state $|0\rangle$ is defined as the state with no photons in any mode:

$$\hat{a}_{\mathbf{k},\lambda}|0\rangle = 0 \quad \forall\mathbf{k},\lambda. \tag{1.261}$$

The energy of the vacuum state is the sum of the zero-point energies of all modes:

$$E_{\text{vacuum}} = \sum_{\mathbf{k},\lambda}\frac{1}{2}\hbar\omega_k. \tag{1.262}$$

1.15.4 THE CASIMIR EFFECT

Modern QFT suggests that the field itself is ubiquitous and permeates all of space, but it localizes into discrete quanta, or photons, when energy is exchanged or interacted with. The field exists everywhere, continuously oscillating with vacuum fluctuations, even in the absence of particles. These fluctuations lead to phenomena such as the Casimir effect and spontaneous emission. The Casimir effect is a physical force arising from the quantum vacuum fluctuations between two closely spaced, uncharged conducting plates. Discovered by Dutch physicist Hendrik Casimir in 1948, this phenomenon demonstrates the reality of vacuum energy in QFT.

In a vacuum, virtual particles constantly appear and annihilate, creating a fluctuating electromagnetic field. When two conducting plates are placed close together, they constrain the wavelengths of the virtual photons that can exist between them, resulting in a measurable force.

To derive the Casimir effect, consider two parallel, uncharged conducting plates separated by a distance a in a vacuum. The boundary conditions impose that the electric field must be zero at the surfaces of the conductors. The allowed modes between the plates are quantized and given by:

$$k_n = \frac{n\pi}{a} \quad \text{for} \quad n = 1,2,3,\ldots \tag{1.263}$$

The energy of each mode is given by:

$$E_n = \frac{\hbar c k_n}{2} = \frac{\hbar c \pi n}{2a} \tag{1.264}$$

The total energy of the modes between the plates can be found by summing over all allowed modes:

$$E = \sum_{n=1}^{\infty} \frac{\hbar c \pi n}{2a} \tag{1.265}$$

This series is divergent, so to find the physical result, we subtract the infinite energy of the modes in the absence of the plates (i.e., the modes in the infinite vacuum), which is given by:

$$E_{\text{vac}} = \int_0^{\infty} \frac{\hbar c k}{2} \frac{dk}{2\pi} \tag{1.266}$$

Regularizing this expression using techniques from QFT, such as zeta function regularization, we obtain the Casimir energy per unit area A between the plates:

$$E_{\text{Casimir}} = \frac{\hbar c \pi^2}{720 a^3} \tag{1.267}$$

The Casimir force F per unit area is then derived from the negative gradient of the energy with respect to the separation a:

$$F = -\frac{\partial E_{\text{Casimir}}}{\partial a} = -\frac{\hbar c \pi^2}{240 a^4} \tag{1.268}$$

This attractive force between the plates is a direct consequence of the quantized nature of the electromagnetic field and vacuum fluctuations, providing a remarkable demonstration of QFT's implications in macroscopic phenomena.

1.15.5 A MODERN PERSPECTIVE

When an interaction occurs, such as an electron transitioning between energy levels in an atom, the field can localize and transfer energy in quantized packets. These packets, or photons, are described by the creation and annihilation operators acting on the vacuum state. This localization is what we perceive as the particle-like behavior of photons.

In this framework, light is not just a collection of particles but rather a quantized excitation of the underlying electromagnetic field. This duality of wave-like and particle-like behavior is a cornerstone of QFT and is fundamental to our understanding of phenomena in quantum optics and beyond.

By quantizing the electromagnetic field, we not only explain the particle nature of light but also gain insights into the continuous wave nature that underpins it, providing a comprehensive picture that bridges classical and quantum descriptions of electromagnetic phenomena.

This derivation leads us to a quantized field Equation that encapsulates the quantized electric field operator in terms of its mode expansion and the creation and annihilation operators:

$$\hat{\mathbf{E}}(\mathbf{r},t) = i \sum_{\mathbf{k},\lambda} \sqrt{\frac{\hbar \omega_k}{2\varepsilon_0 V}} \left(\hat{a}_{\mathbf{k},\lambda} e^{i(\mathbf{k}\cdot\mathbf{r} - \omega_k t)} - \hat{a}^{\dagger}_{\mathbf{k},\lambda} e^{-i(\mathbf{k}\cdot\mathbf{r} - \omega_k t)} \right) \mathbf{e}_{\mathbf{k},\lambda}. \tag{1.269}$$

Which can be simplified to

$$\hat{\mathbf{E}}(\mathbf{r},t) = \mathbf{E}(\mathbf{r},t)\hat{a}_{\mathbf{k},\lambda} - \mathbf{E}^*(\mathbf{r},t)\hat{a}^{\dagger}_{\mathbf{k},\lambda}, \tag{1.270}$$

where the classical field and it's conjugate are:

$$\mathbf{E}(\mathbf{r},t) = i\sum_{\mathbf{k},\lambda}\sqrt{\frac{\hbar\omega_k}{2\varepsilon_0 V}}\mathbf{e}_{\mathbf{k},\lambda}e^{i(\mathbf{k}\cdot\mathbf{r}-\omega_k t)}, \tag{1.271}$$

and

$$\mathbf{E}^*(\mathbf{r},t) = -i\sum_{\mathbf{k},\lambda}\sqrt{\frac{\hbar\omega_k}{2\varepsilon_0 V}}\mathbf{e}_{\mathbf{k},\lambda}e^{-i(\mathbf{k}\cdot\mathbf{r}-\omega_k t)}. \tag{1.272}$$

This equation succinctly represents the quantized nature of the electromagnetic field, showing how classical field components and quantum operators interact to produce observable phenomena such as vacuum fluctuations, spontaneous emission, and the discrete nature of photon exchange.

Equation (1.264) provides a profound perspective on the nature of the electromagnetic field in quantum theory. It suggests that the field exists as a continuous entity throughout space, with the potential to manifest energy in discrete quanta, or photons, localized at specific points in space and time. This conceptualization bridges the classical and quantum views of light, showing that while the field permeates space, the photons are the localized manifestations of energy interactions within this field.

Stadium Simile for Quantized Optical Fields

A helpful simile to understand this quantized field theory is to imagine a football stadium. Consider that the entire set of seats in the stadium represents the electromagnetic field. There is a possibility of finding a quanta, analogous to a person, in any seat, ranging from section 1 to section 50, from rows A to Z, and from seat 1 to 50 within each row. The complete set of possible seat locations represents the field. When a person comes in and sits in a specific seat, for example, section 3, row N, seat 12, there is a localized presence of energy in that particular spot. Similarly, in the quantized field theory, the field is always present throughout space, but the energy (photon) becomes localized when an interaction occurs, manifesting as a discrete packet in a specific location.

In the simile of a football stadium, if there are multiple people, or quanta, present, each person has an equal probability to occupy any seat, which is proportional to $1/\sqrt{N}$, where N is the photon number. This means that before they sit, the probability amplitude for each possible seat is spread uniformly across all seats.

The Dirac notation wavefunction for N people in the stadium before they sit is given by:

$$|\Psi_{\text{before}}\rangle = \frac{1}{\sqrt{N!}}\sum_{i=1}^{N}|i\rangle, \tag{1.273}$$

where i denotes the different possible seats.

After they sit, assuming they occupy specific seats $\{i_1, i_2, \ldots, i_N\}$, the wavefunction collapses to:

$$|\Psi_{\text{after}}\rangle = |i_1, i_2, \ldots, i_N\rangle. \tag{1.274}$$

This analogy highlights how the quantized field theory integrates the continuous nature of fields with the discrete nature of particles, providing a comprehensive understanding of phenomena like spontaneous emission, where photons appear at specific points in space-time as a result of interactions within the pervasive field.

1.16 CHAPTER SUMMARY AND DISCUSSION

This chapter has explored the foundational principles and mathematical framework of quantum mechanics, providing a comprehensive overview of the essential concepts and tools required for understanding and analyzing quantum phenomena. We began with a review of wave-particle duality, quantum states, superposition, and the measurement process, highlighting the core ideas that distinguish quantum mechanics from classical physics.

Wave-particle duality, first introduced through key experiments like Young's double-slit experiment, demonstrates the dual nature of quantum entities. Particles such as electrons and photons exhibit both wave-like and particle-like behaviors, depending on the experimental setup. This duality challenges classical intuition and underscores the complexity of quantum systems.

The concept of quantum states and superposition further distinguishes quantum mechanics. Unlike classical states, which are deterministic, quantum states are described by wave functions or state vectors in a Hilbert space, allowing a system to exist simultaneously in multiple states until measured. This principle is famously illustrated by Schrödingers cat thought experiment and underpins many quantum technologies, including quantum computing.

Measurement in quantum mechanics is not a passive observation but an active process that affects the system's state. The act of measurement causes the wave function to collapse to a single eigenstate, a phenomenon explained by Borns rule. This process introduces intrinsic uncertainties and probabilistic outcomes, diverging from the deterministic predictions of classical mechanics.

We then delved into the mathematical framework of quantum mechanics, focusing on linear algebra and functional analysis. Key concepts include Hilbert spaces, operators, eigenstates, and eigenvalues. Operators such as Hermitian and unitary operators play crucial roles in defining observables and the evolution of quantum states. The time-independent Schrödinger equation was introduced as a fundamental equation governing the dynamics of quantum systems, with applications to simple systems like the particle in a box.

The discussion on quantum harmonic oscillators introduced the creation and annihilation operators, highlighting their role in quantizing the energy levels of the system and explaining the concept of vacuum energy. This formalism is crucial for understanding more complex quantum systems and phenomena such as QFT and the quantization of the electromagnetic field.

Finally, we addressed the quantization of the electromagnetic field and the concept of photons as quanta of the field. Using mode decomposition and field operators, we derived expressions for the quantized field and discussed how this framework leads to an understanding of phenomena such as spontaneous emission and the Casimir effect.

In summary, this chapter provided a thorough introduction to the key principles and mathematical tools of quantum mechanics, laying the groundwork for further exploration of advanced topics. The concepts of wave-particle duality, superposition, measurement, and the mathematical formalism of quantum mechanics are essential for understanding and applying quantum theory to various physical systems and technological applications.

1.17 FINAL COMPREHENSIVE EXAM PROBLEMS

This exam consists of 10 problems. Complete all 10 problems. Clearly show all steps and reasoning for full credit.

PROBLEM 1: TIME-INDEPENDENT PERTURBATION THEORY

Consider a quantum mechanical system with a Hamiltonian given by:

$$\hat{H} = \hat{H}_0 + \lambda \hat{H}', \tag{1.275}$$

where $\hat{H}_0$ has known eigenstates $|n\rangle$ with corresponding eigenvalues $E_n^{(0)}$, and λ is a small parameter.

(a) Calculate the first-order correction to the energy for the ground state $|0\rangle$.

(b) Calculate the first-order correction to the wave function for the ground state $|0\rangle$.

(c) Now consider that the system described is a harmonic oscillator, and the perturbation $\hat{H}'$ is a linear function of position, kx, where k is a constant. The unperturbed ground state wave function of a harmonic oscillator is:

$$\psi_0(x) = \left(\frac{m\omega}{\pi\hbar}\right)^{1/4} e^{-\frac{m\omega x^2}{2\hbar}}, \tag{1.276}$$

where m is the mass of the oscillator, ω is the angular frequency, and $\hbar$ is the reduced Planck's constant.

Solution:

(a) First-Order Correction to the Energy of the Ground State

The first-order correction to the energy is given by:

$$E_0^{(1)} = \langle 0|\hat{H}'|0\rangle. \tag{1.277}$$

(b) First-Order Correction to the Wave Function of the Ground State

The first-order correction to the wave function is given by:

$$|0^{(1)}\rangle = \sum_{n\neq 0} \frac{\langle n|\hat{H}'|0\rangle}{E_0^{(0)} - E_n^{(0)}}|n\rangle. \tag{1.278}$$

(c) Consideration of the Harmonic Oscillator with a Linear Perturbation

The first-order correction to the energy of the ground state with this perturbation is calculated as follows:

$$E_0^{(1)} = \langle 0|kx|0\rangle = k\int_{-\infty}^{\infty} x\psi_0(x)^2\,dx. \tag{1.279}$$

Using the expression for the first-order correction to the energy:

$$E_0^{(1)} = \langle 0|\hat{H}'|0\rangle = \langle 0|kx|0\rangle \tag{1.280}$$

To compute this, substitute the expression for $\psi_0(x)$, the ground state wave function of the harmonic oscillator, into the integral:

$$E_0^{(1)} = \int_{-\infty}^{\infty} \psi_0(x)kx\psi_0(x)\,dx = k\int_{-\infty}^{\infty} x\left(\frac{m\omega}{\pi\hbar}\right)^{1/2} e^{-\frac{m\omega x^2}{\hbar}}\,dx \tag{1.281}$$

Since the integrand $x\left(\frac{m\omega}{\pi\hbar}\right)^{1/2} e^{-\frac{m\omega x^2}{\hbar}}$ is an odd function (because x is odd and the exponential function is even), the integral over symmetric limits from $-\infty$ to ∞ evaluates to zero:

$$E_0^{(1)} = k\times 0 = 0 \tag{1.282}$$

The result $E_0^{(1)} = 0$ demonstrates that the first-order energy correction due to a linear perturbation in a symmetric potential about the origin is zero. This is because the expectation value of x in the symmetric ground state is zero, reflecting the symmetry of the potential and the ground state wave function of the harmonic oscillator. We also see that the linear perturbation does not shift the energy of the ground state of the harmonic oscillator at the first order of perturbation theory.

PROBLEM 2: QUANTUM HARMONIC OSCILLATOR

Consider a one-dimensional quantum harmonic oscillator with Hamiltonian:

$$\hat{H} = \frac{\hat{p}^2}{2m} + \frac{1}{2}m\omega^2\hat{x}^2. \tag{1.283}$$

(a) Write down the creation and annihilation operators $\hat{a}^\dagger$ and $\hat{a}$.
(b) Show that the ground state energy is $\frac{1}{2}\hbar\omega$.
Solution:
(a) The creation and annihilation operators are:

$$\hat{a} = \sqrt{\frac{m\omega}{2\hbar}}\left(\hat{x} + \frac{i\hat{p}}{m\omega}\right), \tag{1.284}$$

$$\hat{a}^\dagger = \sqrt{\frac{m\omega}{2\hbar}}\left(\hat{x} - \frac{i\hat{p}}{m\omega}\right). \tag{1.285}$$

(b) To show that the ground state energy is $\frac{1}{2}\hbar\omega$, start by expressing the Hamiltonian in terms of $\hat{a}$ and $\hat{a}^\dagger$:

$$\hat{x} = \sqrt{\frac{\hbar}{2m\omega}}(\hat{a} + \hat{a}^\dagger), \tag{1.286}$$

$$\hat{p} = i\sqrt{\frac{\hbar m\omega}{2}}(\hat{a}^\dagger - \hat{a}). \tag{1.287}$$

Substitute these into the Hamiltonian formula:

$$\hat{H} = \frac{1}{2m}\left(i\sqrt{\frac{\hbar m\omega}{2}}(\hat{a}^\dagger - \hat{a})\right)^2 + \frac{1}{2}m\omega^2\left(\sqrt{\frac{\hbar}{2m\omega}}(\hat{a} + \hat{a}^\dagger)\right)^2, \tag{1.288}$$

$$\hat{H} = \frac{\hbar\omega}{4}(\hat{a}^\dagger - \hat{a})^2 + \frac{\hbar\omega}{4}(\hat{a} + \hat{a}^\dagger)^2. \tag{1.289}$$

Using the commutation relation $[\hat{a}, \hat{a}^\dagger] = 1$, which implies $\hat{a}\hat{a}^\dagger = \hat{a}^\dagger\hat{a} + 1$ we can simplify:

$$\hat{H} = \frac{\hbar\omega}{4}(\hat{a}^\dagger - \hat{a})^2 + \frac{\hbar\omega}{4}(\hat{a} + \hat{a}^\dagger)^2 \tag{1.290}$$

$$= \frac{\hbar\omega}{4}\left[(\hat{a}^\dagger)^2 + (\hat{a})^2 - 2\hat{a}^\dagger\hat{a} + 2\hat{a}\hat{a}^\dagger\right] \tag{1.291}$$

$$= \frac{\hbar\omega}{4}\left[(\hat{a}^\dagger)^2 + (\hat{a})^2 + 2\hat{a}\hat{a}^\dagger\right] \tag{1.292}$$

$$= \frac{\hbar\omega}{4}\left[\hat{a}^\dagger\hat{a}^\dagger + \hat{a}\hat{a} + 2\hat{a}\hat{a}^\dagger\right] \tag{1.293}$$

$$= \frac{\hbar\omega}{4}\left[\hat{a}^\dagger\hat{a}^\dagger + 2\hat{a}\hat{a}^\dagger + \hat{a}\hat{a}\right] \tag{1.294}$$

$$= \frac{\hbar\omega}{4}\left[\hat{a}^\dagger\hat{a} + \hat{a}\hat{a}^\dagger + \hat{a}\hat{a}\right] \tag{1.295}$$

$$= \frac{\hbar\omega}{4}\left[\hat{a}^\dagger\hat{a} + \hat{a}^\dagger\hat{a} + 1\right] \tag{1.296}$$

$$= \frac{\hbar\omega}{2}\left[\hat{a}^\dagger\hat{a} + \frac{1}{2}\right] \tag{1.297}$$

The ground state energy, corresponding to the vacuum state $|0\rangle$, where $\hat{a}|0\rangle = 0$, is:

$$E_0 = \langle 0|\hat{H}|0\rangle = \frac{1}{2}\hbar\omega. \tag{1.298}$$

PROBLEM 3: TEST FOR ENTANGLEMENT

Consider the following two-particle wavefunctions:

$$|\psi_1\rangle = \frac{1}{\sqrt{2}}(|00\rangle + |11\rangle),$$

$$|\psi_2\rangle = \frac{1}{\sqrt{2}}(|01\rangle + |10\rangle),$$

$$|\psi_3\rangle = \frac{1}{2}(|00\rangle + |01\rangle + |10\rangle + |11\rangle).$$

1. Determine if each wavefunction represents an entangled state.
2. Justify your answer using the mathematical criterion for entanglement.

SOLUTION

1. For $|\psi_1\rangle$:

$$|\psi_1\rangle = \frac{1}{\sqrt{2}}(|00\rangle + |11\rangle) \tag{1.299}$$

To check for entanglement, we calculate the reduced density matrix by tracing out one of the subsystems:

$$\rho_{\psi_1} = |\psi_1\rangle\langle\psi_1| = \frac{1}{2}(|00\rangle + |11\rangle)(\langle 00| + \langle 11|) \tag{1.300}$$

$$= \frac{1}{2}(|00\rangle\langle 00| + |00\rangle\langle 11| + |11\rangle\langle 00| + |11\rangle\langle 11|) \tag{1.301}$$

The partial trace over the second qubit gives:

$$\rho_{\psi_1}^{(1)} = \mathrm{Tr}_2(\rho_{\psi_1}) \tag{1.302}$$

$$= \frac{1}{2}(|0\rangle\langle 0| + |1\rangle\langle 1|) \tag{1.303}$$

$$= \frac{1}{2}I \tag{1.304}$$

Since $\rho_{\psi_1}^{(1)}$ is not a pure state, $|\psi_1\rangle$ is entangled.

2. For $|\psi_2\rangle$:

$$|\psi_2\rangle = \frac{1}{\sqrt{2}}(|01\rangle + |10\rangle) \tag{1.305}$$

Following the same procedure as for $|\psi_1\rangle$:

$$\rho_{\psi_2} = |\psi_2\rangle\langle\psi_2| = \frac{1}{2}(|01\rangle + |10\rangle)(\langle 01| + \langle 10|) \tag{1.306}$$

$$= \frac{1}{2}(|01\rangle\langle 01| + |01\rangle\langle 10| + |10\rangle\langle 01| + |10\rangle\langle 10|) \tag{1.307}$$

The partial trace over the second qubit gives:

$$\rho_{\psi_2}^{(1)} = \mathrm{Tr}_2(\rho_{\psi_2}) \tag{1.308}$$

$$= \frac{1}{2}(|0\rangle\langle 0| + |1\rangle\langle 1|) \tag{1.309}$$

$$= \frac{1}{2}I \tag{1.310}$$

Since $\rho_{\psi_2}^{(1)}$ is not a pure state, $|\psi_2\rangle$ is entangled.

3. For $|\psi_3\rangle$:

$$|\psi_3\rangle = \frac{1}{2}(|00\rangle + |01\rangle + |10\rangle + |11\rangle) \tag{1.311}$$

Again, let's compute the density matrix and the reduced density matrix:

$$\rho_{\psi_3} = |\psi_3\rangle\langle\psi_3| \tag{1.312}$$
$$= \frac{1}{4}(|00\rangle + |01\rangle + |10\rangle + |11\rangle)(\langle 00| + \langle 01| + \langle 10| + \langle 11|) \tag{1.313}$$
$$= \frac{1}{4}(|00\rangle\langle 00| + |00\rangle\langle 01| + |00\rangle\langle 10| + |00\rangle\langle 11| + \tag{1.314}$$
$$|01\rangle\langle 00| + |01\rangle\langle 01| + |01\rangle\langle 10| + |01\rangle\langle 11| + \tag{1.315}$$
$$|10\rangle\langle 00| + |10\rangle\langle 01| + |10\rangle\langle 10| + |10\rangle\langle 11| + \tag{1.316}$$
$$|11\rangle\langle 00| + |11\rangle\langle 01| + |11\rangle\langle 10| + |11\rangle\langle 11|) \tag{1.317}$$

Taking the partial trace over the second qubit:

$$\rho^{(1)}_{\psi_3} = \text{Tr}_2(\rho_{\psi_3}) \tag{1.318}$$
$$= \frac{1}{4}(2|0\rangle\langle 0| + 2|1\rangle\langle 1|) \tag{1.319}$$
$$= \frac{1}{2}(|0\rangle\langle 0| + |1\rangle\langle 1|) \tag{1.320}$$
$$= \frac{1}{2}I \tag{1.321}$$

Like the other two cases, since $\rho^{(1)}_{\psi_3}$ is not a pure state, $|\psi_3\rangle$ is also entangled. Each of these states cannot be factorized into individual particle states, confirming their entangled nature.

PROBLEM 4: MATRIX MANIPULATION IN QUANTUM MECHANICS

Consider the Hamiltonian of a two-level system given by the matrix:

$$H = \begin{pmatrix} 2 & 1 \\ 1 & 3 \end{pmatrix}.$$

1. Verify that H is Hermitian.
2. Find the eigenvalues of H.
3. Find and normalize the eigenvectors of H.
4. Construct a unitary matrix U that diagonalizes H.
5. Find the determinant, trace, and adjoint of H.

SOLUTION

1. To verify that H is Hermitian, we check if $H = H^\dagger$:

$$H^\dagger = \begin{pmatrix} 2 & 1 \\ 1 & 3 \end{pmatrix} = H. \tag{1.322}$$

Therefore, H is Hermitian.

2. To find the eigenvalues, solve the characteristic equation:

$$\det(H - \lambda I) = 0. \tag{1.323}$$

$$\det\begin{pmatrix} 2-\lambda & 1 \\ 1 & 3-\lambda \end{pmatrix} = (2-\lambda)(3-\lambda) - 1 = \lambda^2 - 5\lambda + 5. \tag{1.324}$$

The eigenvalues are the roots of the equation:

$$\lambda = \frac{5 \pm \sqrt{5^2 - 4 \cdot 5}}{2} = \frac{5 \pm \sqrt{5}}{2}. \tag{1.325}$$

3. To find the eigenvectors, substitute each eigenvalue back into $H - \lambda I$:

$$(H - \lambda_1 I)\mathbf{v}_1 = 0, \quad \lambda_1 = \frac{5+\sqrt{5}}{2}, \tag{1.326}$$

$$(H - \lambda_2 I)\mathbf{v}_2 = 0, \quad \lambda_2 = \frac{5-\sqrt{5}}{2}. \tag{1.327}$$

Solving these equations gives the eigenvectors:

$$\mathbf{v}_1 = \begin{pmatrix} 1 \\ \frac{\sqrt{5}-1}{2} \end{pmatrix}, \quad \mathbf{v}_2 = \begin{pmatrix} 1 \\ -\frac{\sqrt{5}+1}{2} \end{pmatrix}. \tag{1.328}$$

Normalize the eigenvectors:

$$\mathbf{v}_1 = \frac{1}{\sqrt{1 + \left(\frac{\sqrt{5}-1}{2}\right)^2}} \begin{pmatrix} 1 \\ \frac{\sqrt{5}-1}{2} \end{pmatrix},$$

$$\mathbf{v}_2 = \frac{1}{\sqrt{1 + \left(\frac{\sqrt{5}+1}{2}\right)^2}} \begin{pmatrix} 1 \\ -\frac{\sqrt{5}+1}{2} \end{pmatrix}. \tag{1.329}$$

4. The unitary matrix U that diagonalizes H is constructed from the normalized eigenvectors:

$$U = \begin{pmatrix} \mathbf{v}_1 & \mathbf{v}_2 \end{pmatrix}. \tag{1.330}$$

$$U = \begin{pmatrix} \frac{2}{\sqrt{5+\sqrt{5}}} & \frac{2}{\sqrt{5-\sqrt{5}}} \\ \frac{\sqrt{5}-1}{\sqrt{5+\sqrt{5}}} & -\frac{\sqrt{5}+1}{\sqrt{5-\sqrt{5}}} \end{pmatrix}. \tag{1.331}$$

5. The determinant, trace, and adjoint of H:

$$\det(H) = \det\begin{pmatrix} 2 & 1 \\ 1 & 3 \end{pmatrix} = (2 \cdot 3) - (1 \cdot 1) = 5. \tag{1.332}$$

$$\mathrm{tr}(H) = 2 + 3 = 5. \tag{1.333}$$

$$\mathrm{adj}(H) = \begin{pmatrix} 3 & -1 \\ -1 & 2 \end{pmatrix}. \tag{1.334}$$

PROBLEM 5: ENERGY OF FERMIONS AND BOSONS IN A ONE-DIMENSIONAL BOX

Consider N fermions and M bosons confined in a one-dimensional box of length L. Assume the particles do not interact with each other, and that fermions and bosons are distinguishable.

Task

1. Calculate the total energy of the system if the particles are fermions. Assume that the fermions fill the lowest available quantum states according to the Pauli exclusion principle.
2. Calculate the total energy of the system if the particles are bosons, assuming all bosons condense into the ground state (Bose-Einstein condensation).
3. Calculate the total energy of the system when both N fermions and M bosons are present in the box.

Given

The energy levels of a particle in a one-dimensional box are given by:

$$E_n = \frac{\hbar^2 \pi^2 n^2}{2mL^2} \tag{1.335}$$

where n is the quantum number associated with the state of the particle ($n = 1, 2, 3, \ldots$), $\hbar$ is the reduced Planck's constant, m is the mass of the particle, and L is the length of the box.

Solution

1. **Fermions:** The total energy $E_{\text{total, fermions}}$ is the sum of the energies of the first N states:

$$E_{\text{total, fermions}} = \sum_{n=1}^{N} \frac{\hbar^2 \pi^2 n^2}{2mL^2} \tag{1.336}$$

2. **Bosons:** Assuming all M bosons condense into the ground state ($n = 1$), the total energy $E_{\text{total, bosons}}$ is:

$$E_{\text{total, bosons}} = M \cdot \frac{\hbar^2 \pi^2}{2mL^2} \tag{1.337}$$

3. **Combined System:** In the combined system, assume the fermions occupy the lowest N energy states. Since bosons do not obey the Pauli exclusion principle, all M bosons can still occupy the ground state, irrespective of the fermions:

$$E_{\text{total, combined}} = E_{\text{total, fermions}} + E_{\text{total, bosons}} \tag{1.338}$$

$$= \sum_{n=1}^{N} \frac{\hbar^2 \pi^2 n^2}{2mL^2} + M \cdot \frac{\hbar^2 \pi^2}{2mL^2} \tag{1.339}$$

Note that this scenario assumes no interaction between fermions and bosons.

PROBLEM 6: SCATTERING PROBLEM

Suppose a scattering potential has the form

$$V(r) = D\frac{e^{-br}}{r}$$

where D and b are constants.

(a) What are the units of D and b?
(b) Find the scattering amplitude.
(c) Calculate the total cross section.

Solution

(a) **Units of D and b**:
The potential $V(r)$ has units of energy. Given that $\frac{e^{-br}}{r}$ is dimensionless, D must have units of energy. The exponential decay parameter b must have units of $\frac{1}{\text{length}}$ to ensure the exponent is dimensionless.

(b) **Scattering Amplitude**:
Using the first Born approximation, the scattering amplitude $f(\theta)$ is given by:

$$f(\theta) = -\frac{2m}{\hbar^2}\int_0^\infty V(r)\frac{\sin(kr\sin\theta)}{kr\sin\theta}r^2 dr \tag{1.340}$$

Substituting $V(r) = D\frac{e^{-br}}{r}$:

$$f(\theta) = -\frac{2mD}{\hbar^2}\int_0^\infty e^{-br}\sin(kr\sin\theta) r dr \tag{1.341}$$

Solving this integral yields:

$$f(\theta) = -\frac{2mD}{\hbar^2}\frac{1}{b^2+(k\sin\theta)^2} \tag{1.342}$$

which is the scattering amplitude for the given potential in the Born approximation.

(c) **Total Cross Section**:
The total cross section is calculated from the scattering amplitude:

$$\begin{aligned}\sigma &= \int |f(\theta)|^2 d\Omega \\ &= \int_0^{2\pi}\int_0^{\pi}\left|-\frac{2mD}{\hbar^2}\frac{1}{b^2+(k\sin\theta)^2}\right|^2 \sin\theta\, d\theta\, d\phi\end{aligned} \tag{1.343}$$

Performing the integration:

$$\sigma = \frac{4\pi m^2 D^2}{\hbar^4}\int_0^{\pi}\frac{\sin\theta}{(b^2+k^2\sin^2\theta)^2}d\theta \tag{1.344}$$

Using a substitution and integrating, we find:

$$\sigma = \frac{4\pi m^2 D^2}{\hbar^4 k b^3} \tag{1.345}$$

which gives the total cross section as a function of the parameters D, b, m, and k.

PROBLEM 7: POTENTIAL AND WAVEFUNCTION

Consider a particle in one dimension subject to the potential

$$V(x) = \begin{cases} -V_0 & \text{if } -a \leq x \leq a \\ 0 & \text{otherwise} \end{cases} \tag{1.346}$$

where $V_0 > 0$ and $a > 0$ are constants.

(a) Consider the wavefunction

$$\psi(x) = \begin{cases} d(x+a) & -a < x < 0 \\ d(a-x) & 0 \leq x < a \\ 0 & \text{otherwise} \end{cases} \tag{1.347}$$

where d is a normalization constant. Find the expectation value of the Hamiltonian in this state.

(b) Use this result to argue that there will always be a bound state for any value of V_0.
(c) Argue that any one-dimensional potential that is (i) zero at $x \to \pm\infty$, (ii) nowhere greater than zero, and (iii) less than zero in some finite interval, always has a bound state.

SOLUTION

(a) **Expectation value of the Hamiltonian:**
First, we normalize the wavefunction $\psi(x)$:

$$\int_{-\infty}^{\infty} |\psi(x)|^2\, dx = \tag{1.348}$$

$$\int_{-a}^{0} d^2(x+a)^2\, dx + \int_{0}^{a} d^2(a-x)^2\, dx = 1 \tag{1.349}$$

$$2d^2 \int_{0}^{a} x^2\, dx = 1 \tag{1.350}$$

$$2d^2 \cdot \frac{a^3}{3} = 1 \tag{1.351}$$

$$d = \sqrt{\frac{3}{2a^3}} \tag{1.352}$$

Now, calculate the expectation value of the Hamiltonian $\langle H \rangle$:

$$\langle H \rangle = \int_{-\infty}^{\infty} \psi^*(x) \left(-\frac{\hbar^2}{2m}\frac{d^2}{dx^2} + V(x) \right) \psi(x)\, dx \tag{1.353}$$

$$= -\frac{\hbar^2}{2m} \int_{-\infty}^{\infty} \psi^*(x) \frac{d^2\psi(x)}{dx^2}\, dx + \int_{-a}^{a} V(x)|\psi(x)|^2\, dx \tag{1.354}$$

$$= -V_0 \int_{-a}^{a} \left(\frac{3}{2a^3}\right)(a^2 - x^2)\, dx \tag{1.355}$$

$$= -V_0 \left(\frac{3}{2a^3}\right)\left(2a^3 - \frac{2a^3}{3}\right) \tag{1.356}$$

$$= -V_0 \left(1 - \frac{1}{3}\right) \tag{1.357}$$

$$= -\frac{2}{3} V_0. \tag{1.358}$$

(b) **Existence of a Bound State:**
The negative expectation value of the Hamiltonian indicates that the particle has a bound state for any $V_0 > 0$, as the energy is lower than the potential outside the well.

(c) **General Condition for a Bound State:**
Any potential that is zero at infinity, negative somewhere, and not positive anywhere will have at least one bound state. This is because the wavefunction can always be localized in the region of negative potential, leading to an overall negative expectation value of the Hamiltonian, indicating binding.

PROBLEM 8: TWO-DIMENSIONAL HILBERT SPACE

Consider a system in a two-dimensional Hilbert space. In the basis consisting of the two energy eigenstates, the Hamiltonian H and some observable B are given by:

$$H = \begin{pmatrix} 2\varepsilon & \varepsilon \\ \varepsilon & 3\varepsilon \end{pmatrix}, \quad B = \begin{pmatrix} 0 & 1 \\ 1 & 0 \end{pmatrix}$$

where $\varepsilon > 0$.

(a) Find the eigenvalues and eigenstates of B.
(b) At $t = 0$, the system is in the lowest energy eigenstate of H, and a measurement of B is made. What are the possible outcomes, and what are their probabilities?
(c) Suppose that this measurement yields the lowest eigenvalue of B. At some later time, what would a measurement of the energy yield?
(d) Write down the time-independent Schrödinger equation for this system.

Solution:
Part (a): To find the eigenvalues of B, solve the characteristic equation $\det(B - \lambda I) = 0$:

$$\det\left(\begin{pmatrix} 0 & 1 \\ 1 & 0 \end{pmatrix} - \lambda \begin{pmatrix} 1 & 0 \\ 0 & 1 \end{pmatrix}\right) = \det\left(\begin{pmatrix} -\lambda & 1 \\ 1 & -\lambda \end{pmatrix}\right) \tag{1.359}$$

$$= \lambda^2 - 1 = 0 \tag{1.360}$$

Thus, the eigenvalues are $\lambda_1 = 1$ and $\lambda_2 = -1$.
The corresponding eigenstates are found by solving $(B - \lambda I)\vec{v} = 0$. For $\lambda_1 = 1$:

$$\begin{pmatrix} -1 & 1 \\ 1 & -1 \end{pmatrix}\begin{pmatrix} x \\ y \end{pmatrix} = 0 \tag{1.361}$$

$$\Rightarrow x = y \tag{1.362}$$

giving the normalized eigenstate $\vec{v}_1 = \frac{1}{\sqrt{2}}\begin{pmatrix} 1 \\ 1 \end{pmatrix}$.

For $\lambda_2 = -1$:

$$\begin{pmatrix} 1 & 1 \\ 1 & 1 \end{pmatrix}\begin{pmatrix} x \\ y \end{pmatrix} = 0 \tag{1.363}$$

$$\Rightarrow x = -y \tag{1.364}$$

giving the normalized eigenstate $\vec{v}_2 = \frac{1}{\sqrt{2}}\begin{pmatrix} 1 \\ -1 \end{pmatrix}$.

Part (b): The lowest energy eigenstate of H can be found by solving the characteristic equation $\det(H - \lambda I) = 0$:

$$\det\left(\begin{pmatrix} 2\varepsilon & \varepsilon \\ \varepsilon & 3\varepsilon \end{pmatrix} - \lambda \begin{pmatrix} 1 & 0 \\ 0 & 1 \end{pmatrix}\right) = \det\left(\begin{pmatrix} 2\varepsilon - \lambda & \varepsilon \\ \varepsilon & 3\varepsilon - \lambda \end{pmatrix}\right) \tag{1.365}$$

$$= (2\varepsilon - \lambda)(3\varepsilon - \lambda) - \varepsilon^2 = \lambda^2 - 5\varepsilon\lambda + 5\varepsilon^2 = 0 \tag{1.366}$$

$$\lambda^2 - 5\varepsilon\lambda + 5\varepsilon^2 = 0 \tag{1.367}$$

Solving this quadratic equation, we find the eigenvalues of H:

$$\lambda = \frac{5\varepsilon \pm \sqrt{(5\varepsilon)^2 - 4 \cdot 5 \cdot \varepsilon^2}}{2} = \frac{5\varepsilon \pm \sqrt{25\varepsilon^2 - 20\varepsilon^2}}{2} = \frac{5\varepsilon \pm \sqrt{5}\varepsilon}{2} \tag{1.368}$$

$$\lambda = \varepsilon\left(\frac{5 \pm \sqrt{5}}{2}\right) \tag{1.369}$$

So, the eigenvalues of H are $\varepsilon\left(\frac{5+\sqrt{5}}{2}\right)$ and $\varepsilon\left(\frac{5-\sqrt{5}}{2}\right)$.

To find the corresponding eigenstates, we substitute each eigenvalue back into the equation $(H - \lambda I)\vec{v} = 0$ and solve for $\vec{v}$.

For $\lambda_1 = \varepsilon\left(\frac{5+\sqrt{5}}{2}\right)$:

$$\begin{pmatrix} 2\varepsilon - \lambda_1 & \varepsilon \\ \varepsilon & 3\varepsilon - \lambda_1 \end{pmatrix}\begin{pmatrix} x \\ y \end{pmatrix} = 0 \tag{1.370}$$

$$\Rightarrow \begin{pmatrix} \varepsilon(\sqrt{5}-1) & \varepsilon \\ \varepsilon & 2\varepsilon(\sqrt{5}-1) \end{pmatrix}\begin{pmatrix} x \\ y \end{pmatrix} = 0 \tag{1.371}$$

$$\Rightarrow \varepsilon(\sqrt{5}-1)x + \varepsilon y = 0, \quad \varepsilon x + 2\varepsilon(\sqrt{5}-1)y = 0 \tag{1.372}$$

$$\Rightarrow y = (\sqrt{5}-1)x, \quad x = (\sqrt{5}-1)y \tag{1.373}$$

This implies that the eigenstate corresponding to λ_1 is proportional to $\begin{pmatrix} 1 \\ \sqrt{5}-1 \end{pmatrix}$. To normalize it, we find the normalization constant N such that $|N\vec{v}_1|^2 = 1$:

$$\begin{aligned} |N\vec{v}_1|^2 &= N^2(1 + (\sqrt{5}-1)^2) \\ &= N^2(5 - 2\sqrt{5} + 1) = N^2(6 - 2\sqrt{5}) = 1 \end{aligned} \tag{1.374}$$

$$\Rightarrow N^2 = \frac{1}{6 - 2\sqrt{5}} \tag{1.375}$$

$$\Rightarrow N = \sqrt{\frac{1}{6 - 2\sqrt{5}}} \tag{1.376}$$

Thus, the normalized eigenstate corresponding to λ_1 is:

$$\vec{v}_1 = \frac{1}{\sqrt{\frac{1}{6-2\sqrt{5}}}}\begin{pmatrix} 1 \\ \sqrt{5}-1 \end{pmatrix} \tag{1.377}$$

For $\lambda_2 = \varepsilon\left(\frac{5-\sqrt{5}}{2}\right)$, similarly, we can find the corresponding normalized eigenstate $\vec{v}_2$.

Part (c): If the measurement of B yields the lowest eigenvalue -1, the system collapses into the eigenstate corresponding to -1, which is $\vec{v}_2$. Therefore, the state of the system after the measurement is $\vec{v}_2 = \frac{1}{\sqrt{2}}\begin{pmatrix} 1 \\ -1 \end{pmatrix}$.

To find the probabilities of measuring each energy eigenvalue at a later time, we need to express the state vector $\vec{v}_2$ in terms of the eigenstates of H. Since $\vec{v}_2$ is already in the eigenbasis of B, we can directly use it to compute the probabilities. The probability of measuring an energy eigenvalue λ_i is given by $|\langle\lambda_i|\vec{v}_2\rangle|^2$, where $|\langle\lambda_i|\vec{v}_2\rangle|^2 = |\vec{v}_i \cdot \vec{v}_2|^2$.

Part (d): The time-independent Schrödinger equation for this system is given by:

$$H|\psi\rangle = E|\psi\rangle \tag{1.378}$$

$$H|\psi\rangle = E|\psi\rangle \tag{1.379}$$

Where:

- $|\psi\rangle$ is the wavefunction representing the state of the system.
- H is the Hamiltonian operator.
- E is the energy eigenvalue associated with the state $|\psi\rangle$.

To solve this equation, we need to find the eigenvalues and eigenvectors of the Hamiltonian matrix H.

The eigenvalues E are obtained by solving the characteristic equation $\det(H-\lambda I)=0$, where λ is an eigenvalue and I is the identity matrix.

Substituting H into the characteristic equation, we have:

$$\det\left(\begin{pmatrix} 2\varepsilon & \varepsilon \\ \varepsilon & 3\varepsilon \end{pmatrix} - \lambda\begin{pmatrix} 1 & 0 \\ 0 & 1 \end{pmatrix}\right) = 0 \tag{1.380}$$

This leads to the characteristic equation:

$$\det\left(\begin{pmatrix} 2\varepsilon-\lambda & \varepsilon \\ \varepsilon & 3\varepsilon-\lambda \end{pmatrix}\right) = (2\varepsilon-\lambda)(3\varepsilon-\lambda)-\varepsilon^2 = 0 \tag{1.381}$$

Solving this equation gives us the eigenvalues E_1 and E_2, which are the possible energy eigenvalues of the system.

Once we have the eigenvalues, we can find the corresponding eigenvectors $|\psi_1\rangle$ and $|\psi_2\rangle$ by solving the equation $(H-E_iI)|\psi_i\rangle = 0$ for $i=1,2$.

The solutions $|\psi_1\rangle$ and $|\psi_2\rangle$ are the wavefunctions associated with the energy eigenvalues E_1 and E_2 respectively.

Finally, we can express the time-independent Schrödinger equation explicitly as:

$$\begin{pmatrix} 2\varepsilon & \varepsilon \\ \varepsilon & 3\varepsilon \end{pmatrix}\begin{pmatrix} \psi_1 \\ \psi_2 \end{pmatrix} = \begin{pmatrix} E_1 & 0 \\ 0 & E_2 \end{pmatrix}\begin{pmatrix} \psi_1 \\ \psi_2 \end{pmatrix} \tag{1.382}$$

Where $\begin{pmatrix} \psi_1 \\ \psi_2 \end{pmatrix}$ represents the wavefunction in the basis of energy eigenstates, and $\begin{pmatrix} E_1 & 0 \\ 0 & E_2 \end{pmatrix}$ is the diagonal matrix containing the energy eigenvalues.

By solving this equation, we obtain the wavefunctions ψ_1 and ψ_2 which describe the state of the system corresponding to the given energy eigenvalues.

PROBLEM 9: HYDROGEN ATOM TRANSITION PROBABILITIES

A hydrogen atom is in the ground state when it is subjected to an electric field $\vec{E}(t) = E_0\exp(-t/\tau)\hat{z}$ for $t \geq 0$.

(a) Find the probability that the atom has made a transition to the $2s$ state as $t \to \infty$.
(b) Find the probability that the atom has made a transition to each of the $2p$ states as $t \to \infty$.

DETAILED SOLUTION FOR HYDROGEN ATOM TRANSITIONS

Given a hydrogen atom in the ground state (1s) subjected to a time-varying electric field $E(t) = E_0e^{-t/\tau}\hat{z}$, we calculate the probability of transitions to the 2s and 2p states.

PART (A): TRANSITION TO THE 2S STATE

The perturbing Hamiltonian is:

$$H'(t) = -eE(t)z = -eE_0e^{-t/\tau}z \tag{1.383}$$

The matrix element $\langle 2s|z|1s\rangle$ is given by:

$$\langle 2s|z|1s\rangle = \int \psi_{2s}^*(r,\theta,\phi)z\psi_{1s}(r,\theta,\phi)d^3r \tag{1.384}$$

Using the hydrogen wavefunctions:

$$\psi_{1s}(r) = \frac{1}{\sqrt{\pi a_0^3}} e^{-r/a_0} \tag{1.385}$$

$$\psi_{2s}(r) = \frac{1}{\sqrt{8\pi a_0^3}} \left(1 - \frac{r}{2a_0}\right) e^{-r/2a_0} \tag{1.386}$$

Since $z = r\cos\theta$, the matrix element becomes:

$$\langle 2s|z|1s\rangle = \frac{1}{\sqrt{8\pi a_0^3 \cdot \pi a_0^3}} \int_0^\infty \int_0^\pi \int_0^{2\pi} \left(1 - \frac{r}{2a_0}\right)$$
$$\times e^{-r/2a_0} r\cos\theta e^{-r/a_0} r^2 \sin\theta d\phi d\theta dr \tag{1.387}$$

Evaluating the integrals:

$$\langle 2s|z|1s\rangle = \frac{1}{8\pi a_0^6} \int_0^\infty r^3 e^{-3r/2a_0} \left(1 - \frac{r}{2a_0}\right) dr$$
$$\int_0^\pi \cos\theta \sin\theta d\theta \int_0^{2\pi} d\phi \tag{1.388}$$

The angular integrals evaluate to:

$$\int_0^\pi \cos\theta \sin\theta d\theta = 0 \tag{1.389}$$

Thus, the matrix element is zero:

$$\langle 2s|z|1s\rangle = 0 \tag{1.390}$$

Hence, the transition probability $P_{1s\to 2s} = 0$.
Hence, the transition probability $P_{1s\to 2s} = 0$.

PART (B): TRANSITION TO THE 2P STATES

The matrix element $\langle 2p_z|z|1s\rangle$ is given by:

$$\langle 2p_z|z|1s\rangle = \int \psi_{2p_z}^*(r,\theta,\phi) z \psi_{1s}(r,\theta,\phi) d^3r \tag{1.391}$$

Using the hydrogen wavefunctions:

$$\psi_{1s}(r) = \frac{1}{\sqrt{\pi a_0^3}} e^{-r/a_0} \tag{1.392}$$

$$\psi_{2p_z}(r,\theta) = \frac{1}{\sqrt{32\pi a_0^3}} r e^{-r/2a_0} \cos\theta \tag{1.393}$$

The matrix element becomes:

$$\langle 2p_z|z|1s\rangle = \frac{1}{\sqrt{\pi a_0^3 \cdot 32\pi a_0^3}}$$
$$\times \int_0^\infty \int_0^\pi \int_0^{2\pi} r^2 e^{-r/2a_0} \cos\theta \cdot r \cdot r^2 e^{-r/a_0} \cos\theta \sin\theta d\phi d\theta dr \tag{1.394}$$

Evaluating the integrals:

$$\langle 2p_z|z|1s\rangle = \frac{1}{32\pi a_0^6}\int_0^\infty r^4 e^{-3r/2a_0}dr \int_0^\pi \cos^2\theta \sin\theta d\theta \int_0^{2\pi} d\phi \tag{1.395}$$

The angular integrals evaluate as follows:

$$\int_0^{2\pi} d\phi = 2\pi \tag{1.396}$$

$$\int_0^\pi \cos^2\theta \sin\theta d\theta = \frac{2}{3} \tag{1.397}$$

So we have:

$$\langle 2p_z|z|1s\rangle = \frac{2\pi}{32\pi a_0^6}\cdot\frac{2}{3}\int_0^\infty r^4 e^{-3r/2a_0}dr \tag{1.398}$$

The radial integral evaluates as:

$$\int_0^\infty r^4 e^{-3r/2a_0}dr = \left(\frac{2a_0}{3}\right)^5 \cdot 4! = \frac{16a_0^5}{243} \tag{1.399}$$

Thus:

$$\langle 2p_z|z|1s\rangle = \frac{2\pi}{32\pi a_0^6}\cdot\frac{2}{3}\cdot\frac{16a_0^5}{243} = \frac{4a_0}{243} \tag{1.400}$$

Next, we consider the time integration for the transition amplitude:

$$c_{1s\to 2p_z}(t\to\infty) = -\frac{i}{\hbar}\langle 2p_z|eE_0z|1s\rangle \int_0^\infty e^{-t'/\tau}e^{i\omega_{21}t'}dt' \tag{1.401}$$

Evaluating the time integral:

$$\int_0^\infty e^{-t'/\tau}e^{i\omega_{21}t'}dt' = \frac{1}{1/\tau - i\omega_{21}} \tag{1.402}$$

So the transition amplitude is:

$$c_{1s\to 2p_z} = -\frac{ieE_0}{\hbar}\cdot\frac{4a_0}{243}\cdot\frac{\tau}{1-i\omega_{21}\tau} \tag{1.403}$$

The transition probability is:

$$P_{1s\to 2p_z} = \left|c_{1s\to 2p_z}\right|^2 = \left(\frac{eE_0 4a_0\tau}{243\hbar}\right)^2 \frac{1}{1+(\omega_{21}\tau)^2} \tag{1.404}$$

Similarly, we can show that transitions to the $2p_x$ and $2p_y$ states will also be non-zero. For each of these states, the matrix element will involve the relevant spherical harmonics, and the transition probability can be calculated in a similar manner.

PROBLEM 10: SPIN-1/2 PARTICLES IN A ONE-DIMENSIONAL INFINITE SQUARE WELL

Consider two identical spin-1/2 particles confined to a one-dimensional infinite square well located between $x = 0$ and $x = a$. They interact via the potential:

$$V(x) = \alpha\delta(x_1 - x_2)S_1\cdot S_2 \tag{1.405}$$

where S_i is the spin angular momentum operator for particle i.

(A) WHAT ARE THE UNITS OF α?

To ensure that $V(x)$ has units of energy (eV, J, etc.), we consider the units of other terms in the equation:

$$\text{Units of } \delta(x_1 - x_2) = \frac{1}{\text{length}} = \frac{1}{\text{m}} \tag{1.406}$$

$$\text{Units of } S_1 \cdot S_2 = (\hbar)^2 \tag{1.407}$$

Since $\hbar$ has units of action (Joule-second or eV-s), the units of α are:

$$\text{Units of } \alpha = \frac{\text{Energy} \times \text{Length}}{(\hbar)^2} \tag{1.408}$$

(B) FIND THE EXACT ENERGY OF THE SYSTEM WHEN THE PARTICLES ARE IN A TRIPLET SPIN STATE.

Assuming the particles are in a symmetric spatial wave function, the interaction energy due to the spin-spin interaction in the triplet state ($S = 1$) is:

$$E_{\text{triplet}} = \alpha\langle\delta(x_1 - x_2)\rangle\langle S_1 \cdot S_2\rangle_{S=1} \tag{1.409}$$

Calculate $\langle S_1 \cdot S_2\rangle_{S=1}$ using:

$$S_1 \cdot S_2 = \frac{1}{2}(S^2 - S_1^2 - S_2^2) = \frac{1}{2}(2\hbar^2 - \frac{3}{4}\hbar^2 - \frac{3}{4}\hbar^2) = \frac{1}{4}\hbar^2 \tag{1.410}$$

(C) FIND THE APPROXIMATE ENERGY (OR ENERGIES) OF THE SYSTEM WHEN THE PARTICLES ARE IN A SINGLET STATE.

For the singlet state ($S = 0$), we expect the spatial wave function to be antisymmetric due to the antisymmetry of the overall wave function in fermionic systems. Therefore, the interaction energy due to the spin-spin interaction in the singlet state ($S = 0$) is given by:

$$E_{\text{singlet}} = \alpha\langle\delta(x_1 - x_2)\rangle\langle S_1 \cdot S_2\rangle_{S=0} \tag{1.411}$$

Using the value calculated for $\langle S_1 \cdot S_2\rangle_{S=0}$:

$$\langle S_1 \cdot S_2\rangle_{S=0} = -\frac{3}{4}\hbar^2 \tag{1.412}$$

Inserting this into the energy equation, we need to calculate $\langle\delta(x_1 - x_2)\rangle$ for an antisymmetric spatial state. For fermions with antisymmetric spatial wave functions, the probability of finding two particles at the same position (as required by the delta function) is zero:

$$\langle\delta(x_1 - x_2)\rangle = 0 \tag{1.413}$$

Thus, the energy contribution from the interaction in the singlet state is zero:

$$E_{\text{singlet}} = \alpha \cdot 0 \cdot -\frac{3}{4}\hbar^2 = 0 \tag{1.414}$$

Therefore, the approximate energy of the system in the singlet state is not influenced by the spin-spin interaction and remains at the kinetic energy defined by the spatial part of the wave function.

EXPECTATION VALUE OF $\delta(X_1 - X_2)$

For a symmetric spatial wave function, $\psi(x_1, x_2) = \psi(x_2, x_1)$, the expectation value is calculated as:

$$\langle \delta(x_1 - x_2) \rangle = \int |\psi(x,x)|^2 \, dx \tag{1.415}$$

This integral evaluates the probability density of finding both particles at the same position.

For an antisymmetric spatial wave function, due to the Pauli exclusion principle, the wave function $\psi(x,x) = 0$ for any x, resulting in:

$$\langle \delta(x_1 - x_2) \rangle = 0 \tag{1.416}$$

Thus, there is no contribution to the potential energy from the interaction term in this case.

2 Lectures 5 & 6: Field Theory

2.1 CLASSICAL FIELDS AND QUANTUM MECHANICS

Classical fields are fundamental entities in physics, describing distributions of physical quantities across space and time. Unlike particles, which are localized to points, fields are extensive and can influence space at large scales. This section introduces the concept of fields and explores different types that appear in classical physics. A classical field is some physical quantity, represented mathematically by a function of space and time. Fields underlie the description of forces and interactions in the universe, from the gravitational pull of planets to the electromagnetic waves emitted by the sun. They can be visualized as maps that fill space and evolve over time. They are essential for describing continuous systems and field theories, such as electromagnetism and general relativity, and provide a more comprehensive description of physical phenomena than point particles alone. Classical fields describe physical quantities distributed across space and time. These fields are essential in both classical and modern physics, providing a framework for understanding various physical phenomena.

2.1.1 OVERVIEW OF SCALAR, VECTOR, AND TENSOR FIELDS

SCALAR FIELDS

A scalar field is defined at every point in space and time by a single scalar value. Common examples include temperature or electric potential. Mathematically, a scalar field $\phi(x,y,z,t)$ represents a scalar function of coordinates and time, providing the value of the field at any given point and time. Scalar fields play a crucial role in various areas of physics, providing simple yet powerful descriptions of physically observed phenomena. Concepts such as gravity potential, electric potentials, and optical fields are examples of scalar fields.

Gravity Potential

The gravitational potential ϕ_g is a scalar field that describes the potential energy per unit mass at a point in a gravitational field. It is defined such that the gravitational force $\vec{F}$ experienced by a mass m is the negative gradient of ϕ_g:

$$\vec{F} = -m\nabla\phi_g \tag{2.1}$$

This equation links the scalar potential to the vector field of gravitational forces, demonstrating the utility of scalar fields in describing force fields.

Electric Potential

The electric potential ϕ, a scalar field, describes the electric field $\vec{E}$ through a gradient relationship:

$$\vec{E} = -\nabla\phi \tag{2.2}$$

This relationship defines the electric field as the negative gradient of the electric potential.

When substituting $\vec{E} = -\nabla\phi$ into Gauss's law, we obtain:

$$\nabla\cdot(-\nabla\phi) = \frac{\rho}{\varepsilon_0} \tag{2.3}$$

DOI: 10.1201/9781003637370-2

which simplifies to:

$$\nabla^2 \phi = -\frac{\rho}{\varepsilon_0} \tag{2.4}$$

This is Poisson's equation for the electric potential, linking the charge distribution ρ to the curvature of the electric potential field.

Optical Fields

In the field of quantum optics, scalar fields are used to describe optical phenomena. For example, the field equation for a simple optical field might be represented as:

$$\hat{E}(z,t) = E_0 \cos(kz - \omega t + \phi) \tag{2.5}$$

This is an example represents the electric field as a function of position and time, and ϕ is a phase shift, a scalar quantity affecting the wave's properties.

Each of these examples demonstrates how scalar fields are integral in simplifying and effectively describing complex physical systems across different areas of physics.

The Higgs Field

The Higgs field is a fundamental field in particle physics, playing a crucial role in the Standard Model of particle physics. It is an example of a scalar field, which means it is described by a scalar value at each point in space and time. The Higgs field is unique because it provides mass to elementary particles through the Higgs mechanism, a process central to our understanding of particle mass. The Higgs field ϕ is represented as a complex scalar field, part of a doublet in the electroweak sector of the Standard Model. Its simplest mathematical representation in the Lagrangian density is given by:

$$\mathscr{L}_{\text{Higgs}} = \left|\partial_\mu \phi\right|^2 - V(\phi) \tag{2.6}$$

where $\left|\partial_\mu \phi\right|^2$ represents the kinetic term involving the spacetime derivative of the field, and $V(\phi)$ is the Higgs potential, which is typically written as:

$$V(\phi) = \mu^2 |\phi|^2 + \lambda |\phi|^4 \tag{2.7}$$

Here, μ^2 and λ are parameters of the model. For the Higgs mechanism to work, μ^2 is taken to be negative, which leads to spontaneous symmetry breaking.

Higgs Mechanism and Mass Generation

The Higgs mechanism occurs when the Higgs field acquires a nonzero vacuum expectation value (VEV). This spontaneous symmetry breaking is due to the shape of the potential $V(\phi)$, which has a non-zero minimum not at $\phi = 0$ but at:

$$|\phi| = \sqrt{-\frac{\mu^2}{2\lambda}} \tag{2.8}$$

This nonzero VEV affects other fields (like the electroweak bosons and fermions) through their interactions with the Higgs field, resulting in their acquiring mass. Specifically, the masses of particles are proportional to the strength of their coupling with the Higgs field. The discovery of the Higgs boson at CERN in 2012 confirmed the existence of the Higgs field and helped solidify the Standard Model's explanation of how particles acquire mass. The Higgs field remains a central element in particle physics and a subject of ongoing research, particularly concerning its potential connections to new physics beyond the Standard Model.

VECTOR FIELDS

Vector fields associate a vector to every point in space, effectively describing vector-valued quantities. For instance, the velocity field in fluid dynamics or the magnetic field in electromagnetism are vector fields. A vector field $\vec{v}(x,y,z,t)$ consists of three components, each being a function of spatial coordinates and time, representing the magnitude and direction of the vector at any point. They associate a vector to every point in space, effectively describing vector-valued quantities. For instance, the velocity field in fluid dynamics or the magnetic field in electromagnetism are vector fields. A vector field $\vec{v}(x,y,z,t)$ consists of three components, each being a function of spatial coordinates and time, representing the magnitude and direction of the vector at any point.

Maxwell's Equations: Scalar or Vector Fields

Maxwell's equations are fundamentally considered a set of vector field equations because they primarily describe the behavior and interactions of two vector fields: the electric field $\vec{E}$ and the magnetic field $\vec{B}$. Maxwell's equations consist of four primary laws that can be expressed in both differential and integral forms, highlighting their nature as vector field equations:

2.1.1.0.1 Gauss's Law for Electricity:

$$\nabla \cdot \vec{E} = \frac{\rho}{\varepsilon_0} \tag{2.9}$$

This law involves the divergence of the electric field $\vec{E}$, which is a vector quantity, and relates it to the scalar quantity of charge density ρ.

2.1.1.0.2 Gauss's Law for Magnetism:

$$\nabla \cdot \vec{B} = 0 \tag{2.10}$$

This states that the magnetic field $\vec{B}$ has zero divergence, indicating no magnetic monopoles.

2.1.1.0.3 Faraday's Law of Electromagnetic Induction:

$$\nabla \times \vec{E} = -\frac{\partial \vec{B}}{\partial t} \tag{2.11}$$

Describes how a changing magnetic field induces a curl in the electric field, both of which are vector fields.

2.1.1.0.4 Ampere's Law with Maxwell's Addition:

$$\nabla \times \vec{B} = \mu_0 \vec{J} + \mu_0 \varepsilon_0 \frac{\partial \vec{E}}{\partial t} \tag{2.12}$$

This relates the curl of the magnetic field to the electric current density and the rate of change of the electric field.

Scalar and Vector Potentials

Maxwell's equations can also be reformulated using scalar (ϕ) and vector ($\vec{A}$) potentials: The electric field $\vec{E}$ is expressed as:

$$\vec{E} = -\nabla\phi - \frac{\partial \vec{A}}{\partial t} \tag{2.13}$$

The magnetic field $\vec{B}$ is defined as:

$$\vec{B} = \nabla \times \vec{A} \tag{2.14}$$

While Maxwell's equations involve both scalar and vector quantities, they are primarily considered vector field equations due to their fundamental role in describing the vector fields of electricity and magnetism.

Electric Field of a Point Charge

The electric field $\vec{E}$ due to a point charge is a classic example of a vector field in electromagnetism. The field at a point in space is given by Coulomb's Law:

$$\vec{E}(\vec{r}) = \frac{1}{4\pi\varepsilon_0}\frac{q}{|\vec{r}-\vec{r}_0|^3}(\vec{r}-\vec{r}_0) \tag{2.15}$$

where q is the charge, $\vec{r}_0$ is the position of the charge, and $\vec{r}$ is the point at which the field is being calculated. This vector field points radially away from (or toward) the charge, depending on the sign of q, and its magnitude decreases with the square of the distance from the charge, illustrating how vector fields can describe force distributions in space.

2.1.1.1 Modern Quantized Electromagnetic Field

As shown in Chapter 1 Section 15, the modern accepted quantized electromagnetic field theory is a vector field representation and is given as:

$$\hat{\mathbf{E}}(\mathbf{r},t) = i\sum_{\mathbf{k},\lambda}\sqrt{\frac{\hbar\omega_k}{2\varepsilon_0 V}}\left(\hat{a}_{\mathbf{k},\lambda}e^{i(\mathbf{k}\cdot\mathbf{r}-\omega_k t)} - \hat{a}^\dagger_{\mathbf{k},\lambda}e^{-i(\mathbf{k}\cdot\mathbf{r}-\omega_k t)}\right)\mathbf{e}_{\mathbf{k},\lambda}. \tag{2.16}$$

Which can be simplified to

$$\hat{\mathbf{E}}(\mathbf{r},t) = \mathbf{E}(\mathbf{r},t)\hat{a}_{\mathbf{k},\lambda} - \mathbf{E}^*(\mathbf{r},t)\hat{a}^\dagger_{\mathbf{k},\lambda}, \tag{2.17}$$

where the classical field and its conjugate are:

$$\mathbf{E}(\mathbf{r},t) = i\sum_{\mathbf{k},\lambda}\sqrt{\frac{\hbar\omega_k}{2\varepsilon_0 V}}\mathbf{e}_{\mathbf{k},\lambda}e^{i(\mathbf{k}\cdot\mathbf{r}-\omega_k t)}, \tag{2.18}$$

and

$$\mathbf{E}^*(\mathbf{r},t) = -i\sum_{\mathbf{k},\lambda}\sqrt{\frac{\hbar\omega_k}{2\varepsilon_0 V}}\mathbf{e}_{\mathbf{k},\lambda}e^{-i(\mathbf{k}\cdot\mathbf{r}-\omega_k t)}. \tag{2.19}$$

TENSOR FIELDS

Tensor fields are a generalization of scalar and vector fields, involving tensors at each point in space and time. These fields are crucial in describing more complex relationships such as the stress tensor in materials or the metric tensor in general relativity. A tensor field $T^{ij}(x,y,z,t)$ at a point provides a tensor of rank defined by its indices i and j, which can vary in form depending on the physical context.

These fields are not just mathematical constructs but are physical entities that influence and describe the behavior of physical systems. Understanding these fields and their dynamics forms the foundation for both classical field theories like electromagnetism and general relativity, as well as quantum field theories.

Examples of Tensor Fields

2.1.1.1.1 Stress Tensor in Continuum Mechanics:

The stress tensor σ_{ij} describes the internal forces experienced by material bodies under stress. It is defined as:

$$\sigma_{ij} = -p\delta_{ij} + \mu \left(\frac{\partial u_i}{\partial x_j} + \frac{\partial u_j}{\partial x_i} \right) \tag{2.20}$$

where p is the pressure, μ is the viscosity (for fluids), δ_{ij} is the Kronecker delta, and u_i represents the velocity field components, indicating how material deformation is related to applied forces.

2.1.1.1.2 Metric Tensor, Einstein Tensor, and the Spacetime Interval:

The metric tensor $g_{\mu\nu}$ defines the spacetime interval ds^2, which is used to measure distances and times in a curved spacetime. The spacetime interval is given by:

$$ds^2 = g_{\mu\nu} dx^\mu dx^\nu \tag{2.21}$$

where dx^μ and dx^ν are differentials of the spacetime coordinates. The metric tensor not only measures distances but also affects the path that matter and light follow in spacetime.

The Einstein tensor $G_{\mu\nu}$, crucial in the field equations of general relativity, is derived from the metric tensor. It is defined as:

$$G_{\mu\nu} = R_{\mu\nu} - \frac{1}{2} R g_{\mu\nu} \tag{2.22}$$

where $R_{\mu\nu}$ is the Ricci curvature tensor and R is the Ricci scalar. The Ricci curvature tensor itself is a contraction of the Riemann curvature tensor, which describes how much spacetime is curved in the vicinity of a point and is given by:

$$R_{\mu\nu} = R^{\lambda}_{\mu\lambda\nu} \tag{2.23}$$

The Ricci scalar R is the trace of the Ricci tensor, computed as:

$$R = g^{\mu\nu} R_{\mu\nu} \tag{2.24}$$

These elements collectively describe the curvature of spacetime due to the mass-energy content encapsulated by the stress-energy tensor $T_{\mu\nu}$ in the Einstein Field Equations:

$$G_{\mu\nu} = \frac{8\pi G}{c^4} T_{\mu\nu} \tag{2.25}$$

This equation directly relates the geometry of spacetime, as described by the Einstein tensor, to the energy and momentum at each point in spacetime, as described by the stress-energy tensor.

This detailed mathematical framework highlights the integral role of the metric tensor in describing the fundamental characteristics of spacetime and gravity according to general relativity, linking it clearly to the physical phenomena such as gravitational forces and the bending of light.

2.1.1.1.3 Permittivity Tensor in Anisotropic Media:

For an anisotropic medium, the permittivity tensor ε_{ij} relates the electric field $\vec{E}$ to the displacement field $\vec{D}$ as follows:

$$D_i = \varepsilon_{ij} E_j \tag{2.26}$$

where the indices run over the spatial dimensions, indicating how the medium's properties affect the electric field.

2.1.1.1.4 Inertia Tensor in Dynamics:

The inertia tensor I_{ij} for a rigid body is defined by:

$$I_{ij} = \int (r^2 \delta_{ij} - x_i x_j) \rho(\vec{r}) \, dV \tag{2.27}$$

where $\rho(\vec{r})$ is the mass density of the body, $\vec{r}$ is the position vector relative to the center of mass, and x_i, x_j are the coordinates. This tensor is crucial for determining the angular momentum $\vec{L}$ in terms of angular velocity $\vec{\omega}$:

$$L_i = I_{ij} \omega_j \tag{2.28}$$

These examples illustrate the diverse applications of tensor fields in various physical theories and highlight their importance in describing complex physical phenomena.

2.1.2 APPLICABLE BASIC PRINCIPLES OF QUANTUM MECHANICS

As reviewed in some detail in Chapter 1, Quantum Mechanics (QM) is a fundamental theory in physics that describes the physical properties of nature at small scales, typically atomic and subatomic levels. It introduces concepts that are markedly different from classical physical theories, including the superposition of states, quantization of physical properties, the uncertainty principle, entanglement, and quantum computing.

Wave-Particle Duality

Quantum mechanics introduces the concept of wave-particle duality, which posits that every particle or quantum entity can exhibit both particle-like and wave-like behavior. This principle is mathematically represented by the de Broglie hypothesis:

$$\lambda = \frac{h}{p} \tag{2.29}$$

where λ is the wavelength, h is Planck's constant, and p is the momentum of the particle. This relationship shows that every particle has a wavelength associated with it, influencing phenomena such as electron diffraction.

Heisenberg Uncertainty Principle

The uncertainty principle, formulated by Werner Heisenberg, is a fundamental theory which states that the more precisely the position of some particle is determined, the less precisely its momentum can be known, and vice versa. This is expressed as:

$$\Delta x \Delta p \geq \frac{\hbar}{2} \tag{2.30}$$

where Δx is the uncertainty in position, Δp is the uncertainty in momentum, and $\hbar$ is the reduced Planck's constant. This principle is a fundamental limit to the precision with which certain pairs of physical properties, like position and momentum, can be simultaneously known.

Schrodinger Equation

The Schrödinger equation is the fundamental equation of motion in quantum mechanics. For a non-relativistic particle, it is typically written as:

$$i\hbar \frac{\partial}{\partial t} \Psi(\vec{r}, t) = \hat{H} \Psi(\vec{r}, t) \tag{2.31}$$

where i is the imaginary unit, $\hbar$ is the reduced Planck constant, $\Psi(\vec{r},t)$ is the wave function of the particle, $\vec{r}$ represents the position, t is time, and $\hat{H}$ is the Hamiltonian operator which represents the total energy of the system. This equation describes how the quantum state of a physical system changes over time.

These principles are crucial for understanding quantum phenomena and provide the theoretical basis for much of modern physics, including quantum field theory.

2.1.3 FERMAT'S PRINCIPLE AS A SPECIAL CASE OF THE PRINCIPLE OF LEAST ACTION

Fermat's Principle, also known as the Principle of Least Time, is a fundamental principle in optics stating that the path taken between two points by a ray of light is the path that can be traversed in the least time. This principle is analogous to the principle of least action in classical mechanics and provides a powerful tool for understanding and deriving the laws of reflection and refraction.

Mathematical Formulation

Fermat's Principle can be mathematically expressed as:

$$\delta \int_A^B \frac{ds}{v} = 0 \tag{2.32}$$

where δ represents the variation, ds is an infinitesimal element of the path, and v is the speed of light along that path segment. The integral is taken from point A to point B, and the variation of the optical path length is set to zero. This expression shows that the path taken by light minimizes the optical path length, which is the product of the distance and the refractive index along the path.

Applications in Optics

Fermat's Principle is central to deriving Snell's Law of refraction, which can be formulated by considering the light path through media with different refractive indices. By applying Fermat's Principle, Snell's Law is obtained as follows:

$$n_1 \sin\theta_1 = n_2 \sin\theta_2 \tag{2.33}$$

where n_1 and n_2 are the refractive indices of the two media, and θ_1 and θ_2 are the angles of incidence and refraction, respectively.

Link to Quantum Mechanics

In quantum mechanics, Fermat's Principle manifests in the principle of stationary phase, which states that quantum mechanical amplitude contributions are most significant where the action is stationary. This is similar to the classical optical path being least, demonstrating the profound connection between classical and quantum theories. The principle of stationary phase is important in explaining the behavior of quantum systems in terms of interference and diffraction, phenomena that are needed to understand wave-like properties of particles. Fermat's Principle not only elucidates its role in classical optics but also connects it to broader physical theories, illustrating its relevance in the foundational principles of quantum mechanics and the nature of light as understood through quantum theory.

2.2 FUNCTIONALS AND VARIATIONAL PRINCIPLES

2.2.1 DEFINITION AND EXAMPLES OF FUNCTIONALS IN PHYSICS

Functionals are integral parts of modern physics, representing mathematical objects that assign a real number to functions, effectively mapping functions to real numbers. In physics, functionals are often used to describe systems in terms of energy, action, or other quantities that depend on the function describing the state of the system.

Mathematical Definition

A functional, denoted as $F[\phi]$, can be defined for a function $\phi(x)$ that maps this function to a scalar. For example, a commonly used functional in physics is the action S, defined in the Lagrangian formalism of mechanics as:

$$S[\phi] = \int L(\phi(x), \phi'(x), x)\, dx \tag{2.34}$$

where L is the Lagrangian of the system, which depends on the function $\phi(x)$, its derivative $\phi'(x)$, and the variable x.

Examples in Physics

1. **Action Functional in Mechanics:** The action functional S is a principal example that takes a path or trajectory $\phi(t)$ and maps it to a number, the action, which is used in the principle of least action to determine the dynamics of a system.

$$S[\phi] = \int_{t_1}^{t_2} L(\phi(t), \dot{\phi}(t), t)\, dt \tag{2.35}$$

where $\dot{\phi}(t)$ is the time derivative of $\phi(t)$.

2. **Energy Functional in Field Theory:** In field theories, particularly in the study of static fields, the energy functional $E[\phi]$ can often be expressed as:

$$E[\phi] = \int \left(\frac{1}{2} (\nabla \phi)^2 + V(\phi) \right) d^3x \tag{2.36}$$

where $V(\phi)$ is the potential energy density as a function of the field ϕ.

Importance in Physics

Functionals play a critical role in the calculus of variations, where one seeks the function that minimizes (or maximizes) a given functional. This mathematical method is fundamental in physics for deriving equations of motion, stability analyses, and in quantifying systems in terms of energies or actions. The extremum (minimum, maximum, or saddle point) of a functional can correspond to the physical state of a system, such as the path taken by a particle in mechanics or the configuration of a field that minimizes energy in field theory.

Functional as a Black Box Analogy

Functionals can be thought of as a type of "black box" in mathematical analysis, especially within the context of physics. This analogy helps in visualizing how functionals operate and their role in various physical theories. Imagine a black box device, where you input a function, and it outputs a specific number. This device does not modify the function itself but evaluates certain characteristics or properties of the function to produce a scalar value. In mathematical terms, a functional is this black box that takes a function ϕ defined over some domain and maps it to a real number.

Mathematical Representation

Mathematically, this can be represented as:

$$F : \{\text{Functions}\} \to \mathbb{R} \tag{2.37}$$

where F is the functional, and $\mathbb{R}$ denotes the set of real numbers. For a given function $\phi(x)$, the functional F might look something like:

$$F[\phi] = \int_a^b f(\phi(x), \phi'(x), x)\, dx \tag{2.38}$$

Here, f is some function that depends on $\phi(x)$, its derivative $\phi'(x)$, and the independent variable x. The functional evaluates this expression over the interval from a to b, producing a single real number.

Physical Interpretation

In physical terms, as shown in Figure 2.1, the black box processes the function (such as the trajectory of a particle or a field configuration) according to the rules defined by the functional, which typically relate to energy, action, or other quantifiable properties of the system. The output is a scalar that often has physical significance, such as the total energy, action, or another conserved quantity.

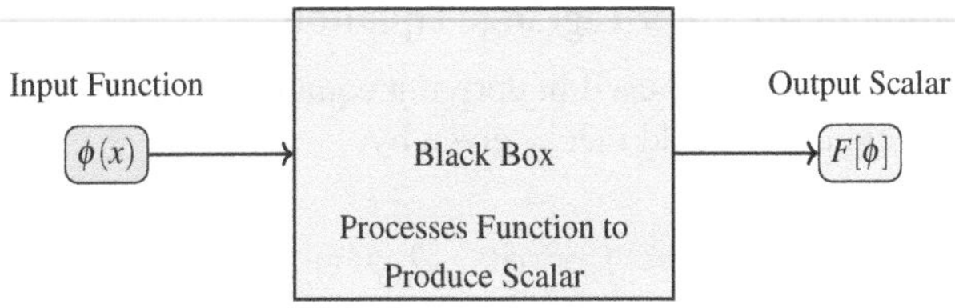

FIGURE 2.1
Functional as a black box: The functional F takes a function $\phi(x)$ as input and produces a scalar $F[\phi]$ as output.

Utility and Applications

The "black box" analogy of functionals is particularly useful in physics as it simplifies the concept of operations on functions, which can otherwise be abstract. This model aids in understanding how various principles, like the principle of least action in classical mechanics or the minimum energy principle in thermodynamics, operate by finding functions that minimize or extremize the output of this black box.

By conceptualizing functionals in this way, students and researchers can more intuitively grasp how to apply these tools in theoretical physics, engineering, and other fields where such mathematical constructs are essential for analyzing and solving complex problems. Functionals are crucial in Quantum Field Theory (QFT) because they allow for the systematic description and calculation of quantum amplitudes through the path integral formulation. This approach involves summing over all possible field configurations, where each configuration is weighted by a functional called "the action." The *action functional*, a key concept in QFT, encodes the dynamics and interactions of fields, enabling the computation of propagators and interaction vertices in a field-theoretic context. Ultimately, functionals provide a framework for understanding the fundamental interactions and symmetries in particle physics, essential for predicting particle behaviors and outcomes in high-energy experiments

2.2.2 INTRODUCTION TO FUNCTIONAL DERIVATIVES AND THE EULER-LAGRANGE EQUATION

Functional derivatives are a fundamental concept in the calculus of variations and field theory. They extend the idea of ordinary derivatives to functionals, which are mappings from functions to real numbers. Functional derivatives are crucial for determining how a small change in a function affects the value of a functional, thus providing a way to obtain equations of motion or field equations from a given action principle.

Definition of Functional Derivative

The functional derivative of a functional $F[\phi]$ with respect to a function $\phi(x)$ at a point x is defined in a way that is analogous to partial derivatives for functions. It measures how F changes in response to a small change in ϕ at x, denoted as $\delta\phi(x)$. Mathematically, the functional derivative $\frac{\delta F}{\delta\phi(x)}$ is defined by the limit:

$$\frac{\delta F}{\delta\phi(x)} = \lim_{\varepsilon\to 0} \frac{F[\phi+\varepsilon\delta_x] - F[\phi]}{\varepsilon} \tag{2.39}$$

where δ_x is a Dirac delta function centered at x, which ensures that the perturbation is localized at x.

Application: Derivation of the Euler-Lagrange Equation

Functional derivatives are extensively used in deriving equations of motion from action principles. Consider the action functional in a field theory given by:

$$S[\phi] = \int \mathscr{L}(\phi(x), \partial_\mu\phi(x), x)\, d^4x \tag{2.40}$$

where $\mathscr{L}$ is the Lagrangian density depending on the field $\phi(x)$, its derivatives $\partial_\mu\phi(x)$, and the spacetime coordinates x.

The Euler-Lagrange equation is derived by requiring that the action $S[\phi]$ be stationary (i.e., $\delta S = 0$) under variations $\delta\phi(x)$ of the field $\phi(x)$ that vanish at the boundaries of the integration domain. Applying the principle of stationary action leads to:

$$\frac{\partial\mathscr{L}}{\partial\phi} - \partial_\mu\left(\frac{\partial\mathscr{L}}{\partial(\partial_\mu\phi)}\right) = 0 \tag{2.41}$$

This is the Euler-Lagrange equation for the field ϕ, a fundamental equation in field theory that specifies the dynamics of the field according to the given Lagrangian.

Significance in Physics

Functional derivatives thus serve as powerful tools in theoretical physics, particularly in the context of Lagrangian and Hamiltonian mechanics, general relativity, and quantum field theory. They provide a systematic approach to deriving dynamic equations that describe how physical systems evolve in time, based on an action principle.

This introduction should help students and researchers understand the importance of functional derivatives and their applications in modern physics, providing the mathematical tools needed to handle variational problems and field equations effectively.

2.2.3 DETAILED WORKED EXAMPLE: EULER-LAGRANGE EQUATION FOR A FREE SCALAR FIELD IN QFT

This example provides a step-by-step derivation of the equation of motion for a free scalar field, demonstrating the use of the Euler-Lagrange equation in Quantum Field Theory. We also explain the underlying mathematical concepts including Minkowski space and the d'Alembertian operator.

Action and Lagrangian Density

Consider a free scalar field ϕ described by the Lagrangian density:

$$\mathscr{L} = \frac{1}{2}\partial^\mu \phi \partial_\mu \phi - \frac{1}{2}m^2\phi^2 \tag{2.42}$$

where:

- $\partial^\mu \phi \partial_\mu \phi$ is the kinetic term involving derivatives of the field. In the notation, μ is a spacetime index that runs over $0,1,2,3$ in four-dimensional spacetime, representing time and spatial components.
- m is the mass of the scalar particles described by the field.

Minkowski Space and Metric

Minkowski space is the setting for special relativity and is characterized by a flat spacetime metric. The metric tensor $g_{\mu\nu}$ in Minkowski space is diag(–1, 1, 1, 1), which means:

$$ds^2 = -dt^2 + dx^2 + dy^2 + dz^2 \tag{2.43}$$

This metric is used to raise and lower indices and to calculate spacetime intervals and derivatives.

Deriving the Euler-Lagrange Equation

Applying the Euler-Lagrange equation:

$$\frac{\partial \mathscr{L}}{\partial \phi} - \partial_\mu \left(\frac{\partial \mathscr{L}}{\partial(\partial_\mu \phi)} \right) = 0 \tag{2.44}$$

we perform the following calculations:

- The partial derivative of the Lagrangian with respect to ϕ is straightforward:

$$\frac{\partial \mathscr{L}}{\partial \phi} = -m^2\phi \tag{2.45}$$

- The derivative with respect to $\partial_\mu \phi$:

$$\frac{\partial \mathscr{L}}{\partial(\partial_\mu \phi)} = \partial^\mu \phi \tag{2.46}$$

  Taking the derivative of this result:

$$\partial_\mu \left(\frac{\partial \mathscr{L}}{\partial(\partial_\mu \phi)} \right) = \partial_\mu \partial^\mu \phi \tag{2.47}$$

This leads to the equation:

$$-m^2\phi - \partial_\mu \partial^\mu \phi = 0 \tag{2.48}$$

The d'Alembertian Operator

The operator $\partial_\mu \partial^\mu$, also known as the d'Alembertian or Box operator, is defined in Minkowski space as:

$$\square = \partial_\mu \partial^\mu = -\frac{\partial^2}{\partial t^2} + \nabla^2 \tag{2.49}$$

where ∇^2 is the Laplacian operator in three spatial dimensions.

Klein-Gordon Equation

The resulting equation is the Klein-Gordon equation for a free scalar field:

$$(\square + m^2)\phi = 0 \tag{2.50}$$

This equation describes how the field evolves over time and space in the absence of interactions, encapsulating the relativistic energy-momentum relationship for scalar fields.

This detailed example not only illustrates the application of the Euler-Lagrange equation in deriving field equations but also provides insights into the mathematical structure of spacetime used in these calculations, critical for understanding the dynamics in Quantum Field Theory.

2.2.4 HAMILTON'S PRINCIPLE OF LEAST ACTION

Hamilton's Principle of Least Action is a fundamental principle in both classical and quantum mechanics, stating that the dynamics of a physical system can be determined by the path that minimizes (or extremizes) the action. This principle provides a unifying framework for the laws of physics, from mechanics to field theory.

Definition of the Action

The action S is a functional of the trajectory or field configuration ϕ that the system takes between two times t_1 and t_2. It is defined as the integral of the Lagrangian L over time:

$$S[\phi] = \int_{t_1}^{t_2} L(\phi(t), \dot{\phi}(t), t)\, dt \tag{2.51}$$

where $\phi(t)$ represents the state of the system at time t, and $\dot{\phi}(t)$ is its time derivative.

Mathematical Formulation of the Principle

Hamilton's principle states that the actual path taken by the system makes the action S stationary (i.e., a local minimum, maximum, or saddle point). This requirement leads to the Euler-Lagrange equation, which provides the equations of motion for the system:

$$\frac{d}{dt}\left(\frac{\partial L}{\partial \dot{\phi}}\right) - \frac{\partial L}{\partial \phi} = 0 \tag{2.52}$$

Significance in Physics

Hamilton's Principle is significant because it offers a powerful and elegant method to derive the laws of motion for any physical system by simply knowing its Lagrangian. This approach is particularly useful in systems where the forces involved are conservative and can be derived from a potential function. Moreover, this principle is not just confined to classical mechanics; it extends to quantum mechanics through the path integral formulation, where paths contribute to the quantum amplitude with a weight given by the exponential of the action.

Applications

Hamilton's Principle is applied across all areas of physics:

- In **classical mechanics**, it simplifies the derivation of motion equations for complex systems, such as celestial mechanics and rigid body dynamics.
- In **optics**, it relates to Fermat's Principle of Least Time, which can be seen as an action principle where the action is time.
- In **quantum mechanics**, the path integral formulation uses an extension of Hamilton's principle, emphasizing the probabilistic nature of path contributions based on the action.

This principle not only underscores the deterministic nature of classical physics but also bridges classical and quantum theories, illustrating the profound conceptual unity underlying physical laws.

2.3 SIMPLE HARMONIC OSCILLATORS AND PHONONS

2.3.1 DISCUSSION OF CLASSICAL AND QUANTUM OSCILLATORS

As we have seen in previous sections, the simple harmonic oscillator (SHO) is a cornerstone in the study of physics, providing a foundation for understanding more complex oscillatory and wave phenomena. The classical description involves a mass m attached to a spring with a spring constant k, where the system exhibits oscillations around an equilibrium position.

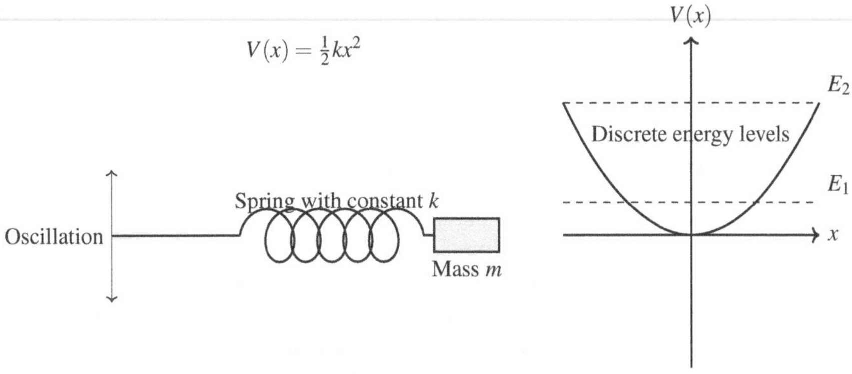

FIGURE 2.2
Classical and Quantum Harmonic Oscillators: (Left) A classical mass-spring system with mass m and spring constant k. (Right) A quantum harmonic oscillator potential well with discrete energy levels E_1 and E_2.

Classical Simple Harmonic Oscillator

The equation of motion for a classical SHO is derived from Newton's second law and Hooke's law:

$$F = -kx = m\ddot{x} \tag{2.53}$$

where x is the displacement from equilibrium, and $\ddot{x}$ represents the acceleration. This leads to the differential equation:

$$\ddot{x} + \omega^2 x = 0 \tag{2.54}$$

with $\omega = \sqrt{\frac{k}{m}}$ being the angular frequency of the oscillator. The general solution is:

$$x(t) = A\cos(\omega t + \phi) \tag{2.55}$$

where A is the amplitude and ϕ is the phase.

Quantum Harmonic Oscillator

In quantum mechanics, the harmonic oscillator is treated by quantizing this classical system. The Hamiltonian for the quantum harmonic oscillator is given by:

$$\hat{H} = \frac{\hat{p}^2}{2m} + \frac{1}{2}k\hat{x}^2 \tag{2.56}$$

where $\hat{p}$ and $\hat{x}$ are the momentum and position operators, respectively. The energy eigenstates $\psi_n(x)$ of the system are solutions to the Schrödinger equation:

$$\hat{H}\psi_n(x) = E_n\psi_n(x) \tag{2.57}$$

with quantized energy levels:

$$E_n = \hbar\omega\left(n + \frac{1}{2}\right) \tag{2.58}$$

for $n = 0, 1, 2, \ldots$. Each level n corresponds to a quantum state of the oscillator with a specific energy and wave function.

2.3.2 PHONONS

Phonons are quantized excitations of the vibrational modes of atoms in a lattice and play a crucial role in understanding thermal and electrical properties of materials. In solid state physics, phonons arise from the quantization of vibrational waves in a crystal lattice. They are bosonic particles, meaning they follow Bose-Einstein statistics and can occupy the same quantum state in large numbers.

Classical Harmonic Oscillators Coupled in a Lattice

To understand phonons, we begin with a classical model of coupled harmonic oscillators representing atoms in a crystalline lattice. This model helps visualize how collective vibrational modes, which are quantized into phonons, emerge from classical mechanics. Consider a one-dimensional lattice of N identical atoms, each of mass m, connected by springs with a spring constant k. The displacement of the n-th atom from its equilibrium position is represented by u_n, and the dynamics of these atoms are governed by Hooke's law and Newton's second law.

Equations of Motion

The force on each atom is due to the displacement relative to its neighbors:

$$F_n = k(u_{n+1} - u_n) - k(u_n - u_{n-1}) \tag{2.59}$$

Using Newtons second law $F = m\ddot{u}$, the equation of motion for each atom becomes:

$$m\ddot{u}_n = k(u_{n+1} + u_{n-1} - 2u_n) \tag{2.60}$$

This equation describes a system of coupled oscillators, where the motion of one oscillator influences its neighbors.

Normal Modes and Frequencies

To solve these equations, we assume solutions of the form:

$$u_n(t) = Ae^{i(\omega t - kna)} \tag{2.61}$$

where A is the amplitude, ω is the angular frequency, k is the wave number, and a is the lattice spacing. Substituting this solution into the equations of motion gives a relation for the frequencies:

$$\omega^2 = \frac{2k}{m}(1 - \cos(ka)) \tag{2.62}$$

This relation, known as the dispersion relation, shows how the frequency of the oscillations depends on the wave number, describing the collective motion of the atoms.

Quantization of Phonons

The quantization of these modes leads to the concept of phonons as particles carrying energy and momentum. The quantum mechanical Hamiltonian for phonons in a crystal is given by:

$$\hat{H} = \sum_q \hbar\omega(q)\left(\hat{a}_q^\dagger \hat{a}_q + \frac{1}{2}\right) \tag{2.63}$$

where $\hat{a}_q^\dagger$ and $\hat{a}_q$ are the creation and annihilation operators for phonons with wave vector q.

This treatment of phonons is essential for understanding various physical properties of materials, such as thermal conductivity, heat capacity, and the interaction of phonons with electrons and photons. This section provides a comprehensive overview of classical and quantum oscillators and introduces the concept of phonons, fundamental for courses on quantum mechanics and solid state physics.

Physical Significance of Phonons

Phonons represent quantized vibrational energy in the lattice. They are bosons and can be used to explain various thermal and electrical properties of materials, including thermal conductivity, specific heat, and sound propagation in solids.

This treatment not only connects classical physics with quantum field theory but also demonstrates the fundamental quantum nature of vibrational energy in solids. By understanding the quantization of lattice vibrations, one gains a deeper insight into material properties and the underlying principles of solid state physics.

2.4 SUMMARY AND DISCUSSION

RECAP OF THE KEY CONCEPTS COVERED IN THE LECTURE

This lecture provided an extensive overview of field theory, emphasizing both its classical and quantum aspects. We explored various types of fields scalar, vector, and tensor and discussed their roles in describing phenomena ranging from gravity to electromagnetism and quantum fields. Key points included:

- **Scalar Fields:** We discussed how scalar fields like the gravitational and electric potentials describe phenomena where a single value at each point in space can represent a physical quantity, such as potential energy or temperature.
- **Vector Fields:** The lecture highlighted the importance of vector fields in physics, exemplified by Maxwell's equations, which govern electromagnetism. These fields associate a vector to every point in space, describing quantities that have magnitude and direction.
- **Tensor Fields:** We examined more complex tensor fields, which include the stress tensor in continuum mechanics and the metric tensor in general relativity, essential for describing the curvature of spacetime and gravitational effects.

- **Quantum Mechanics and Field Theory:** The principles of quantum mechanics were revisited, emphasizing their foundational role in modern physics, including the development of quantum field theory.
- **Harmonic Oscillators and Phonons:** The discussion on classical and quantum harmonic oscillators bridged to the concept of phonons in solid state physics, demonstrating how quantum mechanics describes collective excitations in a lattice.

These topics collectively enhance our understanding of how various fields interact and influence the physical world, from atomic to cosmic scales, and from classical dynamics to quantum interactions.

2.5 QUESTIONS AND PROBLEMS

2.5.1 TRIVIA QUESTIONS

TRIVIA QUESTIONS ON FIELD THEORY

Question 1: What is a scalar field?

(a) A field that associates a vector to every point in space.
(b) A field that associates a scalar value to every point in space and time.
(c) A field that can only describe velocities.
(d) None of the above.

Answer: b. A field that associates a scalar value to every point in space and time.
Question 2: Which equation describes the electric field in terms of the electric potential?

(a) $\vec{E} = \nabla \times \vec{A}$
(b) $\vec{E} = -\nabla\phi - \frac{\partial \vec{A}}{\partial t}$
(c) $\vec{B} = \nabla \times \vec{A}$
(d) $\vec{E} = \nabla \cdot \vec{A}$

Correct Answer: b. $\vec{E} = -\nabla\phi - \frac{\partial \vec{A}}{\partial t}$
Question 3: What physical concept does the Higgs field provide an explanation for?

(a) Electromagnetism
(b) Gravity
(c) Mass of elementary particles
(d) Speed of light

Answer: c. Mass of elementary particles
Question 4: Which of the following is not a vector field?

(a) Magnetic field
(b) Electric field
(c) Gravitational potential
(d) Velocity field

Answer: c. Gravitational potential
Question 5: What mathematical tool is used to derive the motion equations from the Lagrangian?

(a) Fourier Transform
(b) Taylor Series

(c) Euler-Lagrange Equation
(d) Newtons Second Law

Answer: c. Euler-Lagrange Equation
Question 6: What principle states that the path taken by a system between two points is the one for which the action is stationary?

(a) Heisenberg's Uncertainty Principle
(b) Schrödinger's Cat Principle
(c) Hamilton's Principle of Least Action
(d) Pascal's Principle

Answer: c. Hamilton's Principle of Least Action
Question 7: In the context of phonons in a crystal lattice, what are phonons?

(a) Particles that carry electric charge
(b) Quantized excitations of vibrational modes
(c) Subatomic particles in the nucleus
(d) The components of an electric field

Answer: b. Quantized excitations of vibrational modes
Question 8: Which law involves the divergence of the electric field and relates it to charge density?

(a) Gauss's Law for Electricity
(b) Ampere's Law with Maxwell's Addition
(c) Faraday's Law of Induction
(d) Gauss's Law for Magnetism

Answer: a. Gauss's Law for Electricity
Question 9: In quantum mechanics, what fundamental concept suggests that particles exhibit both particle-like and wave-like properties?

(a) Superposition
(b) Entanglement
(c) Wave-particle duality
(d) Quantum tunneling

Answer: c. Wave-particle duality
Question 10: What describes the kinetic term in the Higgs field's Lagrangian density?

(a) $V(\phi)$
(b) $|\phi|^2$
(c) $\left|\partial_\mu \phi\right|^2$
(d) $\lambda\,|\phi|^4$

Answer: c. $\left|\partial_\mu \phi\right|^2$

2.5.2 PROBLEMS

PROBLEM 1: SCALAR FIELD GRADIENT

Problem: Calculate the electric field $\vec{E}$ at a point in space due to a scalar electric potential $\phi(x,y,z) = \frac{k}{\sqrt{x^2+y^2+z^2}}$, where k is a constant.

Solution:

$$\begin{aligned}
\vec{E} &= -\nabla\phi \\
&= -\nabla\left(\frac{k}{\sqrt{x^2+y^2+z^2}}\right) \\
&= -\left(\frac{\partial}{\partial x},\frac{\partial}{\partial y},\frac{\partial}{\partial z}\right)\left(\frac{k}{\sqrt{x^2+y^2+z^2}}\right) \\
&= -\left(\frac{-kx}{(x^2+y^2+z^2)^{3/2}},\frac{-ky}{(x^2+y^2+z^2)^{3/2}},\frac{-kz}{(x^2+y^2+z^2)^{3/2}}\right) \\
&= \frac{k}{(x^2+y^2+z^2)^{3/2}}(x,y,z)
\end{aligned}$$

PROBLEM 2: QUANTUM PARTICLE IN A BOX

Problem: For a quantum particle confined in a one-dimensional box of length L with impenetrable walls, determine the wave function $\psi_n(x)$ for $n = 1$.

Solution:

$$\begin{aligned}
\psi_n(x) &= \sqrt{\frac{2}{L}}\sin\left(\frac{n\pi x}{L}\right) \\
\psi_1(x) &= \sqrt{\frac{2}{L}}\sin\left(\frac{\pi x}{L}\right)
\end{aligned}$$

PROBLEM 3: VECTOR FIELD CURL

Problem: Given the vector field $\vec{A} = (y^2, -x^2, 2xy)$, calculate the curl $\nabla\times\vec{A}$.

Solution:

$$\begin{aligned}
\nabla\times\vec{A} &= \left(\frac{\partial(2xy)}{\partial y}-\frac{\partial(-x^2)}{\partial z},\frac{\partial(y^2)}{\partial z}-\frac{\partial(2xy)}{\partial x},\frac{\partial(-x^2)}{\partial x}-\frac{\partial(y^2)}{\partial y}\right) \\
&= (2x+0, 0-2y, -2-2) \\
&= (2x, -2y, -4)
\end{aligned}$$

PROBLEM 4: PHONONS IN A CRYSTAL LATTICE

Problem: For a simple one-dimensional model of a crystal lattice with atoms of mass m connected by springs of constant k, and spacing a, find the highest frequency mode (ω_{max}).

Solution:

$$\begin{aligned}
\omega_{max} &= \sqrt{\frac{4k}{m}}\sin\left(\frac{\pi a}{2a}\right) \\
&= \sqrt{\frac{4k}{m}}
\end{aligned}$$

PROBLEM 5: KLEIN-GORDON EQUATION

Problem: Solve the Klein-Gordon equation for a free scalar field ϕ with mass m in the absence of external forces: $(\Box + m^2)\phi = 0$ in one spatial dimension.

Solution:

$$\Box\phi + m^2\phi = 0$$
$$-\frac{\partial^2\phi}{\partial t^2} + \frac{\partial^2\phi}{\partial x^2} + m^2\phi = 0$$
$$\text{Assume a solution form: } \phi(x,t) = e^{i(kx-\omega t)}$$
$$-\omega^2 + k^2 + m^2 = 0$$
$$\omega = \sqrt{k^2 + m^2}$$

This is the dispersion relation for a free scalar field, showing the relationship between the energy ω and momentum k.

PROBLEM 6: ELECTRIC FIELD DUE TO A DIPOLE

Problem: Calculate the electric field $\vec{E}$ at a point on the axial line of an electric dipole consisting of charges $+q$ and $-q$ separated by a distance $2a$.

Solution:

$$\begin{aligned}\vec{E} &= \frac{1}{4\pi\varepsilon_0}\left(\frac{q}{(r-a)^2} - \frac{q}{(r+a)^2}\right)\hat{r} \\ &= \frac{q}{4\pi\varepsilon_0}\left(\frac{(r+a)^2-(r-a)^2}{(r^2-a^2)^2}\right)\hat{r} \\ &= \frac{q}{4\pi\varepsilon_0}\frac{4ar}{(r^2-a^2)^2}\hat{r} \\ &= \frac{q}{\pi\varepsilon_0}\frac{ar}{(r^2-a^2)^2}\hat{r}\end{aligned}$$

PROBLEM 7: QUANTUM PARTICLE IN A FINITE POTENTIAL WELL

Problem: Determine the energy levels for a particle of mass m in a one-dimensional box of width L with finite potential V_0 at the walls.

Solution:

$$\text{Inside the well: } \psi''(x) + \frac{2mE}{\hbar^2}\psi(x) = 0$$
$$\text{At the walls: } \psi''(x) + \frac{2m(E-V_0)}{\hbar^2}\psi(x) = 0$$
$$\text{Matching conditions at } x = 0, L : \psi(0) = 0, \psi(L) = 0$$
$$\text{Assuming solutions: } \psi(x) = A\sin(kx) + B\cos(kx),$$
$$k = \sqrt{\frac{2mE}{\hbar^2}}, k' = \sqrt{\frac{2m(E-V_0)}{\hbar^2}}$$
$$\text{Boundary conditions lead to: } kL = n\pi \text{ (for some integer } n)$$
$$\text{Energy levels: } E_n = \frac{\hbar^2k^2}{2m} = \frac{\hbar^2(\frac{n\pi}{L})^2}{2m}$$

PROBLEM 8: MAXWELL'S EQUATIONS IN A VACUUM

Problem: Verify that Maxwells equations reduce to the wave equation for the electric field in a vacuum.

Solution:

$$\nabla \cdot \vec{E} = 0, \quad \nabla \times \vec{E} = -\frac{\partial \vec{B}}{\partial t},$$
$$\nabla \times \vec{B} = \mu_0 \varepsilon_0 \frac{\partial \vec{E}}{\partial t}, \quad \nabla \cdot \vec{B} = 0$$

Taking the curl of both sides of Faradays law:

$$\nabla \times (\nabla \times \vec{E}) = -\nabla \times \frac{\partial \vec{B}}{\partial t}$$

Using the vector identity and Amperes Law:

$$\nabla(\nabla \cdot \vec{E}) - \nabla^2 \vec{E} = -\frac{\partial}{\partial t}(\mu_0 \varepsilon_0 \frac{\partial \vec{E}}{\partial t})$$
$$\nabla^2 \vec{E} - \mu_0 \varepsilon_0 \frac{\partial^2 \vec{E}}{\partial t^2} = 0$$

PROBLEM 9: STRESS TENSOR FOR A FLUID IN MOTION

Problem: Derive the stress tensor for a fluid in motion with velocity field $\vec{v}(x,y,z)$ and pressure $p(x,y,z)$.

Solution:

$$\sigma_{ij} = -p\delta_{ij} + \mu \left(\frac{\partial v_i}{\partial x_j} + \frac{\partial v_j}{\partial x_i} \right)$$

For incompressible fluids with constant viscosity μ:

$$\sigma_{ij} = -p\delta_{ij} + \mu(\nabla\vec{v} + (\nabla\vec{v})^T)$$

PROBLEM 10: KLEIN-GORDON PARTICLE IN A POTENTIAL

Problem: Solve the Klein-Gordon equation for a scalar field ϕ in a potential $V(\phi) = \frac{1}{2}m^2\phi^2$.

Solution: The Klein-Gordon equation for a field ϕ in the presence of a potential V is given by:

$$\Box\phi + \frac{\partial V}{\partial \phi} = 0 \tag{2.64}$$

Given the potential $V(\phi) = \frac{1}{2}m^2\phi^2$, the derivative with respect to ϕ is:

$$\frac{\partial V}{\partial \phi} = m^2\phi \tag{2.65}$$

Substituting this into the Klein-Gordon equation, we get:

$$\Box\phi + m^2\phi = 0 \tag{2.66}$$

where $\Box$ is the d'Alembertian operator in Minkowski space, defined as:

$$\Box = -\frac{\partial^2}{\partial t^2} + \nabla^2 \tag{2.67}$$

Assuming a solution of the form $\phi(x,t) = e^{i(\vec{k}\cdot\vec{x} - \omega t)}$, where $\vec{k}$ is the wave vector and ω is the frequency, we substitute into the equation:

$$-\frac{\partial^2}{\partial t^2}\phi(x,t) = -(-i\omega)^2\phi(x,t) = \omega^2\phi(x,t)$$
$$\nabla^2\phi(x,t) = (ik)^2\phi(x,t) = -k^2\phi(x,t)$$

Thus, the d'Alembertian operator acting on ϕ yields:

$$\Box\phi = (\omega^2 - k^2)\phi \tag{2.68}$$

Plugging this back into the Klein-Gordon equation:

$$(\omega^2 - k^2)\phi + m^2\phi = 0 \tag{2.69}$$

Factoring out ϕ, assuming $\phi \neq 0$, we find:

$$\omega^2 - k^2 + m^2 = 0 \tag{2.70}$$

Solving for ω:

$$\omega = \sqrt{k^2 + m^2} \tag{2.71}$$

This result indicates that the frequency ω depends on both the wave number k and the mass m of the particle described by the field ϕ. The relationship is characteristic of relativistic quantum particles, describing how their energy relates to momentum and mass.

This solution exemplifies the application of the Klein-Gordon equation to a free scalar field with a quadratic potential, a fundamental scenario in quantum field theory for particles with nonzero mass.

BONUS PROBLEM: THREE COUPLED HARMONIC OSCILLATORS WITH PHONON OPERATORS

Problem: Consider three masses, each of mass m, connected in a linear arrangement by springs of spring constant k. The masses are allowed to move only along the x-axis, and the outermost masses are connected to fixed walls by springs. Find the normal modes of oscillation for this system and express the solution using phonon creation and annihilation operators.

Solution: Let x_1, x_2, and x_3 represent the displacements of the first, second, and third mass from their equilibrium positions, respectively.

Equations of Motion: The forces on each mass due to the springs are given by Hooke's Law. The equations of motion for each mass are:

$$m\ddot{x}_1 = -2kx_1 + kx_2, \tag{2.72}$$

$$m\ddot{x}_2 = kx_1 - 2kx_2 + kx_3, \tag{2.73}$$

$$m\ddot{x}_3 = kx_2 - 2kx_3. \tag{2.74}$$

Normal Mode Analysis: Assume solutions of the form $x_i = A_i e^{i\omega t}$ for $i = 1, 2, 3$, where A_i are the amplitudes of the oscillations, and ω is the angular frequency. Substituting these into the equations of motion gives a system of homogeneous equations which can be represented in matrix form:

$$\begin{bmatrix} -2k - m\omega^2 & k & 0 \\ k & -2k - m\omega^2 & k \\ 0 & k & -2k - m\omega^2 \end{bmatrix} \begin{bmatrix} A_1 \\ A_2 \\ A_3 \end{bmatrix} = \begin{bmatrix} 0 \\ 0 \\ 0 \end{bmatrix}. \tag{2.75}$$

Determinant and Characteristic Polynomial: To find the normal modes, we calculate the determinant of the coefficient matrix and set it to zero, yielding a characteristic polynomial in ω^2.

Quantization and Phonon Operators: Upon finding the normal modes, we quantize the system by introducing phonon creation ($a^\dagger$) and annihilation (a) operators for each normal mode. The quantized displacement for each mass can be expressed as:

$$x_i = \sum_{n=1}^{3} \sqrt{\frac{\hbar}{2m\omega_n}} (a_n + a_n^\dagger)\psi_{ni}, \tag{2.76}$$

where ψ_{ni} are the normalized eigenvectors corresponding to each normal mode frequency ω_n, and a_n, $a_n^\dagger$ are the annihilation and creation operators for the phonons in the n-th mode. These operators satisfy the commutation relations:

$$[a_n, a_m^\dagger] = \delta_{nm}, \quad [a_n, a_m] = [a_n^\dagger, a_m^\dagger] = 0. \tag{2.77}$$

Hamiltonian: The Hamiltonian for the system in terms of phonon operators becomes:

$$H = \sum_{n=1}^{3} \hbar\omega_n \left(a_n^\dagger a_n + \frac{1}{2} \right), \tag{2.78}$$

illustrating how the energy of the system is quantized in terms of phonons, where each phonon mode contributes to the total energy.

This approach bridges classical dynamics with quantum field theory, providing a fundamental link between mechanical vibrations and quantum operators, crucial for understanding more complex systems in condensed matter physics.

3 Lectures 6 & 7: Quantization, Path Integrals, Number Operator, and Field Operators

3.1 QUANTUM PARTICLES AND QUANTIZATION

Quantization is one of the fundamental concepts of modern physics, forming the bedrock of quantum mechanics and quantum field theory. This section delves into the nature of quantum particles, the processes that underlie quantization, and how these principles apply to systems ranging from simple particles in a box to complex many-body systems.

Quantum mechanics introduces a radical departure from classical physics, presenting a world where physical quantities such as energy, momentum, and angular momentum take discrete values under certain conditions. This quantization arises from the constraints imposed by wavefunctions and their boundary conditions in quantum systems. For example, a quantum particle confined in a box can only possess specific energy levels, a phenomenon that is both non-intuitive and fundamental to understanding quantum systems.

The concept of quantization extends beyond mere energy levels. It encompasses the entire framework for describing and predicting the behavior of microscopic systems. This includes the first quantization, which applies quantum mechanics to the classical properties of particles, leading to wavefunctions that describe the probability distributions of particle properties. First quantization addresses the need for a quantum description of single particles in potential fields, laying the groundwork for more complex theories.

Second quantization, or field quantization, takes these concepts further by treating fields themselves as quantized. Instead of focusing on particles as the primary entities, second quantization deals with fields, where particles emerge as excitations or quanta of these fields. This approach is particularly powerful in the context of quantum field theory, which unifies and generalizes quantum mechanics and special relativity, and provides the framework for describing fundamental interactions through gauge theories.

In this section, we explore these concepts starting from the basic principles of quantum mechanics, moving through the methodology of first quantization, and culminating in the advanced framework of second quantization. This journey not only illustrates the depth and breadth of quantum theory but also prepares the groundwork for understanding complex phenomena in particle physics, condensed matter physics, and beyond.

Through the detailed discussion of energy levels in quantum boxes, the application of first quantization methods, and an exploration of the profound implications of second quantization, this section aims to provide a thorough understanding of how quantum principles shape our understanding of the physical universe.

DOI: 10.1201/9781003637370-3

3.1.1 ENERGY LEVELS AND WAVE FUNCTIONS FOR A QUANTUM PARTICLE IN A BOX

One of the quintessential problems in quantum mechanics is the quantum particle in a box, which provides profound insights into the behavior of quantum systems confined within boundaries. This model is fundamental not only for its simplicity but also for its ability to illustrate key concepts of quantization and wavefunction behavior.

Theoretical Background

Consider a particle of mass m confined in a one-dimensional box of length L. The potential inside the box is zero, and it is infinite outside the box, effectively confining the particle to the region $0 \leq x \leq L$. The Schrödinger equation for this system in the region of interest (inside the box) is given by:

$$-\frac{\hbar^2}{2m}\frac{d^2\psi}{dx^2} = E\psi, \tag{3.1}$$

where $\hbar$ is the reduced Planck's constant, ψ is the wavefunction of the particle, and E is the energy of the particle.

Boundary Conditions

The boundary conditions require that the wavefunction vanishes at the walls of the box:

$$\psi(0) = \psi(L) = 0. \tag{3.2}$$

These conditions ensure that the particle is found only within the box, as the probability density outside the box is zero.

Solving the Schrödinger Equation

The general solution to the differential equation is:

$$\psi(x) = A\sin(kx) + B\cos(kx), \tag{3.3}$$

where A and B are constants, and k is the wave number related to the energy by $k = \sqrt{\frac{2mE}{\hbar^2}}$. Applying the boundary conditions, B must be zero (since $\cos(0) = 1$ does not satisfy $\psi(0) = 0$), and $\sin(kL) = 0$ which gives:

$$kL = n\pi, \tag{3.4}$$

where n is a positive integer. This quantizes the wave number k as:

$$k = \frac{n\pi}{L}. \tag{3.5}$$

Quantized Energy Levels

Substituting k back into the energy relation provides the quantized energy levels:

$$E_n = \frac{\hbar^2\pi^2}{2mL^2}n^2, \tag{3.6}$$

where $n = 1, 2, 3, \ldots$. These energy levels show that the energy of the particle increases with the square of the quantum number n, and there is no zero-energy ground state, highlighting a fundamental quantum effect of confinement.

Wavefunctions

The normalized wavefunctions for the particle are then given by:

$$\psi_n(x) = \sqrt{\frac{2}{L}} \sin\left(\frac{n\pi x}{L}\right), \tag{3.7}$$

where the normalization factor ensures that the total probability of finding the particle within the box is 1.

Physical Interpretation

The quantized energy levels and corresponding wavefunctions describe a particle that exhibits wave-like properties within the box. Each energy level corresponds to a standing wave pattern of the particle's wavefunction, with nodes at the walls of the box. The lowest energy state ($n = 1$) has no nodes within the box, each higher energy state has more nodes, indicating more rapid fluctuations in the wavefunction and higher kinetic energy.

EXAMPLE: AN ELECTRON IN A NANOMETER BOX

Consider an electron confined in a one-dimensional box of length 1 nanometer (nm). The mass of an electron m_e is approximately 9.109×10^{-31} kg and the reduced Planck's constant $\hbar$ is approximately 1.054×10^{-34} Js. We calculate the ground state energy E_1 using the formula:

$$E_n = \frac{\hbar^2\pi^2}{2mL^2}n^2 \tag{3.8}$$

For the ground state ($n = 1$) and $L = 1$ nm $= 1 \times 10^{-9}$ m:

$$E_1 = \frac{(1.054 \times 10^{-34}\,\text{Js})^2 \cdot \pi^2}{2 \times 9.109 \times 10^{-31}\,\text{kg} \times (1 \times 10^{-9}\,\text{m})^2} \tag{3.9}$$

Calculating this gives:

$$E_1 \approx 6.0247 \times 10^{-20}\,\text{J} \tag{3.10}$$

To convert this energy into electron volts (eV), where 1 eV $\approx 1.60218 \times 10^{-19}$ J:

$$E_1 \approx \frac{6.0247 \times 10^{-20}\,\text{J}}{1.60218 \times 10^{-19}\,\text{J/eV}} \approx 0.376\,\text{eV} \tag{3.11}$$

This result demonstrates the significant quantization effect for an electron in such a confined space.

3.1.2 INTRODUCTION TO FIRST QUANTIZATION AND ITS APPLICATIONS

First quantization is a foundational concept in quantum mechanics, where classical parameters such as position and momentum are replaced with operators that act on wavefunctions. This procedure is fundamental to the quantization of physical systems and serves as a bridge between classical mechanics and quantum theory.

Theoretical Foundation

First quantization starts with the classical phase space variables of a particleits position and momentumand introduces corresponding quantum operators. These operators are defined such that they satisfy the canonical commutation relations:

$$[\hat{x}, \hat{p}] = \hat{x}\hat{p} - \hat{p}\hat{x} = i\hbar \tag{3.12}$$

where $\hat{x}$ is the position operator, $\hat{p}$ is the momentum operator, and i is the imaginary unit. This commutation relation is crucial as it encapsulates the core quantum mechanical property of non-commutativity of measurements.

Wavefunction and Schrödinger Equation

In the framework of first quantization, the state of a physical system is described by a wavefunction $\psi(x)$, which is a function of the position variable x. The wavefunction provides a complete description of the probability amplitude of the position of particles. The dynamics of the wavefunction are governed by the Schrödinger equation:

$$i\hbar\frac{\partial}{\partial t}\psi(x,t) = \hat{H}\psi(x,t) \tag{3.13}$$

where $\hat{H}$ is the Hamiltonian operator of the system, representing the total energy (kinetic + potential) of the system.

3.1.2.1 The Fourier Transform in Quantum Mechanics

The Fourier Transform is a powerful mathematical tool in quantum mechanics that facilitates the transition between position and momentum representations. This technique is fundamental in exploring and understanding the dual aspects of particles as both waves and particles, a core concept in quantum mechanics.

Mathematical Framework

In quantum mechanics, the wavefunction of a particle can be represented in either position space ($\psi(x)$) or momentum space ($\phi(p)$). The Fourier Transform allows for the conversion between these two representations. The position space wavefunction $\psi(x)$ can be transformed to momentum space $\phi(p)$ using the Fourier Transform:

$$\phi(p) = \frac{1}{\sqrt{2\pi\hbar}}\int_{-\infty}^{\infty}\psi(x)e^{-\frac{i}{\hbar}px}\,dx \tag{3.14}$$

Conversely, the inverse Fourier Transform is used to revert from momentum space back to position space:

$$\psi(x) = \frac{1}{\sqrt{2\pi\hbar}}\int_{-\infty}^{\infty}\phi(p)e^{\frac{i}{\hbar}px}\,dp \tag{3.15}$$

Physical Interpretation

The Fourier Transform in quantum mechanics is not merely a mathematical trick; it has deep physical implications:

- **Uncertainty Principle:** The transform pairs $\psi(x)$ and $\phi(p)$ demonstrate the uncertainty principle. A narrow wavefunction in position space, indicating precise knowledge of position, corresponds to a broad wavefunction in momentum space, indicating greater uncertainty in momentum, and vice versa.
- **Wavefunction Analysis:** By analyzing the wavefunction in both position and momentum space, physicists can gain complete information about the quantum state of a system, including how it evolves over time and reacts to different potential fields.
- **Operational Switch:** In many quantum mechanics problems, it is easier to solve the Schrödinger equation in momentum space than in position space, especially when potential energy depends linearly on position.

The Fourier Transform is used extensively across various domains of physics: in spectroscopy for analyzing the frequencies (energy levels) present in a molecular sample, in quantum field theory, to switch between different fields representations, which is crucial for formulating and solving field equations, and in computational physics, to numerically solve quantum mechanics problems more efficiently in the appropriate space.

3.1.2.2 Worked Example: Fourier Transform of a Gaussian Wavefunction

Consider a Gaussian wavefunction $\psi(x)$ centered at zero with width σ, given in position space by:

$$\psi(x) = \frac{1}{\sqrt{\sqrt{2\pi}\sigma}} e^{-\frac{x^2}{2\sigma^2}} \tag{3.16}$$

This form is chosen because it is normalized, and Gaussian wavefunctions are eigenfunctions of the Fourier Transform, making them particularly suitable for such exercises.

Fourier Transform to Momentum Space

The Fourier Transform of $\psi(x)$ to momentum space $\phi(p)$ is given by:

$$\phi(p) = \frac{1}{\sqrt{2\pi\hbar}} \int_{-\infty}^{\infty} \psi(x) e^{-\frac{i}{\hbar}px}\, dx \tag{3.17}$$

Substituting the expression for $\psi(x)$ and integrating, we have:

$$\phi(p) = \frac{1}{\sqrt{2\pi\hbar}} \frac{1}{\sqrt{\sqrt{2\pi}\sigma}} \int_{-\infty}^{\infty} e^{-\frac{x^2}{2\sigma^2}} e^{-\frac{i}{\hbar}px}\, dx \tag{3.18}$$

This integral is a standard form and evaluates to:

$$\phi(p) = \sqrt{\frac{\sigma}{\sqrt{2\pi}\hbar}} e^{-\frac{\sigma^2 p^2}{2\hbar^2}} \tag{3.19}$$

This result shows that the momentum space wavefunction is also Gaussian, with a width inversely proportional to the position space width σ, consistent with the uncertainty principle.

Inverse Fourier Transform to Position Space

To demonstrate the reversibility of the Fourier Transform, we apply the inverse transform to $\phi(p)$:

$$\psi(x) = \frac{1}{\sqrt{2\pi\hbar}} \int_{-\infty}^{\infty} \phi(p) e^{\frac{i}{\hbar}px}\, dp \tag{3.20}$$

Substituting $\phi(p)$:

$$\psi(x) = \frac{1}{\sqrt{2\pi\hbar}} \sqrt{\frac{\sigma}{\sqrt{2\pi}\hbar}} \int_{-\infty}^{\infty} e^{-\frac{\sigma^2 p^2}{2\hbar^2}} e^{\frac{i}{\hbar}px}\, dp \tag{3.21}$$

Solving this integral also yields the original Gaussian form:

$$\psi(x) = \frac{1}{\sqrt{\sqrt{2\pi}\sigma}} e^{-\frac{x^2}{2\sigma^2}} \tag{3.22}$$

This example illustrates the application of the Fourier Transform and its inverse in quantum mechanics, confirming the consistency and symmetry of the transformations between position and momentum spaces. The Gaussian wavefunction serves as an excellent example due to its simple and symmetric form in both spaces, highlighting fundamental quantum mechanical principles like the uncertainty principle.

Applications of the First Quantization

First quantization has been effectively applied in solving a wide range of quantum mechanical problems:

- **Particle in a Potential Well:** It provides a method to solve for the energy states of a particle in potential fields, such as a particle in a box, harmonic oscillator, and the hydrogen atom.
- **Atomic and Molecular Physics:** It allows for the calculation of electronic structures and properties of atoms and molecules, which are essential for understanding chemical bonding and spectra.
- **Tunneling Phenomena:** Quantum tunneling, a phenomenon where particles pass through potential barriers that they classically should not overcome, is described using first quantization.

Significance in Quantum Theory

First quantization lays the groundwork for more advanced theories, including field quantization in quantum field theory. It provides a systematic approach to incorporating quantum effects into classical systems, fundamentally altering our understanding of particles and interactions at microscopic scales. This approach also illustrates the inherent limitations of classical mechanics when applied to atomic and subatomic levels, where quantum effects become dominant.

This subsection elucidates the concept of first quantization, its theoretical basis, and diverse applications across physics, emphasizing its indispensable role in the transition from classical to quantum descriptions of nature.

3.1.3 OVERVIEW OF SECOND QUANTIZATION, ITS BENEFITS, AND PHYSICAL MEANING

Second quantization is a profound reformulation of quantum mechanics that extends the first quantization framework to fields and many-body systems. Unlike first quantization, which quantizes the classical particles' properties (like position and momentum), second quantization quantizes the field itself that describes the quantum state of a system. This approach is especially powerful in the realms of quantum field theory and the study of many-body systems.

Theoretical Framework

In the first quantization the focus is on the uncertainties of a particles position and momentum and is very particle centric. On the other hand, with the second quantization, the focus shifts from particles to fields. A quantum field can be thought of as an operator-valued field, where each point in space has a corresponding operator that creates or annihilates particles at that location. This representation allows for the description of systems with varying particle numbers and interactions more naturally than first quantization.

Mathematical Formulation

The quantum field $\hat{\psi}(x)$ represents the field operator at position x, which satisfies commutation or anticommutation relations depending on whether the particles are bosons or fermions. The Hamiltonian and other physical observables are expressed in terms of these field operators, allowing for a unified description of particles and interactions. For example, the Hamiltonian for a non-relativistic many-body system can be written as:

$$\hat{H} = \int d^3x\, \hat{\psi}^\dagger(x)\left(-\frac{\hbar^2}{2m}\nabla^2 + V(x)\right)\hat{\psi}(x) + \frac{1}{2}\int d^3x d^3x'\, \hat{\psi}^\dagger(x)\hat{\psi}^\dagger(x')U(x,x')\hat{\psi}(x')\hat{\psi}(x) \tag{3.23}$$

where $V(x)$ is the potential energy, $U(x,x')$ is the interaction potential, and $\hat{\psi}^\dagger(x)$, $\hat{\psi}(x)$ are the creation and annihilation operators, respectively.

Understanding Quantum Operators: Field vs. Ladder Operators

In quantum field theory and many-body physics, two types of operators play crucial roles: field operators and ladder operators. Here, we distinguish between $\hat{\psi}^\dagger(x)$ and $\hat{\psi}(x)$, which are field operators, and $\hat{a}^\dagger$ and $\hat{a}$, which are ladder operators.

Field Operators: $\hat{\psi}^\dagger(x)$ and $\hat{\psi}(x)$

- **Role and Definition:** Field operators like $\hat{\psi}^\dagger(x)$ and $\hat{\psi}(x)$ are fundamental in quantum field theory. $\hat{\psi}^\dagger(x)$ is a creation operator that adds a particle at position x, while $\hat{\psi}(x)$ is an annihilation operator that removes a particle from position x.
- **Commutation Relations:** These operators satisfy different commutation relations depending on the nature of the particles:
  - For bosons: $[\hat{\psi}(x), \hat{\psi}^\dagger(y)] = \delta(x-y)$
  - For fermions: $\{\hat{\psi}(x), \hat{\psi}^\dagger(y)\} = \delta(x-y)$
- **Applications:** Field operators are used in the description of quantum fields, such as the electromagnetic field in quantum electrodynamics (QED) and the electron field in condensed matter physics.

Ladder Operators: $\hat{a}^\dagger$ and $\hat{a}$

- **Role and Definition:** Ladder operators, $\hat{a}^\dagger$ and $\hat{a}$, are primarily used in the quantization of the harmonic oscillator and many-body systems. $\hat{a}^\dagger$ raises the energy state of the system (raising operator), while $\hat{a}$ lowers it (lowering operator).
- **Functionality:** These operators are crucial for building the energy eigenstates of the system:

$$\hat{a}^\dagger |n\rangle = \sqrt{n+1}\,|n+1\rangle, \quad \hat{a}|n\rangle = \sqrt{n}\,|n-1\rangle \tag{3.24}$$

- **Applications:** Ladder operators are utilized in systems where the quantization of energy is important, such as in quantum optics for describing the states of photons in a cavity and in solid-state physics for phonons.

These distinctions help clarify the functional roles and applications of quantum operators in different physical contexts, illustrating how quantum mechanics and quantum field theory utilize operators to describe and manipulate quantum systems.

Commutation and Anticommutation Relations in Quantum Mechanics

In quantum mechanics and quantum field theory, the fundamental differences in the statistical behaviors of bosons and fermions are encapsulated by their commutation and anticommutation relations. These relations are crucial for defining the quantum mechanical properties of these particles and have profound implications for their physical behaviors.

Bosons and Commutation Relations

Bosons, such as photons and phonons, obey Bose-Einstein statistics and are described by symmetric wavefunctions. The creation ($\hat{a}^\dagger$) and annihilation ($\hat{a}$) operators for bosons satisfy the following commutation relations:

$$[\hat{a}, \hat{a}^\dagger] = \hat{a}\hat{a}^\dagger - \hat{a}^\dagger\hat{a} = 1 \tag{3.25}$$

Field operators for bosonic fields like $\hat{\psi}(x)$ and $\hat{\psi}^\dagger(x)$ follow similar commutation relations:

$$[\hat{\psi}(x), \hat{\psi}^\dagger(y)] = \delta(x-y) \tag{3.26}$$

These relations allow multiple bosons to occupy the same quantum state, enabling phenomena such as superfluidity and Bose-Einstein condensation.

Fermions and Anticommutation Relations

Fermions, including electrons, protons, and neutrons, follow Fermi-Dirac statistics and are characterized by antisymmetric wavefunctions. Their operators satisfy anticommutation relations:

$$\{\hat{a}, \hat{a}^\dagger\} = \hat{a}\hat{a}^\dagger + \hat{a}^\dagger\hat{a} = 1 \tag{3.27}$$

For fermionic field operators like $\hat{\psi}(x)$ and $\hat{\psi}^\dagger(x)$, the anticommutation relations are:

$$\{\hat{\psi}(x), \hat{\psi}^\dagger(y)\} = \delta(x-y) \tag{3.28}$$

These relations enforce the Pauli Exclusion Principle, which prohibits more than one fermion from occupying the same quantum state.

Mathematical Implications of Commutation and Anticommutation

- **Commutation Relations [ ]:** These relations define the difference between the product of operators applied in opposite orders. For example, if two operators commute, their commutation relation is zero, indicating that the sequence in which the operators are applied does not matter.
- **Anticommutation Relations – ″:** These relations define the sum of the products of operators applied in both possible orders. For fermionic operators, the anticommutation being zero means that applying the same operator twice leads to a nullification, reflecting the exclusion principle.

This framework of commutation and anticommutation is fundamental to the structure of quantum theories, influencing everything from the stability of atoms to the properties of quantum fields.

Benefits and Advantages of Second Quantization

- **Handling of Many-Body Systems:** Second quantization simplifies the treatment of many-body problems significantly. It naturally incorporates the principles of quantum statistics and makes the calculation of many-body states more tractable.
- **Incorporation of Quantum Statistics:** The formalism automatically respects the symmetrization postulate for bosons and the antisymmetrization postulate for fermions due to the commutation relations of the field operators.
- **Flexibility in Particle Number:** It allows the number of particles to vary, facilitating the study of phenomena like particle creation and annihilation, essential in fields like nuclear and high-energy physics.

Physical Meaning and Applications of Second Quantization

Second quantization is not just a mathematical tool; it has deep physical implications and applications:

- **Quantum Field Theory:** It is the backbone of quantum field theory, which describes fundamental forces and particles in the universe as quantum fields interacting with each other.
- **Solid State Physics:** It is crucial in explaining properties of solids, such as conductivity, magnetism, and superconductivity, through the study of electron and phonon fields.
- **Quantum Optics:** It provides a framework for understanding the quantum nature of light and its interaction with matter, explaining phenomena like quantum entanglement and squeezing.

Worked Example: Quantum Field Theory Hamiltonian

Consider a quantum field theory where the field operators are approximated by plane waves and the interactions are simplified to a delta function potential. We also use a harmonic oscillator potential for $V(x)$.

Assumptions

- Field operators:

$$\hat{\psi}(x) = \int \frac{d^3k}{(2\pi)^3} e^{ik\cdot x} \hat{a}_k, \tag{3.29}$$

$$\hat{\psi}^\dagger(x) = \int \frac{d^3k'}{(2\pi)^3} e^{-ik'\cdot x} \hat{a}^\dagger_{k'}. \tag{3.30}$$

- External potential $V(x)$ is the harmonic oscillator potential:

$$V(x) = \frac{1}{2} m\omega^2 x^2. \tag{3.31}$$

- Interaction potential $U(x, x')$ is a delta function:

$$U(x, x') = g\delta(x - x'), \tag{3.32}$$

  where g is the coupling constant.

Hamiltonian Analysis

Inserting these expressions into the Hamiltonian:

$$\hat{H} = \int d^3x\, \hat{\psi}^\dagger(x) \left(-\frac{\hbar^2}{2m} \nabla^2 + \frac{1}{2} m\omega^2 x^2 \right) \hat{\psi}(x) + \frac{1}{2} \int d^3x d^3x'\, \hat{\psi}^\dagger(x) \hat{\psi}^\dagger(x')\, g\delta(x - x')\, \hat{\psi}(x') \hat{\psi}(x). \tag{3.33}$$

Kinetic and Potential Energy Calculation

Using the plane wave expansions of the field operators, the kinetic and potential energy terms are calculated as follows:

3.1.3.0.1 Kinetic Energy:

$$\int d^3x\, \hat{\psi}^\dagger(x) \left(-\frac{\hbar^2}{2m} \nabla^2 \right) \hat{\psi}(x) = \int d^3x \int \frac{d^3k}{(2\pi)^3} \int \frac{d^3k'}{(2\pi)^3} e^{-i\mathbf{k}'\cdot\mathbf{x}} \hat{a}^\dagger_{\mathbf{k}'} \times \left(-\frac{\hbar^2}{2m} \nabla^2 \right) e^{i\mathbf{k}\cdot\mathbf{x}} \hat{a}_{\mathbf{k}}. \tag{3.34}$$

Applying the Laplacian operator ∇^2 to the exponential function yields:

$$\nabla^2 e^{ik\cdot x} = -k^2 e^{ik\cdot x}, \tag{3.35}$$

leading to:

$$\int d^3x \int \frac{d^3k}{(2\pi)^3} \int \frac{d^3k'}{(2\pi)^3} e^{-ik'\cdot x} \hat{a}^\dagger_{k'} \frac{\hbar^2 k^2}{2m} e^{ik\cdot x} \hat{a}_k. \tag{3.36}$$

This integral simplifies due to the orthogonality of the exponential functions, resulting in:

$$\int \frac{d^3k}{(2\pi)^3} \frac{\hbar^2 k^2}{2m} \hat{a}^\dagger_k \hat{a}_k. \tag{3.37}$$

3.1.3.0.2 Potential Energy:

The potential energy term calculation involves similar steps but requires an additional assumption or numerical integration for exact solutions due to the x^2 term.

Interaction Energy Calculation

The interaction term simplifies due to the delta function:

$$\frac{1}{2}\int d^3x d^3x' \, \hat{\psi}^\dagger(x)\hat{\psi}^\dagger(x')g\delta(x-x')\hat{\psi}(x')\hat{\psi}(x)$$
$$= \frac{g}{2}\int d^3x \, \hat{\psi}^\dagger(x)\hat{\psi}^\dagger(x)\hat{\psi}(x)\hat{\psi}(x). \tag{3.38}$$

Final Hamiltonian Expression

After calculating the kinetic, potential, and interaction terms, the Hamiltonian for the quantum field theory with plane wave field operators, a harmonic oscillator potential, and a delta function interaction is given by:

$$\hat{H} = \int \frac{d^3k}{(2\pi)^3}\frac{\hbar^2k^2}{2m}\hat{a}_k^\dagger\hat{a}_k + \int d^3x\left(\frac{1}{2}m\omega^2x^2\hat{\psi}^\dagger(x)\hat{\psi}(x)\right)$$
$$+\frac{g}{2}\int d^3x \, \hat{\psi}^\dagger(x)\hat{\psi}^\dagger(x)\hat{\psi}(x)\hat{\psi}(x). \tag{3.39}$$

3.1.3.0.3 Explanation of Non-Solvability:

The final form of the Hamiltonian includes terms that are challenging to solve analytically:

- The kinetic energy term is expressed in momentum space and is straightforward but highly non-trivial.
  - The kinetic term involves the Laplacian ∇^2, which, when applied to the field operators expressed in terms of plane waves, converts into a multiplication by k^2 in momentum space:

$$\int d^3x \hat{\psi}^\dagger(x)\left(-\frac{\hbar^2}{2m}\nabla^2\right)\hat{\psi}(x) = \int \frac{d^3k}{(2\pi)^3}\frac{\hbar^2k^2}{2m}\hat{a}_k^\dagger\hat{a}_k. \tag{3.40}$$

  - This transformation simplifies the spatial derivatives but requires handling integrals over all momentum states, representing an infinite number of degrees of freedom.
  - The presence of interactions or non-trivial potential landscapes can modify the energy contributions from the kinetic term, necessitating numerical or perturbative methods to evaluate or approximate these contributions effectively.
- The potential energy term involving x^2 requires an assumption about the spatial distribution or a specific representation of $\hat{\psi}(x)$ for further simplification, which often leads to solutions involving special functions or numerical methods. Using the plane wave expansion of the field operators, the potential energy term becomes:
  -

$$\int d^3x \frac{1}{2}m\omega^2x^2\hat{\psi}^\dagger(x)\hat{\psi}(x)$$
$$= \int d^3x \frac{1}{2}m\omega^2x^2\left(\int\frac{d^3k}{(2\pi)^3}\int\frac{d^3k'}{(2\pi)^3}e^{-i(\mathbf{k}-\mathbf{k}')\cdot\mathbf{x}}\hat{a}_k^\dagger\hat{a}_{k'}\right) \tag{3.41}$$
$$= \int\frac{d^3k}{(2\pi)^3}\frac{d^3k'}{(2\pi)^3}\hat{a}_k^\dagger\hat{a}_{k'}\int d^3x \frac{1}{2}m\omega^2x^2e^{-i(\mathbf{k}-\mathbf{k}')\cdot\mathbf{x}}. \tag{3.42}$$

  - This integral is complex and generally requires further assumptions for simplification or numerical methods for practical evaluation.
- The interaction term involves a quartic term in the field operators, typical in many-body quantum field theories, leading to complex many-body interactions that are usually handled through perturbative or numerical techniques rather than direct analytical solutions.

Thus, the complexity of quantum field theoretical problems often goes beyond simple analytical solutions, requiring advanced mathematical techniques and computational approaches.

Physical Interpretation

- The first term in the Hamiltonian, involving $\hat{\psi}^\dagger(x)\hat{\psi}(x)$, effectively counts the number of particles at position x and assigns them kinetic and potential energy.
- The second term describes the pairwise interactions between particles. The factor of $\frac{1}{2}$ prevents double-counting of the interactions.

Realistic Simplification Methods for Quantum Field Theory Hamiltonians

In quantum field theory (QFT), especially when dealing with potential terms like a harmonic oscillator potential within the context of a full field-theoretic model, reducing the Hamiltonian to a more manageable form often requires additional assumptions or approximations. Here are some strategies and considerations that can be used to simplify the Hamiltonian for practical calculations:

1. **Mode Expansion and Normal Modes:** Using normal modes or eigenmodes of the system can simplify the expressions significantly. If the system has translational symmetry or can be confined in a finite volume with boundary conditions, you can expand the field in terms of these modes rather than plane waves. Each mode then behaves like a harmonic oscillator, and the field operators can be rewritten accordingly.
2. **Non-relativistic Limit:** In many practical applications, especially in condensed matter physics or non-relativistic QFT, assuming a non-relativistic limit can simplify the kinetic energy term. This involves expanding around small momenta (compared to the speed of light) and simplifying the kinetic term to its non-relativistic form.
3. **Low-Energy Effective Theories:** Limiting the theory to low energies or low momenta can allow for an effective theory where higher order terms (in terms of energy or momentum) are negligible. This results in a simpler form of the Hamiltonian, focusing only on the most significant terms.
4. **Mean Field Approximation:** The mean field approximation can simplify interactions by replacing field operators with their expectation values in certain states. This reduces the complexity of the interaction terms, effectively decoupling parts of the system.
5. **Symmetry Considerations:** Exploiting symmetries of the potential and interaction terms can lead to simplifications. For example, if the system is isotropic, you can reduce three-dimensional integrals to one-dimensional radial integrals.
6. **Perturbative Techniques:** If the interaction term is small compared to other terms in the Hamiltonian, perturbative methods can be used. Here, the Hamiltonian is split into a solvable part and a perturbation, and the problem is solved iteratively.
7. **Numerical Methods:** When analytical solutions are unfeasible, numerical methods such as lattice field theory or Monte Carlo simulations can provide approximate solutions. These methods discretize space and time and numerically solve the resulting finite-dimensional problems.

Example: Harmonic Oscillator Potential Simplification

Consider a system confined in a box with length L and impose periodic boundary conditions. This confines the possible wave vectors to discrete values, and the field operators can be expanded as:

$$\hat{\psi}(x) = \sum_n \phi_n(x)\hat{a}_n, \tag{3.43}$$

$$\hat{\psi}^\dagger(x) = \sum_n \phi_n^*(x)\hat{a}_n^\dagger \tag{3.44}$$

where $\phi_n(x)$ are the eigenfunctions of the harmonic oscillator potential.

Derivation of the Hamiltonian from Field Operators

Given the field operators for a quantum system confined in a box with harmonic oscillator potential, we express the field operators in terms of the eigenfunctions of the Hamiltonian. Let's consider a one-dimensional quantum box of length L with periodic boundary conditions for simplicity.

Step 1: Express the field operators

$$\hat{\psi}(x) = \int \frac{d^3k}{(2\pi)^3} e^{ik\cdot x}\hat{a}_k, \tag{3.45}$$

$$\hat{\psi}^\dagger(x) = \int \frac{d^3k}{(2\pi)^3} e^{-ik\cdot x}\hat{a}_k^\dagger. \tag{3.46}$$

Step 2: Expand the field operators in eigenfunctions

$$\hat{\psi}(x) = \sum_n \phi_n(x)\hat{a}_n, \tag{3.47}$$

$$\hat{\psi}^\dagger(x) = \sum_n \phi_n^*(x)\hat{a}_n^\dagger, \tag{3.48}$$

where $\phi_n(x)$ are the eigenfunctions of the Hamiltonian with harmonic oscillator potential and $\hat{a}_n$, $\hat{a}_n^\dagger$ are the annihilation and creation operators for mode n.

Step 3: Substitute into the Hamiltonian The Hamiltonian of a quantum field theory with a harmonic oscillator potential and interaction term can be written as:

$$\hat{H} = \int d^3x \left(\hat{\psi}^\dagger(x) \left(-\frac{\hbar^2}{2m}\nabla^2 + \frac{1}{2}m\omega^2 x^2 \right) \hat{\psi}(x) \right) + \frac{1}{2}\int d^3x d^3x' \, \hat{\psi}^\dagger(x)\hat{\psi}^\dagger(x')U(x,x')\hat{\psi}(x')\hat{\psi}(x), \tag{3.49}$$

where $U(x,x')$ is the interaction potential.

Step 4: Simplify the kinetic and potential energy terms

$$\int d^3x \, \hat{\psi}^\dagger(x) \left(-\frac{\hbar^2}{2m}\nabla^2 \right) \hat{\psi}(x) = \sum_n \frac{\hbar^2 k_n^2}{2m} \hat{a}_n^\dagger \hat{a}_n, \tag{3.50}$$

$$\int d^3x \, \hat{\psi}^\dagger(x) \frac{1}{2} m\omega^2 x^2 \hat{\psi}(x) = \sum_n \frac{1}{2} m\omega^2 \langle x^2 \rangle_n \hat{a}_n^\dagger \hat{a}_n, \tag{3.51}$$

where $\langle x^2 \rangle_n$ is the expectation value of x^2 in state n.

Step 5: Simplify the interaction term

$$\frac{1}{2}\int d^3x d^3x' \, \hat{\psi}^\dagger(x)\hat{\psi}^\dagger(x') g\delta(x-x')\hat{\psi}(x')\hat{\psi}(x) = \frac{g}{2}\sum_n \hat{a}_n^\dagger \hat{a}_n^\dagger \hat{a}_n \hat{a}_n, \tag{3.52}$$

assuming a simplified interaction potential.

Final Hamiltonian:

$$\hat{H} = \sum_n \left(\frac{\hbar^2 k_n^2}{2m} + \frac{1}{2} m\omega^2 \langle x^2 \rangle_n \right) \hat{a}_n^\dagger \hat{a}_n + \frac{g}{2} \sum_n \hat{a}_n^\dagger \hat{a}_n^\dagger \hat{a}_n \hat{a}_n. \tag{3.53}$$

This derivation clearly outlines how the Hamiltonian can be expressed in terms of the mode expansions of the field operators, focusing on a realistic case with a harmonic oscillator potential. The approach highlights the practical methods to handle and simplify quantum field theoretical models for physical systems.

Next we need to express this in terms of the general operator and interaction matrix elements.

Step 1: Simplify the non-interacting part We define the non-interacting part of the Hamiltonian (the first term in the sum) as the eigenenergies ε_n:

$$\varepsilon_n = \frac{\hbar^2 k_n^2}{2m} + \frac{1}{2} m\omega^2 \langle x^2 \rangle_n, \tag{3.54}$$

which accounts for the kinetic energy and potential energy of each mode.

Step 2: Transform the interaction term The interaction term (second part of the sum) needs to be expressed in terms of the interaction matrix elements U_{nmlk}. Assuming that the interaction potential can be decomposed into the eigenbasis used for the field operators, this is done as follows:

$$U_{nmlk} = \int d^3x\, d^3x'\, \phi_n^*(x) \phi_m^*(x') U(x,x') \phi_l(x') \phi_k(x), \tag{3.55}$$

$$\hat{H}_{\text{interaction}} = \frac{1}{2} \sum_{n,m,l,k} U_{nmlk} \hat{a}_n^\dagger \hat{a}_m^\dagger \hat{a}_l \hat{a}_k. \tag{3.56}$$

Here, $U(x,x')$ is the spatial representation of the interaction potential, typically a function like the delta function or a screened Coulomb potential, and $\phi_n(x)$ are the eigenfunctions of the non-interacting part of the Hamiltonian.

Final Hamiltonian: The Hamiltonian then becomes:

$$\hat{H} = \sum_n \varepsilon_n \hat{a}_n^\dagger \hat{a}_n + \frac{1}{2} \sum_{n,m,l,k} U_{nmlk} \hat{a}_n^\dagger \hat{a}_m^\dagger \hat{a}_l \hat{a}_k, \tag{3.57}$$

where ε_n are the eigenenergies for each mode, and U_{nmlk} are the interaction matrix elements representing the strength and nature of the interaction between different modes.

Explanation: This formulation allows for an explicit calculation of interactions within a quantized field framework, adapting the model to handle both non-interacting and interacting particles in a quantum system. The transformation of the interaction term is crucial for incorporating real physical interactions and making the theory applicable to realistic systems, such as those found in condensed matter physics or quantum electrodynamics.

3.2 PATH INTEGRAL FORMULATION

The path integral formulation of quantum mechanics, developed by Richard Feynman, offers a unique perspective on quantum theory that is particularly powerful in the field of quantum field theory. This formulation complements the standard operator method, providing a more intuitive approach based on the summation over histories concept.

3.2.1 FUNDAMENTALS OF PATH INTEGRAL FORMULATION

The path integral formulation considers all possible paths that a particle can take between initial and final states. The probability amplitude for a transition from a state $|x_i, t_i\rangle$ to a state $|x_f, t_f\rangle$ is given by summing over all possible paths, $x(t)$, connecting these points:

$$\langle x_f, t_f | x_i, t_i \rangle = \int \mathscr{D}[x(t)]\, e^{\frac{i}{\hbar} S[x(t)]}, \tag{3.58}$$

where $\mathscr{D}[x(t)]$ denotes the path integral measure, and $S[x(t)]$ is the action integral, evaluated along the path $x(t)$:

$$S[x(t)] = \int_{t_i}^{t_f} L(x, \dot{x}, t)\, dt, \tag{3.59}$$

with L representing the Lagrangian of the system.

3.2.2 QUANTIZATION OF FIELDS

In quantum field theory, the path integral formulation extends to fields. For a scalar field ϕ, the transition amplitude, Z, between two field configurations $\phi_i(x)$ and $\phi_f(x)$ is:

$$Z = \int \mathscr{D}[\phi]\, e^{iS[\phi]}, \tag{3.60}$$

where $S[\phi]$ is the action expressed as a functional of the field $\phi(x)$:

$$S[\phi] = \int d^4x\, \mathscr{L}(\phi, \partial_\mu \phi), \tag{3.61}$$

and $\mathscr{L}$ is the Lagrangian density, which depends on the field and its derivatives.

Significance of the Partition Function Z

In quantum field theory, Z represents the partition function or generating functional, playing a crucial role in the formulation and analysis of the theory. Below are the key aspects of its significance:

1. **Statistical Mechanics Analogy:** In statistical mechanics, the partition function Z encapsulates all thermodynamic information of a system, summing over all possible states, each weighted by its Boltzmann factor $e^{-\beta E}$. In quantum field theory, Z serves a similar purpose, summing over all possible field configurations:
$$Z = \int \mathscr{D}[\phi]\, e^{iS[\phi]}, \tag{3.62}$$
where $S[\phi]$ is the action functional.
2. **Generating Functional for Correlation Functions:** Z generates all correlation functions of the field theory. Functional derivatives of Z with respect to field sources yield the correlation functions:
$$\left.\frac{\delta^n Z[J]}{\delta J(x_1)\delta J(x_2)\ldots\delta J(x_n)}\right|_{J=0} = \langle 0|T\hat{\phi}(x_1)\hat{\phi}(x_2)\ldots\hat{\phi}(x_n)|0\rangle, \tag{3.63}$$
where J is a source coupled to the field ϕ, and T denotes time ordering.
3. **Perturbative Expansions:** Z facilitates the perturbative expansion in terms of power series in the coupling constants, leading to Feynman diagrams (discussed in a later chapter):
$$Z[J] = \sum(\text{Feynman diagrams}), \tag{3.64}$$
each diagram corresponding to a term in the series expansion of the interactions.
4. **Vacuum Properties:** The value of Z in the absence of external sources ($J = 0$) relates to the vacuum energy of the theory, impacting the stability and ground state properties.

5. **Connection to Experiment:** Correlation functions derived from Z are related to measurable quantities in experiments, such as cross-sections and decay rates, bridging theory with experimental outcomes.

These aspects underscore Z as not merely a mathematical construct but as a central component in linking the dynamics, symmetries, and interactions of quantum fields to physical, observable phenomena.

3.2.3 SIGNIFICANCE AND ADVANTAGES

One of the significant advantages of the path integral approach is its natural accommodation of symmetries and conservation laws, such as gauge invariance, through the action principle. It also simplifies the treatment of systems with a large number of degrees of freedom, such as quantum fields, and provides a direct method for statistical mechanics.

3.3 QUANTUM FIELDS AND OPERATORS

3.3.1 BASIC CONCEPTS OF QUANTUM FIELD THEORY

Quantum Field Theory (QFT) extends the framework of quantum mechanics to fields, treating particles as excited states of an underlying field, each characterized by a quantum number. In this theory, the classical fields are replaced by operator-valued fields that act on the quantum state space. QFT combines classical field theory, special relativity, and quantum mechanics; it is the foundation for understanding fundamental interactions in particle physics, particularly within the Standard Model.

The central concept in QFT is the field operator, which replaces the concept of a particle trajectory. Fields are functions of space and time that obey quantum mechanical principles. Each point in space and time has an associated field operator, which can create or annihilate particles at that point. This approach allows the description of processes where the number of particles is not conserved, such as particle decay or particle creation in collisions.

3.3.2 NUMBER OPERATOR

In quantum field theory, the number operator and ladder operators (creation and annihilation operators) are essential tools for quantizing fields. The number operator, denoted as $\hat{N}$, measures the number of particles in a given quantum state. It is defined for a mode of the field with quantum number k by:

$$\hat{N}_k = \hat{a}_k^\dagger \hat{a}_k, \tag{3.65}$$

where $\hat{a}_k^\dagger$ and $\hat{a}_k$ are the creation and annihilation operators for the mode k, respectively. These operators satisfy the commutation relations for bosons and anticommutation relations for fermions.

Example: Calculating the Number Operator

Consider a harmonic oscillator in quantum mechanics, which is analogous to a single mode of a quantum field. The number operator $\hat{N}$ is defined as:

$$\hat{N} = \hat{a}^\dagger \hat{a}, \tag{3.66}$$

where $\hat{a}^\dagger$ and $\hat{a}$ are the creation and annihilation operators, respectively.

3.3.2.0.1 Action on a Quantum State:

To see how the number operator acts on a state, consider a state $|n\rangle$ which represents the oscillator being in its n-th excited state. This state is an eigenstate of the number operator:

$$\hat{N}|n\rangle = \hat{a}^\dagger \hat{a}|n\rangle. \tag{3.67}$$

Applying the annihilation operator, $\hat{a}$, reduces the state by one quantum of energy:

$$\hat{a}|n\rangle = \sqrt{n}|n-1\rangle, \tag{3.68}$$

where $\sqrt{n}$ is the normalization factor arising from the definition of the ladder operators.

Then, applying the creation operator, $\hat{a}^\dagger$, to $\hat{a}|n\rangle$:

$$\hat{a}^\dagger \hat{a}|n\rangle = \hat{a}^\dagger \sqrt{n}|n-1\rangle = \sqrt{n}\hat{a}^\dagger|n-1\rangle = \sqrt{n}\sqrt{n}|n\rangle = n|n\rangle. \tag{3.69}$$

3.3.2.0.2 Result:

Therefore, we find that:

$$\hat{N}|n\rangle = n|n\rangle. \tag{3.70}$$

This shows that the number operator $\hat{N}$ measures the number of quanta in the state $|n\rangle$, and the eigenvalue n corresponds to the number of quanta or particles in that state.

3.3.2.0.3 Physical Interpretation:

In the context of quantum field theory, this calculation demonstrates how the number operator can be used to determine the number of particles present in a given quantum field mode. This is crucial for understanding phenomena such as particle creation and annihilation in fields like Quantum Electrodynamics (QED) and for calculating the energy levels of systems described by quantum fields.

3.3.3 FIELD OPERATORS AND COMMUTATION RELATIONS

Field operators play a crucial role in the formulation of quantum field theory. For each point in space and time, there is a field operator $\hat{\psi}(x)$ and its Hermitian conjugate $\hat{\psi}^\dagger(x)$, which create and annihilate particles at the position x. The commutation relations for these operators depend on whether the particles are bosons or fermions:

Bosons: For bosonic fields, such as the electromagnetic field or scalar fields, the field operators obey the commutation relations:

$$[\hat{\psi}(x), \hat{\psi}^\dagger(y)] = \delta(x-y), \tag{3.71}$$

where $\delta(x-y)$ is the Dirac delta function, ensuring that the commutation is local in space.

3.3.3.0.1 Example Calculation for Bosons:

Suppose we have a bosonic field described by $\hat{\psi}(x)$ and $\hat{\psi}^\dagger(x)$. Consider two states at points x and y:

$$\hat{\psi}^\dagger(x)\hat{\psi}(x)|0\rangle = |x\rangle, \tag{3.72}$$

where $|0\rangle$ is the vacuum state and $|x\rangle$ represents the state with a particle at position x. Applying the commutation relation, we see:

$$[\hat{\psi}(x), \hat{\psi}^\dagger(x)]|0\rangle = (\hat{\psi}(x)\hat{\psi}^\dagger(x) - \hat{\psi}^\dagger(x)\hat{\psi}(x))|0\rangle = \delta(0)|0\rangle. \tag{3.73}$$

Fermions: For fermionic fields, such as the electron field in quantum electrodynamics, the operators obey anticommutation relations:

$$\{\hat{\psi}(x), \hat{\psi}^\dagger(y)\} = \delta(x-y), \tag{3.74}$$

reflecting the Pauli exclusion principle, which states that no two fermions can occupy the same quantum state.

3.3.3.0.2 Example Calculation for Fermions:

Using fermionic field operators, if we apply $\hat{\psi}^\dagger(x)$ to the vacuum state:

$$\hat{\psi}^\dagger(x)|0\rangle = |x\rangle, \tag{3.75}$$

then applying the anticommutation relation leads to:

$$\hat{\psi}(x)\hat{\psi}^\dagger(x)|0\rangle + \hat{\psi}^\dagger(x)\hat{\psi}(x)|0\rangle = \delta(0)|0\rangle, \tag{3.76}$$

$$\hat{\psi}(x)|x\rangle = 0, \tag{3.77}$$

demonstrating that applying $\hat{\psi}(x)$ to a state already containing a fermion at x results in annihilation (zero), consistent with the exclusion principle.

3.3.3.1 Physical Implications

These commutation and anticommutation relations are fundamental for describing the statistics and dynamics of quantum fields. They dictate how particles can be created or annihilated and how they interact with each other in quantum field theory, providing a framework for everything from the formulation of particle interactions in the Standard Model to the properties of quantum gases. The commutation or anticommutation relations are crucial for determining the behavior of the quantum system, particularly in how quantum states evolve over time and how particles scatter or interact.

3.4 SUMMARY AND DISCUSSION

This lecture series has explored a range of foundational and advanced topics in quantum mechanics and quantum field theory, emphasizing the development and applications of quantum fields, quantization techniques, and the mathematical formalisms underlying modern theoretical physics.

- **Quantization Techniques:** We began with a discussion on quantization, starting from the basic principles in quantum mechanics to more complex formulations in quantum field theory. This included first quantization's role in transitioning classical properties into quantum observables and second quantization's expansion into quantizing fields rather than just particles.
- **Quantum Particles and Quantum Fields:** The lecture detailed the theoretical treatment of quantum particles trapped in potential wells, illustrating how their quantized nature emerges from boundary conditions and the Schrödinger equation. For quantum fields, we extended these concepts to fields characterized by quantum numbers, highlighting their role in describing particle interactions and field dynamics.
- **Path Integral Formulation:** The path integral approach provided a unique perspective on quantum mechanics, emphasizing its utility in quantizing fields and calculating amplitudes through a summation over all possible histories.
- **Number Operator and Field Operators:** The use of number and ladder operators was elaborated to explain their fundamental roles in quantum field theory, particularly in counting quantum states and facilitating particle interactions within a given field.
- **Symmetries and Statistics:** The discussion on symmetries and statistical behaviors underpinned much of quantum field theory, detailing how particles like bosons and fermions behave under different quantum statistical rules.
- **Practical Computational Techniques:** Throughout the lectures, various computational strategies and simplifications were discussed to tackle the complexity of quantum field theoretical problems, such as mode expansions, perturbative techniques, and numerical methods.

3.5 QUESTIONS AND PROBLEMS

3.5.1 TRIVIA QUESTIONS

Question 1: What is the main focus of second quantization in quantum mechanics?

(a) Quantizing classical trajectories of particles.
(b) Quantizing fields themselves, not just particles.
(c) Quantizing only the gravitational interactions.
(d) Quantizing time-space interactions.

Answer: b. Quantizing fields themselves, not just particles.
Question 2: Which mathematical tool is crucial for transitioning between position and momentum space in quantum mechanics?

(a) Algebraic topology.
(b) The Fourier Transform.
(c) Differential calculus.
(d) Matrix algebra.

Answer: b. The Fourier Transform.
Question 3: What does the number operator measure in quantum field theory?

(a) The energy of the field.
(b) The number of particles in a given quantum state.
(c) The momentum of the field.
(d) The position of particles.

Answer: b. The number of particles in a given quantum state.
Question 4: In the context of quantum particles, what does quantization primarily refer to?

(a) Conversion of classical fields into quantum fields.
(b) Making quantum mechanics non-deterministic.
(c) Discretization of certain physical properties like energy.
(d) Creation of particles from fields.

Answer: c. Discretization of certain physical properties like energy.
Question 5: Which equation is central to the theoretical foundation of quantum mechanics?

(a) Einstein's field equations.
(b) The Schrödinger equation.
(c) Newton's law of universal gravitation.
(d) Maxwell's equations.

Answer: b. The Schrödinger equation.
Question 6: What role do commutation relations play in quantum mechanics?

(a) They determine the classical properties of particles.
(b) They define the non-commutative nature of quantum measurements.
(c) They solve the equations of motion for particles.
(d) They provide a method for integrating quantum equations.

Answer: b. They define the non-commutative nature of quantum measurements.
Question 7: Which concept in quantum mechanics describes the probability amplitude of finding a particle at a certain position?

(a) Hamiltonian dynamics.
(b) The wavefunction.
(c) The uncertainty principle.
(d) Quantum tunneling.

Answer: b. The wavefunction.
Question 8: What principle prevents two identical fermions from occupying the same quantum state?

(a) Heisenberg's uncertainty principle.
(b) Pauli's exclusion principle.
(c) Newton's third law.
(d) Conservation of energy.

Answer: b. Pauli's exclusion principle.
Question 9: Which type of statistics do bosons follow?

(a) Maxwell-Boltzmann statistics.
(b) Fermi-Dirac statistics.
(c) Bose-Einstein statistics.
(d) None of the above.

Answer: c. Bose-Einstein statistics.
Question 10: What is the significance of the path integral formulation in quantum field theory?

(a) It calculates classical trajectories only.
(b) It sums over all possible histories to calculate amplitudes.
(c) It is used only for solving statistical mechanics problems.
(d) It describes non-quantum phenomena.

Answer: b. It sums over all possible histories to calculate amplitudes.

3.5.2 PROBLEMS

PROBLEM 1: QUANTUM PARTICLE IN A BOX

Calculate the first three quantized energy levels for a particle of mass m in a one-dimensional box of length L.

Solution: The energy levels for a particle in a box are given by:

$$E_n = \frac{\hbar^2 \pi^2 n^2}{2mL^2} \tag{3.78}$$

Substituting $n = 1, 2, 3$ into the equation gives:

$$E_1 = \frac{\hbar^2 \pi^2}{2mL^2}$$
$$E_2 = \frac{4\hbar^2 \pi^2}{2mL^2} = 4E_1$$
$$E_3 = \frac{9\hbar^2 \pi^2}{2mL^2} = 9E_1$$

These values show the quantized nature of energy levels in a confined system.

PROBLEM 2: FOURIER TRANSFORM OF A GAUSSIAN WAVEFUNCTION

Calculate the Fourier transform of a Gaussian wavefunction $\psi(x) = \frac{1}{\sqrt{\sqrt{2\pi}\sigma}} e^{-\frac{x^2}{2\sigma^2}}$ to find its representation in momentum space.

Solution:

1. **Wavefunction in Position Space:** The given Gaussian wavefunction is:

$$\psi(x) = \frac{1}{\sqrt{\sqrt{2\pi}\sigma}} e^{-\frac{x^2}{2\sigma^2}} \tag{3.79}$$

where σ is the standard deviation of the Gaussian distribution.

2. **Fourier Transform Definition:** The Fourier transform $\phi(p)$ of the wavefunction $\psi(x)$ is given by:

$$\phi(p) = \frac{1}{\sqrt{2\pi\hbar}} \int_{-\infty}^{\infty} \psi(x) e^{-\frac{i}{\hbar}px} dx \tag{3.80}$$

3. **Substitute $\psi(x)$ into the Fourier Transform:**

$$\phi(p) = \frac{1}{\sqrt{2\pi\hbar}} \int_{-\infty}^{\infty} \frac{1}{\sqrt{\sqrt{2\pi}\sigma}} e^{-\frac{x^2}{2\sigma^2}} e^{-\frac{i}{\hbar}px} dx \tag{3.81}$$

4. **Combine the Exponents:**

$$\phi(p) = \frac{1}{\sqrt{2\pi\hbar\sqrt{2\pi}\sigma}} \int_{-\infty}^{\infty} e^{-\frac{x^2}{2\sigma^2} - \frac{i}{\hbar}px} dx \tag{3.82}$$

$$= \frac{1}{\sqrt{2\pi\hbar\sqrt{2\pi}\sigma}} \int_{-\infty}^{\infty} e^{-\frac{1}{2\sigma^2}(x^2 + 2i\frac{\sigma^2}{\hbar}px)} dx \tag{3.83}$$

5. **Complete the Square:** Completing the square in the exponent:

$$x^2 + 2i\frac{\sigma^2}{\hbar}px = \left(x + i\frac{\sigma^2}{\hbar}p\right)^2 - \left(i\frac{\sigma^2}{\hbar}p\right)^2 \tag{3.84}$$

$$= \left(x + i\frac{\sigma^2}{\hbar}p\right)^2 - \frac{\sigma^4}{\hbar^2}p^2 \tag{3.85}$$

6. **Integrate the Gaussian:**

$$\phi(p) = \frac{1}{\sqrt{2\pi\hbar\sqrt{2\pi}\sigma}} e^{\frac{\sigma^4}{2\hbar^2\sigma^2}p^2} \int_{-\infty}^{\infty} e^{-\frac{1}{2\sigma^2}\left(x + i\frac{\sigma^2}{\hbar}p\right)^2} dx \tag{3.86}$$

$$= \frac{1}{\sqrt{2\pi\hbar}} e^{-\frac{\sigma^2 p^2}{2\hbar^2}} \tag{3.87}$$

where the integral of a Gaussian function over all space is $\sqrt{2\pi}\sigma$, and the integral shift does not change the limits or result due to the properties of the exponential function.

Result: The Fourier transform $\phi(p)$ of the Gaussian wavefunction $\psi(x)$ is another Gaussian function in momentum space, demonstrating the duality and symmetry properties inherent in quantum mechanical descriptions. The width of the momentum space Gaussian is inversely proportional to the position space width σ, consistent with the uncertainty principle.

PROBLEM 3: COMMUTATION RELATION VALIDATION

Verify the commutation relation $[\hat{x}, \hat{p}] = i\hbar$ using the position operator $\hat{x}$ and the momentum operator $\hat{p} = -i\hbar\frac{d}{dx}$.

Solution: Applying the operators to a function $f(x)$, we have:

$$\hat{x}\hat{p}f(x) = \hat{x}(-i\hbar\frac{d}{dx}f(x)) = -i\hbar x\frac{d}{dx}f(x)$$
$$\hat{p}\hat{x}f(x) = -i\hbar\frac{d}{dx}(xf(x)) = -i\hbar(\frac{dx}{dx}f(x) + x\frac{df}{dx})$$
$$[\hat{x}, \hat{p}]f(x) = -i\hbar x\frac{d}{dx}f(x) + i\hbar f(x) + i\hbar x\frac{df}{dx} = i\hbar f(x)$$

Thus, $[\hat{x}, \hat{p}] = i\hbar$, confirming the canonical commutation relation.

PROBLEM 4: PATH INTEGRAL CALCULATION

Calculate the path integral for a free particle moving from point x_i at time t_i to point x_f at time t_f.

Solution: For a free particle, the action $S[x(t)]$ is given by $\int_{t_i}^{t_f} \frac{1}{2}m\dot{x}^2 dt$. The path integral is:

$$\langle x_f, t_f | x_i, t_i \rangle = \int \mathscr{D}[x(t)] e^{\frac{i}{\hbar}\int_{t_i}^{t_f} \frac{1}{2}m\dot{x}^2 dt} \tag{3.88}$$

This integral results in a propagator for a free particle, illustrating the basic principle of the path integral approach in quantum mechanics.

PROBLEM 5: FIELD OPERATOR ACTION

Consider a quantum field $\hat{\psi}(x)$ described by the field operators $\hat{\psi}^\dagger(x)$ and $\hat{\psi}(x)$. If the field is initially in the vacuum state $|0\rangle$, calculate the state resulting from applying $\hat{\psi}^\dagger(x)$ at position x_0.

Solution: The quantum field $\hat{\psi}(x)$ and its Hermitian conjugate $\hat{\psi}^\dagger(x)$ are operators that annihilate and create particles at position x, respectively. The vacuum state $|0\rangle$ is defined as the state with no particles, and it is annihilated by the annihilation operator:

$$\hat{\psi}(x)|0\rangle = 0 \tag{3.89}$$

Now, applying the creation operator $\hat{\psi}^\dagger(x)$ at position x_0 to the vacuum state does the following:

$$\hat{\psi}^\dagger(x_0)|0\rangle = |x_0\rangle \tag{3.90}$$

This operation creates a particle at position x_0. The state $|x_0\rangle$ represents a quantum state where a single particle is localized at position x_0. This state is often referred to as a one-particle state or a single-particle excitation at x_0.

BONUS PROBLEM: EVALUATING THE COMMUTATOR OF FIELD OPERATORS IN QUANTUM FIELD THEORY

Consider a quantum field theory in one spatial dimension where the field operators $\hat{\psi}(x)$ and $\hat{\psi}^\dagger(x)$ describe the annihilation and creation of particles at position x, respectively. Calculate the commutator $[\hat{\psi}(x), \hat{\psi}^\dagger(y)]$ for bosonic particles and evaluate it at $x = y$.

Solution:

1. **Definition of Field Operators:** The field operators in quantum field theory are defined as follows:

$$\hat{\psi}(x) = \int \frac{dk}{\sqrt{2\pi}} e^{ikx} \hat{a}_k, \tag{3.91}$$

$$\hat{\psi}^\dagger(x) = \int \frac{dk}{\sqrt{2\pi}} e^{-ikx} \hat{a}_k^\dagger. \tag{3.92}$$

where $\hat{a}_k$ and $\hat{a}_k^\dagger$ are the annihilation and creation operators in momentum space.

2. **Calculate the Commutator:** Using the integral expressions for $\hat{\psi}(x)$ and $\hat{\psi}^\dagger(y)$, we can express the commutator as follows:

$$[\hat{\psi}(x), \hat{\psi}^\dagger(y)] = \left[\int \frac{dk}{\sqrt{2\pi}} e^{ikx} \hat{a}_k, \int \frac{dk'}{\sqrt{2\pi}} e^{-ik'y} \hat{a}_{k'}^\dagger\right] \tag{3.93}$$

$$= \int \frac{dk}{\sqrt{2\pi}} \int \frac{dk'}{\sqrt{2\pi}} e^{ikx} e^{-ik'y} [\hat{a}_k, \hat{a}_{k'}^\dagger]. \tag{3.94}$$

3. **Using the Bosonic Commutation Relations:** For bosons, the commutation relation is given by:

$$[\hat{a}_k, \hat{a}_{k'}^\dagger] = \delta(k - k'). \tag{3.95}$$

Substituting this into the integral:

$$[\hat{\psi}(x), \hat{\psi}^\dagger(y)] = \int \frac{dk}{\sqrt{2\pi}} \int \frac{dk'}{\sqrt{2\pi}} e^{ikx} e^{-ik'y} \delta(k - k') \tag{3.96}$$

$$= \int \frac{dk}{2\pi} e^{ikx} e^{-iky} \tag{3.97}$$

$$= \delta(x - y). \tag{3.98}$$

4. **Evaluation at $x = y$:** When $x = y$, the commutator simplifies to:

$$[\hat{\psi}(x), \hat{\psi}^\dagger(x)] = \delta(0), \tag{3.99}$$

which is an ill-defined expression in the continuum but represents the infinite density at a point, characteristic of quantum field theoretical descriptions.

This problem exemplifies the non-trivial nature of field theory calculations and the care needed when handling delta functions and commutators in QFT.

4 Lecture 8: Symmetries, Indistinguishability, Statistics, Renormalization, and Gauge Theories

4.1 SYMMETRIES

Symmetries and indistinguishability are pivotal concepts in quantum mechanics and quantum field theory, shaping the fundamental laws of physics. This section explores how these principles underpin the behaviors of particles in field theories and contribute to our understanding of quantum statistics.

4.1.1 IMPORTANCE OF SYMMETRY IN QUANTUM FIELD THEORY

Symmetries in quantum field theory lead to conservation laws and dictate interaction forms through Noether's theorem. In physics, a **symmetry transformation** refers to any change that can be made to a system that leaves certain fundamental properties or the form of physical laws invariant. This concept is fundamental in modern theoretical physics, particularly in fields like quantum mechanics and field theories such as Quantum Field Theory (QFT) and General Relativity.

KEY POINTS ABOUT SYMMETRY TRANSFORMATIONS

1. **Invariance of Physical Laws**: A symmetry transformation preserves the form of the physical laws governing a system. For instance, laws that are invariant under rotation demonstrate that outcomes are the same, irrespective of the system's orientation in space.
2. **Types of Symmetries**:
   - *Spatial Symmetries*: These include translations, rotations, and reflections, affecting how physical systems are oriented or positioned in space.
   - *Temporal Symmetries*: These involve changes in the time coordinate, suggesting the constancy of physical laws over time.
   - *Internal Symmetries*: These involve transformations of internal degrees of freedom, such as phase transformations in quantum mechanics.
3. **Mathematical Formulation**: In mathematical terms, symmetry transformations are operations that apply to fields or states within a theory. For example, a field $\phi(x)$ may transform under a spatial translation to $\phi'(x) = \phi(x+a)$, indicating translational symmetry.
4. **Noether's Theorem**: This theorem connects differentiable symmetries of the action of a physical system with corresponding conservation laws. For example, time translation symmetry correlates with energy conservation.
5. **Role in Fundamental Theories**: Symmetry transformations underpin theories like the Standard Model, which relies on gauge symmetries, and General Relativity, where principles like the equivalence principle reflect symmetry considerations.

Symmetry transformations reflect profound insights into the nature of reality, guiding the development of theoretical physics and influencing experimental designs and predictions.

DOI: 10.1201/9781003637370-4

TENSOR REPRESENTATION OF SYMMETRIES

Tensors provide a mathematical framework to represent symmetries, especially useful in physics for describing transformations that preserve certain properties of physical systems.

4.1.1.1 Tensor Transformation Law

A tensor's transformation under a change of coordinates, which can reflect symmetry transformations, is given by:

$$T'^{\mu\nu} = \frac{\partial x'^{\mu}}{\partial x^{\alpha}} \frac{\partial x'^{\nu}}{\partial x^{\beta}} T^{\alpha\beta} \tag{4.1}$$

where $T^{\mu\nu}$ is a tensor, and x^{μ} and x'^{μ} represent the old and new coordinates, respectively. This transformation law ensures that the tensor equation holds true in all coordinate systems if it holds in any one coordinate system, highlighting the intrinsic symmetry.

4.1.2 EXAMPLES OF SYMMETRY

- **Lorentz Symmetry**: In the theory of relativity, tensors must be invariant under Lorentz transformations, which describe how physical quantities transform between two coordinate frames that move at constant velocity relative to each other.
- **Gauge Symmetry**: In quantum field theory, gauge symmetries can be expressed using tensor fields that transform under local gauge transformations to maintain the form of the physical laws.

WORKED EXAMPLE: TENSOR SYMMETRY UNDER 2D ROTATIONS

This example demonstrates how a two-dimensional stress tensor remains invariant under a rotation transformation, showcasing the tensor's symmetry.

Rotation Transformation

Consider a rotation by an angle θ in the plane. The rotation matrix R is given by:

$$R = \begin{pmatrix} \cos\theta & -\sin\theta \\ \sin\theta & \cos\theta \end{pmatrix} \tag{4.2}$$

Transformation of a Symmetric Tensor

Let T be a symmetric stress tensor represented as:

$$T = \begin{pmatrix} \sigma_{xx} & \sigma_{xy} \\ \sigma_{xy} & \sigma_{yy} \end{pmatrix} \tag{4.3}$$

where σ_{xx}, σ_{yy}, and σ_{xy} are the components of the stress tensor.

The transformed tensor T' under the rotation is given by:

$$T' = RTR^T \tag{4.4}$$

where R^T is the transpose of R.

Calculation

The transformed tensor is computed as follows:

$$T' = \begin{pmatrix} \cos\theta & -\sin\theta \\ \sin\theta & \cos\theta \end{pmatrix} \begin{pmatrix} \sigma_{xx} & \sigma_{xy} \\ \sigma_{xy} & \sigma_{yy} \end{pmatrix}$$
$$\times \begin{pmatrix} \cos\theta & \sin\theta \\ -\sin\theta & \cos\theta \end{pmatrix}$$
$$= \begin{pmatrix} \cos\theta\sigma_{xx} + \sin\theta\sigma_{xy} & \cos\theta\sigma_{xy} + \sin\theta\sigma_{yy} \\ -\sin\theta\sigma_{xx} + \cos\theta\sigma_{xy} & -\sin\theta\sigma_{xy} + \cos\theta\sigma_{yy} \end{pmatrix}$$
$$[3pt] \times \begin{pmatrix} \cos\theta & \sin\theta \\ -\sin\theta & \cos\theta \end{pmatrix}$$
$$= \begin{pmatrix} \begin{array}{l} \sigma_{xx}\cos^2\theta + \\ 2\sigma_{xy}\sin\theta\cos\theta + \\ \sigma_{yy}\sin^2\theta \end{array} & * \\ * & \begin{array}{l} \sigma_{xx}\sin^2\theta - \\ 2\sigma_{xy}\sin\theta\cos\theta + \\ \sigma_{yy}\cos^2\theta \end{array} \end{pmatrix} \tag{4.5}$$

The final expression for T' shows that the tensor components transform in a way that preserves the tensor's symmetry properties under rotation. This conservation is crucial for maintaining the physical interpretations of stresses in a material under rotation.

WORKED EXAMPLE: TENSOR INVARIANCE UNDER LORENTZ TRANSFORMATIONS

Lorentz transformations are a fundamental symmetry in the theory of special relativity. They describe how measurements of space and time by two observers are related to each other in different inertial frames. A key feature of these transformations is the invariance of the spacetime interval, represented by the metric tensor $\eta_{\mu\nu}$.

Lorentz Transformation

The Lorentz transformation for coordinates is given by:

$$x'^{\mu} = \Lambda^{\mu}{}_{\nu} x^{\nu} \tag{4.6}$$

where $\Lambda^{\mu}{}_{\nu}$ are the components of the Lorentz transformation matrix, and x^{μ} and x'^{μ} are the spacetime coordinates in the original and transformed frames, respectively.

Transformation of the Metric Tensor

The metric tensor in the Minkowski spacetime (used in special relativity) is invariant under Lorentz transformations. This means:

$$\eta'_{\mu\nu} = \Lambda^{\alpha}{}_{\mu} \Lambda^{\beta}{}_{\nu} \eta_{\alpha\beta} \tag{4.7}$$

The invariance of the metric tensor $\eta_{\mu\nu}$ under Lorentz transformations can be shown as follows:

$$\eta'_{\mu\nu} = \Lambda^{\alpha}_{\ \mu}\Lambda^{\beta}_{\ \nu}\eta_{\alpha\beta} \tag{4.8}$$

$$= \Lambda^{\alpha}_{\ \mu}\Lambda^{\beta}_{\ \nu}\begin{pmatrix} -1 & 0 & 0 & 0 \\ 0 & 1 & 0 & 0 \\ 0 & 0 & 1 & 0 \\ 0 & 0 & 0 & 1 \end{pmatrix} \tag{4.9}$$

$$= \begin{pmatrix} -1 & 0 & 0 & 0 \\ 0 & 1 & 0 & 0 \\ 0 & 0 & 1 & 0 \\ 0 & 0 & 0 & 1 \end{pmatrix} \tag{4.10}$$

$$= \eta_{\mu\nu} \tag{4.11}$$

The calculation confirms that the metric tensor, which encodes the geometry of spacetime, is invariant under Lorentz transformations. This invariance is a direct consequence of the symmetry properties of spacetime under these transformations, crucial for the consistency of the theory of relativity.

4.1.3 SYMMETRIZATION POSTULATE FOR BOSONS

The Symmetrization Postulate is a fundamental principle in quantum mechanics that dictates the behavior of identical bosons. According to this postulate, the wavefunction describing a system of identical bosons must be symmetric under the exchange of any two particles. Mathematically, for a wavefunction $\Phi(\vec{x}_1, \vec{x}_2, \ldots, \vec{x}_N)$ describing N identical bosons, the symmetry condition is given by:

$$\Phi(\vec{x}_1, \vec{x}_2, \ldots, \vec{x}_N) = \Phi(\vec{x}_{\sigma(1)}, \vec{x}_{\sigma(2)}, \ldots, \vec{x}_{\sigma(N)}), \tag{4.12}$$

where σ is any permutation of the particle indices. This symmetric property ensures that swapping any two bosons does not change the wavefunction, reflecting their indistinguishable nature and leading to the characteristic statistical behavior described by Bose-Einstein statistics.

4.1.4 NOETHER'S THEOREM

Noether's Theorem establishes a fundamental connection between symmetries and conservation laws in physics. It states that every differentiable symmetry of the action of a physical system corresponds to a particular conservation law. Consider a field ϕ described by a Lagrangian $\mathscr{L}(\phi, \partial_\mu \phi)$, where ∂_μ represents the partial derivative with respect to the spacetime coordinates. Suppose the action S is invariant under a continuous symmetry transformation of the field:

$$\phi(x) \to \phi'(x) = \phi(x) + \delta\phi(x) \tag{4.13}$$

where $\delta\phi(x)$ is a small variation of the field. By the principle of least action, the action S given by

$$S = \int d^4x \mathscr{L}(\phi, \partial_\mu \phi) \tag{4.14}$$

remains invariant under the transformation. The variation of the Lagrangian under this transformation is zero:

$$\delta\mathscr{L} = \frac{\partial \mathscr{L}}{\partial \phi}\delta\phi + \frac{\partial \mathscr{L}}{\partial(\partial_\mu \phi)}\delta(\partial_\mu \phi) = 0 \tag{4.15}$$

Using the product rule, $\delta(\partial_\mu \phi) = \partial_\mu(\delta\phi)$, and integrating by parts, we get:

$$\int \delta\mathscr{L}\, d^4x = \int \left(\frac{\partial \mathscr{L}}{\partial \phi}\delta\phi + \frac{\partial \mathscr{L}}{\partial(\partial_\mu \phi)}\partial_\mu(\delta\phi) \right) d^4x \tag{4.16}$$

Applying integration by parts to the second term:

$$\int \frac{\partial \mathcal{L}}{\partial(\partial_\mu \phi)} \partial_\mu (\delta\phi)\, d^4x = -\int \partial_\mu \left(\frac{\partial \mathcal{L}}{\partial(\partial_\mu \phi)} \right) \delta\phi\, d^4x + \text{surface term} \tag{4.17}$$

Assuming the surface term vanishes (common in field theory under suitable boundary conditions), we get:

$$\int \left(\frac{\partial \mathcal{L}}{\partial \phi} - \partial_\mu \frac{\partial \mathcal{L}}{\partial(\partial_\mu \phi)} \right) \delta\phi\, d^4x = 0 \tag{4.18}$$

This step leads directly to the Euler-Lagrange equation:

$$\frac{\partial \mathcal{L}}{\partial \phi} - \partial_\mu \frac{\partial \mathcal{L}}{\partial(\partial_\mu \phi)} = 0 \tag{4.19}$$

$$\delta \mathcal{L} = \left(\frac{\partial \mathcal{L}}{\partial \phi} - \partial_\mu \frac{\partial \mathcal{L}}{\partial(\partial_\mu \phi)} \right) \delta\phi + \partial_\mu \left(\frac{\partial \mathcal{L}}{\partial(\partial_\mu \phi)} \delta\phi \right) = 0 \tag{4.20}$$

The first term must vanish independently for the action to be stationary, leading to the Euler-Lagrange equations:

$$\frac{\partial \mathcal{L}}{\partial \phi} - \partial_\mu \frac{\partial \mathcal{L}}{\partial(\partial_\mu \phi)} = 0 \tag{4.21}$$

The second term leads to the conservation law associated with the symmetry. Defining the Noether current J^μ by:

$$J^\mu = \frac{\partial \mathcal{L}}{\partial(\partial_\mu \phi)} \delta\phi \tag{4.22}$$

the conservation of J^μ follows from the condition of vanishing divergence, derived from the symmetry of the action:

$$\partial_\mu J^\mu = 0 \tag{4.23}$$

This equation implies that the Noether current J^μ is conserved, representing the conservation law corresponding to the continuous symmetry of the action.

Understanding the Non-Zero Variation in Noether's Theorem

In the context of Noether's Theorem, the variation $\delta\phi$ represents a small change in the field ϕ due to a symmetry transformation. The statement $\delta\phi \neq 0$ is not an arbitrary assumption but is based on the premise that we are considering a genuine symmetry of the action.

Here's why $\delta\phi \neq 0$ is a meaningful condition:

1. **Symmetry Transformation**: The variation $\delta\phi$ is derived from a specific symmetry transformation applicable to the field ϕ. For example, in the case of a spatial translation symmetry, $\delta\phi$ might represent the change in the field due to a small shift in position. The transformation must affect the field in some nontrivial way for it to be considered a symmetry rather than an identity operation.
2. **Action Invariance**: Noether's Theorem is specifically concerned with continuous symmetries that leave the action of the system invariant. If $\delta\phi$ were zero, it would imply that the field does not change under the transformation, contradicting the idea that we are examining the effects of a symmetry. A nonzero $\delta\phi$ ensures that the symmetry has a real, observable impact on the system's physical configuration.

3. **Physical Relevance**: In practical scenarios, such as in field theory or mechanics, $\delta\phi$ represents actual physical changes like displacements, rotations, or changes in field configurations due to gauge transformations. These changes are essential for understanding how physical systems respond to symmetries, and thus $\delta\phi = 0$ would imply no response, which is generally not the case.
4. **Derivation of Conserved Quantities**: The core of Noether's theorem is that each symmetry leads to a corresponding conserved quantity. The conservation laws (like conservation of energy, momentum, or charge) derive from the fact that $\delta\phi$ modifies the fields in a way dictated by the symmetry, leading to specific forms of the Euler-Lagrange equations. If $\delta\phi$ were zero, these conservation laws would not manifest, as there would be no perturbation to track through the system's dynamics.

In summary, assuming $\delta\phi \neq 0$ is essential for the application of Noether's Theorem, as it ensures that the symmetry under consideration actively influences the physical system, leading to meaningful and observable physical laws.

4.1.5 SYMMETRY AND ANTISYMMETRY

Symmetry principles in quantum field theory dictate that the laws of physics are invariant under a set of transformations, leading to conservation laws according to Noether's theorem. An example is the invariance under time translation, which leads to conservation of energy.

$$\mathscr{L}(\phi, \partial_\mu \phi) = \mathscr{L}(\phi', \partial_\mu \phi') \tag{4.24}$$

$$\text{where } \phi'(x) = \phi(x-a), \quad a \text{ is a constant vector.} \tag{4.25}$$

This invariance under translation in space and time signifies conservation of momentum and energy, respectively, central to the formulation and solutions of field equations in quantum mechanics and field theory.

Antisymmetry is a crucial principle in quantum mechanics, especially when dealing with fermionic particles such as electrons, protons, and neutrons. This principle originates from the Pauli Exclusion Principle, which asserts that no two identical fermions can occupy the same quantum state within a quantum system at the same time. This requirement leads to wavefunctions for systems of identical fermions being antisymmetric with respect to the exchange of any two particles.

In quantum mechanics, the wavefunction for a system of identical fermions must be antisymmetric with respect to the exchange of any two particles. This means that if the spatial coordinates or any other quantum numbers of two fermions are exchanged, the overall wavefunction changes its sign. The mathematical expression for this property is given by:

$$\Psi(x_1, x_2, ..., x_N) = -\Psi(x_2, x_1, ..., x_N) \tag{4.26}$$

where $\Psi(x_1, x_2, ..., x_N)$ is the wavefunction of the system and x_i represents the set of all quantum numbers associated with the i-th fermion. This antisymmetric property ensures that the wavefunction becomes zero if any two fermions are in the same state, reflecting the Pauli exclusion principle.

This principle has profound implications, affecting the structure of atoms, the properties of electrons in metals, and the stability of matter itself.

Worked Example: Noether's Theorem and the Conservation of Momentum

Consider a free particle of mass m moving in one-dimensional space. The Lagrangian $\mathscr{L}$ of this system is given by the kinetic energy of the particle:

$$\mathscr{L} = \frac{1}{2} m \dot{x}^2 \tag{4.27}$$

where $\dot{x}$ is the velocity of the particle. According to Noether's Theorem, if the Lagrangian does not explicitly depend on the position x, which indicates translational symmetry, there is a corresponding conservation law. Determine this law.

Solution:

The symmetry transformation in question is $x \to x + \varepsilon$, where ε is a small constant. Under this transformation, the Lagrangian remains unchanged:

$$\mathscr{L}(x,\dot{x}) = \mathscr{L}(x+\varepsilon,\dot{x}) \tag{4.28}$$

This shows that the Lagrangian is symmetric under translations in space.

Applying Noether's Theorem

The translational symmetry implies that the Lagrangian does not explicitly depend on the position x. For a small displacement ε, the transformation $x \to x + \varepsilon$ keeps the Lagrangian invariant because it depends only on $\dot{x}$, not on x:

$$\mathscr{L}(x+\varepsilon,\dot{x}) = \frac{1}{2}m\dot{x}^2 = \mathscr{L}(x,\dot{x}) \tag{4.29}$$

Thus, $\mathscr{L}$ is symmetric under translation in space.

Noether's theorem relates symmetries of the action (integral of the Lagrangian) to conserved quantities. The theorem states that if the action does not change under a continuous symmetry transformation, there exists a conserved current. The conserved current j for translational symmetry is derived as follows:

$$j = \frac{\partial \mathscr{L}}{\partial \dot{x}}\frac{dx}{d\varepsilon} \tag{4.30}$$

Given the transformation $x \to x + \varepsilon$, we have $\frac{dx}{d\varepsilon} = 1$.

Compute the derivative of $\mathscr{L}$ with respect to $\dot{x}$:

$$\frac{\partial \mathscr{L}}{\partial \dot{x}} = \frac{\partial}{\partial \dot{x}}\left(\frac{1}{2}m\dot{x}^2\right) = m\dot{x} \tag{4.31}$$

Now substituting back into the expression for j:

$$j = m\dot{x} \cdot 1 = m\dot{x} \tag{4.32}$$

This result, $m\dot{x}$, is the linear momentum, which Noether's theorem confirms as conserved due to the translational symmetry of the Lagrangian.

4.2 INDISTINGUISHABILITY

In quantum field theory, indistinguishability implies that particles of a specific type are fundamentally identical and interchangeable, influencing the statistical behavior of systems. This principle is crucial in determining the symmetrical properties of wavefunctions for bosons and fermions.

For bosons, the wavefunction must be symmetric:

$$\Phi(\vec{x}_1,\vec{x}_2,\ldots,\vec{x}_N) = \Phi(\vec{x}_{\sigma(1)},\vec{x}_{\sigma(2)},\ldots,\vec{x}_{\sigma(N)}), \tag{4.33}$$

For fermions, the wavefunction must be antisymmetric:

$$\Psi(\vec{x}_1,\vec{x}_2,\ldots,\vec{x}_N) = \mathrm{sgn}(\sigma)\Psi(\vec{x}_{\sigma(1)},\vec{x}_{\sigma(2)},\ldots,\vec{x}_{\sigma(N)}), \tag{4.34}$$

where σ represents a permutation of particle indices and $\mathrm{sgn}(\sigma)$ is the sign of the permutation, which is $+1$ for even permutations and -1 for odd permutations.

4.2.1 IMPLICATIONS OF INDISTINGUISHABILITY IN QFT

Indistinguishability in QFT leads to significant theoretical and practical outcomes. For example, the concept of identical particles necessitates the use of symmetric (bosonic) or antisymmetric (fermionic) wavefunctions, as shown above. This has profound implications:

1. **Statistical Mechanics and Thermodynamics**: The behavior of gases, plasmas, and even the vacuum itself are deeply influenced by the statistics that these wavefunctions dictate. Bosons can condense into the same energy state, allowing phenomena like Bose-Einstein condensation, while fermions fill up states due to the Pauli exclusion principle, creating the structure of the periodic table and defining the properties of metals and insulators.

2. **Quantum Field Operators**: The creation and annihilation operators in QFT must also obey symmetric or antisymmetric commutation relations:

$$\begin{aligned} \text{Bosons:} \quad & [\hat{a}_p, \hat{a}_q^\dagger] = \delta_{pq}, \\ \text{Fermions:} \quad & \{\hat{b}_p, \hat{b}_q^\dagger\} = \delta_{pq}, \end{aligned} \tag{4.35}$$

where $\hat{a}_p$ and $\hat{b}_p$ are the annihilation operators for bosons and fermions, respectively, and δ_{pq} is the Kronecker delta, which is 1 if $p = q$ and 0 otherwise.

These operators follow from the indistinguishable nature of quantum particles, enforcing that physical observables in QFT must be invariant under particle exchange, which is a cornerstone in the formulation and calculation of quantum processes.

This symmetry requirement leads to fundamental constraints on the allowed interactions and the form of the Hamiltonian, which governs the dynamics of quantum fields, ensuring that physical laws are consistent with observed quantum statistics.

4.3 STATISTICS

In QFT, the indistinguishability of particles leads to specific statistical distributions for fermions and bosons, governed by Fermi-Dirac and Bose-Einstein statistics, respectively. These statistics describe how identical particles distribute themselves among various energy states.

$$\begin{aligned} & \text{Fermi-Dirac Distribution:} \\ & f_{FD}(E) = \frac{1}{e^{(E-\mu)/k_B T} + 1}, \end{aligned} \tag{4.36}$$

$$\begin{aligned} & \text{Bose-Einstein Distribution:} \\ & f_{BE}(E) = \frac{1}{e^{(E-\mu)/k_B T} - 1}, \end{aligned} \tag{4.37}$$

where E is the energy of a quantum state, μ is the chemical potential, k_B is the Boltzmann constant, and T is the temperature.

4.3.1 MATHEMATICAL EXAMPLE OF STATISTICS IN QFT:

Consider a system of non-interacting bosons in a quantum field theory framework. The number of bosons N in a particular state of energy E can be modeled by the Bose-Einstein distribution. For example, the average number of photons (bosons) in a mode of the electromagnetic field at thermal equilibrium is given by:

$$\langle n \rangle = \frac{1}{e^{\hbar\omega/k_B T} - 1}, \tag{4.38}$$

where $\hbar\omega$ represents the energy of the photons corresponding to the mode with frequency ω. This equation illustrates how quantum statistics directly results from the symmetrical properties of the bosonic wavefunction, which allows multiple identical bosons to occupy the same quantum state.

These statistical principles are fundamental in predicting the properties of quantum systems and are pivotal in many areas of physics, including condensed matter physics and quantum thermodynamics.

Worked Example: Calculating the Number of Photons in an Electromagnetic Field

Assuming a quantized electromagnetic field in a cavity of volume V at a temperature T, the average number of photons $\langle n \rangle$ in a mode of the field with frequency ω can be calculated using the Bose-Einstein distribution:

$$\langle n \rangle = \frac{1}{e^{\frac{\hbar\omega}{k_B T}} - 1} \tag{4.39}$$

where $\hbar$ is the reduced Planck constant, k_B is the Boltzmann constant, and T is the temperature in Kelvin. If we assume that the cavity supports N such modes uniformly distributed in frequency, the total number of photons in the electromagnetic field is:

$$N_{\text{photons}} = \sum_{i=1}^{N} \frac{1}{e^{\frac{\hbar\omega_i}{k_B T}} - 1} \tag{4.40}$$

Worked Example: Calculating the Number of Allowed Electron States in a Quantum Field

For a system of N electrons confined in a potential well or a quantum dot with discrete energy levels, the number of allowed quantum states can be computed assuming that each electron occupies a unique quantum state due to the Pauli Exclusion Principle. If the electrons can occupy states in an energy band with M levels (where $M \geq N$):

$$N_{\text{states}} = \binom{M}{N} = \frac{M!}{N!(M-N)!} \tag{4.41}$$

This equation calculates the number of ways N electrons can be distributed among M energy levels, adhering to the exclusion principle that no two electrons can occupy the same state.

4.4 RENORMALIZATION

Renormalization addresses the infinities that arise in field theory calculations, ensuring theories are predictive and consistent. This section will delve into the role and implementation of renormalization in QFT. It modifies the parameters of a theory to absorb the infinities into definable physical constants. It is essential for making the predictions of quantum field theory match observed physical results.

$$\hat{\phi}_{\text{ren}}(x) = Z^{1/2}\hat{\phi}(x), \tag{4.42}$$

where Z is the renormalization constant adjusting the field's normalization.

4.4.1 RENORMALIZATION OF THE ELECTRON MASS IN QED

Consider the renormalization of the electron mass in Quantum Electrodynamics (QED) as an example. In QED, the bare mass m_0 of the electron is modified by quantum corrections due to interactions with the vacuum. These corrections are represented by self-energy diagrams. The physical, observed mass m differs from m_0 due to these interactions.

Calculating Self-Energy of the Electron

The self-energy $\Sigma(p)$ at one-loop level can be expressed as an integral that typically diverges. This divergence is handled through renormalization. The corrected mass of the electron can be calculated using the relation:

$$m = m_0 + \Sigma(m), \tag{4.43}$$

where $\Sigma(m)$ is the self-energy evaluated at the physical mass m.

Renormalization of the Electron Mass

To make the theory predictive and consistent with observations, the parameters of the theory (like the electron mass and charge) are renormalized. The renormalization of the mass involves adjusting the bare mass m_0 and the field renormalization constant Z to absorb the infinities from $\Sigma(m)$:

$$\hat{\phi}_{\text{ren}}(x) = Z^{1/2}\hat{\phi}(x), \tag{4.44}$$

where Z is the renormalization constant that adjusts the normalization of the field. The renormalized mass m is defined in such a way that it cancels the divergences in $\Sigma(m)$, leading to:

$$m = m_0 + \text{Re}[\Sigma(m)], \tag{4.45}$$

and

$$m_0 = m - \text{Re}[\Sigma(m)]. \tag{4.46}$$

This renormalization ensures that m remains finite and matches the experimentally observed mass of the electron.

4.4.2 HANDLING INFINITIES IN QUANTUM FIELD THEORY

In quantum field theory, the calculation of certain physical quantities, like the self-energy of particles, often leads to mathematical expressions that diverge, meaning they yield infinite results. These infinities arise primarily from the integrals over all possible states of the field, including states with arbitrarily high energy or momentum, which are an inherent part of the quantum field formulations.

Nature of Infinities

These divergences can typically be categorized as:

- **Ultraviolet (UV) divergences**: These occur due to the integration over very high energy (short wavelength) states, leading to infinite results. They are associated with the high-energy behavior of field theories.
- **Infrared (IR) divergences**: These arise from the integration over very low energy (long wavelength) states and are often related to the properties of massless particles, like photons in Quantum Electrodynamics (QED).

Absorbing the Infinities

Renormalization is a method to deal with these divergences by absorbing them into the parameters of the theory, such as masses and coupling constants. This is done through several steps:

1. **Regularization**: Introducing a cutoff or modifying the theory in a way (like dimensional regularization) that makes the divergent integrals finite temporarily. This allows the manipulation of these expressions mathematically.

2. **Renormalization**: Adjusting the parameters of the theory (such as the bare mass and charge in QED) so that the predictions of the theory match the observed physical quantities. The original, 'bare' parameters are replaced with physical, 'renormalized' parameters.
3. **Renormalization Group Equations**: These equations describe how the parameters of the theory change with the energy scale, providing a deeper understanding of the physical processes at different scales.

$$\begin{aligned} m_{\text{ren}} &= m_{\text{bare}} + \delta m, \\ e_{\text{ren}} &= e_{\text{bare}} + \delta e, \end{aligned} \tag{4.47}$$

where δm and δe represent the adjustments needed to cancel the infinities in the theory. The renormalized parameters m_{ren} and e_{ren} are finite and correspond to the physically measured values.

The process ensures that despite the presence of theoretical infinities, the predictions of quantum field theory remain accurate, reliable, and consistent with experimental data. This method not only preserves the predictive power of the theory but also offers insights into the behavior of quantum systems at different energy scales.

4.5 GAUGE THEORIES AND THEIR SIGNIFICANCE IN QFT

Gauge theories form the backbone of our understanding of fundamental interactions, based on the principle of gauge invariance. Quantum Electrodynamics (QED) and Quantum Chromodynamics (QCD) apply gauge symmetry to describe forces.

4.5.1 MATHEMATICAL FRAMEWORK FOR GAUGE THEORIES

Gauge transformations in QED are described by:

$$\psi(x) \to e^{ie\theta(x)}\psi(x), \quad A_\mu(x) \to A_\mu(x) + \frac{1}{e}\partial_\mu\theta(x), \tag{4.48}$$

where $\psi(x)$ represents the electron field, $A_\mu(x)$ the electromagnetic potential, and $\theta(x)$ the gauge function.

4.5.2 UNDERSTANDING THE GAUGE FUNCTION

The gauge function in gauge theories, such as Quantum Electrodynamics (QED), plays a crucial role in defining the transformations that preserve the form of the theory's equations despite changes in the fields' configurations. This function, denoted typically as $\theta(x)$, is a scalar function of spacetime coordinates and is integral to maintaining local gauge invariance.

Role of the Gauge Function

The gauge function $\theta(x)$ allows for local transformations of the field variables, which means that its value can vary from one point in spacetime to another. This flexibility is key to the gauge symmetry characteristic of fundamental interactions, allowing for the definition of physically equivalent but locally distinct field configurations. Consider a simple electromagnetic field described by the vector potential $A_\mu(x)$ in QED. Under a gauge transformation, the form of the vector potential is altered by the gradient of $\theta(x)$. The transformation is given by:

$$A_\mu(x) \to A_\mu(x) + \frac{1}{e}\partial_\mu\theta(x), \tag{4.49}$$

To see the effect of this transformation, let's calculate the change in the electromagnetic field tensor $F_{\mu\nu}$, which is defined as:

$$F_{\mu\nu} = \partial_\mu A_\nu - \partial_\nu A_\mu. \tag{4.50}$$

After the gauge transformation, $F_{\mu\nu}$ transforms as follows:

$$F_{\mu\nu} \to \partial_\mu(A_\nu + \frac{1}{e}\partial_\nu\theta) - \partial_\nu(A_\mu + \frac{1}{e}\partial_\mu\theta) \tag{4.51}$$

$$= \partial_\mu A_\nu - \partial_\nu A_\mu + \frac{1}{e}(\partial_\mu\partial_\nu\theta - \partial_\nu\partial_\mu\theta) \tag{4.52}$$

$$= F_{\mu\nu} + \frac{1}{e}(0) \tag{4.53}$$

$$= F_{\mu\nu}. \tag{4.54}$$

This calculation shows that the electromagnetic field tensor $F_{\mu\nu}$, which determines the observable electromagnetic fields, is invariant under the gauge transformation. The gauge function $\theta(x)$, despite altering the potential $A_\mu(x)$, does not change the physical electromagnetic fields, demonstrating the principle of gauge invariance.

This example highlights how gauge functions enable different descriptions of the same physical situation, reflecting the profound symmetry underlying the laws of electromagnetism and other fundamental forces.

4.5.3 GAUGE FUNCTION AND ITS PHYSICAL IMPLICATIONS

The gauge function $\theta(x)$ and its gradient $\partial_\mu\theta(x)$ in gauge theories, such as Quantum Electrodynamics (QED), serve as mathematical tools to ensure gauge invariance. Physically, they allow for the flexible choice of gauge fields' configurations that, while different locally, yield equivalent physical outcomes. This can be likened to choosing different reference frames in classical physics.

PRACTICAL SCENARIOS INFLUENCING THE GRADIENT $\theta(X)$

Electromagnetic Calibration:

- *Local Adjustments:* Different instruments in electromagnetic measurements might require calibration adjustments to accurately measure fields or potentials. The gauge function $\theta(x)$ facilitates these local adjustments by modifying the gauge field $A_\mu(x)$, compensating for offsets or variations specific to each instrument's calibration.

Interference Effects:

- *Phase Adjustments in Interferometry:* Environmental factors affecting the phase of electromagnetic waves can necessitate local phase adjustments in interferometry setups. The gauge function $\theta(x)$ mathematically represents these local phase shifts to maintain coherent interference patterns.

Theoretical Simulations:

- *Boundary Conditions:* In simulations involving electromagnetic fields, boundary conditions can influence Maxwells equations' solutions. Modifying the vector potential via $\theta(x)$ helps satisfy specific boundary conditions without altering the physical content of the electromagnetic field tensor $F_{\mu\nu}$.

These examples illustrate that the gauge function $\theta(x)$ is not merely a theoretical construct but has practical applications in experimental and theoretical physics, ensuring that local modifications to gauge fields do not alter the observable physics and maintain the theory's gauge invariance.

4.6 APPLICATIONS AND PRACTICAL IMPLICATIONS

This section explores how the theories discussed apply to phenomena such as the quantum Hall effect and other condensed matter systems, demonstrating the relevance of QFT in both fundamental and applied physics contexts.

4.6.1 QUANTUM HALL EFFECT

The Quantum Hall Effect is a quantum phenomenon observed in two-dimensional electron systems subjected to low temperatures and strong magnetic fields. It is characterized by the quantization of the Hall conductivity.

4.6.1.0.1 Mathematical Description:

The Hall conductivity σ_{xy} in the quantum Hall effect is quantized and given by:

$$\sigma_{xy} = \frac{e^2}{h}\nu, \tag{4.55}$$

where e is the electron charge, h is Planck's constant, and ν is the filling factor, which can be an integer or fraction depending on the electron density and magnetic field.

4.6.1.0.2 Example Calculation:

For a two-dimensional electron gas, the energy levels (Landau levels) in a magnetic field B are quantized as:

$$E_n = \hbar\omega_c\left(n + \frac{1}{2}\right), \tag{4.56}$$

where $\omega_c = \frac{eB}{m}$ is the cyclotron frequency, and n is an integer. The degeneracy of each Landau level per unit area, D, is given by:

$$D = \frac{eB}{h}. \tag{4.57}$$

When a Landau level is completely filled, the filling factor ν equals the number of filled Landau levels. The Hall conductivity then becomes:

$$\sigma_{xy} = \frac{e^2}{h} \times \text{integer}, \tag{4.58}$$

reflecting the quantized values observed experimentally.

4.6.2 BCS THEORY AND SUPERCONDUCTIVITY IN CONDENSED MATTER PHYSICS

BCS theory, or Bardeen-Cooper-Schrieffer theory, is a fundamental theoretical framework in condensed matter physics that explains superconductivity in certain materials at low temperatures. Named after its developers John Bardeen, Leon Cooper, and Robert Schrieffer, who received the Nobel Prize in Physics in 1972 for their work, BCS theory describes how electrons in a superconductor pair up into Cooper pairs, leading to the phenomenon of superconductivity. Superconductivity provides a compelling example of quantum field theory applications, particularly through the BCS theory.

KEY POINTS OF BCS THEORY

- **Cooper Pairs:**
  - Electrons in a conductor typically repel each other due to their negative charge. However, in the BCS theory, two electrons can indirectly attract each other by interacting with phonons, which are quanta of lattice vibrations in the material.
  - This interaction leads to the formation of bound pairs of electrons called Cooper pairs. These pairs have opposite momenta and spins, effectively forming a boson-like entity.
- **Energy Gap:**
  - The formation of Cooper pairs leads to an energy gap in the electronic excitation spectrum. This gap requires energy to break the pairs apart, contributing to the material's superconducting properties, as the paired state is energetically favorable.
- **Macroscopic Quantum State:**
  - Below a certain critical temperature, these Cooper pairs condense into a single quantum mechanical state, described by a macroscopic wave function. This state exhibits quantum coherence over macroscopic distances, a hallmark of superconductivity.
- **Meissner Effect:**
  - Superconductors expel magnetic fields, a phenomenon known as the Meissner effect. BCS theory explains this by showing how the superconducting state reacts to magnetic fields, maintaining zero electrical resistance and expelling magnetic fields from the interior of the superconductor.
- **Predictions and Applications:**
  - BCS theory successfully predicts the critical temperature below which superconductivity occurs and explains the temperature dependence of the energy gap. It also describes how thermal energy and magnetic fields can destroy superconductivity.

4.6.2.0.1 Mathematical Framework:

The BCS ground state wavefunction is described by:

$$|\text{BCS}\rangle = \prod_{\mathbf{k}}(u_{\mathbf{k}} + v_{\mathbf{k}} c^{\dagger}_{\mathbf{k}\uparrow} c^{\dagger}_{-\mathbf{k}\downarrow})|0\rangle, \tag{4.59}$$

where $u_{\mathbf{k}}$ and $v_{\mathbf{k}}$ are coherence factors that satisfy $u_{\mathbf{k}}^2 + v_{\mathbf{k}}^2 = 1$, and $c^{\dagger}_{\mathbf{k}\sigma}$ is the creation operator for an electron with momentum $\mathbf{k}$ and spin σ.

EXAMPLE: CALCULATING COHERENCE FACTORS AND THE GAP EQUATION

Energy Dispersion and Gap Equation

In BCS theory, the energy of an electron pair with momentum $\mathbf{k}$ is given by:

$$E_{\mathbf{k}} = \sqrt{\xi_{\mathbf{k}}^2 + \Delta^2} \tag{4.60}$$

where $\xi_{\mathbf{k}}$ is the normal state energy relative to the Fermi energy, and Δ is the superconducting gap parameter.

BCS Gap Equation

The gap Δ is determined by the self-consistent solution of the BCS gap equation:

$$\Delta = V \sum_{\mathbf{k}} \frac{\Delta}{2E_{\mathbf{k}}} \tanh\left(\frac{E_{\mathbf{k}}}{2k_B T}\right), \tag{4.61}$$

where V is the pairing potential, k_B is the Boltzmann constant, and T is the temperature.

Coherence Factors

The coherence factors $u_{\mathbf{k}}$ and $v_{\mathbf{k}}$ are given by:

$$u_{\mathbf{k}}^2 = \frac{1}{2}\left(1 + \frac{\xi_{\mathbf{k}}}{E_{\mathbf{k}}}\right), \quad v_{\mathbf{k}}^2 = \frac{1}{2}\left(1 - \frac{\xi_{\mathbf{k}}}{E_{\mathbf{k}}}\right). \tag{4.62}$$

Probability of Finding an Electron in the Cooper Pair State

The probability of finding an electron in the Cooper pair state can be calculated using the coherence factors. For instance, the probability P that a state with momentum $\mathbf{k}$ is occupied is:

$$P = v_{\mathbf{k}}^2 = \frac{1}{2}\left(1 - \frac{\xi_{\mathbf{k}}}{E_{\mathbf{k}}}\right). \tag{4.63}$$

This example illustrates how the BCS theory mathematically describes the superconducting state, providing insights into the behavior of Cooper pairs and the energy gap that defines superconductivity.

4.6.3 COOPER PAIRS AND BOSE-EINSTEIN STATISTICS

While electrons are fermions and follow Fermi-Dirac statistics, Cooper pairs, formed by two electrons with opposite momenta and spins, behave differently. When two electrons pair up to form a Cooper pair, they effectively create a composite particle with an integer spin (spin-0 or spin-1). As a result, these Cooper pairs act as bosons, which obey Bose-Einstein statistics.

This bosonic nature of Cooper pairs allows them to condense into the same quantum ground state, leading to the phenomenon of superconductivity. Below a critical temperature, a large fraction of these Cooper pairs occupy the lowest energy state, forming a macroscopic quantum state that exhibits superconducting properties. The condensation of Cooper pairs into a single quantum state with long-range phase coherence is a key feature of the superconducting state.

4.6.3.0.1 Implications:

The bosonic behavior of Cooper pairs means that they can occupy the same quantum state without restriction, unlike individual electrons which must obey the Pauli exclusion principle. This collective behavior leads to the unique properties of superconductors, such as zero electrical resistance and the expulsion of magnetic fields (Meissner effect).

Thus, the transformation of fermionic electrons into bosonic Cooper pairs is crucial for the emergence of superconductivity, as it allows for the collective quantum behavior that underpins this phenomenon.

4.6.4 COOPER PAIRS AND BOSE-EINSTEIN STATISTICS FROM A QFT PERSPECTIVE

From a Quantum Field Theory (QFT) perspective, the formation of Cooper pairs and their behavior can be understood in terms of field operators and symmetry breaking. In QFT, fields represent particles, and interactions between these fields can lead to the emergence of new collective phenomena.

4.6.4.0.1 Field Operators and Pairing:

In the context of superconductivity, the electron field operators $\psi_\uparrow(\mathbf{r})$ and $\psi_\downarrow(\mathbf{r})$ describe the creation and annihilation of electrons with spin up and spin down, respectively. The formation of Cooper pairs can be represented by a pairing field operator $\Delta(\mathbf{r})$, which is related to the electron fields by:

$$\Delta(\mathbf{r}) = \langle \psi_\uparrow(\mathbf{r}) \psi_\downarrow(\mathbf{r}) \rangle. \tag{4.64}$$

This pairing field $\Delta(\mathbf{r})$ describes the condensation of electron pairs into a coherent state. The expectation value $\langle\cdot\rangle$ indicates a ground state expectation, reflecting the formation of Cooper pairs in the superconducting state.

4.6.4.0.2 Spontaneous Symmetry Breaking:

The condensation of Cooper pairs corresponds to a spontaneous breaking of the $U(1)$ gauge symmetry associated with the electromagnetic field. In the normal (non-superconducting) state, the gauge symmetry is unbroken, and the phase of the Cooper pair wavefunction can vary freely. However, in the superconducting state, the gauge symmetry is broken, and the phase of the Cooper pair wavefunction becomes fixed, leading to a macroscopic quantum coherence.

4.6.4.0.3 Effective Field Theory:

In the effective field theory of superconductivity, the dynamics of the Cooper pairs can be described by a complex scalar field $\Phi(\mathbf{r},t)$ representing the order parameter of the superconducting phase. This field follows Bose-Einstein statistics and its behavior can be captured by a Ginzburg-Landau-type Lagrangian:

$$\mathscr{L} = \alpha|\Phi|^2 + \frac{\beta}{2}|\Phi|^4 + \gamma|\nabla\Phi|^2, \tag{4.65}$$

where α, β, and γ are phenomenological parameters. The term $\alpha|\Phi|^2$ determines the phase transition, while $\beta|\Phi|^4$ ensures the stability of the superconducting state, and $\gamma|\nabla\Phi|^2$ describes the spatial variations of the order parameter.

4.6.4.0.4 Implications in QFT:

The bosonic nature of Cooper pairs in QFT highlights the importance of symmetry and collective excitations in condensed matter systems. The transition from fermionic electrons to bosonic Cooper pairs and the associated spontaneous symmetry breaking are central themes in QFT, illustrating how macroscopic quantum phenomena emerge from microscopic interactions. This perspective not only deepens our understanding of superconductivity but also connects it to broader concepts in field theory, such as phase transitions and the Higgs mechanism.

4.7 SUMMARY AND DISCUSSION

In this chapter, we explored the fundamental principles of symmetries, indistinguishability, statistics, renormalization, and gauge theories within the context of quantum field theory (QFT).

SYMMETRIES AND NOETHER'S THEOREM

We began with the importance of symmetries in QFT, where symmetries lead to conservation laws via Noether's theorem. Symmetry transformations, such as spatial and temporal symmetries, play a crucial role in maintaining the invariance of physical laws. Noether's theorem provides a powerful tool to derive conservation laws from these symmetries, highlighting the deep connection between symmetry and physical observables.

INDISTINGUISHABILITY AND QUANTUM STATISTICS

The concept of indistinguishability was discussed, emphasizing how identical particles, whether bosons or fermions, must adhere to symmetric or antisymmetric wavefunctions. This leads to distinct statistical behaviors: Bose-Einstein statistics for bosons, allowing multiple particles to occupy the same state, and Fermi-Dirac statistics for fermions, enforcing the Pauli exclusion principle.

These statistical properties are fundamental in determining the macroscopic behavior of quantum systems.

RENORMALIZATION IN QUANTUM FIELD THEORY

Renormalization was introduced as a method to handle the infinities that arise in field theory calculations. By adjusting the parameters of the theory, such as masses and coupling constants, renormalization absorbs these infinities into finite, physically meaningful quantities. This ensures that QFT remains predictive and consistent with observed experimental results.

GAUGE THEORIES AND THEIR APPLICATIONS

Gauge theories, foundational to our understanding of fundamental interactions, were explored with a focus on QED. The role of the gauge function and its implications for maintaining gauge invariance were discussed, illustrating how different gauge configurations can describe the same physical situation.

APPLICATIONS IN CONDENSED MATTER PHYSICS

Finally, we examined practical applications of these theories in condensed matter physics. The Quantum Hall Effect was described as a manifestation of quantum statistics in two-dimensional electron systems under strong magnetic fields. BCS theory was introduced to explain superconductivity, where Cooper pairs of electrons form a bosonic state, leading to macroscopic quantum coherence and zero electrical resistance. This highlights the profound impact of QFT concepts on understanding and predicting condensed matter phenomena.

CONCLUSION

The integration of symmetries, indistinguishability, statistics, renormalization, and gauge theories forms the backbone of modern theoretical physics, providing a coherent framework to describe and predict the behavior of quantum systems. These principles not only enhance our understanding of fundamental particles and interactions but also have wide-ranging applications in various fields of physics, from particle physics to condensed matter physics.

This chapter serves as a foundation for further exploration into more advanced topics in quantum field theory and their implications for both theoretical research and practical applications.

4.8 QUESTIONS AND PROBLEMS

4.8.1 TRIVIA QUESTIONS

Question 1: Which principle explains the symmetry of the wavefunction for bosons?

(a) Pauli Exclusion Principle
(b) Fermi-Dirac Principle
(c) Symmetrization Postulate
(d) Uncertainty Principle

Answer: c. Symmetrization Postulate

Question 2: Which theorem states that each differentiable symmetry of the action of a physical system has a corresponding conservation law?

(a) Heisenberg's Theorem

(b) Noether's Theorem
(c) Gauss's Theorem
(d) Stokes' Theorem

Answer: b. Noether's Theorem

Question 3: In Quantum Field Theory, what are the commutation relations for bosonic field operators?

(a) $- \hat{\psi}(x), \hat{\psi}^{\dagger}(y) '' = \delta(x-y)$
(b) $- \hat{\psi}(x), \hat{\psi}^{\dagger}(y) '' = \delta(x-y)$
(c) $- \hat{\psi}(x), \hat{\psi}(y) '' = \delta(x-y)$
(d) $- \hat{\psi}(x), \hat{\psi}(y) '' = \delta(x-y)$

Answer: b. $- \hat{\psi}(x), \hat{\psi}^{\dagger}(y) '' = \delta(x-y)$
Question 4: In the context of renormalization, what does the constant Z represent?

(a) Charge renormalization
(b) Mass renormalization
(c) Field strength renormalization
(d) Coupling constant renormalization

Answer: c. Field strength renormalization
Question 5: Which particles follow Bose-Einstein statistics?

(a) Electrons
(b) Protons
(c) Neutrons
(d) Photons

Answer: d. Photons
Question 6: What kind of symmetry is responsible for the conservation of electric charge in QED?

(a) Rotational symmetry
(b) Translational symmetry
(c) Gauge symmetry
(d) Reflection symmetry

Answer: c. Gauge symmetry
Question 7: In BCS theory, what are Cooper pairs composed of?

(a) Two electrons with parallel spins
(b) Two electrons with opposite spins
(c) Two protons with parallel spins
(d) Two protons with opposite spins

Answer: b. Two electrons with opposite spins
Question 8: What is the filling factor in the context of the quantum Hall effect?

(a) The ratio of the number of filled Landau levels to the total number of electrons
(b) The number of electrons per unit volume
(c) The ratio of the number of filled Landau levels to the number of flux quanta
(d) The number of flux quanta per unit area

Answer: c. The ratio of the number of filled Landau levels to the number of flux quanta

Question 9: In the renormalization of the electron mass, what does the self-energy $\Sigma(p)$ represent?

(a) The bare mass of the electron
(b) The interaction energy with the vacuum
(c) The observed mass of the electron
(d) The charge of the electron

Answer: b. The interaction energy with the vacuum

Question 10: What is the physical significance of the Meissner effect in superconductors?

(a) Complete expulsion of magnetic fields from the interior
(b) Complete expulsion of electric fields from the interior
(c) Generation of magnetic fields inside the superconductor
(d) Generation of electric fields inside the superconductor

Answer: a. Complete expulsion of magnetic fields from the interior

4.8.2 PROBLEMS

Problem 1:

Show that the wavefunction for two identical fermions is antisymmetric under particle exchange.

Solution:

Consider two identical fermions with coordinates $\vec{x}_1$ and $\vec{x}_2$. The total wavefunction $\Psi(\vec{x}_1, \vec{x}_2)$ must satisfy the antisymmetry condition:

$$\Psi(\vec{x}_1, \vec{x}_2) = -\Psi(\vec{x}_2, \vec{x}_1) \tag{4.66}$$

To verify this, assume a simple form of the wavefunction where the fermions occupy single-particle states ψ_a and ψ_b:

$$\Psi(\vec{x}_1, \vec{x}_2) = \frac{1}{\sqrt{2}}[\psi_a(\vec{x}_1)\psi_b(\vec{x}_2) - \psi_b(\vec{x}_1)\psi_a(\vec{x}_2)] \tag{4.67}$$

Now, exchange the particles:

$$\Psi(\vec{x}_2, \vec{x}_1) = \frac{1}{\sqrt{2}}[\psi_a(\vec{x}_2)\psi_b(\vec{x}_1) - \psi_b(\vec{x}_2)\psi_a(\vec{x}_1)] \tag{4.68}$$

$$= -\frac{1}{\sqrt{2}}[\psi_a(\vec{x}_1)\psi_b(\vec{x}_2) - \psi_b(\vec{x}_1)\psi_a(\vec{x}_2)] \tag{4.69}$$

$$= -\Psi(\vec{x}_1, \vec{x}_2) \tag{4.70}$$

Thus, the wavefunction is antisymmetric under the exchange of two identical fermions.

Problem 2:

Prove Noether's theorem for a Lagrangian that is invariant under time translation symmetry.

Solution:

Consider a Lagrangian $\mathscr{L}(q, \dot{q}, t)$ that is invariant under time translations $t \to t + \varepsilon$. The action S is given by:

$$S = \int_{t_1}^{t_2} \mathscr{L}(q, \dot{q}, t)\, dt \tag{4.71}$$

If $\mathscr{L}$ is invariant under time translation, then:

$$\frac{d}{d\varepsilon}\mathscr{L}(q,\dot{q},t+\varepsilon)\Big|_{\varepsilon=0}=0 \tag{4.72}$$

Using the chain rule:

$$\frac{\partial\mathscr{L}}{\partial t}+\frac{\partial\mathscr{L}}{\partial q}\frac{dq}{d\varepsilon}+\frac{\partial\mathscr{L}}{\partial\dot{q}}\frac{d\dot{q}}{d\varepsilon}=0 \tag{4.73}$$

Since q and $\dot{q}$ do not explicitly depend on ε, the second and third terms vanish, leading to:

$$\frac{\partial\mathscr{L}}{\partial t}=0 \tag{4.74}$$

This implies that $\mathscr{L}$ does not explicitly depend on t, and thus the Hamiltonian $H=\dot{q}\frac{\partial\mathscr{L}}{\partial\dot{q}}-\mathscr{L}$ is conserved. Therefore, Noether's theorem shows that time translation symmetry corresponds to the conservation of energy.

Problem 3:

Calculate the Bose-Einstein distribution for a photon gas at temperature, T.

Solution:

The Bose-Einstein distribution function for particles, such as photons, is given by:

$$f_{BE}(E)=\frac{1}{e^{(E-\mu)/k_BT}-1} \tag{4.75}$$

For photons, the chemical potential $\mu=0$, so the distribution simplifies to:

$$f_{BE}(E)=\frac{1}{e^{E/k_BT}-1} \tag{4.76}$$

This function describes the average number of photons in a state with energy E at temperature T.

Problem 4:

Show that the commutation relation $[\hat{a}_p,\hat{a}_q^\dagger]=\delta_{pq}$ holds for bosonic creation and annihilation operators.

Solution:

Consider the bosonic annihilation $\hat{a}_p$ and creation $\hat{a}_p^\dagger$ operators. The vacuum state $|0\rangle$ satisfies $\hat{a}_p|0\rangle=0$. The commutation relation is defined by:

$$[\hat{a}_p,\hat{a}_q^\dagger]=\hat{a}_p\hat{a}_q^\dagger-\hat{a}_q^\dagger\hat{a}_p \tag{4.77}$$

To prove this, consider the action on the vacuum state:

$$\hat{a}_p\hat{a}_q^\dagger|0\rangle=\hat{a}_p|q\rangle=\delta_{pq}|0\rangle \tag{4.78}$$

$$\hat{a}_q^\dagger\hat{a}_p|0\rangle=\hat{a}_q^\dagger\cdot 0=0 \tag{4.79}$$

Therefore:

$$[\hat{a}_p,\hat{a}_q^\dagger]|0\rangle=\delta_{pq}|0\rangle \tag{4.80}$$

Since this holds for any state, the commutation relation is:

$$[\hat{a}_p, \hat{a}_q^\dagger] = \delta_{pq} \tag{4.81}$$

Problem 5:

Derive the expression for the renormalized electron mass in QED.

Solution:

In Quantum Electrodynamics (QED), the renormalized electron mass m is related to the bare mass m_0 and the self-energy correction $\Sigma(m)$:

$$m = m_0 + \Sigma(m) \tag{4.82}$$

To calculate $\Sigma(m)$, consider the self-energy diagram at one-loop level. The self-energy $\Sigma(p)$ can be expressed as:

$$\Sigma(p) = e^2 \int \frac{d^4k}{(2\pi)^4} \frac{\gamma^\mu(\not{p} - \not{k} + m_0)\gamma_\mu}{((p-k)^2 - m_0^2 + i\varepsilon)(k^2 + i\varepsilon)} \tag{4.83}$$

This integral is divergent and requires regularization and renormalization. Introducing a cutoff Λ, we obtain:

$$\Sigma(p) = \frac{e^2}{8\pi^2}\left[\not{p}\log\left(\frac{\Lambda^2}{m_0^2}\right) - 2m_0\log\left(\frac{\Lambda^2}{m_0^2}\right)\right] \tag{4.84}$$

Renormalizing the mass, we redefine m_0 such that the divergence is absorbed:

$$m_{\text{ren}} = m_0 + \frac{e^2}{4\pi^2} m_0 \log\left(\frac{\Lambda^2}{m_0^2}\right) \tag{4.85}$$

Hence, the renormalized mass m_{ren} is finite and matches the experimentally observed mass of the electron.

BONUS PROBLEM

Using Noether's theorem, derive the conservation of angular momentum for a Lagrangian invariant under rotations.

Solution:

Consider a Lagrangian $\mathscr{L}$ that describes a system of particles and is invariant under rotations. The rotation symmetry implies that the Lagrangian does not change under infinitesimal rotations.

An infinitesimal rotation around the z-axis by an angle $\delta\theta$ can be described by the transformation of the coordinates $\vec{r} = (x, y, z)$:

$$\begin{aligned} x' &= x - y\delta\theta, \\ y' &= y + x\delta\theta, \\ z' &= z. \end{aligned} \tag{4.86}$$

The infinitesimal change in the coordinates can be written as:

$$\delta x = -y\delta\theta, \quad \delta y = x\delta\theta, \quad \delta z = 0. \tag{4.87}$$

The corresponding change in the Lagrangian is given by:

$$\delta\mathscr{L} = \frac{\partial\mathscr{L}}{\partial x}\delta x + \frac{\partial\mathscr{L}}{\partial y}\delta y + \frac{\partial\mathscr{L}}{\partial z}\delta z + \frac{\partial\mathscr{L}}{\partial \dot{x}}\delta\dot{x} + \frac{\partial\mathscr{L}}{\partial \dot{y}}\delta\dot{y} + \frac{\partial\mathscr{L}}{\partial \dot{z}}\delta\dot{z}. \tag{4.88}$$

Since the Lagrangian is invariant under rotation, $\delta\mathscr{L} = 0$:

$$0 = \frac{\partial \mathscr{L}}{\partial x}(-y\delta\theta) + \frac{\partial \mathscr{L}}{\partial y}(x\delta\theta) + \frac{\partial \mathscr{L}}{\partial \dot{x}}(-\dot{y}\delta\theta) + \frac{\partial \mathscr{L}}{\partial \dot{y}}(\dot{x}\delta\theta). \tag{4.89}$$

Simplifying and factoring out $\delta\theta$, we get:

$$0 = -y\frac{\partial \mathscr{L}}{\partial x} + x\frac{\partial \mathscr{L}}{\partial y} - \dot{y}\frac{\partial \mathscr{L}}{\partial \dot{x}} + \dot{x}\frac{\partial \mathscr{L}}{\partial \dot{y}}. \tag{4.90}$$

By the Euler-Lagrange equations, $\frac{\partial \mathscr{L}}{\partial x} = \frac{d}{dt}\left(\frac{\partial \mathscr{L}}{\partial \dot{x}}\right)$ and $\frac{\partial \mathscr{L}}{\partial y} = \frac{d}{dt}\left(\frac{\partial \mathscr{L}}{\partial \dot{y}}\right)$, the equation simplifies to:

$$\frac{d}{dt}\left(x\frac{\partial \mathscr{L}}{\partial \dot{y}} - y\frac{\partial \mathscr{L}}{\partial \dot{x}}\right) = 0. \tag{4.91}$$

This indicates that the quantity

$$L_z = x\frac{\partial \mathscr{L}}{\partial \dot{y}} - y\frac{\partial \mathscr{L}}{\partial \dot{x}} \tag{4.92}$$

is conserved. This quantity is the z-component of the angular momentum. Therefore, we have derived the conservation of angular momentum from the rotational invariance of the Lagrangian using Noether's theorem.

5 Lecture 9: Second Quantization, Fock Space, the Tight Binding Hamiltonian, Two and N Particle Operators, and the Hubbard Model

5.1 INTRODUCTION

In this chapter, we explore the advanced formalism of second quantization, which provides a powerful framework for describing quantum systems with variable particle numbers. We begin with an overview of how operators are second quantized, introducing the essential tools of creation and annihilation operators. These operators form the foundation of Fock space, a versatile representation that accommodates the quantum states of systems with differing particle counts.

We then delve into the tight binding Hamiltonian, a key model in condensed matter physics that simplifies the description of electrons in a lattice by considering only nearest-neighbor interactions. This model serves as a stepping stone to understanding more complex interactions within multiparticle systems.

Our discussion extends to two and N particle operators, which allow for the examination of interactions between multiple particles within the second quantization framework. These operators are crucial for modeling and analyzing the collective behavior in many-body systems.

Finally, we introduce the Hubbard model, a cornerstone in the study of strongly correlated electron systems. The Hubbard model captures the competition between kinetic energy and electron-electron interactions, providing insights into phenomena such as magnetism and superconductivity.

By mastering these concepts, you will gain a deeper understanding of the theoretical tools used to describe and analyze many-body quantum systems, paving the way for further exploration in fields such as condensed matter physics and quantum information science.

5.2 HOW TO SECOND QUANTIZE OPERATORS

5.2.1 FIRST QUANTIZATION REVIEW

Recall from previous discussions that in first quantization, the fundamental principles of quantum mechanics are applied to describe the behavior of individual particles. This formalism uses wavefunctions to represent the quantum states of particles and operators to represent physical observables.

5.2.1.0.1 Wavefunctions

The state of a single particle in first quantization is described by a wavefunction $\psi(\vec{r},t)$, which is a complex-valued function of the particle's position $\vec{r}$ and time t. The probability density of finding the particle at a position $\vec{r}$ at time t is given by $|\psi(\vec{r},t)|^2$.

DOI: 10.1201/9781003637370-5

5.2.1.0.2 Operators

Operators are mathematical entities that act on wavefunctions to extract physical information. Key operators include:

- **Position Operator:** $\hat{\vec{r}}\psi(\vec{r}) = \vec{r}\psi(\vec{r})$
- **Momentum Operator:** $\hat{\vec{p}} = -i\hbar\nabla$
- **Hamiltonian Operator:** The Hamiltonian $\hat{H}$ represents the total energy of the system. For a particle in a potential $V(\vec{r})$, the Hamiltonian is given by:

$$\hat{H} = -\frac{\hbar^2}{2m}\nabla^2 + V(\vec{r}) \tag{5.1}$$

5.2.1.0.3 Schrödinger Equation

The time evolution of the wavefunction is governed by the Schrödinger equation:

$$i\hbar\frac{\partial\psi(\vec{r},t)}{\partial t} = \hat{H}\psi(\vec{r},t) \tag{5.2}$$

For a time-independent Hamiltonian, the wavefunction can be separated into spatial and temporal parts: $\psi(\vec{r},t) = \phi(\vec{r})e^{-iEt/\hbar}$, where $\phi(\vec{r})$ satisfies the time-independent Schrödinger equation:

$$\hat{H}\phi(\vec{r}) = E\phi(\vec{r}) \tag{5.3}$$

5.2.1.0.4 Review Example: Particle in a Box

As we have discussed in detail in previous lectures a particle confined in a one-dimensional box of length L with infinitely high potential walls has a potential $V(x)$ that is zero inside the box $(0 < x < L)$ and infinite outside. The time-independent Schrödinger equation inside the box is:

$$-\frac{\hbar^2}{2m}\frac{d^2\phi(x)}{dx^2} = E\phi(x) \tag{5.4}$$

Solving this, we obtain the eigenfunctions and eigenvalues:

$$\phi_n(x) = \sqrt{\frac{2}{L}}\sin\left(\frac{n\pi x}{L}\right), \quad E_n = \frac{n^2\pi^2\hbar^2}{2mL^2}, \quad n = 1,2,3,\ldots \tag{5.5}$$

This brief review of first quantization sets the stage for understanding second quantization, where we generalize these principles to systems with varying particle numbers.

5.2.2 TRANSITION TO SECOND QUANTIZATION

Second quantization is a formalism used to describe quantum systems with varying particle numbers, such as systems of identical particles where the number of particles can change due to interactions. This approach extends the principles of quantum mechanics to many-body systems, providing a more flexible framework for dealing with particles that are created or annihilated.

5.2.2.0.1 Need for Second Quantization

In first quantization, the wavefunction $\psi(\vec{r},t)$ describes a fixed number of particles. However, many physical systems, such as those in quantum field theory or condensed matter physics, involve interactions that can change the particle number. Second quantization addresses this by introducing operators that act on a state space with an indefinite number of particles.

5.2.2.0.2 Creation and Annihilation Operators

Central to the formalism of second quantization are the creation and annihilation operators. The creation operator $\hat{a}^\dagger$ adds a particle to a given state, while the annihilation operator $\hat{a}$ removes a particle from a state. These operators satisfy specific commutation or anticommutation relations, depending on the statistics of the particles.

For bosons (particles obeying Bose-Einstein statistics), the operators satisfy the commutation relations:

$$[\hat{a}_i, \hat{a}_j^\dagger] = \delta_{ij}, \quad [\hat{a}_i, \hat{a}_j] = [\hat{a}_i^\dagger, \hat{a}_j^\dagger] = 0 \tag{5.6}$$

For fermions (particles obeying Fermi-Dirac statistics), the operators satisfy the anticommutation relations:

$$\{\hat{a}_i, \hat{a}_j^\dagger\} = \delta_{ij}, \quad \{\hat{a}_i, \hat{a}_j\} = \{\hat{a}_i^\dagger, \hat{a}_j^\dagger\} = 0 \tag{5.7}$$

5.2.2.0.3 Field Operators

Field operators provide a continuous representation of the creation and annihilation processes. The field operator $\hat{\psi}(x)$ annihilates a particle at position x, and the conjugate field operator $\hat{\psi}^\dagger(x)$ creates a particle at position x. These operators are expressed in terms of the discrete creation and annihilation operators as:

$$\hat{\psi}(x) = \sum_i \phi_i(x)\hat{a}_i, \quad \hat{\psi}^\dagger(x) = \sum_i \phi_i^*(x)\hat{a}_i^\dagger \tag{5.8}$$

where $\phi_i(x)$ are the single-particle wavefunctions.

The field operators obey commutation or anticommutation relations, ensuring the correct quantum statistics for the particles:

$$[\hat{\psi}(x), \hat{\psi}^\dagger(y)] = \delta(x-y) \quad \text{(for bosons)}, \tag{5.9}$$

$$\{\hat{\psi}(x), \hat{\psi}^\dagger(y)\} = \delta(x-y) \quad \text{(for fermions)}. \tag{5.10}$$

5.2.2.0.4 Hamiltonian in Second Quantization

In the second quantization formalism, the Hamiltonian of a many-body system can be expressed in terms of the creation and annihilation operators. For example, the Hamiltonian for a system of non-interacting particles is given by:

$$\hat{H} = \sum_{i,j} \langle i|\hat{h}|j\rangle \hat{a}_i^\dagger \hat{a}_j, \tag{5.11}$$

where $\hat{h}$ is the single-particle Hamiltonian and $\langle i|\hat{h}|j\rangle$ are its matrix elements.

For interacting particles, additional terms involving products of four operators can be included to account for interactions.

5.2.2.0.5 Example: Second Quantization of a Harmonic Oscillator

Consider a one-dimensional harmonic oscillator with the Hamiltonian:

$$\hat{H} = \frac{\hat{p}^2}{2m} + \frac{1}{2}m\omega^2\hat{x}^2. \tag{5.12}$$

In second quantization, we express $\hat{H}$ using the creation and annihilation operators:

$$\hat{H} = \hbar\omega\left(\hat{a}^\dagger\hat{a} + \frac{1}{2}\right), \tag{5.13}$$

where $\hat{a}$ and $\hat{a}^\dagger$ are the annihilation and creation operators, respectively. This formulation highlights the ease with which second quantization handles systems with variable particle numbers and interactions.

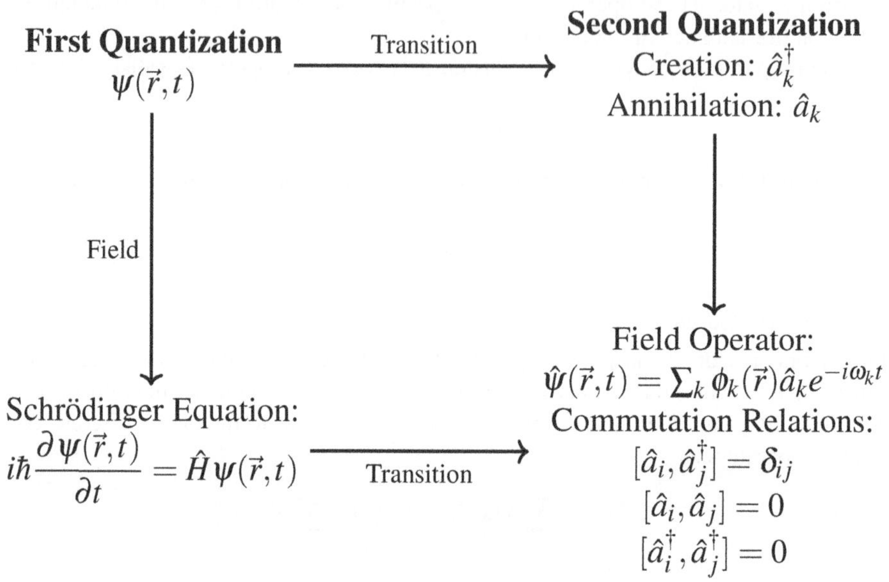

FIGURE 5.1
Visual representation of the transition from first to second quantization, showing how wavefunctions are replaced by creation and annihilation operators.

The transition from first to second quantization is visually depicted in Figure 5.1. This figure illustrates the fundamental shift in quantum mechanics from describing particles with wavefunctions, as seen in first quantization, to employing creation and annihilation operators in second quantization. In first quantization, the Schrödinger equation governs the time evolution of wavefunctions, representing the quantum state of a single particle. Second quantization, however, introduces a more versatile framework where field operators, built from creation and annihilation operators, describe systems with variable particle numbers. This transition is crucial for understanding many-body quantum systems, where the commutation relations between these operators encapsulate the statistical properties of bosons or fermions. The figure encapsulates these concepts, highlighting the replacement of wavefunctions with operators and demonstrating how second quantization provides a powerful tool for analyzing complex quantum systems.

5.2.3 SECOND QUANTIZATION AND REAL-WORLD EXAMPLE: LASER EMISSION

Second quantization provides a powerful framework for describing systems where particle numbers can change, such as in the process of laser emission. Lasers operate based on the principles of quantum mechanics and quantum optics, where photons are created and annihilated in a controlled manner to produce coherent light.

5.2.3.0.1 Principles of Laser Emission

Laser emission involves three key processes:

1. **Absorption**: An atom in a lower energy state absorbs a photon and transitions to a higher energy state.

2. **Spontaneous Emission**: An excited atom spontaneously decays to a lower energy state, emitting a photon in the process.
3. **Stimulated Emission**: An excited atom, when struck by a photon of the same energy, is stimulated to decay to a lower energy state, emitting a photon that is coherent with the stimulating photon.

5.2.3.0.2 *Second Quantization Formalism*

In the second quantization framework, we describe the electromagnetic field in terms of photon creation and annihilation operators. Let $\hat{a}_k^\dagger$ and $\hat{a}_k$ be the creation and annihilation operators for a photon in mode k. These operators satisfy the bosonic commutation relations:

$$[\hat{a}_k, \hat{a}_{k'}^\dagger] = \delta_{kk'}, \quad [\hat{a}_k, \hat{a}_{k'}] = [\hat{a}_k^\dagger, \hat{a}_{k'}^\dagger] = 0. \tag{5.14}$$

The Hamiltonian of the electromagnetic field can be written as:

$$\hat{H}_{\text{field}} = \sum_k \hbar\omega_k \left(\hat{a}_k^\dagger \hat{a}_k + \frac{1}{2} \right), \tag{5.15}$$

where ω_k is the frequency of mode k.

5.2.3.0.3 *Interaction with Atoms*

Consider a two-level atom with ground state $|g\rangle$ and excited state $|e\rangle$, separated by an energy $\hbar\omega_0$. The interaction between the atom and the electromagnetic field is described by the Jaynes-Cummings model. The Jaynes-Cummings model is a fundamental theoretical framework in quantum optics that describes the interaction between a two-level atom and a single mode of the electromagnetic field, capturing essential quantum phenomena such as Rabi oscillations and quantum entanglement. The interaction Hamiltonian is given as:

$$\hat{H}_{\text{int}} = \hbar g(\hat{\sigma}_+ \hat{a} + \hat{\sigma}_- \hat{a}^\dagger), \tag{5.16}$$

where g is the coupling constant, $\hat{\sigma}_+ = |e\rangle\langle g|$ and $\hat{\sigma}_- = |g\rangle\langle e|$ are the raising and lowering operators for the atom, and $\hat{a}$ and $\hat{a}^\dagger$ are the annihilation and creation operators for the photon field.

5.2.3.0.4 *Stimulated Emission*

In the process of stimulated emission, an incoming photon stimulates the transition of the atom from the excited state $|e\rangle$ to the ground state $|g\rangle$, accompanied by the emission of an additional photon. This results in two coherent photons.

The transition rate for stimulated emission can be derived using Fermi's golden rule. The transition probability per unit time is given by:

$$W_{i\to f} = \frac{2\pi}{\hbar} |\langle f|\hat{H}_{\text{int}}|i\rangle|^2 \delta(E_f - E_i). \tag{5.17}$$

For an initial state $|i\rangle = |e,n\rangle$ (an atom in the excited state with n photons) and a final state $|f\rangle = |g,n+1\rangle$ (an atom in the ground state with $n+1$ photons), the matrix element is:

$$\langle g,n+1|\hat{H}_{\text{int}}|e,n\rangle = \hbar g\sqrt{n+1}. \tag{5.18}$$

The transition rate becomes:

$$\begin{aligned} W_{e\to g} &= \frac{2\pi}{\hbar} |\hbar g\sqrt{n+1}|^2 \delta(E_{g,n+1} - E_{e,n}) \\ &= 2\pi g^2 (n+1) \delta(\hbar\omega - \hbar\omega_0). \end{aligned} \tag{5.19}$$

5.2.3.0.5 Laser Threshold Condition

For laser operation, the gain due to stimulated emission must exceed the losses in the laser cavity. The condition for achieving this is known as the laser threshold condition. It can be expressed as:

$$R_{\text{stimulated}} > R_{\text{loss}}, \tag{5.20}$$

where $R_{\text{stimulated}}$ is the rate of stimulated emission and R_{loss} represents various loss mechanisms, such as absorption and scattering.

5.2.4 SPONTANEOUS EMISSION AND VACUUM ENERGY

Spontaneous emission, traditionally thought of as an inherently random process where an excited atom emits a photon and transitions to a lower energy state, is not truly "spontaneous" in the strictest sense. Rather, it is driven by the vacuum fluctuations of the electromagnetic field, which are omnipresent even in a perfect vacuum. According to quantum electrodynamics, these vacuum fluctuations provide the perturbations necessary to stimulate the emission of photons from excited atoms. This process, known as vacuum-stimulated emission, highlights the significant role of the zero-point energy of the vacuum in facilitating atomic transitions. Thus, the emission is not entirely intrinsic to the atom but is influenced by the surrounding quantum vacuum field, giving a more nuanced understanding of the nature of spontaneous emission.

To illustrate this using ladder operators, consider the interaction Hamiltonian for an atom interacting with the quantized electromagnetic field:

$$\hat{H}_{\text{int}} = \hbar g(\hat{a}\hat{\sigma}^{+} + \hat{a}^{\dagger}\hat{\sigma}^{-}), \tag{5.21}$$

where $\hat{a}$ and $\hat{a}^{\dagger}$ are the annihilation and creation operators for the photon field, and $\hat{\sigma}^{+}$ and $\hat{\sigma}^{-}$ are the atomic raising and lowering operators, respectively.

The annihilation operator $\hat{a}$ and creation operator $\hat{a}^{\dagger}$ satisfy the commutation relations:

$$[\hat{a}, \hat{a}^{\dagger}] = 1, \tag{5.22}$$

which reflect the quantized nature of the electromagnetic field. The atomic operators $\hat{\sigma}^{+}$ and $\hat{\sigma}^{-}$ describe transitions between the excited state $|e\rangle$ and the ground state $|g\rangle$ of the atom:

$$\hat{\sigma}^{+} = |e\rangle\langle g|, \quad \hat{\sigma}^{-} = |g\rangle\langle e|. \tag{5.23}$$

The interaction Hamiltonian shows that the presence of the vacuum field, represented by $\hat{a}$ and $\hat{a}^{\dagger}$, can stimulate the emission or absorption of a photon by the atom, thereby illustrating how spontaneous emission is influenced by vacuum fluctuations. Second quantization provides a natural and powerful language to describe the quantum dynamics of laser emission. By using creation and annihilation operators, we can elegantly model the processes of absorption, spontaneous emission, and stimulated emission, which are fundamental to the operation of lasers. This formalism not only simplifies the theoretical description but also enables precise calculations of transition rates and thresholds, essential for designing and understanding laser systems.

5.3 FOCK SPACE

5.3.1 DEFINITION AND STRUCTURE

Fock space is the Hilbert space that encompasses all possible quantum states of a system with a variable number of particles. It is constructed as a direct sum of the n-particle Hilbert spaces, ranging from the vacuum state (no particles) to states with an arbitrary number of particles.

Mathematically, the Fock space $\mathscr{F}$ is defined as:

$$\mathscr{F} = \bigoplus_{n=0}^{\infty} \mathscr{H}^{(n)}, \tag{5.24}$$

where $\mathscr{H}^{(n)}$ represents the n-particle Hilbert space.

The operator $\oplus$, sometimes called "oplus," is used to denote the direct sum of vector spaces or Hilbert spaces. The direct sum is a way of combining several vector spaces into a new, larger vector space.

In the context of Fock space, $\oplus$ signifies that the Fock space is constructed as a direct sum of subspaces corresponding to different particle numbers. Each subspace $\mathscr{H}^{(n)}$ contains the states with exactly n particles.

Heres a brief explanation of how the direct sum works:

Direct Sum of Vector Spaces: If V and W are vector spaces, their direct sum $V \oplus W$ is the vector space consisting of all ordered pairs (v, w), where $v \in V$ and $w \in W$. The operations of addition and scalar multiplication are defined component-wise:

$$(v_1, w_1) + (v_2, w_2) = (v_1 + v_2, w_1 + w_2), \tag{5.25}$$

$$\alpha(v, w) = (\alpha v, \alpha w). \tag{5.26}$$

Direct Sum of Hilbert Spaces: For Hilbert spaces $\mathscr{H}_1$ and $\mathscr{H}_2$, their direct sum $\mathscr{H}_1 \oplus \mathscr{H}_2$ is the Hilbert space of pairs (h_1, h_2) with $h_1 \in \mathscr{H}_1$ and $h_2 \in \mathscr{H}_2$, equipped with the inner product:

$$\langle (h_1, h_2), (h_1', h_2') \rangle = \langle h_1, h_1' \rangle_{\mathscr{H}_1} + \langle h_2, h_2' \rangle_{\mathscr{H}_2}. \tag{5.27}$$

Worked Example: Direct Sum Inner Product in Hilbert Spaces

Let's consider two Hilbert spaces $\mathscr{H}_1$ and $\mathscr{H}_2$:

1. $\mathscr{H}_1$ is a 2-dimensional complex vector space with basis vectors $|e_1\rangle$ and $|e_2\rangle$. 2. $\mathscr{H}_2$ is a 2-dimensional complex vector space with basis vectors $|f_1\rangle$ and $|f_2\rangle$.

For simplicity, let's use the following vectors in these Hilbert spaces:

$$h_1 = 2|e_1\rangle + 3|e_2\rangle, \tag{5.28}$$

$$h_1' = |e_1\rangle + 4|e_2\rangle, \tag{5.29}$$

$$h_2 = -|f_1\rangle + 5|f_2\rangle, \tag{5.30}$$

$$h_2' = 3|f_1\rangle - 2|f_2\rangle. \tag{5.31}$$

The inner products in $\mathscr{H}_1$ and $\mathscr{H}_2$ are given by:

$$\langle h_1, h_1' \rangle_{\mathscr{H}_1} = (2^*) \cdot 1 + (3^*) \cdot 4 = 2 + 12 = 14, \tag{5.32}$$

$$\langle h_2, h_2' \rangle_{\mathscr{H}_2} = (-1^*) \cdot 3 + (5^*) \cdot (-2) = -3 - 10 = -13. \tag{5.33}$$

Using these results, the inner product in the direct sum space $\mathscr{H}_1 \oplus \mathscr{H}_2$ is:

$$\langle (h_1, h_2), (h_1', h_2') \rangle = \langle h_1, h_1' \rangle_{\mathscr{H}_1} + \langle h_2, h_2' \rangle_{\mathscr{H}_2}, \tag{5.34}$$

$$\langle (h_1, h_2), (h_1', h_2') \rangle = 14 + (-13) = 1. \tag{5.35}$$

Thus, the inner product of the vectors (h_1, h_2) and (h_1', h_2') in the direct sum Hilbert space $\mathscr{H}_1 \oplus \mathscr{H}_2$ is 1.

Direct Sum in Fock Space: When constructing the Fock space $\mathscr{F}$, each $\mathscr{H}^{(n)}$ represents the Hilbert space for n-particle states. The Fock space is the direct sum of these n-particle spaces:

$$\mathscr{F} = \bigoplus_{n=0}^{\infty} \mathscr{H}^{(n)}. \tag{5.36}$$

This notation captures the idea that the total Fock space includes all possible configurations of different numbers of particles, each configuration forming its own subspace within the larger Fock space.

For example, the total Fock space for a single species of particles can be written as:

$$\mathscr{F} = \mathbb{C} \oplus \mathscr{H} \oplus (\mathscr{H} \otimes \mathscr{H})_S \oplus (\mathscr{H} \otimes \mathscr{H} \otimes \mathscr{H})_S \oplus \cdots, \tag{5.37}$$

where $\mathbb{C}$ represents the vacuum state, $\mathscr{H}$ is the single-particle Hilbert space, and the subscripts S denote symmetrization (for bosons) or antisymmetrization (for fermions) of the tensor products. This process is illustrated in Figure 5.2.

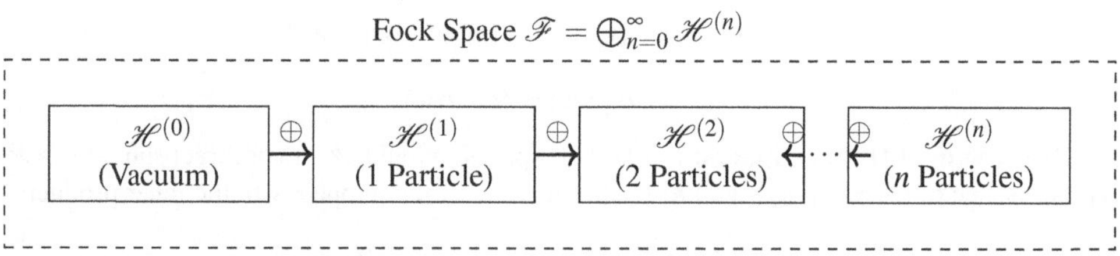

FIGURE 5.2
Schematic of Fock space showing the direct sum of subspaces corresponding to different particle numbers.

Worked Example: Direct Sum in Fock Space

Let's consider the Fock space constructed from a single-particle Hilbert space $\mathscr{H}$. The Fock space $\mathscr{F}$ is the direct sum of the n-particle Hilbert spaces $\mathscr{H}^{(n)}$:

$$\mathscr{F} = \bigoplus_{n=0}^{\infty} \mathscr{H}^{(n)}. \tag{5.38}$$

For this example, we will limit our consideration to the 0-particle, 1-particle, and 2-particle subspaces.

1. The 0-particle subspace $\mathscr{H}^{(0)}$ is spanned by the vacuum state $|0\rangle$. 2. The 1-particle subspace $\mathscr{H}^{(1)}$ is spanned by the states $|1\rangle$ and $|2\rangle$. 3. The 2-particle subspace $\mathscr{H}^{(2)}$ is spanned by the states $|1,1\rangle$, $|1,2\rangle$, and $|2,2\rangle$.

We can construct a state in the Fock space $\mathscr{F}$ as a direct sum of states from these subspaces. For example, consider the following states:

$$\psi_0 = |0\rangle \in \mathscr{H}^{(0)}, \tag{5.39}$$

$$\psi_1 = \alpha|1\rangle + \beta|2\rangle \in \mathscr{H}^{(1)}, \tag{5.40}$$

$$\psi_2 = \gamma|1,1\rangle + \delta|1,2\rangle + \varepsilon|2,2\rangle \in \mathscr{H}^{(2)}, \tag{5.41}$$

where $\alpha, \beta, \gamma, \delta$, and ε are complex coefficients.

The state in the Fock space $\mathscr{F}$ is then:

$$\Psi = \psi_0 \oplus \psi_1 \oplus \psi_2. \tag{5.42}$$

For numerical values, lets take:

$$\alpha = 1, \quad \beta = 2, \quad \gamma = 1, \quad \delta = -1, \quad \varepsilon = 0. \tag{5.43}$$

Then the specific states are:

$$\psi_1 = |1\rangle + 2|2\rangle, \tag{5.44}$$

$$\psi_2 = |1,1\rangle - |1,2\rangle. \tag{5.45}$$

The full Fock space state Ψ is:

$$\Psi = |0\rangle \oplus (|1\rangle + 2|2\rangle) \oplus (|1,1\rangle - |1,2\rangle). \tag{5.46}$$

This represents a state in the Fock space $\mathscr{F}$ that combines contributions from the 0-particle, 1-particle, and 2-particle subspaces.

5.3.2 BASIS STATES

The basis states in Fock space are constructed using creation operators. For a given mode k, the creation operator $\hat{a}_k^\dagger$ adds a particle to that mode, while the annihilation operator $\hat{a}_k$ removes a particle.

The vacuum state $|0\rangle$ is the state with no particles:

$$\hat{a}_k |0\rangle = 0 \quad \text{for all } k. \tag{5.47}$$

Applying the creation operator $\hat{a}_k^\dagger$ to the vacuum state generates the single-particle states:

$$|1_k\rangle = \hat{a}_k^\dagger |0\rangle. \tag{5.48}$$

Multi-particle states are generated by applying multiple creation operators:

$$|n_{k_1}, n_{k_2}, \ldots, n_{k_m}\rangle = \frac{(\hat{a}_{k_1}^\dagger)^{n_{k1}} (\hat{a}_{k_2}^\dagger)^{n_{k2}} \cdots (\hat{a}_{k_m}^\dagger)^{n_{km}}}{\sqrt{n_{k_1}! n_{k_2}! \cdots n_{k_m}!}} |0\rangle, \tag{5.49}$$

where n_{k_i} is the number of particles in mode k_i.

For example, the two-particle state with one particle in mode k_1 and one particle in mode k_2 is:

$$|1_{k_1}, 1_{k_2}\rangle = \hat{a}_{k_1}^\dagger \hat{a}_{k_2}^\dagger |0\rangle. \tag{5.50}$$

5.3.3 OCCUPATION NUMBER REPRESENTATION

The occupation number representation is a convenient way to describe states in Fock space by specifying the number of particles in each mode. In this representation, a state is labeled by the set of occupation numbers $\{n_k\}$.

A general state in Fock space can be written as:

$$|\{n_k\}\rangle = |n_{k_1}, n_{k_2}, \ldots, n_{k_m}\rangle. \tag{5.51}$$

For example, the vacuum state is:

$$|0\rangle = |0, 0, \ldots, 0\rangle. \tag{5.52}$$

A state with one particle in mode k_1 and two particles in mode k_2 is:

$$|1_{k_1}, 2_{k_2}, 0, \ldots\rangle = \hat{a}^\dagger_{k_1} (\hat{a}^\dagger_{k_2})^2 |0\rangle . \tag{5.53}$$

In this representation, the action of the creation and annihilation operators is straightforward:

$$\hat{a}^\dagger_k |n_{k_1}, \ldots, n_k, \ldots\rangle = \sqrt{n_k + 1}\, |n_{k_1}, \ldots, n_k + 1, \ldots\rangle , \tag{5.54}$$

$$\hat{a}_k |n_{k_1}, \ldots, n_k, \ldots\rangle = \sqrt{n_k}\, |n_{k_1}, \ldots, n_k - 1, \ldots\rangle . \tag{5.55}$$

The occupation number representation is significant because it simplifies the description of many-body systems and facilitates the calculation of physical properties in quantum field theory.

5.4 THE TIGHT BINDING HAMILTONIAN

5.4.1 LATTICE MODELS

Lattice models are used to describe systems where particles, such as electrons, move through a regular array of potential wells, typically representing atoms in a crystal lattice. In the tight binding approximation, it is assumed that the wavefunctions of the electrons are strongly localized around the lattice sites, and the hopping of electrons between neighboring sites is treated perturbatively. Figure 5.3 shows that in the tight binding model in a one-dimensional lattice electrons can hop between neighboring sites.

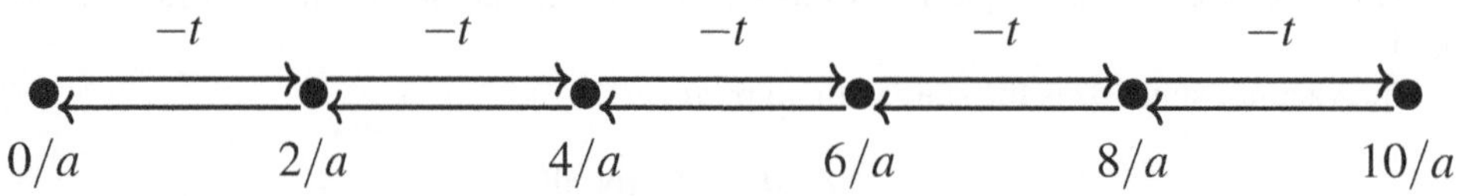

FIGURE 5.3
Tight binding model in a one-dimensional lattice. Electrons can hop between neighboring sites with a hopping parameter t.

5.4.2 FORMULATION OF THE TIGHT BINDING HAMILTONIAN

To derive the tight binding Hamiltonian, consider a one-
dimensional lattice with lattice constant a. The lattice sites are labeled by an integer n. The tight binding Hamiltonian for an electron hopping between nearest neighbor sites is given by:

$$\hat{H} = -t \sum_n \left(\hat{c}^\dagger_{n+1} \hat{c}_n + \hat{c}^\dagger_n \hat{c}_{n+1} \right) + \varepsilon_0 \sum_n \hat{c}^\dagger_n \hat{c}_n , \tag{5.56}$$

where:

- $\hat{c}^\dagger_n$ and $\hat{c}_n$ are the creation and annihilation operators at site n,
- t is the hopping parameter representing the amplitude for an electron to hop from one site to an adjacent site,
- ε_0 is the on-site energy.

The first term represents the hopping of electrons between neighboring sites, while the second term represents the energy of an electron at a particular site.

In the case of a periodic lattice with periodic boundary conditions, the Hamiltonian can be diagonalized by transforming to momentum space. Define the Fourier transform of the annihilation operator:

$$\hat{c}_k = \frac{1}{\sqrt{N}} \sum_n \hat{c}_n e^{-ikna} , \tag{5.57}$$

where N is the number of lattice sites and k is the wavevector. The inverse Fourier transform is given by:

$$\hat{c}_n = \frac{1}{\sqrt{N}} \sum_k \hat{c}_k e^{ikna}. \tag{5.58}$$

Substituting these into the Hamiltonian, we get:

$$\hat{H} = \sum_k (-2t\cos(ka) + \varepsilon_0)\, \hat{c}_k^\dagger \hat{c}_k. \tag{5.59}$$

This shows that the energy eigenvalues in the tight binding model form a band:

$$E(k) = \varepsilon_0 - 2t\cos(ka). \tag{5.60}$$

5.4.3 APPLICATIONS

The tight binding model has wide applications in condensed matter physics, including:

- **Electronic band structure:** The tight binding model is used to calculate the electronic band structure of materials, providing insight into their conductive properties.
- **Graphene:** The unique electronic properties of graphene can be described using a tight binding model on a honeycomb lattice.
- **Topological insulators:** The tight binding model helps in understanding the edge states and topological properties of materials.

5.4.3.1 Electronic Band Structure

To understand the electronic band structure, consider a simple cubic lattice in three dimensions. The tight binding Hamiltonian for this lattice is:

$$\hat{H} = -t \sum_{\langle i,j \rangle} \left(\hat{c}_i^\dagger \hat{c}_j + \hat{c}_j^\dagger \hat{c}_i \right), \tag{5.61}$$

where $\langle i, j \rangle$ denotes summation over nearest neighbor pairs. Applying periodic boundary conditions and transforming to momentum space, we obtain:

$$\hat{H} = \sum_{\mathbf{k}} \varepsilon(\mathbf{k}) \hat{c}_{\mathbf{k}}^\dagger \hat{c}_{\mathbf{k}}, \tag{5.62}$$

where the energy dispersion relation is:

$$\varepsilon(\mathbf{k}) = -2t\left(\cos(k_x a) + \cos(k_y a) + \cos(k_z a)\right). \tag{5.63}$$

This dispersion relation describes the energy bands in the material, which are critical for understanding its electrical conductivity.

5.4.3.2 Graphene

Graphene's electronic properties can be described using a tight binding model on a honeycomb lattice. Each unit cell of graphene contains two atoms, which we label as A and B. The tight binding Hamiltonian is:

$$\hat{H} = -t \sum_{\langle i,j \rangle} \left(\hat{a}_i^\dagger \hat{b}_j + \hat{b}_j^\dagger \hat{a}_i \right), \tag{5.64}$$

where $\hat{a}_i^\dagger$ and $\hat{b}_j^\dagger$ are creation operators for A and B atoms, respectively. Transforming to momentum space, we get:

$$\hat{H} = \sum_{\mathbf{k}} \begin{pmatrix} \hat{a}_{\mathbf{k}}^\dagger & \hat{b}_{\mathbf{k}}^\dagger \end{pmatrix} \begin{pmatrix} 0 & -tf(\mathbf{k}) \\ -tf^*(\mathbf{k}) & 0 \end{pmatrix} \begin{pmatrix} \hat{a}_{\mathbf{k}} \\ \hat{b}_{\mathbf{k}} \end{pmatrix}, \tag{5.65}$$

where $f(\mathbf{k}) = 1 + e^{ik_y a} + e^{i(k_x a + k_y a)}$. Diagonalizing this Hamiltonian yields the energy bands:

$$E(\mathbf{k}) = \pm t|f(\mathbf{k})|. \tag{5.66}$$

These bands include the famous Dirac cones at the K points in the Brillouin zone, shown in Figure 5.4, explaining graphene's high conductivity and unique electronic properties.

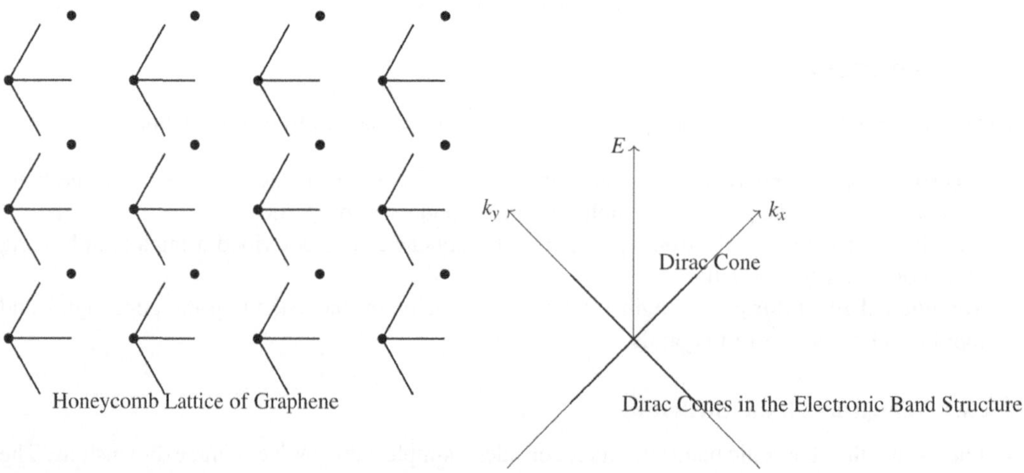

FIGURE 5.4
Representation of the honeycomb lattice of graphene and its Dirac cones in the electronic band structure.

5.4.3.3 Topological Insulators

Topological insulators are materials with insulating bulk and conducting edge states. A simple model to describe such systems is the Bernevig-Hughes-Zhang (BHZ) model. The tight binding Hamiltonian on a two-dimensional square lattice is given by:

$$\hat{H} = \sum_{\mathbf{k}} \Psi_{\mathbf{k}}^\dagger \mathscr{H}(\mathbf{k}) \Psi_{\mathbf{k}}, \tag{5.67}$$

where

$$\mathscr{H}(\mathbf{k}) = \begin{pmatrix} M - 2B(2-\cos(k_x a) - \cos(k_y a)) & A(\sin(k_x a) - i\sin(k_y a)) \\ A(\sin(k_x a) + i\sin(k_y a)) & -[M - 2B(2-\cos(k_x a) - \cos(k_y a))] \end{pmatrix} \tag{5.68}$$

and $\Psi_{\mathbf{k}}^\dagger = (\hat{c}_{\mathbf{k},\uparrow}^\dagger, \hat{c}_{\mathbf{k},\downarrow}^\dagger)$ is a spinor of creation operators for spin-up and spin-down electrons. Diagonalizing this Hamiltonian reveals edge states that are protected by the topological nature of the system.

To diagonalize this Hamiltonian, we solve the eigenvalue problem:

$$\mathscr{H}(\mathbf{k})\begin{pmatrix}u_{\mathbf{k}}\\ v_{\mathbf{k}}\end{pmatrix}=E_{\mathbf{k}}\begin{pmatrix}u_{\mathbf{k}}\\ v_{\mathbf{k}}\end{pmatrix}. \tag{5.69}$$

Solving this yields the energy eigenvalues:

$$E_{\mathbf{k}}^{\pm}=\pm\left[(M-2B(2-\cos(k_xa)-\cos(k_ya)))^2 +A^2\left(\sin^2(k_xa)+\sin^2(k_ya)\right)\right]^{1/2} \tag{5.70}$$

These eigenvalues describe the bulk bands of the topological insulator. However, if we consider a finite system with edges, additional states appear within the bulk energy gap. These states are localized at the edges and are protected by the topological nature of the system. They cross the bulk gap, connecting the valence and conduction bands, and are robust against perturbations that do not close the bulk gap.

This diagonalization and the resulting edge states illustrate how the tight binding model can describe the unique properties of topological insulators, including their conducting edge states, which are protected by the topological invariants of the system.

5.5 TWO AND N PARTICLE OPERATORS

5.5.1 TWO-PARTICLE OPERATORS

In second quantization, two-particle operators are crucial for describing interactions between pairs of particles. These operators act on the Fock space and can be written in terms of creation and annihilation operators. A common example is the two-particle interaction operator, which is used to describe interactions such as the Coulomb interaction in an electron gas.

Consider a two-particle interaction described by a potential $V(\mathbf{r}_1,\mathbf{r}_2)$. The corresponding operator in second quantization is:

$$\hat{V}=\frac{1}{2}\sum_{ijkl}\langle ij|V|kl\rangle\hat{a}_i^{\dagger}\hat{a}_j^{\dagger}\hat{a}_k\hat{a}_l, \tag{5.71}$$

where:

- $\hat{a}_i^{\dagger}$ and $\hat{a}_i$ are the creation and annihilation operators for the i-th single-particle state,
- $\langle ij|V|kl\rangle$ are the matrix elements of the interaction potential V in the single-particle basis.

For a specific example, consider the Coulomb interaction between two electrons:

$$V(\mathbf{r}_1,\mathbf{r}_2)=\frac{e^2}{|\mathbf{r}_1-\mathbf{r}_2|}. \tag{5.72}$$

The matrix elements in a given basis (e.g., plane waves) are:

$$\langle\mathbf{k}_1\mathbf{k}_2|V|\mathbf{k}_3\mathbf{k}_4\rangle=\frac{e^2}{V}\delta_{\mathbf{k}_1+\mathbf{k}_2,\mathbf{k}_3+\mathbf{k}_4}\frac{4\pi}{|\mathbf{k}_1-\mathbf{k}_3|^2}, \tag{5.73}$$

where V is the volume of the system. The two-particle operator in second quantization then becomes:

$$\hat{V}=\frac{1}{2}\sum_{\mathbf{k}_1\mathbf{k}_2\mathbf{k}_3\mathbf{k}_4}\frac{e^2}{V}\delta_{\mathbf{k}_1+\mathbf{k}_2,\mathbf{k}_3+\mathbf{k}_4}\frac{4\pi}{|\mathbf{k}_1-\mathbf{k}_3|^2}\hat{a}_{\mathbf{k}_1}^{\dagger}\hat{a}_{\mathbf{k}_2}^{\dagger}\hat{a}_{\mathbf{k}_3}\hat{a}_{\mathbf{k}_4}. \tag{5.74}$$

5.5.2 N-PARTICLE OPERATORS

The concept of two-particle operators can be generalized to N-particle operators, which are used to describe interactions involving more than two particles. These operators are essential in the study of many-body quantum systems, where correlations between large numbers of particles must be considered.

An N-particle operator can be written as:

$$\hat{W} = \frac{1}{N!} \sum_{i_1 \cdots i_N} \sum_{j_1 \cdots j_N} \langle i_1 i_2 \cdots i_N | W | j_1 j_2 \cdots j_N \rangle \\ \times \hat{a}^\dagger_{i_1} \hat{a}^\dagger_{i_2} \cdots \hat{a}^\dagger_{i_N} \hat{a}_{j_N} \cdots \hat{a}_{j_2} \hat{a}_{j_1} \tag{5.75}$$

where:

- $\hat{a}^\dagger_i$ and $\hat{a}_i$ are the creation and annihilation operators for the i-th single-particle state,
- $\langle i_1 i_2 \cdots i_N | W | j_1 j_2 \cdots j_N \rangle$ are the matrix elements of the N-particle interaction potential W.

For instance, consider a three-particle interaction with potential $W(\mathbf{r}_1, \mathbf{r}_2, \mathbf{r}_3)$. The corresponding operator in second quantization is:

$$\hat{W} = \frac{1}{3!} \sum_{i_1 i_2 i_3 j_1 j_2 j_3} \langle i_1 i_2 i_3 | W | j_1 j_2 j_3 \rangle \hat{a}^\dagger_{i_1} \hat{a}^\dagger_{i_2} \hat{a}^\dagger_{i_3} \hat{a}_{j_3} \hat{a}_{j_2} \hat{a}_{j_1}. \tag{5.76}$$

This formalism is particularly useful in describing complex systems such as nuclei, where interactions among multiple nucleons play a critical role.

5.5.2.1 Example: Three-Particle Interaction

Consider a simplified model where the three-particle interaction potential is given by:

$$W(\mathbf{r}_1, \mathbf{r}_2, \mathbf{r}_3) = g\delta(\mathbf{r}_1 - \mathbf{r}_2)\delta(\mathbf{r}_2 - \mathbf{r}_3), \tag{5.77}$$

where g is a coupling constant. The matrix elements in a basis of plane waves are:

$$\langle \mathbf{k}_1 \mathbf{k}_2 \mathbf{k}_3 | W | \mathbf{k}_4 \mathbf{k}_5 \mathbf{k}_6 \rangle = g\delta_{\mathbf{k}_1+\mathbf{k}_2+\mathbf{k}_3, \mathbf{k}_4+\mathbf{k}_5+\mathbf{k}_6}. \tag{5.78}$$

The three-particle operator in second quantization becomes:

$$\hat{W} = \frac{g}{3!} \sum_{\mathbf{k}_1 \mathbf{k}_2 \mathbf{k}_3 \mathbf{k}_4 \mathbf{k}_5 \mathbf{k}_6} \delta_{\mathbf{k}_1+\mathbf{k}_2+\mathbf{k}_3, \mathbf{k}_4+\mathbf{k}_5+\mathbf{k}_6} \hat{a}^\dagger_{\mathbf{k}_1} \hat{a}^\dagger_{\mathbf{k}_2} \hat{a}^\dagger_{\mathbf{k}_3} \hat{a}_{\mathbf{k}_6} \hat{a}_{\mathbf{k}_5} \hat{a}_{\mathbf{k}_4}. \tag{5.79}$$

This example illustrates the complexity and richness of many-body interactions in quantum systems, highlighting the importance of N-particle operators in theoretical and computational studies. This formalism can be used to describe three entangled photons. In the case of entangled photons, the creation and annihilation operators will act on the photonic states, and the interaction terms will reflect the specific entanglement and correlations among the three photons. For example, if the photons are entangled in polarization or momentum states, the corresponding interaction terms in the N-particle operator will capture these correlations.

5.5.3 ENTANGLED PHOTONS IN YOUNG'S TWO-SLIT EXPERIMENT

In Young's two-slit experiment, photons are emitted from a source, pass through two slits, and interfere on a detection screen. We assume the photons are polarized and can be in either horizontal ($|H\rangle$) or vertical ($|V\rangle$) polarization states. The goal is to measure the entanglement between photons based on their polarization states.

5.5.4 PHOTON STATES BEFORE THE SLITS

Initially, consider a single photon in a superposition state passing through both slits:

$$|\psi\rangle = \frac{1}{\sqrt{2}}(|s_1\rangle + |s_2\rangle), \tag{5.80}$$

where $|s_1\rangle$ and $|s_2\rangle$ represent the photon passing through slit 1 and slit 2, respectively.

Now, introduce polarization states. Assume the photon can have horizontal ($|H\rangle$) or vertical ($|V\rangle$) polarization. The state of the photon becomes:

$$|\psi\rangle = \frac{1}{\sqrt{2}}(|s_1\rangle \otimes |H\rangle + |s_2\rangle \otimes |V\rangle). \tag{5.81}$$

5.5.5 TWO-PHOTON ENTANGLED STATE

Consider two photons emitted from the source. The combined state of the photons can be written as:

$$|\Psi\rangle = \frac{1}{2}(|s_1\rangle \otimes |H\rangle_1 + |s_2\rangle \otimes |V\rangle_1) \otimes (|s_1\rangle \otimes |H\rangle_2 + |s_2\rangle \otimes |V\rangle_2). \tag{5.82}$$

Expanding this, we get:

$$\begin{aligned}|\Psi\rangle = \\ &\frac{1}{2}(|s_1\rangle_1 \otimes |H\rangle_1 \otimes |s_1\rangle_2 \otimes |H\rangle_2 + |s_1\rangle_1 \otimes |H\rangle_1 \otimes |s_2\rangle_2 \otimes |V\rangle_2 \\ &+ |s_2\rangle_1 \otimes |V\rangle_1 \otimes |s_1\rangle_2 \otimes |H\rangle_2 + |s_2\rangle_1 \otimes |V\rangle_1 \otimes |s_2\rangle_2 \otimes |V\rangle_2).\end{aligned} \tag{5.83}$$

5.5.6 TWO-PHOTON INTERFERENCE

When the photons interfere on the detection screen, the probability amplitude for detecting both photons at specific points on the screen depends on their polarization states and paths. The interference pattern is affected by the polarization correlation.

The state of the photons after passing through the slits can be written as:

$$\begin{aligned}|\Psi_{\text{screen}}\rangle = &\frac{1}{\sqrt{2}}(|D_1\rangle \otimes |H\rangle_1 + |D_2\rangle \\ &\otimes |V\rangle_1) \otimes \frac{1}{\sqrt{2}}(|D_1\rangle \otimes |H\rangle_2 + |D_2\rangle \otimes |V\rangle_2),\end{aligned} \tag{5.84}$$

where $|D_1\rangle$ and $|D_2\rangle$ represent the detection points corresponding to slit 1 and slit 2, respectively.

5.5.7 MEASUREMENT AND ENTANGLEMENT

To measure the entanglement, we can analyze the correlation between the polarization states of the photons detected at different points. The joint probability of detecting photon 1 with polarization $|H\rangle$ and photon 2 with polarization $|V\rangle$ is given by:

$$P(H_1, V_2) = |\langle H_1, V_2|\Psi_{\text{screen}}\rangle|^2. \tag{5.85}$$

Calculating this for the given state:

$$
\begin{aligned}
|\Psi_{\text{screen}}\rangle &= \frac{1}{2}(|D_1\rangle \otimes |H\rangle_1 + |D_2\rangle \otimes |V\rangle_1) \otimes (|D_1\rangle \otimes |H\rangle_2 \\
&+ |D_2\rangle \otimes |V\rangle_2) \quad (5.86)\\
&= \frac{1}{2}(|D_1\rangle \otimes |H\rangle_1 \otimes |D_1\rangle \otimes |H\rangle_2 \\
&+ |D_1\rangle \otimes |H\rangle_1 \otimes |D_2\rangle \otimes |V\rangle_2 \\
&+ |D_2\rangle \otimes |V\rangle_1 \otimes |D_1\rangle \otimes |H\rangle_2 \\
&+ |D_2\rangle \otimes |V\rangle_1 \otimes |D_2\rangle \otimes |V\rangle_2). \quad (5.87)
\end{aligned}
$$

The probability amplitude for detecting photon 1 with polarization $|H\rangle$ at $|D_1\rangle$ and photon 2 with polarization $|V\rangle$ at $|D_2\rangle$ is:

$$
\begin{aligned}
&\langle D_1, H_1| \otimes \langle D_2, V_2|\Psi_{\text{screen}}\rangle = \\
&\qquad \frac{1}{2}(\langle D_1|D_1\rangle \otimes \langle H_1|H_1\rangle \otimes \langle D_2|D_2\rangle \otimes \langle V_2|V_2\rangle). \quad (5.88)
\end{aligned}
$$

Given the orthonormality of the states:

$$
P(H_1, V_2) = \left|\frac{1}{2}\right|^2 = \frac{1}{4}. \quad (5.89)
$$

This shows that the probability of detecting the entangled photons at the given polarizations is non-zero, indicating the presence of entanglement.

In this example, we demonstrated how to use the second quantization formalism to describe the behavior of indistinguishable photons passing through Young's two-slits and how polarization states can be used to measure the entanglement between them. The entanglement arises from the superposition and interference of the polarization states, leading to measurable correlations that reflect the underlying quantum nature of the system.

5.5.8 ENTANGLEMENT IN YOUNG'S TWO-SLIT EXPERIMENT USING N-PARTICLE OPERATORS

To describe the entanglement of photons in Young's two-slit experiment using N-particle operators, we extend the formalism to multiple photons. We use creation and annihilation operators to represent the state of the system and the polarization of the photons.

5.5.8.0.1 State Representation with N-Particle Operators

Consider an ensemble of N indistinguishable photons, each of which can pass through either slit and can be in one of two polarization states, horizontal ($|H\rangle$) or vertical ($|V\rangle$).

The state of the system can be written using the N-particle creation operator:

$$
|\Psi\rangle = \frac{1}{\sqrt{2^N}} \left(\prod_{i=1}^{N} \hat{a}^\dagger_{s_1,H} + \prod_{i=1}^{N} \hat{a}^\dagger_{s_2,V} \right) |0\rangle, \quad (5.90)
$$

where $\hat{a}^\dagger_{s_1,H}$ and $\hat{a}^\dagger_{s_2,V}$ are the creation operators for photons passing through slit 1 with horizontal polarization and slit 2 with vertical polarization, respectively.

5.5.8.0.2 Two-Photon Case

For $N = 2$, the state can be explicitly written as:

$$|\Psi\rangle = \frac{1}{2}\left(\hat{a}^{\dagger}_{s_1,H}\hat{a}^{\dagger}_{s_1,H} + \hat{a}^{\dagger}_{s_1,H}\hat{a}^{\dagger}_{s_2,V} + \hat{a}^{\dagger}_{s_2,V}\hat{a}^{\dagger}_{s_1,H} + \hat{a}^{\dagger}_{s_2,V}\hat{a}^{\dagger}_{s_2,V}\right)|0\rangle. \tag{5.91}$$

5.5.8.0.3 Measurement and Entanglement

To measure the entanglement, we consider the joint detection probabilities. The state of the system after passing through the slits and reaching the detection screen is:

$$|\Psi_{\text{screen}}\rangle = \frac{1}{2}\left(\hat{a}^{\dagger}_{D_1,H}\hat{a}^{\dagger}_{D_1,H} + \hat{a}^{\dagger}_{D_1,H}\hat{a}^{\dagger}_{D_2,V} + \hat{a}^{\dagger}_{D_2,V}\hat{a}^{\dagger}_{D_1,H} + \hat{a}^{\dagger}_{D_2,V}\hat{a}^{\dagger}_{D_2,V}\right)|0\rangle. \tag{5.92}$$

The probability amplitude for detecting one photon with horizontal polarization at D_1 and another with vertical polarization at D_2 is given by:

$$\begin{aligned}\langle 0|\hat{a}_{D_1,H}\hat{a}_{D_2,V}|\Psi_{\text{screen}}\rangle &= \\ &\frac{1}{2}(\langle 0|\hat{a}_{D_1,H}\hat{a}_{D_2,V}(\hat{a}^{\dagger}_{D_1,H}\hat{a}^{\dagger}_{D_1,H} \\ &+ \hat{a}^{\dagger}_{D_1,H}\hat{a}^{\dagger}_{D_2,V} + \hat{a}^{\dagger}_{D_2,V}\hat{a}^{\dagger}_{D_1,H} + \hat{a}^{\dagger}_{D_2,V}\hat{a}^{\dagger}_{D_2,V}))|0\rangle.\end{aligned} \tag{5.93}$$

Using the commutation relations for bosons:

$$[\hat{a}_i, \hat{a}^{\dagger}_j] = \delta_{ij}, \tag{5.94}$$

we can simplify the probability amplitude:

$$\langle 0|\hat{a}_{D_1,H}\hat{a}_{D_2,V}\hat{a}^{\dagger}_{D_1,H}\hat{a}^{\dagger}_{D_2,V}|0\rangle = \langle 0|\delta_{D_1,D_1}\delta_{D_2,D_2}|0\rangle = 1. \tag{5.95}$$

The corresponding probability is:

$$P(H_1, V_2) = \left|\frac{1}{2}\right|^2 = \frac{1}{4}. \tag{5.96}$$

5.5.8.0.4 Generalization to N-Photon Case

For N photons, the state and measurements follow similarly, with the creation and annihilation operators extending to N-particle states. The general form of the state becomes:

$$|\Psi\rangle = \frac{1}{\sqrt{2^N}}\left(\prod_{i=1}^{N}\hat{a}^{\dagger}_{s_1,H} + \prod_{i=1}^{N}\hat{a}^{\dagger}_{s_2,V}\right)|0\rangle, \tag{5.97}$$

and the measurement probabilities can be calculated by extending the above approach using commutation relations and the structure of N-particle states.

By using the N-particle operators, we can effectively describe and analyze the entanglement properties of photons in Young's two-slit experiment. The formalism not only provides a clear framework for calculations but also highlights the fundamental quantum mechanical principles underlying photon behavior and entanglement.

5.5.9 SECOND-ORDER CORRELATION FUNCTION

The second-order correlation function, $g^{(2)}(\tau)$, provides a measure of the intensity correlations between two points in time or space. It is defined as:

$$g^{(2)}(\tau) = \frac{\langle \hat{E}^{(-)}(t)\hat{E}^{(-)}(t+\tau)\hat{E}^{(+)}(t+\tau)\hat{E}^{(+)}(t)\rangle}{\langle \hat{E}^{(-)}(t)\hat{E}^{(+)}(t)\rangle\langle \hat{E}^{(-)}(t+\tau)\hat{E}^{(+)}(t+\tau)\rangle}, \tag{5.98}$$

where $\hat{E}^{(+)}$ and $\hat{E}^{(-)}$ are the positive and negative frequency parts of the electric field operator, respectively.

At zero time delay ($\tau = 0$), the second-order correlation function is:

$$g^{(2)}(0) = \frac{\langle \hat{E}^{(-)}(t)\hat{E}^{(-)}(t)\hat{E}^{(+)}(t)\hat{E}^{(+)}(t)\rangle}{\langle \hat{E}^{(-)}(t)\hat{E}^{(+)}(t)\rangle^2}. \tag{5.99}$$

5.5.9.0.1 *Calculation of $g^{(2)}(0)$ for Photons Through Young's Two-Slit*

Consider photons passing through Young's two slits and interfering on a detection screen. The state of the photons can be written using creation and annihilation operators. Assume the electric field operators for the photons are given by:

$$\hat{E}^{(+)} = \hat{a}_1 + \hat{a}_2, \tag{5.100}$$

$$\hat{E}^{(-)} = \hat{a}_1^\dagger + \hat{a}_2^\dagger, \tag{5.101}$$

where $\hat{a}_1$ and $\hat{a}_2$ are the annihilation operators for photons passing through slit 1 and slit 2, respectively.

The numerator of $g^{(2)}(0)$ is:

$$\begin{aligned}
&\langle \hat{E}^{(-)}\hat{E}^{(-)}\hat{E}^{(+)}\hat{E}^{(+)}\rangle \\
&= \langle (\hat{a}_1^\dagger + \hat{a}_2^\dagger)(\hat{a}_1^\dagger + \hat{a}_2^\dagger)(\hat{a}_1 + \hat{a}_2)(\hat{a}_1 + \hat{a}_2)\rangle \\
&= \langle \hat{a}_1^\dagger\hat{a}_1^\dagger\hat{a}_1\hat{a}_1\rangle + \langle \hat{a}_1^\dagger\hat{a}_1^\dagger\hat{a}_2\hat{a}_2\rangle + \langle \hat{a}_2^\dagger\hat{a}_2^\dagger\hat{a}_1\hat{a}_1\rangle + \langle \hat{a}_2^\dagger\hat{a}_2^\dagger\hat{a}_2\hat{a}_2\rangle \\
&+ 2\langle \hat{a}_1^\dagger\hat{a}_1^\dagger\hat{a}_1\hat{a}_2\rangle + 2\langle \hat{a}_1^\dagger\hat{a}_2^\dagger\hat{a}_1\hat{a}_2\rangle + 2\langle \hat{a}_2^\dagger\hat{a}_2^\dagger\hat{a}_1\hat{a}_2\rangle \\
&+ 4\langle \hat{a}_1^\dagger\hat{a}_1\rangle\langle \hat{a}_2^\dagger\hat{a}_2\rangle.
\end{aligned} \tag{5.102}$$

For simplicity, assume the photons are in a coherent state, where the operators satisfy:

$$\langle \hat{a}_i^\dagger\hat{a}_j\rangle = \delta_{ij}\bar{n}, \tag{5.103}$$

$$\langle \hat{a}_i^\dagger\hat{a}_j^\dagger\hat{a}_k\hat{a}_l\rangle = \bar{n}^2(\delta_{ij}\delta_{kl} + \delta_{ik}\delta_{jl} + \delta_{il}\delta_{jk}). \tag{5.104}$$

Substituting these into the expression for the numerator:

$$\langle \hat{E}^{(-)}\hat{E}^{(-)}\hat{E}^{(+)}\hat{E}^{(+)}\rangle = \bar{n}^2(1+1+1+1) + 2\bar{n}^2(1+1) + 4\bar{n}^2 \tag{5.105}$$

$$= 8\bar{n}^2. \tag{5.106}$$

The denominator of $g^{(2)}(0)$ is:

$$\langle \hat{E}^{(-)}\hat{E}^{(+)}\rangle^2 = (2\bar{n})^2 = 4\bar{n}^2. \tag{5.107}$$

Thus, the second-order correlation function is:

$$g^{(2)}(0) = \frac{8\bar{n}^2}{4\bar{n}^2} = 2. \tag{5.108}$$

This result indicates photon bunching, which is characteristic of thermal light. The second-order correlation function $g^{(2)}(0) = 2$ shows that photons are more likely to be detected together than independently, illustrating the quantum statistical properties of the light passing through Young's two slits.

5.5.9.1 Second-Order Correlation Function in QFT

In Quantum Field Theory (QFT), the second-order correlation function, $g^{(2)}$, provides insight into the statistical properties of quantum fields, particularly the behavior of particles such as photons in a quantum state.

5.5.9.1.1 Field Operators and Correlations

In QFT, field operators $\hat{\psi}(\mathbf{r},t)$ and their conjugates $\hat{\psi}^\dagger(\mathbf{r},t)$ describe the creation and annihilation of particles at a point in space-time. Correlation functions measure the probability amplitudes of finding particles at different points in space and time, providing insight into the spatial and temporal coherence of the field.

5.5.9.1.2 Definition of $g^{(2)}$

The second-order correlation function is defined as:

$$g^{(2)}(\mathbf{r}_1,t_1;\mathbf{r}_2,t_2) = \frac{\langle \hat{\psi}^\dagger(\mathbf{r}_1,t_1)\hat{\psi}^\dagger(\mathbf{r}_2,t_2)\hat{\psi}(\mathbf{r}_2,t_2)\hat{\psi}(\mathbf{r}_1,t_1)\rangle}{\langle \hat{\psi}^\dagger(\mathbf{r}_1,t_1)\hat{\psi}(\mathbf{r}_1,t_1)\rangle\langle \hat{\psi}^\dagger(\mathbf{r}_2,t_2)\hat{\psi}(\mathbf{r}_2,t_2)\rangle}. \tag{5.109}$$

5.5.9.1.3 Interpretation of $g^{(2)}$ Values

- $g^{(2)} = 1$: Indicates a coherent state, such as a laser beam, where the field has Poissonian statistics.
- $g^{(2)} > 1$: Indicates bunching, where particles tend to cluster together. This is characteristic of thermal or chaotic light.
- $g^{(2)} < 1$: Indicates anti-bunching, where particles tend to avoid each other. This is a signature of quantum states with sub-Poissonian statistics, such as single-photon sources.

5.5.9.1.4 Example Calculation for Photons Through Slits

Consider photons passing through a double-slit setup. The second-order correlation function $g^{(2)}(\tau)$ at zero time delay ($\tau = 0$) can be calculated for the photon field.

- For indistinguishable photons, if the initial state is a coherent state, $g^{(2)}(0) = 1$.
- For thermal light, $g^{(2)}(0) = 2$.
- For a single-photon state, $g^{(2)}(0) = 0$.

In summary, the second-order correlation function $g^{(2)}$ in QFT provides crucial information about the statistical properties and coherence of quantum fields. It is a powerful tool for distinguishing between different types of quantum states and understanding the underlying particle interactions and distributions.

5.6 THE HUBBARD MODEL

The Hubbard model is a simplified representation of interacting electrons in a lattice. It is crucial for studying strongly correlated electron systems, where electron-electron interactions play a significant role. The model captures essential physics of phenomena such as magnetism, metal-insulator transitions, and high-temperature superconductivity. By focusing on the competition between kinetic energy and interaction energy, the Hubbard model provides insights into the collective behavior of electrons in materials.

5.6.1 HUBBARD HAMILTONIAN

The Hubbard Hamiltonian describes the behavior of electrons on a lattice and is given by:

$$\hat{H} = -t \sum_{\langle i,j \rangle,\sigma} \left(\hat{c}_{i\sigma}^{\dagger} \hat{c}_{j\sigma} + \hat{c}_{j\sigma}^{\dagger} \hat{c}_{i\sigma} \right) + U \sum_{i} \hat{n}_{i\uparrow} \hat{n}_{i\downarrow}, \tag{5.110}$$

where:

- $\hat{c}_{i\sigma}^{\dagger}$ and $\hat{c}_{i\sigma}$ are the creation and annihilation operators for an electron with spin σ (up or down) at site i.
- t is the hopping parameter, representing the kinetic energy associated with an electron moving from one site to a neighboring site.
- $\langle i, j \rangle$ denotes that the sum is over nearest-neighbor pairs.
- U is the on-site Coulomb repulsion energy, representing the potential energy due to two electrons occupying the same site.
- $\hat{n}_{i\sigma} = \hat{c}_{i\sigma}^{\dagger} \hat{c}_{i\sigma}$ is the number operator for electrons with spin σ at site i.

Hopping Term The hopping term $-t \sum_{\langle i,j \rangle,\sigma} \left(\hat{c}_{i\sigma}^{\dagger} \hat{c}_{j\sigma} + \hat{c}_{j\sigma}^{\dagger} \hat{c}_{i\sigma} \right)$ describes the kinetic energy of electrons moving between neighboring sites. This term favors the delocalization of electrons over the lattice.

Interaction Term The interaction term $U \sum_{i} \hat{n}_{i\uparrow} \hat{n}_{i\downarrow}$ describes the repulsive interaction energy when two electrons of opposite spin occupy the same site. This term favors localization of electrons to minimize double occupancy.

5.6.2 PHYSICAL IMPLICATIONS

The Hubbard model captures key physical phenomena observed in strongly correlated materials.

Metal-Insulator Transition At half-filling (one electron per site on average), the competition between the kinetic energy (hopping term) and the interaction energy (repulsion term) can lead to a metal-insulator transition. For small U, the kinetic energy dominates, and the system behaves as a metal. For large U, the interaction energy dominates, and the system becomes a Mott insulator, where electron localization due to strong repulsion prevents electrical conduction.

Mott Insulator

A Mott insulator is a type of insulator that arises due to strong electron-electron interactions, which prevent electrons from conducting despite an odd number of electrons per unit cell, which would typically imply metallic behavior. The concept is named after Sir Nevill Francis Mott, who first explained this phenomenon.

Mathematical Example of a Mott Insulator

Consider the one-dimensional Hubbard model at half-filling. The Hamiltonian is given by:

$$\hat{H} = -t \sum_{i,\sigma} \left(\hat{c}_{i\sigma}^{\dagger} \hat{c}_{i+1\sigma} + \hat{c}_{i+1\sigma}^{\dagger} \hat{c}_{i\sigma} \right) + U \sum_{i} \hat{n}_{i\uparrow} \hat{n}_{i\downarrow}, \tag{5.111}$$

where t is the hopping parameter, U is the on-site Coulomb repulsion, $\hat{c}_{i\sigma}^{\dagger}$ and $\hat{c}_{i\sigma}$ are the creation and annihilation operators for an electron with spin σ at site i, and $\hat{n}_{i\sigma} = \hat{c}_{i\sigma}^{\dagger} \hat{c}_{i\sigma}$ is the number operator.

Half-Filled Band Case At half-filling, the number of electrons is equal to the number of lattice sites. In the absence of interactions ($U = 0$), the system would be metallic. However, when U is large compared to t, the electrons experience strong repulsion, leading to an insulating state.

Strong Coupling Limit In the limit where $U \gg t$, the kinetic energy term (hopping) is much smaller than the interaction term (repulsion). In this regime, we can treat the hopping term as a perturbation to the interaction term.

To understand the insulating behavior, consider the energy cost of having two electrons on the same site. The on-site repulsion U creates an energy gap that prevents electrons from moving freely, leading to an insulating state. The effective Hamiltonian in the large U limit can be mapped onto the Heisenberg model, which describes spin interactions:

$$\hat{H}_{\text{eff}} = J \sum_{\langle i,j \rangle} \hat{\mathbf{S}}_i \cdot \hat{\mathbf{S}}_j, \tag{5.112}$$

where $J = \frac{4t^2}{U}$ is the exchange interaction, and $\hat{\mathbf{S}}_i$ is the spin operator at site i.

Energy Gap The Mott insulating state is characterized by an energy gap Δ for charge excitations. This gap can be approximated by the energy difference between the ground state (no double occupancy) and the state with one doubly occupied site:

$$\Delta \approx U - 2zt, \tag{5.113}$$

where z is the coordination number (number of nearest neighbors).

Worked Example: 1D Lattice

Consider a simple 1D lattice with $t = 1$ and $U = 10$. The effective exchange interaction J is given by:

$$J = \frac{4t^2}{U} = \frac{4 \times 1^2}{10} = 0.4. \tag{5.114}$$

The energy gap Δ can be approximated by:

$$\Delta \approx U - 2zt = 10 - 2 \times 1 \times 1 = 8. \tag{5.115}$$

Thus, the system exhibits an energy gap of 8 units, indicating insulating behavior due to strong electron-electron interactions.

In summary, a Mott insulator is an insulating state driven by strong interactions rather than band theory. The energy gap arises from the repulsion between electrons, which prevents them from conducting electricity, even in a half-filled band.

Magnetism In the limit of large U, the Hubbard model can be mapped onto the Heisenberg model, which describes magnetic interactions. The effective Hamiltonian for large U is given by:

$$\hat{H}_{\text{eff}} = J \sum_{\langle i,j \rangle} \hat{\mathbf{S}}_i \cdot \hat{\mathbf{S}}_j, \tag{5.116}$$

where $J = \frac{4t^2}{U}$ is the exchange interaction and $\hat{\mathbf{S}}_i$ is the spin operator at site i. This effective Hamiltonian shows that the Hubbard model can describe antiferromagnetic order in materials.

Superconductivity The Hubbard model is also used to study high-temperature superconductivity. In particular, the two-dimensional Hubbard model on a square lattice is relevant for understanding the cuprate superconductors. Pairing mechanisms and the formation of Cooper pairs can be explored within this framework, although a complete theoretical description of high-temperature superconductivity remains an open question.

5.6.2.1 Example: Half-Filled One-Dimensional Hubbard Model

Consider a one-dimensional Hubbard model at half-filling ($n = 1$). In this case, the Hamiltonian is:

$$\hat{H}_{1D} = -t\sum_{i,\sigma}\left(\hat{c}^{\dagger}_{i\sigma}\hat{c}_{i+1\sigma} + \hat{c}^{\dagger}_{i+1\sigma}\hat{c}_{i\sigma}\right) + U\sum_{i}\hat{n}_{i\uparrow}\hat{n}_{i\downarrow}. \tag{5.117}$$

For $U = 0$, the system is a metal with a gapless excitation spectrum. For $U \neq 0$, the system can undergo a transition to a Mott insulator with a charge gap.

From the above example we can now see that the Hubbard model serves as a powerful theoretical tool for studying a wide range of phenomena in strongly correlated electron systems. By capturing the essential competition between kinetic and interaction energies, it provides valuable insights into the emergent properties of materials.

5.7 SUMMARY AND DISCUSSION

In this chapter, we have explored several advanced concepts in quantum mechanics and condensed matter physics, focusing on the formalism of second quantization, the structure and applications of Fock space, the tight binding Hamiltonian, multi-particle operators, and the Hubbard model.

5.7.0.1 Second Quantization

We began by reviewing the principles of first quantization, emphasizing wavefunctions and operators. Transitioning to second quantization, we introduced creation and annihilation operators, which are fundamental for describing systems with variable particle numbers. These operators facilitate the formulation of Hamiltonians for many-body systems, allowing for a more comprehensive analysis of quantum dynamics.

5.7.0.2 Fock Space

We defined Fock space as the Hilbert space that encompasses all possible quantum states of a system with varying particle numbers. We discussed the construction of basis states using creation operators and the significance of the occupation number representation. By employing the direct sum of subspaces corresponding to different particle numbers, Fock space provides a versatile framework for studying quantum systems.

5.7.0.3 Tight Binding Hamiltonian

We derived the tight binding Hamiltonian for a simple lattice, highlighting its utility in condensed matter physics. This model simplifies the description of electrons in a lattice by considering only nearest-neighbor interactions. We applied the tight binding model to various physical systems, including the calculation of electronic band structures, the unique properties of graphene, and the edge states in topological insulators. These applications underscore the model's relevance to understanding the electronic properties of materials.

5.7.0.4 Two and N Particle Operators

We extended the formalism of second quantization to describe interactions between multiple particles using two-particle and N-particle operators. These operators are crucial for modeling and analyzing collective behavior in many-body systems. Through examples, we demonstrated how to construct and use these operators to describe complex interactions, such as three-particle interactions.

5.7.0.5 Hubbard Model

We introduced the Hubbard model, emphasizing its importance in studying strongly correlated electron systems. The Hubbard Hamiltonian captures the competition between kinetic energy and electron-electron interactions, providing insights into phenomena such as metal-insulator transitions, magnetism, and superconductivity. We discussed the physical implications of the Hubbard model, including the description of Mott insulators and the effective spin interactions in the strong coupling limit.

5.7.0.6 Discussion

The formalism and models presented in this chapter are foundational for understanding the behavior of many-body quantum systems. Second quantization and Fock space offer powerful tools for describing quantum systems with varying particle numbers, facilitating the study of complex interactions and emergent phenomena. The tight binding model and the Hubbard model provide essential frameworks for analyzing the electronic properties of materials, from simple lattices to strongly correlated systems.

The concepts explored in this chapter have wide-ranging applications in condensed matter physics, quantum information science, and beyond. By mastering these tools, you are equipped to tackle advanced topics in quantum mechanics and to contribute to the ongoing research in understanding the fundamental properties of quantum materials.

In the following chapters, we will build on these concepts to delve deeper into specific applications and explore more complex systems, further enriching our understanding of the quantum world.

5.8 QUESTIONS AND PROBLEMS

Provide a set of problems and exercises related to the topics covered in this chapter to reinforce understanding and application of the concepts.

5.8.1 TRIVIA QUESTIONS

Question 1: What does the creation operator $\hat{a}^\dagger$ do in second quantization?

(a) Removes a particle from a state
(b) Adds a particle to a state
(c) Measures the number of particles in a state
(d) Commutes with the annihilation operator

Answer: b. Adds a particle to a state
Question 2: Which of the following represents the Fock space?

(a) $\mathscr{H} = \bigoplus_{n=0}^{\infty} \mathscr{H}^{(n)}$
(b) $\mathscr{H} = \bigotimes_{n=0}^{\infty} \mathscr{H}^{(n)}$
(c) $\mathscr{H} = \sum_{n=0}^{\infty} \mathscr{H}^{(n)}$
(d) $\mathscr{H} = \int_0^{\infty} \mathscr{H}^{(n)}\, dn$

Answer: a. $\mathscr{H} = \bigoplus_{n=0}^{\infty} \mathscr{H}^{(n)}$
Question 3: In the tight binding model, what does the hopping parameter t represent?

(a) The on-site energy of electrons
(b) The interaction energy between electrons
(c) The amplitude for an electron to hop between neighboring sites

(d) The energy gap in the electronic band structure

Answer: c. The amplitude for an electron to hop between neighboring sites
Question 4: What is the significance of the term U in the Hubbard model?

(a) It represents the kinetic energy of electrons
(b) It represents the on-site Coulomb repulsion energy
(c) It describes the hopping of electrons between sites
(d) It is the total energy of the system

Answer: b. It represents the on-site Coulomb repulsion energy
Question 5: In second quantization, the field operator $\hat{\psi}(x)$ can be expressed as:

(a) $\hat{\psi}(x) = \sum_i \phi_i(x)\hat{a}_i$
(b) $\hat{\psi}(x) = \prod_i \phi_i(x)\hat{a}_i$
(c) $\hat{\psi}(x) = \int \phi_i(x)\hat{a}_i\, dx$
(d) $\hat{\psi}(x) = \frac{1}{N}\sum_i \phi_i(x)\hat{a}_i$

Answer: a. $\hat{\psi}(x) = \sum_i \phi_i(x)\hat{a}_i$
Question 6: What does the occupation number representation describe in Fock space?

(a) The spatial position of particles
(b) The number of particles in each mode
(c) The energy of each particle
(d) The spin state of particles

Answer: b. The number of particles in each mode
Question 7: Which equation represents the time-independent Schrödinger equation?

(a) $i\hbar\frac{\partial\psi(\vec{r},t)}{\partial t} = \hat{H}\psi(\vec{r},t)$
(b) $\hat{H}\phi(\vec{r}) = E\phi(\vec{r})$
(c) $\hat{p} = -i\hbar\nabla$
(d) $\hat{a}_k = \frac{1}{\sqrt{N}}\sum_n \hat{c}_n e^{-ikna}$

Answer: b. $\hat{H}\phi(\vec{r}) = E\phi(\vec{r})$
Question 8: In the context of the Hubbard model, what physical phenomenon does the parameter t help to describe?

(a) Superconductivity
(b) Ferromagnetism
(c) Electron hopping between sites
(d) Dielectric constant

Answer: c. Electron hopping between sites
Question 9: What is the direct sum notation $\oplus$ used for in Fock space?

(a) Combining different energy levels
(b) Summing over all possible spin states
(c) Representing the total Fock space as a sum of subspaces
(d) Describing the Hamiltonian of a many-body system

Answer: c. Representing the total Fock space as a sum of subspaces
Question 10: What is the purpose of using creation and annihilation operators in second quantization?

(a) To describe fixed particle numbers in a system
(b) To allow the number of particles to vary in a quantum system
(c) To measure the energy levels of particles
(d) To calculate the momentum of particles

Answer: b. To allow the number of particles to vary in a quantum system

5.8.2 PROBLEMS

Problem 1: Creation and Annihilation Operators

Consider a single mode of a quantum harmonic oscillator. The creation ($\hat{a}^\dagger$) and annihilation ($\hat{a}$) operators satisfy the commutation relation:

$$[\hat{a}, \hat{a}^\dagger] = 1$$

Given the ground state $|0\rangle$ such that $\hat{a}|0\rangle = 0$, find the normalized first excited state $|1\rangle$.

Solution:

The first excited state $|1\rangle$ is obtained by applying the creation operator $\hat{a}^\dagger$ to the ground state $|0\rangle$:

$$|1\rangle = \hat{a}^\dagger |0\rangle \tag{5.118}$$

To normalize $|1\rangle$, we calculate its norm:

$$\langle 1|1\rangle = \langle 0|\hat{a}\hat{a}^\dagger|0\rangle \tag{5.119}$$

Using the commutation relation:

$$\hat{a}\hat{a}^\dagger = 1 + \hat{a}^\dagger\hat{a} \tag{5.120}$$

Therefore,

$$\langle 0|\hat{a}\hat{a}^\dagger|0\rangle = \langle 0|1 + \hat{a}^\dagger\hat{a}|0\rangle = \langle 0|0\rangle + \langle 0|\hat{a}^\dagger\hat{a}|0\rangle = 1 \tag{5.121}$$

Hence, $|1\rangle$ is already normalized:

$$|1\rangle = \hat{a}^\dagger |0\rangle \tag{5.122}$$

Problem 2: Fock Space and Direct Sum

Define the Fock space as the direct sum of n-particle Hilbert spaces. Given $\mathscr{H}^{(0)}$ as the vacuum state, $\mathscr{H}^{(1)}$ as the single-particle state, and $\mathscr{H}^{(2)}$ as the two-particle state, write a state in the Fock space combining these subspaces.

Solution:

The Fock space $\mathscr{F}$ is the direct sum of n-particle Hilbert spaces:

$$\mathscr{F} = \bigoplus_{n=0}^{\infty} \mathscr{H}^{(n)} \tag{5.123}$$

Let's construct a state combining the vacuum state $|0\rangle$, a single-particle state $|1\rangle$, and a two-particle state $|2\rangle$:

$$\psi_0 = |0\rangle \in \mathscr{H}^{(0)} \tag{5.124}$$

$$\psi_1 = \alpha|1_1\rangle + \beta|1_2\rangle \in \mathscr{H}^{(1)} \tag{5.125}$$

$$\psi_2 = \gamma|2_1\rangle + \delta|2_2\rangle \in \mathscr{H}^{(2)} \tag{5.126}$$

The combined state in Fock space is:

$$\Psi = \psi_0 \oplus \psi_1 \oplus \psi_2 = |0\rangle \oplus (\alpha|1_1\rangle + \beta|1_2\rangle) \oplus (\gamma|2_1\rangle + \delta|2_2\rangle) \tag{5.127}$$

Problem 3: Tight Binding Hamiltonian

Derive the energy dispersion relation for electrons in a one-dimensional tight binding model with nearest-neighbor hopping.

Solution:

The tight binding Hamiltonian in one dimension is:

$$\hat{H} = -t\sum_{n}\left(\hat{c}_{n+1}^{\dagger}\hat{c}_{n} + \hat{c}_{n}^{\dagger}\hat{c}_{n+1}\right) \tag{5.128}$$

We use the Fourier transform of the annihilation operator:

$$\hat{c}_k = \frac{1}{\sqrt{N}}\sum_{n}\hat{c}_n e^{-ikna} \tag{5.129}$$

Substituting this into the Hamiltonian:

$$\hat{H} = -t\sum_{k}\left(\frac{1}{\sqrt{N}}\sum_{n}\hat{c}_n e^{-ik(n+1)a}\right)^{\dagger}\left(\frac{1}{\sqrt{N}}\sum_{m}\hat{c}_m e^{-ikma}\right) + \text{h.c.} \tag{5.130}$$

Simplifying and recognizing the Kronecker delta function:

$$\hat{H} = -t\sum_{k}\left(e^{-ika}\hat{c}_k^{\dagger}\hat{c}_k + e^{ika}\hat{c}_k^{\dagger}\hat{c}_k\right) \tag{5.131}$$

Thus, the energy dispersion relation is:

$$E(k) = -2t\cos(ka) \tag{5.132}$$

Problem 4: Two-Particle Interaction

Consider a two-particle system with a Coulomb interaction potential $V(\mathbf{r}_1,\mathbf{r}_2) = \frac{e^2}{|\mathbf{r}_1-\mathbf{r}_2|}$. Write the second quantized form of the interaction Hamiltonian.

Solution:

The second quantized form of the two-particle interaction Hamiltonian is:

$$\hat{V} = \frac{1}{2}\sum_{ijkl}\langle ij|V|kl\rangle \hat{a}_i^{\dagger}\hat{a}_j^{\dagger}\hat{a}_k\hat{a}_l \tag{5.133}$$

For the Coulomb interaction:

$$V(\mathbf{r}_1,\mathbf{r}_2) = \frac{e^2}{|\mathbf{r}_1-\mathbf{r}_2|} \tag{5.134}$$

The matrix elements in a plane wave basis are:

$$\langle \mathbf{k}_1\mathbf{k}_2|V|\mathbf{k}_3\mathbf{k}_4\rangle = \frac{e^2}{V}\delta_{\mathbf{k}_1+\mathbf{k}_2,\mathbf{k}_3+\mathbf{k}_4}\frac{4\pi}{|\mathbf{k}_1-\mathbf{k}_3|^2} \tag{5.135}$$

Thus, the interaction Hamiltonian in second quantization is:

$$\hat{V} = \frac{1}{2}\sum_{\mathbf{k}_1\mathbf{k}_2\mathbf{k}_3\mathbf{k}_4}\frac{e^2}{V}\delta_{\mathbf{k}_1+\mathbf{k}_2,\mathbf{k}_3+\mathbf{k}_4}\frac{4\pi}{|\mathbf{k}_1-\mathbf{k}_3|^2}\hat{a}_{\mathbf{k}_1}^{\dagger}\hat{a}_{\mathbf{k}_2}^{\dagger}\hat{a}_{\mathbf{k}_3}\hat{a}_{\mathbf{k}_4} \tag{5.136}$$

Problem 5: Hubbard Model

Consider the Hubbard model on a one-dimensional lattice at half-filling with $t = 1$ and $U = 4$. Determine the effective Hamiltonian in the large U limit.

Solution:

The Hubbard Hamiltonian is:

$$\hat{H} = -t\sum_{i,\sigma}\left(\hat{c}_{i\sigma}^{\dagger}\hat{c}_{i+1\sigma} + \hat{c}_{i+1\sigma}^{\dagger}\hat{c}_{i\sigma}\right) + U\sum_{i}\hat{n}_{i\uparrow}\hat{n}_{i\downarrow} \tag{5.137}$$

In the large U limit ($U \gg t$), we treat the hopping term as a perturbation to the interaction term. The effective Hamiltonian describes spin interactions in the half-filled case. It is given by the Heisenberg model:

$$\hat{H}_{\text{eff}} = J\sum_{\langle i,j\rangle}\hat{\mathbf{S}}_i \cdot \hat{\mathbf{S}}_j \tag{5.138}$$

where $J = \frac{4t^2}{U}$ is the exchange interaction.

For $t = 1$ and $U = 4$:

$$J = \frac{4 \times 1^2}{4} = 1 \tag{5.139}$$

Thus, the effective Hamiltonian is:

$$\hat{H}_{\text{eff}} = \sum_{\langle i,j\rangle}\hat{\mathbf{S}}_i \cdot \hat{\mathbf{S}}_j \tag{5.140}$$

BONUS PROBLEM: HANBURY BROWN-TWISS INTERFEROMETER

The Hanbury Brown-Twiss (HBT) interferometer is used to measure the statistical properties of photons, particularly their correlations. The HBT interferometer measures the second-order correlation function to distinguish between coherent and thermal states of light. Using the second quantization formalism, we can describe how different quantum states of light exhibit different correlation properties, which are detected by the HBT interferometer. Describe the HBT effect using the formalism of second quantization and N-particle operators. Show how the second-order correlation function $g^{(2)}(0)$ is calculated for a coherent state and a thermal state of light.

Solution:

In second quantization, the state of the electromagnetic field is described using creation and annihilation operators. The HBT interferometer measures the intensity correlations of photons, which can be analyzed using the second-order correlation function $g^{(2)}(0)$.

5.8.2.0.1 Second-Order Correlation Function

The second-order correlation function is defined as:

$$g^{(2)}(\tau) = \frac{\langle \hat{E}^{(-)}(t)\hat{E}^{(-)}(t+\tau)\hat{E}^{(+)}(t+\tau)\hat{E}^{(+)}(t)\rangle}{\langle \hat{E}^{(-)}(t)\hat{E}^{(+)}(t)\rangle\langle \hat{E}^{(-)}(t+\tau)\hat{E}^{(+)}(t+\tau)\rangle}, \tag{5.141}$$

where $\hat{E}^{(+)}$ and $\hat{E}^{(-)}$ are the positive and negative frequency parts of the electric field operator, respectively.

At zero time delay ($\tau = 0$):

$$g^{(2)}(0) = \frac{\langle \hat{E}^{(-)}(t)\hat{E}^{(-)}(t)\hat{E}^{(+)}(t)\hat{E}^{(+)}(t)\rangle}{\langle \hat{E}^{(-)}(t)\hat{E}^{(+)}(t)\rangle^2}. \tag{5.142}$$

5.8.2.0.2 Field Operators in Terms of Creation and Annihilation Operators

The positive and negative frequency parts of the electric field operator can be written as:

$$\hat{E}^{(+)}(t) = \sum_k \mathscr{E}_k \hat{a}_k e^{-i\omega_k t}, \tag{5.143}$$

$$\hat{E}^{(-)}(t) = \sum_k \mathscr{E}_k^* \hat{a}_k^\dagger e^{i\omega_k t}, \tag{5.144}$$

where $\mathscr{E}_k$ is the amplitude of the mode k and $\hat{a}_k^\dagger$ and $\hat{a}_k$ are the creation and annihilation operators.

5.8.2.0.3 Coherent State

A coherent state $|\alpha\rangle$ is an eigenstate of the annihilation operator $\hat{a}$:

$$\hat{a}\,|\alpha\rangle = \alpha\,|\alpha\rangle\,. \tag{5.145}$$

For a coherent state, the second-order correlation function $g^{(2)}(0)$ is:

$$g^{(2)}(0) = \frac{\langle\alpha|\hat{a}^\dagger\hat{a}^\dagger\hat{a}\hat{a}|\alpha\rangle}{\langle\alpha|\hat{a}^\dagger\hat{a}|\alpha\rangle^2} = \frac{|\alpha|^4}{|\alpha|^4} = 1. \tag{5.146}$$

This result shows that for a coherent state, $g^{(2)}(0) = 1$, indicating no photon bunching or antibunching.

5.8.2.0.4 Thermal State

A thermal state ρ_{th} is described by a density matrix with a diagonal form in the number basis:

$$\rho_{\text{th}} = \sum_n \frac{\bar{n}^n}{(\bar{n}+1)^{n+1}}\,|n\rangle\,\langle n|\,, \tag{5.147}$$

where $\bar{n}$ is the average photon number.

For a thermal state, the second-order correlation function $g^{(2)}(0)$ is:

$$g^{(2)}(0) = \frac{\langle\hat{a}^\dagger\hat{a}^\dagger\hat{a}\hat{a}\rangle_{\text{th}}}{\langle\hat{a}^\dagger\hat{a}\rangle_{\text{th}}^2}. \tag{5.148}$$

Using the properties of the thermal state:

$$\langle\hat{a}^\dagger\hat{a}\rangle_{\text{th}} = \bar{n}, \tag{5.149}$$

$$\langle\hat{a}^\dagger\hat{a}^\dagger\hat{a}\hat{a}\rangle_{\text{th}} = \bar{n}^2 + \bar{n}. \tag{5.150}$$

Thus,

$$g^{(2)}(0) = \frac{\bar{n}^2+\bar{n}}{\bar{n}^2} = 1 + \frac{1}{\bar{n}}. \tag{5.151}$$

For a thermal state, $g^{(2)}(0) > 1$, indicating photon bunching.

5.8.2.0.5 Hanbury Brown-Twiss Effect

The N-particle operators and second quantization provide a powerful framework for analyzing the statistical properties of light in experiments like the HBT interferometer, revealing the underlying quantum nature of photon correlations.

6 Lectures 10 & 11: Relativistic Quantum Mechanics, the Klein-Gordon Equation, Contour Integrals, the Scalar Feynman Propagator, and Green's Functions

6.1 INTRODUCTION

In modern physics, the fusion of quantum mechanics and special relativity has created a very profound understanding of the microscopic world. This chapter delves into the intricacies of relativistic quantum mechanics, focusing on key concepts and equations that form the foundation of this advanced theoretical framework. We begin with the motivation for extending quantum mechanics to accommodate relativistic principles, addressing the need for equations that remain invariant under Lorentz transformations.

We will explore the Klein-Gordon equation, the cornerstone of relativistic quantum mechanics for spin-0 particles. We will derive this equation from the relativistic energy-momentum relationship and examine its properties, solutions, and physical interpretations. Despite its limitations, the Klein-Gordon equation provides a crucial stepping stone toward a fully relativistic quantum theory.

Following this, we will review the mathematical techniques of contour integrals, an essential tool in the arsenal of theoretical physicists. A brief review of complex analysis sets the stage for understanding how contour integration is employed in quantum mechanics, particularly in the evaluation of propagators and Green's functions.

An in-depth discussion of the scalar Feynman propagator, a fundamental concept in quantum field theory (QFT) is then covered. We will derive the Feynman propagator using contour integration and connect it to the Klein-Gordon equation. By exploring its properties and applications, we demonstrate how the scalar Feynman propagator facilitates the calculation of particle interactions and propagations in QFT.

We then introduce Green's functions, a powerful method for solving inhomogeneous differential equations that arise in many areas of physics, including quantum field theory. We will explore their definition, properties, and how they are used to describe the response of a system to external forces.

Throughout this chapter, we highlight the interplay between these foundational elements and their collective significance in quantum field theory. By understanding relativistic quantum mechanics, contour integrals, the Feynman propagator, and Green's functions, we lay the groundwork for exploring the rich and complex landscape of QFT, where particles are created and annihilated, and interactions are mediated by quantum fields.

DOI: 10.1201/9781003637370-6

6.2 RELATIVISTIC QUANTUM MECHANICS

6.2.1 MOTIVATION AND HISTORICAL BACKGROUND

The unification of quantum mechanics with special relativity led to the development of relativistic quantum mechanics, a crucial step toward the formulation of quantum field theory (QFT). Classical mechanics failed to describe phenomena at high velocities and small scales, while non-relativistic quantum mechanics could not account for the effects of special relativity. The need to reconcile these two fundamental theories became apparent with the discovery of the electron's spin, the prediction of antiparticles, and the explanation of fine structure in atomic spectra.

The Klein-Gordon and Dirac equations were pioneering efforts in this direction, aiming to describe particles with spin-0 and spin-$\frac{1}{2}$, respectively. These equations paved the way for understanding particle interactions at relativistic speeds and laid the groundwork for quantum electrodynamics (QED) and QFT.

6.2.2 THE DIRAC EQUATION

The Dirac equation is a relativistic wave equation that provides a description of spin-$\frac{1}{2}$ particles, such as electrons, in the framework of quantum mechanics. It was formulated by Paul Dirac in 1928 and successfully unified quantum mechanics with special relativity, addressing the limitations of the Klein-Gordon equation for spin-$\frac{1}{2}$ particles.

6.2.2.1 Derivation of the Dirac Equation

The starting point for deriving the Dirac equation is the relativistic energy-momentum relation:

$$E^2 = p^2c^2 + m^2c^4. \tag{6.1}$$

Dirac proposed a linearization of this equation by introducing matrices α_i and β such that:

$$E = \alpha \cdot pc + \beta mc^2. \tag{6.2}$$

By replacing the energy and momentum with their corresponding quantum mechanical operators, we get the Dirac equation in the form:

$$i\hbar\frac{\partial \psi}{\partial t} = \left(-i\hbar c\alpha \cdot \nabla + \beta mc^2\right)\psi, \tag{6.3}$$

where ψ is a four-component spinor, and α and β are 4×4 matrices satisfying the following anti-commutation relations:

$$\begin{aligned} \alpha_i\alpha_j + \alpha_j\alpha_i &= 2\delta_{ij}I, \\ \alpha_i\beta + \beta\alpha_i &= 0, \\ \beta^2 &= I. \end{aligned} \tag{6.4}$$

6.2.2.2 Properties of the Dirac matrices

The Dirac matrices α and β are constructed to ensure the Lorentz invariance of the equation. They can be represented in terms of the Pauli matrices σ_i and the identity matrix I as follows:

$$\alpha_i = \begin{pmatrix} 0 & \sigma_i \\ \sigma_i & 0 \end{pmatrix}, \quad \beta = \begin{pmatrix} I & 0 \\ 0 & -I \end{pmatrix}. \tag{6.5}$$

These matrices satisfy the required algebraic properties and ensure the correct relativistic behavior of the spinor field ψ.

6.2.2.3 Plane Wave Solutions

To find the solutions of the Dirac equation, consider a plane wave of the form:

$$\psi(\mathbf{r},t) = u(\mathbf{p})e^{i(\mathbf{p}\cdot\mathbf{r}-Et)/\hbar}, \tag{6.6}$$

where $u(\mathbf{p})$ is a four-component spinor. Substituting this into the Dirac equation gives the following algebraic equation:

$$(E - \alpha\cdot pc - \beta mc^2)u(\mathbf{p}) = 0. \tag{6.7}$$

This equation has nontrivial solutions if the determinant of the matrix is zero, leading to the energy eigenvalues:

$$E = \pm\sqrt{p^2c^2 + m^2c^4}, \tag{6.8}$$

corresponding to positive and negative energy solutions. The negative energy solutions are interpreted as antiparticles, such as positrons.

6.2.2.4 Importance in quantum field theory

The Dirac equation's introduction of spin and antiparticles was a significant advancement in quantum mechanics and provided a natural framework for the creation and annihilation of particles, which are fundamental concepts in quantum field theory (QFT). In QFT, the Dirac field is quantized, leading to the development of quantum electrodynamics (QED) and other gauge theories. The Dirac equation's treatment of spin-$\frac{1}{2}$ particles and their interactions with electromagnetic fields underpins much of modern particle physics and QFT.

6.2.3 LORENTZ TRANSFORMATIONS AND INVARIANCE

A cornerstone of special relativity is the Lorentz transformation, which relates the coordinates of events as observed in different inertial frames. For a boost in the x-direction, the Lorentz transformation is given by:

$$\begin{aligned} t' &= \gamma(t - \frac{vx}{c^2}), \\ x' &= \gamma(x - vt), \\ y' &= y, \\ z' &= z, \end{aligned} \tag{6.9}$$

where $\gamma = \frac{1}{\sqrt{1-\frac{v^2}{c^2}}}$ is the Lorentz factor and v is the relative velocity between the two frames.

In relativistic quantum mechanics, the invariance of physical laws under Lorentz transformations is a fundamental requirement. This invariance ensures that the equations describing quantum states and observables are consistent across all inertial frames. For a scalar field $\phi(x)$, this requirement translates to:

$$\phi'(x') = \phi(x). \tag{6.10}$$

The Klein-Gordon equation, for instance, must remain invariant under Lorentz transformations. The Klein-Gordon equation is derived from the relativistic energy-momentum relation:

$$E^2 = p^2c^2 + m^2c^4. \tag{6.11}$$

Replacing the energy and momentum with their quantum mechanical operators, $E \to i\hbar\frac{\partial}{\partial t}$ and $\mathbf{p} \to -i\hbar\nabla$, we obtain:

$$\left(\frac{1}{c^2}\frac{\partial^2}{\partial t^2} - \nabla^2 + \frac{m^2c^2}{\hbar^2}\right)\phi(x) = 0, \tag{6.12}$$

where $\Box = \frac{1}{c^2}\frac{\partial^2}{\partial t^2} - \nabla^2$ is the d'Alembertian operator. This equation is manifestly Lorentz invariant, making it a suitable candidate for describing relativistic quantum fields.

The concept of Lorentz invariance extends to the formulation of Lagrangians in QFT, ensuring that the action remains invariant under Lorentz transformations. This invariance is crucial for the consistency and predictability of physical theories in QFT.

Overall, Lorentz transformations and the requirement of invariance form the bedrock of relativistic quantum mechanics, ensuring that the laws of physics are the same in all inertial frames. This principle is fundamental to the development of QFT, where fields and their interactions must adhere to relativistic constraints to accurately describe the behavior of particles at high energies.

6.2.3.1 Lorentz boost

A Lorentz boost is a transformation in special relativity that changes the velocity of an inertial reference frame along a particular direction without altering the orientation of the spatial axes. It is used to describe how physical quantities, such as time, position, and spinor fields, transform when moving from one inertial frame to another at a constant relative velocity

WORKED EXAMPLE: LORENTZ TRANSFORMATION IN QFT

To illustrate a Lorentz transformation relevant to quantum field theory, consider a Lorentz boost along the x-axis with velocity v. The Lorentz boost matrix Λ in the x-direction is given by:

$$\Lambda = \begin{pmatrix} \gamma & -\gamma\beta & 0 & 0 \\ -\gamma\beta & \gamma & 0 & 0 \\ 0 & 0 & 1 & 0 \\ 0 & 0 & 0 & 1 \end{pmatrix}, \tag{6.13}$$

where $\beta = \frac{v}{c}$ and $\gamma = \frac{1}{\sqrt{1-\beta^2}}$ is the Lorentz factor.

Consider a four-vector $x^\mu = (ct, x, y, z)$. Under the Lorentz transformation, the transformed coordinates x'^μ are given by:

$$x'^\mu = \Lambda^\mu{}_\nu x^\nu. \tag{6.14}$$

Applying the Lorentz boost to the four-vector x^μ:

$$\begin{pmatrix} ct' \\ x' \\ y' \\ z' \end{pmatrix} = \begin{pmatrix} \gamma & -\gamma\beta & 0 & 0 \\ -\gamma\beta & \gamma & 0 & 0 \\ 0 & 0 & 1 & 0 \\ 0 & 0 & 0 & 1 \end{pmatrix} \begin{pmatrix} ct \\ x \\ y \\ z \end{pmatrix}. \tag{6.15}$$

Performing the matrix multiplication, we get:

$$ct' = \gamma(ct - \beta x), \tag{6.16}$$

$$x' = \gamma(x - \beta ct), \tag{6.17}$$

$$y' = y, \tag{6.18}$$

$$z' = z. \tag{6.19}$$

This example shows how the time and space coordinates transform under a Lorentz boost. In the context of QFT, Lorentz transformations are crucial for ensuring that the equations governing the fields remain invariant under changes in the reference frame, preserving the principle of relativity.

6.2.3.2 Worked Example: Lorentz Transformation of a Dirac spinor

In quantum field theory, understanding how fields transform under Lorentz transformations is crucial. Here, we provide an example of how a Dirac spinor transforms under a Lorentz boost.

Consider a Lorentz boost in the x-direction. The Lorentz transformation matrix for a boost with velocity v is given by:

$$\Lambda^{\mu}_{\ \nu} = \begin{pmatrix} \gamma & -\gamma\beta & 0 & 0 \\ -\gamma\beta & \gamma & 0 & 0 \\ 0 & 0 & 1 & 0 \\ 0 & 0 & 0 & 1 \end{pmatrix}, \tag{6.20}$$

where $\beta = \frac{v}{c}$ and $\gamma = \frac{1}{\sqrt{1-\beta^2}}$ is the Lorentz factor.

The Dirac spinor ψ transforms under this boost according to:

$$\psi'(x') = S(\Lambda)\psi(x), \tag{6.21}$$

where $S(\Lambda)$ is the spinor representation of the Lorentz transformation. For a boost in the x-direction, $S(\Lambda)$ is given by:

$$S(\Lambda) = e^{-\frac{i}{2}\omega_{\mu\nu}\Sigma^{\mu\nu}}, \tag{6.22}$$

where $\omega_{\mu\nu}$ are the parameters of the Lorentz transformation and $\Sigma^{\mu\nu}$ are the generators of the Lorentz group in the spinor representation. Specifically, for a boost in the x-direction:

$$\omega^{01} = -\omega^{10} = \eta, \quad \text{and all other} \quad \omega^{\mu\nu} = 0, \tag{6.23}$$

where η is the rapidity, related to the velocity by $\tanh\eta = \beta$.

The generators $\Sigma^{\mu\nu}$ for boosts are given by:

$$\Sigma^{01} = \frac{i}{2}\begin{pmatrix} 0 & \sigma_x \\ \sigma_x & 0 \end{pmatrix}, \tag{6.24}$$

where σ_x is the Pauli matrix.

For a boost in the x-direction, the spinor transformation matrix becomes:

$$S(\Lambda) = e^{-\eta\Sigma^{01}} = \cosh\left(\frac{\eta}{2}\right)I - \sinh\left(\frac{\eta}{2}\right)\Sigma^{01}. \tag{6.25}$$

Explicitly, this is:

$$S(\Lambda) = \cosh\left(\frac{\eta}{2}\right)I - \sinh\left(\frac{\eta}{2}\right)\frac{i}{2}\begin{pmatrix} 0 & \sigma_x \\ \sigma_x & 0 \end{pmatrix}. \tag{6.26}$$

Since η is related to the velocity by $\tanh\eta = \beta$, and $\gamma = \cosh\eta$, we have:

$$S(\Lambda) = \cosh\left(\frac{\eta}{2}\right)I - \sinh\left(\frac{\eta}{2}\right)\frac{i}{2}\begin{pmatrix} 0 & \sigma_x \\ \sigma_x & 0 \end{pmatrix}, \tag{6.27}$$

$$S(\Lambda) = \frac{1}{\sqrt{2(1+\gamma)}}\begin{pmatrix} \sqrt{1+\gamma} & 0 & 0 & \frac{\beta\sqrt{1+\gamma}}{\gamma} \\ 0 & \sqrt{1+\gamma} & \frac{\beta\sqrt{1+\gamma}}{\gamma} & 0 \\ 0 & \frac{\beta\sqrt{1+\gamma}}{\gamma} & \sqrt{1+\gamma} & 0 \\ \frac{\beta\sqrt{1+\gamma}}{\gamma} & 0 & 0 & \sqrt{1+\gamma} \end{pmatrix}. \tag{6.28}$$

Applying this transformation to a Dirac spinor $\psi(x)$, we obtain the transformed spinor $\psi'(x')$:

$$\psi'(x') = S(\Lambda)\psi(x). \tag{6.29}$$

This example illustrates how Lorentz transformations are applied to Dirac spinors, ensuring the relativistic invariance of the theory. Understanding these transformations is crucial in quantum field theory for analyzing the behavior of particles under changes in reference frames.

6.3 THE KLEIN-GORDON EQUATION

6.3.1 DERIVATION FROM THE ENERGY-MOMENTUM RELATIONSHIP

The Klein-Gordon equation is derived from the relativistic energy-momentum relationship. In special relativity, the energy and momentum of a free particle are related by the equation:

$$E^2 = p^2c^2 + m^2c^4, \tag{6.30}$$

where E is the energy, p is the momentum, m is the rest mass of the particle, and c is the speed of light.

In quantum mechanics, energy and momentum operators are given by:

$$E \to i\hbar\frac{\partial}{\partial t}, \quad \mathbf{p} \to -i\hbar\nabla. \tag{6.31}$$

Substituting these operators into the relativistic energy-momentum relationship, we get:

$$\left(i\hbar\frac{\partial}{\partial t}\right)^2 = (-i\hbar\nabla)^2 c^2 + m^2c^4. \tag{6.32}$$

Simplifying, we obtain:

$$-\hbar^2\frac{\partial^2}{\partial t^2} = -\hbar^2c^2\nabla^2 + m^2c^4. \tag{6.33}$$

Rearranging terms, we arrive at the Klein-Gordon equation:

$$\left(\frac{1}{c^2}\frac{\partial^2}{\partial t^2} - \nabla^2 + \frac{m^2c^2}{\hbar^2}\right)\phi = 0, \tag{6.34}$$

where ϕ is the scalar field.

6.3.2 PROPERTIES AND SOLUTIONS

The Klein-Gordon equation is a second-order partial differential equation. It is Lorentz invariant, meaning it retains its form under Lorentz transformations, making it suitable for relativistic quantum mechanics.

6.3.2.0.1 Plane Wave Solutions

A common solution to the Klein-Gordon equation is the plane wave solution. Assume:

$$\phi(\mathbf{r},t) = e^{i(\mathbf{k}\cdot\mathbf{r} - \omega t)}, \tag{6.35}$$

where $\mathbf{k}$ is the wave vector and ω is the angular frequency.

Substituting this into the Klein-Gordon equation, we get:

$$\left(\frac{1}{c^2}\omega^2 - k^2 - \frac{m^2c^2}{\hbar^2}\right)e^{i(\mathbf{k}\cdot\mathbf{r} - \omega t)} = 0. \tag{6.36}$$

This implies:

$$\frac{\omega^2}{c^2} - k^2 = \frac{m^2c^2}{\hbar^2}. \tag{6.37}$$

Rearranging, we obtain the dispersion relation:

$$\omega^2 = c^2k^2 + \frac{m^2c^4}{\hbar^2}. \tag{6.38}$$

6.3.3 PHYSICAL INTERPRETATIONS AND LIMITATIONS

The Klein-Gordon equation describes scalar particles in a relativistic framework. However, it has several limitations:

6.3.3.0.1 negative probability density

The probability density associated with the Klein-Gordon equation is not always positive definite, leading to issues with the interpretation of probabilities. The probability density is given by:

$$\rho = \frac{i\hbar}{2mc^2}(\phi^* \partial_t \phi - \phi \partial_t \phi^*). \tag{6.39}$$

Negative probability densities suggest that the Klein-Gordon equation does not provide a consistent probabilistic interpretation on its own. This limitation led to the development of quantum field theory (QFT), where the Klein-Gordon equation is reinterpreted in the context of fields rather than single-particle wavefunctions. In QFT, the fields ϕ and ϕ^* are treated as operators that create and annihilate particles, and the concept of probability density is replaced by more general field-theoretic constructs.

6.3.3.0.2 Second-Order Nature

The Klein-Gordon equation is second-order in both space and time, unlike the first-order Schrödinger equation. This makes it more challenging to interpret the solutions in terms of quantum mechanics' probabilistic framework.

Despite these limitations, the Klein-Gordon equation is crucial in quantum field theory (QFT), where it describes spin-0 particles, such as mesons. It provides a foundation for understanding more complex fields and interactions in relativistic quantum theories.

WORKED EXAMPLE: LORENTZ INVARIANCE OF THE KLEIN-GORDON EQUATION

Show that the Klein-Gordon equation is Lorentz invariant, meaning it retains its form under Lorentz transformations. To demonstrate this, we need to show that the equation remains unchanged when we transform to a new reference frame using Lorentz transformations.

The Klein-Gordon equation is:

$$\left(\frac{1}{c^2}\frac{\partial^2}{\partial t^2} - \nabla^2 + \frac{m^2c^2}{\hbar^2}\right)\phi = 0. \tag{6.40}$$

Consider a Lorentz transformation from coordinates $(t, \mathbf{r})$ to new coordinates $(t', \mathbf{r}')$. For simplicity, let's consider a boost in the x-direction with velocity v. The Lorentz transformation equations are:

$$t' = \gamma\left(t - \frac{vx}{c^2}\right), \quad x' = \gamma(x - vt), \quad y' = y, \quad z' = z, \tag{6.41}$$

where $\gamma = \frac{1}{\sqrt{1-\frac{v^2}{c^2}}}$ is the Lorentz factor.

The partial derivatives transform as follows:

$$\frac{\partial}{\partial t} = \gamma\left(\frac{\partial}{\partial t'} + v\frac{\partial}{\partial x'}\right), \quad \frac{\partial}{\partial x} = \gamma\left(\frac{\partial}{\partial x'} + \frac{v}{c^2}\frac{\partial}{\partial t'}\right). \tag{6.42}$$

For the second-order partial derivatives, we get:

$$\frac{\partial^2}{\partial t^2} = \gamma^2\left(\frac{\partial^2}{\partial t'^2} + 2v\frac{\partial^2}{\partial t'\partial x'} + v^2\frac{\partial^2}{\partial x'^2}\right), \tag{6.43}$$

$$\frac{\partial^2}{\partial x^2} = \gamma^2 \left(\frac{\partial^2}{\partial x'^2} + 2\frac{v}{c^2}\frac{\partial^2}{\partial x' \partial t'} + \frac{v^2}{c^4}\frac{\partial^2}{\partial t'^2} \right). \tag{6.44}$$

Substitute these into the Klein-Gordon equation:

$$\left[\frac{\gamma^2}{c^2}\left(\frac{\partial^2}{\partial t'^2} + 2v\frac{\partial^2}{\partial t' \partial x'} + v^2\frac{\partial^2}{\partial x'^2} \right) \right.$$
$$\left. -\gamma^2\left(\frac{\partial^2}{\partial x'^2} + \frac{2v}{c^2}\frac{\partial^2}{\partial x' \partial t'} + \frac{v^2}{c^4}\frac{\partial^2}{\partial t'^2} \right) + \frac{m^2c^2}{\hbar^2} \right] \phi = 0 \tag{6.45}$$

Factor out γ^2:

$$\gamma^2\left(\frac{1}{c^2}\frac{\partial^2}{\partial t'^2} + \frac{2v}{c^2}\frac{\partial^2}{\partial t' \partial x'} + \frac{v^2}{c^2}\frac{\partial^2}{\partial x'^2} \right)$$
$$-\gamma^2\left(\frac{\partial^2}{\partial x'^2} + \frac{2v}{c^2}\frac{\partial^2}{\partial x' \partial t'} + \frac{v^2}{c^4}\frac{\partial^2}{\partial t'^2} \right) + \frac{m^2c^2}{\hbar^2}\phi = 0 \tag{6.46}$$

Simplify the terms inside the parentheses:

$$\gamma^2\left(\left(\frac{1}{c^2} - \frac{v^2}{c^4} \right)\frac{\partial^2}{\partial t'^2} + \left(\frac{v^2}{c^2} - 1 \right)\frac{\partial^2}{\partial x'^2} \right)\phi = 0. \tag{6.47}$$

Since $\gamma^2(1 - \frac{v^2}{c^2}) = 1$, we have:

$$\left(\frac{1}{c^2}\frac{\partial^2}{\partial t'^2} - \nabla'^2 + \frac{m^2c^2}{\hbar^2} \right)\phi = 0. \tag{6.48}$$

This shows that the Klein-Gordon equation retains its form under Lorentz transformations, confirming its Lorentz invariance.

6.4 CONTOUR INTEGRALS

6.4.1 REVIEW OF COMPLEX ANALYSIS

Contour integrals are a fundamental concept in complex analysis, which deals with functions of a complex variable. As shown in Figure 6.1 contour integral involves integrating a complex function along a specific path, or contour, in the complex plane.

6.4.1.0.1 Complex Functions and Analyticity

A complex function $f(z)$ of a complex variable $z = x + iy$ is said to be analytic if it is differentiable at every point in its domain. Analytic functions have many remarkable properties, including the fact that their integral around a closed contour depends only on the singularities enclosed by the contour.

6.4.1.0.2 Cauchy's integral theorem

Cauchy's Integral Theorem states that if $f(z)$ is analytic within and on a simple closed contour C, then:

$$\oint_C f(z)\,dz = 0. \tag{6.49}$$

6.4.1.0.3 *Cauchy's Integral Formula*

Cauchy's Integral Formula provides a method for evaluating integrals of analytic functions. If $f(z)$ is analytic inside and on a simple closed contour C and z_0 is a point inside C, then:

$$f(z_0) = \frac{1}{2\pi i} \oint_C \frac{f(z)}{z - z_0} dz. \tag{6.50}$$

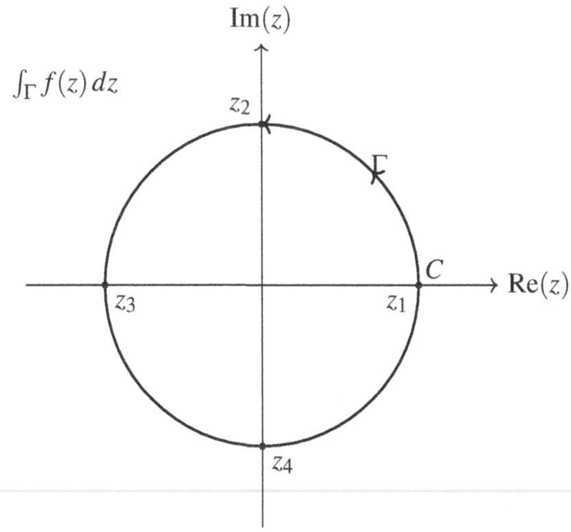

FIGURE 6.1
A contour integral $\int_\Gamma f(z)\,dz$ along the contour Γ in the complex plane. The contour is a circle centered at the origin with marked points z_1, z_2, z_3, and z_4.

6.4.1.0.4 *Residue Theorem*

The Residue Theorem is a powerful tool for evaluating contour integrals. It states that if $f(z)$ is analytic inside and on a simple closed contour C except for isolated singularities $z_1, z_2, \ldots, z_n$ inside C, then:

$$\oint_C f(z)\,dz = 2\pi i \sum_{k=1}^{n} \mathrm{Res}(f, z_k), \tag{6.51}$$

where $\mathrm{Res}(f, z_k)$ is the residue of $f(z)$ at the singularity z_k.

6.4.2 APPLICATION TO QUANTUM MECHANICS

In quantum mechanics, contour integrals are used in various contexts, such as evaluating Green's functions and propagators. These integrals often involve complex functions that represent physical quantities, and their evaluation provides insights into the behavior of quantum systems.

6.4.2.0.1 *Example: Evaluating a Green's function*

Consider the Green's function $G(z)$ for a one-dimensional free particle:

$$G(z) = \frac{1}{z^2 + k^2}. \tag{6.52}$$

To evaluate the integral of $G(z)$ along a contour that encloses the singularities at $z = \pm ik$, we use the Residue Theorem.

First, we identify the residues at the singularities:

$$\text{Res}(G, ik) = \lim_{z \to ik} (z - ik) G(z) = \frac{1}{2ik}, \tag{6.53}$$

$$\text{Res}(G, -ik) = \lim_{z \to -ik} (z + ik) G(z) = -\frac{1}{2ik}. \tag{6.54}$$

Then, we use the Residue Theorem to evaluate the integral:

$$\oint_C G(z)\, dz = 2\pi i \left(\frac{1}{2ik} - \frac{1}{2ik} \right) = 0. \tag{6.55}$$

In summary, contour integrals play a significant role in quantum mechanics and quantum field theory, particularly in evaluating Green's functions and propagators. Understanding these techniques is essential for analyzing the behavior of quantum systems and predicting experimental outcomes.

6.5 THE SCALAR FEYNMAN PROPAGATOR

The scalar Feynman propagator is a crucial concept in quantum field theory, representing the probability amplitude for a scalar particle to propagate from one spacetime point to another. This section discusses the derivation, properties, and applications of the scalar Feynman propagator and its connection to the Klein-Gordon equation.

6.5.1 DERIVATION USING CONTOUR INTEGRATION

The scalar Feynman propagator $\Delta_F(x)$ for a scalar field is derived using contour integration in the complex plane. It is defined as the time-ordered product of field operators:

$$\Delta_F(x - y) = \langle 0 | T\{\phi(x)\phi(y)\} | 0 \rangle. \tag{6.56}$$

In momentum space, the propagator is given by:

$$\Delta_F(p) = \frac{i}{p^2 - m^2 + i\varepsilon}, \tag{6.57}$$

where ε is a small positive number ensuring the correct boundary conditions for causal propagation.

To derive the position-space propagator, we perform a Fourier transform:

$$\Delta_F(x) = \int \frac{d^4p}{(2\pi)^4} \frac{e^{-ip\cdot x}}{p^2 - m^2 + i\varepsilon}. \tag{6.58}$$

We evaluate the integral over p^0 using contour integration. The integrand has poles at $p^0 = \pm\sqrt{\mathbf{p}^2 + m^2} - i\varepsilon$.

$$I = \int_{-\infty}^{\infty} \frac{dp^0}{2\pi} \frac{e^{-ip^0 t}}{(p^0)^2 - (\omega_{\mathbf{p}}^2 - i\varepsilon)}, \tag{6.59}$$

where $\omega_{\mathbf{p}} = \sqrt{\mathbf{p}^2 + m^2}$.

For $t > 0$, we close the contour in the lower half-plane, picking up the residue at $p^0 = \omega_{\mathbf{p}} - i\varepsilon$:

$$I = 2\pi i \text{Res}\left(\frac{e^{-ip^0 t}}{(p^0)^2 - (\omega_{\mathbf{p}}^2 - i\varepsilon)}, \omega_{\mathbf{p}} - i\varepsilon \right). \tag{6.60}$$

The residue is:

$$\text{Res}\left(\frac{e^{-ip^0t}}{(p^0)^2-(\omega_{\mathbf{p}}^2-i\varepsilon)},\omega_{\mathbf{p}}-i\varepsilon\right)=\frac{e^{-i\omega_{\mathbf{p}}t}}{-2\omega_{\mathbf{p}}}. \tag{6.61}$$

Thus, the integral is:

$$I=\frac{e^{-i\omega_{\mathbf{p}}t}}{2\omega_{\mathbf{p}}}. \tag{6.62}$$

Combining this with the integral over **p**, we obtain the Feynman propagator:

$$\Delta_F(x)=\int\frac{d^3p}{(2\pi)^3}\frac{e^{-i\mathbf{p}\cdot\mathbf{x}}e^{-i\omega_{\mathbf{p}}t}}{2\omega_{\mathbf{p}}}. \tag{6.63}$$

For $x^0>0$, we close the contour in the lower half-plane and for $x^0<0$, we close the contour in the upper half-plane. The resulting integrals give:

$$\Delta_F(x)=\int\frac{d^3p}{(2\pi)^3}\frac{e^{i\mathbf{p}\cdot\mathbf{x}}}{2\omega_{\mathbf{p}}}\left(\theta(x^0)e^{-i\omega_{\mathbf{p}}x^0}+\theta(-x^0)e^{i\omega_{\mathbf{p}}x^0}\right), \tag{6.64}$$

where $\omega_{\mathbf{p}}=\sqrt{\mathbf{p}^2+m^2}$ and $\theta(x^0)$ is the Heaviside step function.

This evaluation of the Feynman propagator using contour integration highlights its importance in quantum field theory. The propagator describes the propagation of particles and is a crucial component in calculating scattering amplitudes and other physical observables.

6.5.2 PROPERTIES AND APPLICATIONS

The scalar Feynman propagator has several important properties:

- **Causality:** The propagator respects causality, meaning that the effect cannot precede the cause. This is ensured by the $i\varepsilon$ prescription, which shifts the poles slightly off the real axis.
- **Lorentz Invariance:** The propagator is Lorentz invariant, meaning it has the same form in all inertial reference frames. This invariance ensures that the laws of physics are the same for all observers.
- **Green's Function:** The propagator acts as a Green's function for the Klein-Gordon equation, meaning it satisfies the inhomogeneous Klein-Gordon equation with a delta function source:

$$(\partial^2+m^2)\Delta_F(x-y)=-\delta^4(x-y). \tag{6.65}$$

6.5.2.0.1 Applications:

The scalar Feynman propagator is widely used in quantum field theory, including:

- **Perturbation Theory:** It forms the building block for Feynman diagrams, which are used to calculate scattering amplitudes and other physical observables in perturbation theory.
- **Virtual Particles:** The propagator describes the propagation of virtual particles between interaction vertices in Feynman diagrams.
- **Field Correlations:** It is used to compute correlations between field operators at different spacetime points, providing insights into the structure of quantum fields.

6.5.3 CONNECTION TO THE KLEIN-GORDON EQUATION

The Klein-Gordon equation is a relativistic wave equation for scalar fields. The connection between the scalar Feynman propagator and the Klein-Gordon equation is fundamental to understanding the dynamics of scalar fields.

The Klein-Gordon equation in position space is:

$$(\partial^2 + m^2)\phi(x) = 0, \tag{6.66}$$

where $\partial^2 = \partial_\mu \partial^\mu$ is the d'Alembertian operator.

The propagator $\Delta_F(x-y)$ satisfies the inhomogeneous Klein-Gordon equation with a delta function source:

$$(\partial^2 + m^2)\Delta_F(x-y) = -\delta^4(x-y). \tag{6.67}$$

This relationship shows that the propagator acts as a Green's function for the Klein-Gordon equation, providing the solution to the equation for a given source. In this way, the propagator encodes the response of the field to a point source, capturing the causal and relativistic nature of the field's evolution.

The connection between the scalar Feynman propagator and the Klein-Gordon equation is central to quantum field theory, as it provides a consistent framework for describing the propagation of scalar particles and their interactions.

WORKED EXAMPLE: LORENTZ INVARIANCE OF THE FEYNMAN PROPAGATOR

To demonstrate the Lorentz invariance of the Feynman propagator, we need to show that the propagator remains unchanged under a Lorentz transformation. The Feynman propagator in momentum space is given by:

$$\Delta_F(p) = \frac{i}{p^2 - m^2 + i\varepsilon}, \tag{6.68}$$

where $p^2 = p^\mu p_\mu = p_0^2 - \mathbf{p}^2$.

Under a Lorentz transformation Λ, the four-momentum p^μ transforms as:

$$p'^\mu = \Lambda^\mu{}_\nu p^\nu. \tag{6.69}$$

The Lorentz invariance of the propagator can be checked by transforming p and showing that the form of $\Delta_F(p)$ remains the same.

6.5.3.0.1 *Step 1: Transforming the Propagator*

Under the Lorentz transformation, the momentum squared p^2 transforms as:

$$p'^2 = \Lambda^\mu{}_\alpha p^\alpha \Lambda_{\mu\beta} p^\beta = \Lambda^\mu{}_\alpha \Lambda_{\mu\beta} p^\alpha p^\beta. \tag{6.70}$$

Since the metric tensor $\eta_{\mu\nu}$ is invariant under Lorentz transformations, we have:

$$\eta_{\mu\nu}\Lambda^\mu{}_\alpha \Lambda^\nu{}_\beta = \eta_{\alpha\beta}. \tag{6.71}$$

Thus,

$$p'^2 = p^\mu p_\mu = p^2. \tag{6.72}$$

This shows that p^2 is invariant under Lorentz transformations.

6.5.3.0.2 *Step 2: Invariance of the Propagator Form*

The Feynman propagator in momentum space can be rewritten in terms of p':

$$\Delta_F(p') = \frac{i}{p'^2 - m^2 + i\varepsilon}. \tag{6.73}$$

Given that $p'^2 = p^2$, we have:

$$\Delta_F(p') = \frac{i}{p^2 - m^2 + i\varepsilon} = \Delta_F(p). \tag{6.74}$$

This demonstrates that the form of the Feynman propagator remains unchanged under a Lorentz transformation, proving that the Feynman propagator is Lorentz invariant.

6.5.3.0.3 *Step 3: Position-Space Propagator*

To further illustrate the Lorentz invariance, consider the Feynman propagator in position space, given by:

$$\Delta_F(x-y) = \int \frac{d^4p}{(2\pi)^4} \frac{e^{-ip\cdot(x-y)}}{p^2 - m^2 + i\varepsilon}. \tag{6.75}$$

Under a Lorentz transformation, the spacetime coordinates x^μ and y^μ transform as:

$$x'^\mu = \Lambda^\mu{}_\nu x^\nu, \quad y'^\mu = \Lambda^\mu{}_\nu y^\nu. \tag{6.76}$$

The difference $x - y$ transforms as:

$$(x-y)'^\mu = \Lambda^\mu_\nu (x^\nu - y^\nu). \tag{6.77}$$

The exponent in the Fourier transform becomes:

$$e^{-ip\cdot(x-y)} \to e^{-ip\cdot\Lambda(x-y)} = e^{-i(\Lambda^{-1}p)\cdot(x-y)}. \tag{6.78}$$

Since the measure d^4p is invariant under Lorentz transformations, we have:

$$\Delta_F(x'-y') = \int \frac{d^4p}{(2\pi)^4} \frac{e^{-i(\Lambda^{-1}p)\cdot(x-y)}}{p^2 - m^2 + i\varepsilon} = \Delta_F(x-y). \tag{6.79}$$

This shows that the position-space propagator $\Delta_F(x-y)$ is also Lorentz invariant, confirming that the Feynman propagator maintains its form under Lorentz transformations both in momentum space and in position space.

6.6 GREEN'S FUNCTIONS

Green's functions are powerful mathematical tools used to solve differential equations with boundary conditions. They play a crucial role in many areas of physics, including quantum field theory (QFT), where they are used to describe propagators and correlation functions.

6.6.1 DEFINITION AND PROPERTIES

A Green's function $G(x,x')$ is defined as the solution to an inhomogeneous differential equation of the form:

$$\mathscr{L}G(x,x') = \delta(x-x'), \tag{6.80}$$

where $\mathscr{L}$ is a linear differential operator, x and x' are points in the domain of the operator, and $\delta(x-x')$ is the Dirac delta function.

Properties of Green's functions

Linearity: If $G_1(x,x')$ and $G_2(x,x')$ are Green's functions corresponding to differential operators $\mathscr{L}_1$ and $\mathscr{L}_2$, respectively, then for any constants a and b:

$$G(x,x') = aG_1(x,x') + bG_2(x,x') \tag{6.81}$$

is the Green's function for the operator $a\mathscr{L}_1 + b\mathscr{L}_2$.
Symmetry: In many cases, Green's functions are symmetric:

$$G(x,x') = G(x',x). \tag{6.82}$$

Boundary Conditions: The specific form of the Green's function depends on the boundary conditions of the problem.

6.6.2 GREEN'S FUNCTIONS FOR DIFFERENTIAL EQUATIONS

Green's functions provide a method to solve inhomogeneous differential equations. Consider the differential equation:

$$\mathscr{L}\phi(x) = f(x), \tag{6.83}$$

where $\mathscr{L}$ is a linear differential operator and $f(x)$ is a source term. The solution $\phi(x)$ can be expressed in terms of the Green's function $G(x,x')$ as:

$$\phi(x) = \int G(x,x') f(x')\, dx'. \tag{6.84}$$

Example: Green's Function for the Poisson Equation

Consider the Poisson equation in one dimension:

$$-\frac{d^2\phi(x)}{dx^2} = f(x). \tag{6.85}$$

The Green's function $G(x,x')$ satisfies:

$$-\frac{d^2G(x,x')}{dx^2} = \delta(x-x'). \tag{6.86}$$

The solution to this equation is:

$$G(x,x') = \begin{cases} x'-x, & x \leq x' \\ x-x', & x > x' \end{cases}, \tag{6.87}$$

which ensures continuity and the appropriate jump in the first derivative at $x = x'$.
The solution $\phi(x)$ to the Poisson equation is then given by:

$$\phi(x) = \int_{-\infty}^{\infty} G(x,x') f(x')\, dx'. \tag{6.88}$$

6.6.3 APPLICATIONS IN QUANTUM FIELD THEORY

In quantum field theory, Green's functions are used to describe the propagation of particles and fields. The Green's function for a field operator provides information about the probability amplitude for a particle to propagate from one point to another.

Feynman propagator as a Green's Function

The Feynman propagator $\Delta_F(x-y)$ is the Green's function for the Klein-Gordon equation:

$$(\Box + m^2)\Delta_F(x-y) = \delta^4(x-y), \tag{6.89}$$

where $\Box$ is the d'Alembertian operator and m is the mass of the particle.

Correlation Functions

In QFT, Green's functions are also used to compute correlation functions, which are essential for calculating physical observables. The two-point correlation function, for example, is given by the expectation value of the time-ordered product of field operators:

$$G(x,y) = \langle 0|T\{\phi(x)\phi(y)\}|0\rangle. \tag{6.90}$$

This correlation function describes the propagation of a particle between points x and y.

Example: Scalar Field Propagator

For a free scalar field ϕ, the propagator $\Delta_F(x-y)$ in momentum space is given by:

$$\Delta_F(p) = \frac{i}{p^2 - m^2 + i\varepsilon}. \tag{6.91}$$

Transforming back to position space using Fourier transform, we get:

$$\Delta_F(x-y) = \int \frac{d^4p}{(2\pi)^4} \frac{e^{-ip\cdot(x-y)}}{p^2 - m^2 + i\varepsilon}. \tag{6.92}$$

The Green's function approach in QFT provides a unified framework for understanding particle propagation, interactions, and scattering processes, making it a fundamental tool in theoretical physics.

6.7 SUMMARY AND DISCUSSION

In this chapter, we explored the integration of special relativity with quantum mechanics, laying the groundwork for understanding relativistic quantum mechanics and its applications in quantum field theory (QFT). We began by motivating the need for relativistic formulations in quantum mechanics and discussing the historical context that led to the development of key equations like the Klein-Gordon and Dirac equations.

The Klein-Gordon equation, derived from the relativistic energy-momentum relationship, was presented as the fundamental equation for spin-0 particles. We discussed its properties, solutions, and physical interpretations, highlighting its limitations, such as negative probability densities, which led to the development of QFT where fields are quantized, and particle creation and annihilation are naturally described.

We also covered the mathematical technique of contour integration, crucial for evaluating complex integrals in quantum mechanics. This included a review of complex analysis and the application of contour integrals to evaluate propagators and Green's functions, essential tools in theoretical physics.

The scalar Feynman propagator, a central concept in QFT, was derived using contour integration. We explored its properties, applications, and its connection to the Klein-Gordon equation, demonstrating its role in describing particle propagation and interactions in QFT. We showed the Lorentz invariance of the Feynman propagator, reinforcing the consistency of relativistic formulations in QFT.

Finally, we introduced Green's functions, defining their properties and discussing their role in solving differential equations and their applications in QFT. We demonstrated how Green's functions, including the Feynman propagator, are used to describe the propagation of particles and fields, providing a unified framework for understanding interactions in QFT.

This chapter has laid a comprehensive foundation for understanding relativistic quantum mechanics and its extensions to quantum field theory. By integrating these concepts, we have provided the tools and insights necessary to explore the rich and complex landscape of QFT, where particles and fields interact in a relativistic framework.

6.8 QUESTIONS AND PROBLEMS

To reinforce the concepts covered in this chapter, the following questions and problems are provided:

6.8.1 TRIVIA QUESTIONS

Question 1: Which equation is considered the cornerstone of relativistic quantum mechanics for spin-0 particles?

(a) Dirac Equation
(b) Schrödinger Equation
(c) Klein-Gordon Equation
(d) Pauli Equation

Answer: c. Klein-Gordon Equation
Question 2: What is the primary motivation for extending quantum mechanics to accommodate relativistic principles?

(a) To explain blackbody radiation
(b) To account for high-velocity phenomena
(c) To understand atomic structure
(d) To describe superconductivity

Answer: b. To account for high-velocity phenomena
Question 3: What is the Lorentz factor, γ, for a Lorentz transformation in the x-direction?

(a) $\gamma = \frac{1}{\sqrt{1-\frac{v^2}{c^2}}}$
(b) $\gamma = \frac{v}{c}$
(c) $\gamma = \frac{1}{v-c}$
(d) $\gamma = \frac{c}{v}$

Answer: a. $\gamma = \frac{1}{\sqrt{1-\frac{v^2}{c^2}}}$

Question 4: What is the form of the Dirac equation in terms of the matrices α and β?

(a) $i\hbar\frac{\partial\psi}{\partial t} = \left(-i\hbar c\alpha\cdot\nabla + \beta mc^2\right)\psi$
(b) $-\hbar^2\nabla^2\psi = E\psi$
(c) $\left(\frac{1}{c^2}\frac{\partial^2}{\partial t^2} - \nabla^2 + \frac{m^2c^2}{\hbar^2}\right)\phi = 0$
(d) $i\hbar\frac{\partial\psi}{\partial t} = H\psi$

Answer: a. $i\hbar\frac{\partial\psi}{\partial t} = \left(-i\hbar c\alpha\cdot\nabla + \beta mc^2\right)\psi$
Question 5: Which of the following is a limitation of the Klein-Gordon equation?

(a) It only applies to non-relativistic particles.
(b) It predicts negative probability densities.
(c) It does not account for quantum effects.
(d) It cannot describe spin-$\frac{1}{2}$ particles.

Answer: b. It predicts negative probability densities
Question 6: What does the Feynman propagator describe in quantum field theory?

(a) The wave function of a particle
(b) The probability amplitude for a particle to propagate from one spacetime point to another
(c) The energy levels of an atom
(d) The statistical distribution of particles in a gas

Answer: b. The probability amplitude for a particle to propagate from one spacetime point to another
Question 7: What does a Green's function solve in the context of differential equations?

(a) Homogeneous differential equations
(b) Algebraic equations
(c) Inhomogeneous differential equations
(d) Integral equations

Answer: c. Inhomogeneous differential equations
Question 8: Which of the following is a key property of the scalar Feynman propagator?

(a) It is not Lorentz invariant.
(b) It only applies to non-relativistic particles.
(c) It respects causality.
(d) It describes the energy levels of bound states.

Answer: c. It respects causality
Question 9: The scalar Feynman propagator is the Green's function for which equation?

(a) Schrödinger Equation
(b) Dirac Equation
(c) Klein-Gordon Equation
(d) Pauli Equation

Answer: c. Klein-Gordon Equation
Question 10: What role do contour integrals play in quantum mechanics?

(a) Describing the energy levels of atoms
(b) Evaluating integrals involving complex functions
(c) Calculating statistical distributions
(d) Solving algebraic equations

Answer: b. Evaluating integrals involving complex functions

PROBLEMS

6.8.1.1 Problem 1:

Derive the Klein-Gordon equation from the relativistic energy-momentum relationship and discuss its physical significance.

Solution:

Starting with the relativistic energy-momentum relationship:

$$E^2 = p^2c^2 + m^2c^4, \tag{6.93}$$

we replace the energy and momentum with their corresponding quantum mechanical operators:

$$E \to i\hbar\frac{\partial}{\partial t}, \quad \mathbf{p} \to -i\hbar\nabla. \tag{6.94}$$

Substituting these into the energy-momentum relationship, we get:

$$\left(i\hbar\frac{\partial}{\partial t}\right)^2 = (-i\hbar\nabla)^2 c^2 + m^2c^4. \tag{6.95}$$

Simplifying, we obtain:

$$-\hbar^2\frac{\partial^2}{\partial t^2} = -\hbar^2c^2\nabla^2 + m^2c^4. \tag{6.96}$$

Rearranging terms, we arrive at the Klein-Gordon equation:

$$\left(\frac{1}{c^2}\frac{\partial^2}{\partial t^2} - \nabla^2 + \frac{m^2c^2}{\hbar^2}\right)\phi = 0, \tag{6.97}$$

where ϕ is the scalar field.

The Klein-Gordon equation is significant because it describes scalar particles in a relativistic framework and is Lorentz invariant, meaning it retains its form under Lorentz transformations.

6.8.1.2 Problem 2:

Solve the Klein-Gordon equation for a free particle and interpret the solutions.

Solution:

Assume a plane wave solution:

$$\phi(\mathbf{r},t) = e^{i(\mathbf{k}\cdot\mathbf{r}-\omega t)}, \tag{6.98}$$

where $\mathbf{k}$ is the wave vector and ω is the angular frequency.

Substituting this into the Klein-Gordon equation, we get:

$$\left(\frac{1}{c^2}\omega^2 - k^2 - \frac{m^2c^2}{\hbar^2}\right)e^{i(\mathbf{k}\cdot\mathbf{r}-\omega t)} = 0. \tag{6.99}$$

This implies:

$$\frac{\omega^2}{c^2} - k^2 = \frac{m^2c^2}{\hbar^2}. \tag{6.100}$$

Rearranging, we obtain the dispersion relation:

$$\omega^2 = c^2k^2 + \frac{m^2c^4}{\hbar^2}. \tag{6.101}$$

The solutions $\phi(\mathbf{r},t) = e^{i(\mathbf{k}\cdot\mathbf{r}-\omega t)}$ represent plane waves with positive and negative energy solutions corresponding to particles and antiparticles.

6.8.1.3 Problem 3:

Show that the Klein-Gordon equation is Lorentz invariant.

Solution:

The Klein-Gordon equation is:

$$\left(\frac{1}{c^2}\frac{\partial^2}{\partial t^2} - \nabla^2 + \frac{m^2c^2}{\hbar^2}\right)\phi = 0. \tag{6.102}$$

Consider a Lorentz transformation from coordinates $(t, \mathbf{r})$ to new coordinates $(t', \mathbf{r}')$. For simplicity, let's consider a boost in the x-direction with velocity v. The Lorentz transformation equations are:

$$t' = \gamma\left(t - \frac{vx}{c^2}\right), \quad x' = \gamma(x - vt), \quad y' = y, \quad z' = z, \tag{6.103}$$

where $\gamma = \frac{1}{\sqrt{1-\frac{v^2}{c^2}}}$ is the Lorentz factor.

The partial derivatives transform as follows:

$$\frac{\partial}{\partial t} = \gamma\left(\frac{\partial}{\partial t'} + v\frac{\partial}{\partial x'}\right), \quad \frac{\partial}{\partial x} = \gamma\left(\frac{\partial}{\partial x'} + \frac{v}{c^2}\frac{\partial}{\partial t'}\right). \tag{6.104}$$

For the second-order partial derivatives, we get:

$$\frac{\partial^2}{\partial t^2} = \gamma^2\left(\frac{\partial^2}{\partial t'^2} + 2v\frac{\partial^2}{\partial t'\partial x'} + v^2\frac{\partial^2}{\partial x'^2}\right), \tag{6.105}$$

$$\frac{\partial^2}{\partial x^2} = \gamma^2\left(\frac{\partial^2}{\partial x'^2} + 2\frac{v}{c^2}\frac{\partial^2}{\partial x'\partial t'} + \frac{v^2}{c^4}\frac{\partial^2}{\partial t'^2}\right). \tag{6.106}$$

Substitute these into the Klein-Gordon equation:

$$\left(\frac{\gamma^2}{c^2}\left(\frac{\partial^2}{\partial t'^2} + 2v\frac{\partial^2}{\partial t'\partial x'} + v^2\frac{\partial^2}{\partial x'^2}\right)\right.$$
$$\left. - \gamma^2\left(\frac{\partial^2}{\partial x'^2} + 2\frac{v}{c^2}\frac{\partial^2}{\partial x'\partial t'} + \frac{v^2}{c^4}\frac{\partial^2}{\partial t'^2}\right) + \frac{m^2c^2}{\hbar^2}\right)\phi = 0 \tag{6.107}$$

Factor out γ^2:

$$\gamma^2\left(\left(\frac{1}{c^2} - \frac{v^2}{c^4}\right)\frac{\partial^2}{\partial t'^2} + \left(\frac{v^2}{c^2} - 1\right)\frac{\partial^2}{\partial x'^2}\right)\phi = 0. \tag{6.108}$$

Since $\gamma^2(1 - \frac{v^2}{c^2}) = 1$, we have:

$$\left(\frac{1}{c^2}\frac{\partial^2}{\partial t'^2} - \nabla'^2 + \frac{m^2c^2}{\hbar^2}\right)\phi = 0. \tag{6.109}$$

This shows that the Klein-Gordon equation retains its form under Lorentz transformations, confirming its Lorentz invariance.

6.8.1.4 Problem 4:

Derive the Dirac equation and discuss its implications for spin-$\frac{1}{2}$ particles.

Solution:

Starting with the relativistic energy-momentum relationship:

$$E^2 = p^2c^2 + m^2c^4, \tag{6.110}$$

Dirac proposed a linearization of this equation by introducing matrices α_i and β such that:

$$E = \alpha \cdot pc + \beta mc^2. \tag{6.111}$$

By replacing the energy and momentum with their corresponding quantum mechanical operators, we get the Dirac equation in the form:

$$i\hbar \frac{\partial \psi}{\partial t} = \left(-i\hbar c \alpha \cdot \nabla + \beta mc^2\right) \psi, \tag{6.112}$$

where ψ is a four-component spinor, and α and β are 4×4 matrices satisfying the following anti-commutation relations:

$$\begin{aligned} \alpha_i \alpha_j + \alpha_j \alpha_i &= 2\delta_{ij} I, \\ \alpha_i \beta + \beta \alpha_i &= 0, \\ \beta^2 &= I. \end{aligned} \tag{6.113}$$

The Dirac equation describes spin-$\frac{1}{2}$ particles, such as electrons, in a relativistic framework. It predicts the existence of antiparticles and provides a natural explanation for the electron's spin and magnetic moment.

6.8.1.5 Problem 5:

Evaluate a contour integral for a given complex function and discuss its relevance to quantum mechanics.

Solution:

Consider the contour integral of the function $f(z) = \frac{1}{z-a}$ around a closed contour C that encloses the point $z = a$:

$$\oint_C \frac{1}{z-a} dz. \tag{6.114}$$

Using the Residue Theorem, we know that the integral of a function around a closed contour is $2\pi i$ times the sum of the residues of the function inside the contour. The residue of $f(z) = \frac{1}{z-a}$ at $z = a$ is 1, so:

$$\oint_C \frac{1}{z-a} dz = 2\pi i. \tag{6.115}$$

In quantum mechanics, contour integrals are used to evaluate integrals in the complex plane, which appear in the calculation of propagators and Green's functions. These integrals often involve singularities, and the Residue Theorem provides a powerful method for evaluating them.

6.8.1.6 Problem 6

Derive the scalar Feynman propagator using contour integration and discuss its properties.

Solution:

The scalar Feynman propagator $\Delta_F(x)$ for a scalar field is given by:

$$\Delta_F(x-y) = \langle 0|T\{\phi(x)\phi(y)\}|0\rangle. \tag{6.116}$$

In momentum space, the propagator is:

$$\Delta_F(p) = \frac{i}{p^2 - m^2 + i\varepsilon}, \tag{6.117}$$

To derive the position-space propagator, we perform a Fourier transform:

$$\Delta_F(x) = \int \frac{d^4p}{(2\pi)^4} \frac{e^{-ip\cdot x}}{p^2 - m^2 + i\varepsilon}. \tag{6.118}$$

We evaluate the integral over p^0 using contour integration. The integrand has poles at $p^0 = \pm\sqrt{\mathbf{p}^2 + m^2} - i\varepsilon$.

For $t > 0$, we close the contour in the lower half-plane, picking up the residue at $p^0 = \omega_{\mathbf{p}} - i\varepsilon$:

$$I = 2\pi i \operatorname{Res}\left(\frac{e^{-ip^0 t}}{(p^0)^2 - (\omega_{\mathbf{p}}^2 - i\varepsilon)}, \omega_{\mathbf{p}} - i\varepsilon \right). \tag{6.119}$$

The residue is:

$$\operatorname{Res}\left(\frac{e^{-ip^0 t}}{(p^0)^2 - (\omega_{\mathbf{p}}^2 - i\varepsilon)}, \omega_{\mathbf{p}} - i\varepsilon \right) = \frac{e^{-i\omega_{\mathbf{p}} t}}{-2\omega_{\mathbf{p}}}. \tag{6.120}$$

Thus, the integral is:

$$I = \frac{e^{-i\omega_{\mathbf{p}} t}}{2\omega_{\mathbf{p}}}. \tag{6.121}$$

Combining this with the integral over $\mathbf{p}$, we obtain the Feynman propagator:

$$\Delta_F(x) = \int \frac{d^3p}{(2\pi)^3} \frac{e^{-i\mathbf{p}\cdot\mathbf{x}} e^{-i\omega_{\mathbf{p}} t}}{2\omega_{\mathbf{p}}}. \tag{6.122}$$

The Feynman propagator has several important properties:

- **Causality:** The propagator respects causality, meaning that the effect cannot precede the cause.
- **Lorentz Invariance:** The propagator is Lorentz invariant, ensuring that the laws of physics are the same for all observers.
- **Green's Function:** The propagator acts as a Green's function for the Klein-Gordon equation, satisfying the inhomogeneous Klein-Gordon equation with a delta function source.

6.8.1.7 Problem 7:

Show that the scalar Feynman propagator is Lorentz invariant.

Solution:

The Feynman propagator in momentum space is given by:

$$\Delta_F(p) = \frac{i}{p^2 - m^2 + i\varepsilon}, \tag{6.123}$$

where $p^2 = p^\mu p_\mu = p_0^2 - \mathbf{p}^2$.

Under a Lorentz transformation Λ, the four-momentum p^μ transforms as:

$$p'^\mu = \Lambda^\mu{}_\nu p^\nu. \tag{6.124}$$

The Lorentz invariance of the propagator can be checked by transforming p and showing that the form of $\Delta_F(p)$ remains the same.

Under the Lorentz transformation, the momentum squared p^2 transforms as:

$$p'^2 = \Lambda^\mu{}_\alpha p^\alpha \Lambda_{\mu\beta} p^\beta = \Lambda^\mu{}_\alpha \Lambda_{\mu\beta} p^\alpha p^\beta. \tag{6.125}$$

Since the metric tensor $\eta_{\mu\nu}$ is invariant under Lorentz transformations, we have:

$$\eta_{\mu\nu} \Lambda^\mu{}_\alpha \Lambda^\nu{}_\beta = \eta_{\alpha\beta}. \tag{6.126}$$

Thus,

$$p'^2 = p^\mu p_\mu = p^2. \tag{6.127}$$

This shows that p^2 is invariant under Lorentz transformations.

The Feynman propagator in momentum space can be rewritten in terms of p':

$$\Delta_F(p') = \frac{i}{p'^2 - m^2 + i\varepsilon}. \tag{6.128}$$

Given that $p'^2 = p^2$, we have:

$$\Delta_F(p') = \frac{i}{p^2 - m^2 + i\varepsilon} = \Delta_F(p). \tag{6.129}$$

This demonstrates that the form of the Feynman propagator remains unchanged under a , proving that the Feynman propagator is Lorentz invariant.

6.8.1.8 Problem 8

Discuss the role of Green's functions in solving differential equations and their applications in QFT.

Solution:

A Green's function $G(x,x')$ is defined as the solution to an inhomogeneous differential equation of the form:

$$\mathscr{L} G(x,x') = \delta(x - x'), \tag{6.130}$$

where $\mathscr{L}$ is a linear differential operator, x and x' are points in the domain of the operator, and $\delta(x - x')$ is the Dirac delta function.

Green's functions provide a method to solve inhomogeneous differential equations. Consider the differential equation:

$$\mathscr{L} \phi(x) = f(x), \tag{6.131}$$

The solution $\phi(x)$ can be expressed in terms of the Green's function $G(x,x')$ as:

$$\phi(x) = \int G(x,x') f(x') \, dx'. \tag{6.132}$$

In quantum field theory, Green's functions are used to describe the propagation of particles and fields. The Green's function for a field operator provides information about the probability amplitude for a particle to propagate from one point to another.

The Feynman propagator $\Delta_F(x - y)$ is the Green's function for the Klein-Gordon equation:

$$(\Box + m^2) \Delta_F(x - y) = \delta^4(x - y). \tag{6.133}$$

In QFT, Green's functions are used to compute correlation functions, which are essential for calculating physical observables. The two-point correlation function, for example, is given by the expectation value of the time-ordered product of field operators:

$$G(x,y) = \langle 0 | T\{\phi(x)\phi(y)\} | 0 \rangle. \tag{6.134}$$

This correlation function describes the propagation of a particle between points x and y.

6.8.1.9 Problem 9:

Compute the Feynman propagator for a scalar field and interpret the results.

Solution:

The Feynman propagator in momentum space is given by:

$$\Delta_F(p) = \frac{i}{p^2 - m^2 + i\varepsilon}, \tag{6.135}$$

Transforming back to position space using Fourier transform, we get:

$$\Delta_F(x-y) = \int \frac{d^4p}{(2\pi)^4} \frac{e^{-ip\cdot(x-y)}}{p^2 - m^2 + i\varepsilon}. \tag{6.136}$$

We evaluate the integral over p^0 using contour integration. The integrand has poles at $p^0 = \pm\sqrt{\mathbf{p}^2 + m^2} - i\varepsilon$.

For $t > 0$, we close the contour in the lower half-plane, picking up the residue at $p^0 = \omega_\mathbf{p} - i\varepsilon$:

$$I = 2\pi i \operatorname{Res}\left(\frac{e^{-ip^0 t}}{(p^0)^2 - (\omega_\mathbf{p}^2 - i\varepsilon)}, \omega_\mathbf{p} - i\varepsilon\right). \tag{6.137}$$

The residue is:

$$\operatorname{Res}\left(\frac{e^{-ip^0 t}}{(p^0)^2 - (\omega_\mathbf{p}^2 - i\varepsilon)}, \omega_\mathbf{p} - i\varepsilon\right) = \frac{e^{-i\omega_\mathbf{p} t}}{-2\omega_\mathbf{p}}. \tag{6.138}$$

Thus, the integral is:

$$I = \frac{e^{-i\omega_\mathbf{p} t}}{2\omega_\mathbf{p}}. \tag{6.139}$$

Combining this with the integral over $\mathbf{p}$, we obtain the Feynman propagator:

$$\Delta_F(x) = \int \frac{d^3p}{(2\pi)^3} \frac{e^{-i\mathbf{p}\cdot\mathbf{x}} e^{-i\omega_\mathbf{p} t}}{2\omega_\mathbf{p}}. \tag{6.140}$$

The Feynman propagator describes the probability amplitude for a particle to propagate from one spacetime point to another. It is a fundamental concept in quantum field theory, used to calculate scattering amplitudes and other physical observables.

6.8.1.10 Problem 10

Explain the significance of negative probability densities in the Klein-Gordon equation and how this issue is resolved in QFT.

Solution:

The probability density associated with the Klein-Gordon equation is given by:

$$\rho = \frac{i\hbar}{2mc^2}(\phi^* \partial_t \phi - \phi \partial_t \phi^*). \tag{6.141}$$

This probability density can be negative, which poses a problem for the interpretation of probabilities in quantum mechanics. Negative probability densities suggest that the Klein-Gordon equation does not provide a consistent probabilistic interpretation on its own.

In quantum field theory (QFT), this issue is resolved by reinterpreting the Klein-Gordon equation in the context of fields rather than single-particle wavefunctions. In QFT, the fields ϕ and ϕ^* are treated as operators that create and annihilate particles, and the concept of probability density is replaced by more general field-theoretic constructs.

By treating the Klein-Gordon field as a quantum field, we can describe the creation and annihilation of particles and antiparticles, providing a consistent framework for interpreting the solutions to the Klein-Gordon equation. This approach resolves the issue of negative probability densities and allows for a consistent probabilistic interpretation in the context of quantum field theory.

Bonus Problem: Klein-Gordon Equation with an External Electromagnetic Field

Consider a scalar particle described by the Klein-Gordon equation in the presence of an external electromagnetic field. The Klein-Gordon equation in this context is given by:

$$\left(\frac{1}{c^2}\left(\frac{\partial}{\partial t} - i\frac{e}{\hbar}A_0\right)^2 - \left(\nabla - i\frac{e}{\hbar c}\mathbf{A}\right)^2 + \frac{m^2c^2}{\hbar^2}\right)\phi = 0, \tag{6.142}$$

where A_0 is the scalar potential, $\mathbf{A}$ is the vector potential, e is the charge of the particle, and $\hbar$ is the reduced Planck constant. Solve this equation for a uniform magnetic field $\mathbf{B} = B\hat{z}$ using the gauge $\mathbf{A} = \frac{1}{2}\mathbf{B} \times \mathbf{r}$.

Solution:

1. **Setting Up the Problem:**

We begin by choosing a gauge for the vector potential $\mathbf{A}$ such that it corresponds to a uniform magnetic field $\mathbf{B} = B\hat{z}$. A convenient choice is the symmetric gauge:

$$\mathbf{A} = \frac{1}{2}\mathbf{B} \times \mathbf{r}. \tag{6.143}$$

For $\mathbf{B} = B\hat{z}$, this becomes:

$$\mathbf{A} = \frac{B}{2}(-y, x, 0). \tag{6.144}$$

2. **Substituting the Vector Potential:**

The Klein-Gordon equation in the presence of the electromagnetic field is:

$$\left(\frac{1}{c^2}\left(\frac{\partial}{\partial t} - i\frac{e}{\hbar}A_0\right)^2 - \left(\nabla - i\frac{e}{\hbar c}\mathbf{A}\right)^2 + \frac{m^2c^2}{\hbar^2}\right)\phi = 0. \tag{6.145}$$

For a static magnetic field, $A_0 = 0$, so the equation simplifies to:

$$\left(\frac{1}{c^2}\frac{\partial^2}{\partial t^2} - \left(\nabla - i\frac{e}{\hbar c}\mathbf{A}\right)^2 + \frac{m^2c^2}{\hbar^2}\right)\phi = 0. \tag{6.146}$$

3. **Expressing the Covariant Derivative:**

The covariant derivative in the presence of the vector potential $\mathbf{A}$ is:

$$\left(\nabla - i\frac{e}{\hbar c}\mathbf{A}\right)^2 = \nabla^2 - i\frac{e}{\hbar c}(\nabla \cdot \mathbf{A}) - \frac{e^2}{\hbar^2 c^2}\mathbf{A}^2 - 2i\frac{e}{\hbar c}\mathbf{A} \cdot \nabla. \tag{6.147}$$

Since $\nabla \cdot \mathbf{A} = 0$ in the chosen gauge, the equation reduces to:

$$\left(\nabla - i\frac{e}{\hbar c}\mathbf{A}\right)^2 = \nabla^2 - \frac{e^2}{\hbar^2 c^2}\mathbf{A}^2 - 2i\frac{e}{\hbar c}\mathbf{A} \cdot \nabla. \tag{6.148}$$

4. **Substituting $\mathbf{A}$ into the Equation:**

Substitute $\mathbf{A} = \frac{B}{2}(-y, x, 0)$ into the equation:

$$\mathbf{A}^2 = \left(\frac{B}{2}\right)^2 (y^2 + x^2). \tag{6.149}$$

Therefore,

$$-\frac{e^2}{\hbar^2 c^2}\mathbf{A}^2 = -\frac{e^2 B^2}{4\hbar^2 c^2}(x^2+y^2). \tag{6.150}$$

The term $-2i\frac{e}{\hbar c}\mathbf{A}\cdot\nabla$ becomes:

$$-2i\frac{e}{\hbar c}\left(\frac{B}{2}(-y,x,0)\right)\cdot\nabla = -i\frac{eB}{\hbar c}(-y\partial_x + x\partial_y). \tag{6.151}$$

5. **Simplified Klein-Gordon Equation:**
The Klein-Gordon equation now reads:

$$\left(\frac{1}{c^2}\frac{\partial^2}{\partial t^2} - \nabla^2 + \frac{e^2B^2}{4\hbar^2c^2}(x^2+y^2) + i\frac{eB}{\hbar c}(y\partial_x - x\partial_y) + \frac{m^2c^2}{\hbar^2}\right)\phi = 0. \tag{6.152}$$

6. **Solving the Equation:**
Assume a separable solution of the form:

$$\phi(x,y,z,t) = e^{-iEt/\hbar}\psi(x,y,z). \tag{6.153}$$

The equation for $\psi(x,y,z)$ becomes:

$$\left(-E^2 + \hbar^2c^2\nabla^2 - \frac{e^2B^2}{4}(x^2+y^2) - i\hbar eB(y\partial_x - x\partial_y) - m^2c^4\right)\psi = 0. \tag{6.154}$$

7. **Landau levels:**
This equation describes the quantum harmonic oscillator in the presence of a magnetic field, leading to quantized energy levels known as Landau levels. The energy eigenvalues are:

$$E_n = \hbar\omega_c\left(n+\frac{1}{2}\right) + p_z^2c^2 + m^2c^4, \tag{6.155}$$

where $\omega_c = \frac{eB}{m}$ is the cyclotron frequency and n is a non-negative integer representing the Landau level.

This example illustrates how the Klein-Gordon equation is modified in the presence of an electromagnetic field and highlights the appearance of Landau levels, which are significant in various physical contexts such as quantum Hall effect and magnetoresistance.

7 Lecture 12: Advanced Quantum Field Theory: Beyond Single Particle Quantum Theory

7.1 INTRODUCTION

In this chapter, we delve into the limitations of single-particle quantum theory and explore the framework of quantum field theory (QFT) as a natural extension. We will discuss different representations of quantum mechanics, including the interaction picture and the Schrödinger and Heisenberg pictures, with a focus on their time evolution operators. We will also examine interaction terms in QFT, specifically the ϕ^4 theory, the sum of fields $\phi_1 + \phi_2$, and the interaction with an external source field. This chapter aims to provide a comprehensive understanding of the transition from single-particle quantum mechanics to the many-body description of QFT.

7.2 THE LIMITATIONS OF SINGLE-PARTICLE QUANTUM THEORY

7.2.1 HISTORICAL CONTEXT AND MOTIVATION

The early development of quantum mechanics in the 1920s and 1930s primarily focused on the behavior of single particles or small systems of particles. Key equations such as the Schrödinger equation provided a powerful framework for understanding the wave-like properties of particles and their probabilistic nature. However, as experimental techniques advanced, it became clear that single-particle quantum mechanics could not adequately describe all observed phenomena, especially in the realms of high-energy physics and condensed matter physics.

For instance, the discovery of the positron by Carl Anderson in 1932, which was predicted by Paul Dirac's relativistic quantum mechanics, pointed to the existence of antimatter. Moreover, processes involving the creation and annihilation of particles, such as electron-positron pair production, could not be fully explained by single-particle theory. These limitations necessitated the development of a more comprehensive frameworkquantum field theory (QFT).

7.2.2 DERIVATION FROM DIRAC'S EQUATION: ELECTRON AND POSITRON

Paul Dirac's formulation of the relativistic wave equation for spin-$\frac{1}{2}$ particles marked a significant advancement in quantum mechanics, providing insight into the existence of antimatter. The Dirac equation is given by:

$$(i\gamma^\mu \partial_\mu - m)\psi = 0, \tag{7.1}$$

where γ^μ are the Dirac gamma matrices, ∂_μ represents the partial derivative with respect to space-time coordinates, m is the mass of the particle, and ψ is the Dirac spinor.

DOI: 10.1201/9781003637370-7

Energy-Momentum Relation

To derive the possible energy solutions, we start with the plane wave solution for the Dirac spinor:

$$\psi(x) = u(p)e^{-ip\cdot x}, \tag{7.2}$$

where $u(p)$ is a spinor and $p\cdot x = p^{\mu}x_{\mu}$.
Substituting this into the Dirac equation, we obtain:

$$(i\gamma^{\mu} p_{\mu} - m)u(p)e^{-ip\cdot x} = 0. \tag{7.3}$$

Simplifying, we get the algebraic equation for $u(p)$:

$$(\gamma^{\mu} p_{\mu} - m)u(p) = 0. \tag{7.4}$$

Multiplying both sides by $(\gamma^{\nu} p_{\nu} + m)$ to eliminate the gamma matrices, we obtain:

$$(\gamma^{\mu} p_{\mu} + m)(\gamma^{\nu} p_{\nu} - m)u(p) = 0. \tag{7.5}$$

Using Clifford Algebra to Derive Energy-Momentum Relation

The Clifford algebra of the gamma matrices is given by the anticommutation relation:

$$\{\gamma^{\mu}, \gamma^{\nu}\} = \gamma^{\mu}\gamma^{\nu} + \gamma^{\nu}\gamma^{\mu} = 2g^{\mu\nu}I, \tag{7.6}$$

where $g^{\mu\nu}$ is the metric tensor of spacetime, and I is the identity matrix. Using this relation, we can derive the energy-momentum relation from the Dirac equation.

Starting from the Dirac equation:

$$(\gamma^{\mu} p_{\mu} - m)u(p) = 0, \tag{7.7}$$

we multiply both sides by $(\gamma^{\nu} p_{\nu} + m)$:

$$(\gamma^{\nu} p_{\nu} + m)(\gamma^{\mu} p_{\mu} - m)u(p) = 0. \tag{7.8}$$

Expanding the product inside the parentheses, we get:

$$(\gamma^{\nu} p_{\nu}\gamma^{\mu} p_{\mu} + \gamma^{\nu} p_{\nu} m - m\gamma^{\mu} p_{\mu} - m^2)u(p) = 0. \tag{7.9}$$

Rearranging the terms, we have:

$$(\gamma^{\nu} p_{\nu}\gamma^{\mu} p_{\mu} - m^2)u(p) = 0. \tag{7.10}$$

To simplify $\gamma^{\nu} p_{\nu}\gamma^{\mu} p_{\mu}$, we use the anticommutation relation of the gamma matrices:

$$\gamma^{\nu}\gamma^{\mu} = \frac{1}{2}(\gamma^{\nu}\gamma^{\mu} + \gamma^{\mu}\gamma^{\nu}) + \frac{1}{2}(\gamma^{\nu}\gamma^{\mu} - \gamma^{\mu}\gamma^{\nu}). \tag{7.11}$$

The first term is the symmetric part, and the second term is the antisymmetric part. Using the Clifford algebra, we get:

$$\gamma^{\nu}\gamma^{\mu} = g^{\nu\mu}I + \Sigma^{\nu\mu}, \tag{7.12}$$

where $\Sigma^{\nu\mu} = \frac{1}{2}[\gamma^{\nu}, \gamma^{\mu}]$ is the commutator of the gamma matrices.
Substituting this into our expression, we get:

$$\gamma^{\nu} p_{\nu}\gamma^{\mu} p_{\mu} = g^{\nu\mu} p_{\nu} p_{\mu} I + \Sigma^{\nu\mu} p_{\nu} p_{\mu}. \tag{7.13}$$

Since the commutator term vanishes for the symmetric part $g^{\nu\mu} p_\nu p_\mu$, we have:

$$\gamma^\nu p_\nu \gamma^\mu p_\mu = g^{\nu\mu} p_\nu p_\mu I = p^\nu p_\nu I = p^2 I, \tag{7.14}$$

where $p^2 = p_\mu p^\mu$ is the invariant length of the four-momentum.
Substituting this back into our original equation, we obtain:

$$(p^2 I - m^2 I)u(p) = 0. \tag{7.15}$$

Simplifying, we find:

$$(p^2 - m^2)u(p) = 0. \tag{7.16}$$

For nontrivial solutions $u(p) \neq 0$, this implies:

$$p^2 = m^2. \tag{7.17}$$

Thus, we have derived the relativistic energy-momentum relation:

$$E^2 = \mathbf{p}^2 + m^2. \tag{7.18}$$

This relation shows that the energy E of a particle is related to its momentum $\mathbf{p}$ and mass m, accounting for the relativistic effects. The solutions to this equation include both positive and negative energy states, corresponding to particles and antiparticles.

Interpretation of Solutions

The positive energy solution, $E = +\sqrt{\mathbf{p}^2 + m^2}$, corresponds to particles with energy E. However, the negative energy solution, $E = -\sqrt{\mathbf{p}^2 + m^2}$, initially posed a problem as it suggested the existence of particles with negative energy, which seemed unphysical.

Dirac's ingenious interpretation was to propose that these negative energy states were actually states of a new type of particle, the positron, which has the same mass as the electron but opposite charge. In this interpretation, all negative energy states are filled in a "Dirac sea," and holes in this sea behave like positively charged particles. This led to the prediction and subsequent discovery of the positron, confirming the validity of Dirac's theory.

Mathematical Representation of Electron and Positron

In the presence of an external field, transitions can occur between the filled negative energy states and the positive energy states. An electron can be excited from a negative energy state to a positive energy state, leaving behind a hole interpreted as a positron. This is mathematically represented by the creation and annihilation operators in quantum field theory:

$$a^\dagger(p) \text{ creates an electron with momentum } p, \tag{7.19}$$

$$b^\dagger(p) \text{ creates a positron with momentum } p. \tag{7.20}$$

This formalism naturally incorporates the creation and annihilation processes, explaining phenomena such as electron-positron pair production and annihilation. This derivation from the Dirac equation not only provided a theoretical foundation for the existence of antimatter but also demonstrated the necessity of a quantum field theory framework to describe particle creation and annihilation processes.

7.2.3 EMERGENCE OF QUANTUM FIELD THEORY

Quantum Field Theory emerged as a natural extension of quantum mechanics and special relativity. It treats particles as excitations of underlying fields, allowing for the creation and annihilation of particles. This framework was initially developed in the late 1920s and early 1930s by pioneers such as Dirac, Heisenberg, and Pauli, but it was not until the development of quantum electrodynamics (QED) in the 1940s by Feynman, Schwinger, and Tomonaga that QFT gained its modern form.

QFT elegantly combines quantum mechanics with special relativity by describing fields rather than individual particles. The quantization of these fields leads to the concept of particles as quanta of the field. This approach not only explains particle creation and annihilation but also provides a consistent method for calculating interactions using perturbation theory and Feynman diagrams.

7.2.4 PARTICLE CREATION AND ANNIHILATION

One of the most profound implications of QFT is the ability to describe processes where particles are created and annihilated. In single-particle quantum mechanics, the number of particles is fixed, but in QFT, the particle number can change. This is crucial for understanding high-energy particle collisions and decay processes.

To illustrate this, consider the quantization of a scalar field $\phi(x)$. The field can be expanded in terms of creation and annihilation operators $a^\dagger(\mathbf{k})$ and $a(\mathbf{k})$:

$$\phi(x) = \int \frac{d^3k}{(2\pi)^3} \frac{1}{\sqrt{2\omega_{\mathbf{k}}}} \left(a(\mathbf{k})e^{-ik\cdot x} + a^\dagger(\mathbf{k})e^{ik\cdot x} \right), \tag{7.21}$$

where $\omega_{\mathbf{k}} = \sqrt{\mathbf{k}^2 + m^2}$ is the energy of the field mode with wave vector $\mathbf{k}$. The operators $a(\mathbf{k})$ and $a^\dagger(\mathbf{k})$ satisfy the commutation relations:

$$[a(\mathbf{k}), a^\dagger(\mathbf{k}')] = (2\pi)^3\delta^3(\mathbf{k} - \mathbf{k}'), \tag{7.22}$$

$$[a(\mathbf{k}), a(\mathbf{k}')] = [a^\dagger(\mathbf{k}), a^\dagger(\mathbf{k}')] = 0. \tag{7.23}$$

The vacuum state $|0\rangle$ is defined by $a(\mathbf{k})|0\rangle = 0$. Acting on the vacuum with $a^\dagger(\mathbf{k})$ creates a particle with momentum $\mathbf{k}$:

$$|\mathbf{k}\rangle = a^\dagger(\mathbf{k})|0\rangle. \tag{7.24}$$

This formalism allows for the description of processes where particles are created or annihilated, which is essential for understanding phenomena such as pair production and scattering events.

7.2.4.0.1 *Example: Electron-Positron Pair Production*

Consider the process of electron-positron pair production, where a high-energy photon converts into an electron and a positron. In the framework of quantum field theory (QFT), this process can be described by the interaction of the quantized electromagnetic field with the electron field. The Hamiltonian for this interaction includes terms that allow for the creation and annihilation of electron-positron pairs from the vacuum.

The Hamiltonian H for the interaction of the electron-positron field with the electromagnetic field is given by:

$$H = H_0 + H_{\text{int}}, \tag{7.25}$$

where H_0 is the free Hamiltonian, and H_{int} is the interaction Hamiltonian.

7.2.4.0.1.1 Free Hamiltonian H_0: The free Hamiltonian H_0 consists of two parts: the Hamiltonian for the free electron field H_ψ and the Hamiltonian for the free electromagnetic field H_A:

$$H_0 = H_\psi + H_A. \tag{7.26}$$

The Hamiltonian for the free electron field is given by:

$$H_\psi = \int d^3x\, \psi^\dagger(\mathbf{x}) \left(-i\hbar c\, \alpha \cdot \nabla + \beta mc^2\right) \psi(\mathbf{x}), \tag{7.27}$$

where $\psi(\mathbf{x})$ is the electron field operator, α and β are the Dirac matrices, and m is the mass of the electron.

The Hamiltonian for the free electromagnetic field is given by:

$$H_A = \int d^3x \left(\frac{1}{2}\mathbf{E}^2 + \frac{1}{2}\mathbf{B}^2\right), \tag{7.28}$$

where $\mathbf{E}$ and $\mathbf{B}$ are the electric and magnetic field operators, respectively.

7.2.4.0.1.2 Interaction Hamiltonian H_{int} The interaction Hamiltonian H_{int} describes the interaction between the electron field and the electromagnetic field. In the minimal coupling scheme, this interaction is introduced by replacing the partial derivatives in the Dirac equation with covariant derivatives. The interaction Hamiltonian is given by:

$$H_{\text{int}} = e \int d^3x\, \psi^\dagger(\mathbf{x}) \gamma^\mu A_\mu(\mathbf{x}) \psi(\mathbf{x}), \tag{7.29}$$

where e is the electric charge, γ^μ are the gamma matrices, $A_\mu(\mathbf{x})$ is the electromagnetic field operator, and $\psi(\mathbf{x})$ is the electron field operator.

7.2.4.0.1.3 Terms in the Interaction Hamiltonian: The interaction Hamiltonian H_{int} can be expanded into creation and annihilation operators for the electron, positron, and photon fields. The key terms include:

1. **Electron-Photon Interaction:**

$$H_{\text{int}} \supset e \int d^3x \left(\psi^\dagger \gamma^\mu \psi\right) A_\mu, \tag{7.30}$$

which describes the interaction between the electron field and the electromagnetic field.

2. **Pair Creation and Annihilation:**

$$H_{\text{int}} \supset e \int d^3x \left(\psi^\dagger \gamma^\mu \psi\right) A_\mu, \tag{7.31}$$

where $\psi^\dagger$ and ψ represent the creation and annihilation operators for the electron and positron fields. This term describes processes such as pair creation (photon converting into an electron-positron pair) and pair annihilation (electron and positron annihilating to produce a photon).

In the context of electron-positron pair production, the relevant term in H_{int} is the one that involves the creation of an electron-positron pair from the vacuum due to the interaction with the photon field. This process can be represented by Feynman diagrams and calculated using perturbation theory in quantum electrodynamics (QED).

By understanding the structure and components of the Hamiltonian in QFT, we can describe and predict the outcomes of high-energy interactions, such as electron-positron pair production, which are fundamental to particle physics and the Standard Model.

7.3 THE SCHRÖDINGER VERSUS HEISENBERG PICTURES

In quantum mechanics, there are different formulations to describe the time evolution of a system. The two primary pictures are the Schrödinger picture and the Heisenberg picture. Both pictures are mathematically equivalent and provide the same physical predictions, but they approach the problem of time evolution from different perspectives.

7.3.1 THE SCHRÖDINGER PICTURE

7.3.1.1 State Vectors and Operators

In the Schrödinger picture, the state vectors (or wave functions) evolve with time, while the operators representing observables remain fixed. The state of a quantum system at time t is described by the state vector $|\psi(t)\rangle$. The time evolution of the state vector is governed by the Schrödinger equation:

$$i\hbar\frac{\partial}{\partial t}|\psi(t)\rangle = \hat{H}|\psi(t)\rangle, \tag{7.32}$$

where $\hat{H}$ is the Hamiltonian operator of the system.

7.3.1.2 Time Evolution Operator

The solution to the time-dependent Schrödinger equation can be expressed in terms of the time evolution operator $\hat{U}(t,t_0)$, which evolves the state from an initial time t_0 to a later time t:

$$|\psi(t)\rangle = \hat{U}(t,t_0)|\psi(t_0)\rangle. \tag{7.33}$$

The time evolution operator $\hat{U}(t,t_0)$ is given by:

$$\hat{U}(t,t_0) = e^{-\frac{i}{\hbar}\hat{H}(t-t_0)}. \tag{7.34}$$

For a time-independent Hamiltonian, the operator $\hat{U}(t,t_0)$ satisfies the following properties:

- Unitarity: $\hat{U}^\dagger(t,t_0)\hat{U}(t,t_0) = \hat{I}$
- Initial condition: $\hat{U}(t_0,t_0) = \hat{I}$
- Composition: $\hat{U}(t,t_0) = \hat{U}(t,t_1)\hat{U}(t_1,t_0)$ for any intermediate time t_1

7.3.2 THE HEISENBERG PICTURE

In the Heisenberg picture, the state vectors are considered time-independent, while the operators evolve with time. The state vector $|\psi\rangle$ remains constant, and the time evolution is transferred to the operators.

7.3.2.1 Operator Dynamics

The time evolution of an operator $\hat{O}$ in the Heisenberg picture is given by:

$$\hat{O}_H(t) = \hat{U}^\dagger(t,t_0)\hat{O}\hat{U}(t,t_0), \tag{7.35}$$

where $\hat{O}_H(t)$ is the Heisenberg picture operator and $\hat{U}(t,t_0)$ is the time evolution operator from the Schrödinger picture.

7.3.2.2 Heisenberg Equation of Motion

The Heisenberg equation of motion describes the time evolution of operators in the Heisenberg picture. It is derived by differentiating $\hat{O}_H(t)$ with respect to time:

$$\frac{d\hat{O}_H(t)}{dt} = \frac{1}{i\hbar}[\hat{O}_H(t), \hat{H}] + \left(\frac{\partial \hat{O}_H(t)}{\partial t}\right)_{\text{explicit}}, \tag{7.36}$$

where $[\hat{O}_H(t), \hat{H}]$ is the commutator of $\hat{O}_H(t)$ and the Hamiltonian $\hat{H}$, and the second term accounts for any explicit time dependence of $\hat{O}_H(t)$.

For time-independent operators, the Heisenberg equation of motion simplifies to:

$$\frac{d\hat{O}_H(t)}{dt} = \frac{1}{i\hbar}[\hat{O}_H(t), \hat{H}]. \tag{7.37}$$

7.3.3 COMPARISON AND EQUIVALENCE OF THE PICTURES

Both the Schrödinger and Heisenberg pictures provide a complete description of the quantum system's time evolution. Although the formalisms appear different, they are mathematically equivalent.

7.3.3.1 Equivalence of Expectations Values

The equivalence of the two pictures can be shown by comparing the expectation values of observables. In the Schrödinger picture, the expectation value of an observable $\hat{O}$ is given by:

$$\langle \hat{O} \rangle_S = \langle \psi(t) | \hat{O} | \psi(t) \rangle. \tag{7.38}$$

In the Heisenberg picture, the expectation value is:

$$\langle \hat{O} \rangle_H = \langle \psi | \hat{O}_H(t) | \psi \rangle. \tag{7.39}$$

Using the definition of $\hat{O}_H(t)$:

$$\langle \hat{O} \rangle_H = \langle \psi | \hat{U}^\dagger(t,t_0) \hat{O} \hat{U}(t,t_0) | \psi \rangle. \tag{7.40}$$

Since $|\psi(t)\rangle = \hat{U}(t,t_0)|\psi\rangle$, we have:

$$\langle \hat{O} \rangle_H = \langle \psi(t) | \hat{O} | \psi(t) \rangle = \langle \hat{O} \rangle_S. \tag{7.41}$$

This demonstrates that both pictures yield the same physical predictions.

7.3.3.2 Example: Time Evolution of a Simple Harmonic Oscillator

Consider a simple harmonic oscillator with the Hamiltonian:

$$\hat{H} = \frac{\hat{p}^2}{2m} + \frac{1}{2} m\omega^2 \hat{x}^2. \tag{7.42}$$

In the Schrödinger picture, the time evolution of the state vector is given by:

$$|\psi(t)\rangle = e^{-\frac{i}{\hbar}\hat{H}t} |\psi(0)\rangle. \tag{7.43}$$

In the Heisenberg picture, the operators evolve with time while the state vectors remain constant. The position and momentum operators evolve as:

$$\hat{x}_H(t) = \hat{U}^\dagger(t) \hat{x} \hat{U}(t), \tag{7.44}$$

$$\hat{p}_H(t) = \hat{U}^\dagger(t)\hat{p}\hat{U}(t). \tag{7.45}$$

where $\hat{U}(t) = e^{-\frac{i}{\hbar}\hat{H}t}$.

Using the Heisenberg equation of motion, we find the time derivatives of the operators:

$$\frac{d\hat{x}_H(t)}{dt} = \frac{1}{i\hbar}[\hat{x}_H(t), \hat{H}] = \frac{\hat{p}_H(t)}{m}, \tag{7.46}$$

$$\frac{d\hat{p}_H(t)}{dt} = \frac{1}{i\hbar}[\hat{p}_H(t), \hat{H}] = -m\omega^2\hat{x}_H(t). \tag{7.47}$$

These are coupled differential equations for $\hat{x}_H(t)$ and $\hat{p}_H(t)$. Solving these equations, we obtain:

$$\hat{x}_H(t) = \hat{x}(0)\cos(\omega t) + \frac{\hat{p}(0)}{m\omega}\sin(\omega t), \tag{7.48}$$

$$\hat{p}_H(t) = \hat{p}(0)\cos(\omega t) - m\omega\hat{x}(0)\sin(\omega t). \tag{7.49}$$

To show that the Schrödinger picture solution is equivalent to the Heisenberg picture solution, we examine the expectation values of $\hat{x}$ and $\hat{p}$. In the Schrödinger picture, the expectation value of $\hat{x}$ is:

$$\langle\psi(t)|\hat{x}|\psi(t)\rangle = \langle\psi(0)|\, e^{\frac{i}{\hbar}\hat{H}t}\hat{x}e^{-\frac{i}{\hbar}\hat{H}t}\,|\psi(0)\rangle. \tag{7.50}$$

Using the unitary property of $e^{-\frac{i}{\hbar}\hat{H}t}$, we can write:

$$\langle\psi(t)|\hat{x}|\psi(t)\rangle = \langle\psi(0)|\, \hat{x}_H(t)\,|\psi(0)\rangle, \tag{7.51}$$

where $\hat{x}_H(t) = e^{\frac{i}{\hbar}\hat{H}t}\hat{x}e^{-\frac{i}{\hbar}\hat{H}t}$ as previously derived.

The same procedure applies for $\hat{p}$:

$$\langle\psi(t)|\hat{p}|\psi(t)\rangle = \langle\psi(0)|\, e^{\frac{i}{\hbar}\hat{H}t}\hat{p}e^{-\frac{i}{\hbar}\hat{H}t}\,|\psi(0)\rangle = \langle\psi(0)|\, \hat{p}_H(t)\,|\psi(0)\rangle. \tag{7.52}$$

Therefore, the time evolution in the Schrödinger picture given by $|\psi(t)\rangle = e^{-\frac{i}{\hbar}\hat{H}t}\,|\psi(0)\rangle$ is equivalent to the time evolution of operators in the Heisenberg picture where the operators $\hat{x}_H(t)$ and $\hat{p}_H(t)$ evolve as:

$$\hat{x}_H(t) = \hat{x}(0)\cos(\omega t) + \frac{\hat{p}(0)}{m\omega}\sin(\omega t), \tag{7.53}$$

$$\hat{p}_H(t) = \hat{p}(0)\cos(\omega t) - m\omega\hat{x}(0)\sin(\omega t). \tag{7.54}$$

These results illustrate how the Heisenberg picture describes the time evolution of operators, while the Schrödinger picture focuses on the time evolution of state vectors.

In summary, both the Schrödinger and Heisenberg pictures are essential tools in quantum mechanics. They offer complementary views of the time evolution of quantum systems and highlight the rich structure of quantum theory.

7.4 INTERACTION TERMS IN QUANTUM FIELD THEORY

7.4.1 ϕ^4 THEORY

7.4.1.1 Lagrangian and Field Equations

In ϕ^4 theory, the Lagrangian for a scalar field ϕ with a quartic interaction term is given by:

$$\mathcal{L} = \frac{1}{2}\partial_\mu\phi\partial^\mu\phi - \frac{1}{2}m^2\phi^2 - \frac{\lambda}{4!}\phi^4, \tag{7.55}$$

where m is the mass of the scalar particle and λ is the coupling constant for the ϕ^4 interaction.

The Euler-Lagrange equation for the field ϕ is derived from the Lagrangian:

$$\partial_\mu \left(\frac{\partial \mathscr{L}}{\partial(\partial_\mu \phi)} \right) - \frac{\partial \mathscr{L}}{\partial \phi} = 0. \tag{7.56}$$

Substituting the Lagrangian, we obtain the field equation:

$$\partial_\mu \partial^\mu \phi + m^2 \phi + \frac{\lambda}{3!} \phi^3 = 0, \tag{7.57}$$

which is a nonlinear Klein-Gordon equation with a ϕ^3 term.

7.4.1.2 Renormalization in ϕ^4 Theory

Renormalization is a process used to remove infinities that arise in quantum field theory calculations. In ϕ^4 theory, renormalization involves redefining the parameters m, λ, and ϕ to absorb the infinities into redefined (renormalized) quantities.

The renormalized Lagrangian is written as:

$$\mathscr{L}_R = \frac{1}{2} Z_\phi \partial_\mu \phi \partial^\mu \phi - \frac{1}{2} Z_m m^2 \phi^2 - \frac{Z_\lambda \lambda}{4!} \phi^4, \tag{7.58}$$

where Z_ϕ, Z_m, and Z_λ are the renormalization constants.

The renormalized quantities are related to the bare quantities by:

$$\phi_B = \sqrt{Z_\phi} \phi, \quad m_B^2 = Z_m m^2, \quad \lambda_B = Z_\lambda \lambda. \tag{7.59}$$

Renormalization ensures that the predictions of ϕ^4 theory remain finite and physically meaningful.

7.4.2 SUM OF FIELDS: $\phi_1 + \phi_2$

7.4.2.1 Interaction Lagrangian

When considering two interacting scalar fields ϕ_1 and ϕ_2, the interaction Lagrangian can be written as:

$$\mathscr{L} = \mathscr{L}_1 + \mathscr{L}_2 + \mathscr{L}_{\text{int}}, \tag{7.60}$$

where $\mathscr{L}_1$ and $\mathscr{L}_2$ are the free Lagrangians for ϕ_1 and ϕ_2, respectively, and $\mathscr{L}_{\text{int}}$ describes the interaction between the fields.

An example of an interaction term is:

$$\mathscr{L}_{\text{int}} = -g \phi_1^2 \phi_2^2, \tag{7.61}$$

where g is the coupling constant.

7.4.2.2 Coupling and Mixing of Fields

The interaction term $\mathscr{L}_{\text{int}} = -g \phi_1^2 \phi_2^2$ leads to coupling and mixing of the fields ϕ_1 and ϕ_2. This interaction affects the dynamics of the fields.

The coupled field equations are:

$$\partial_\mu \partial^\mu \phi_1 + m_1^2 \phi_1 + 2g \phi_1 \phi_2^2 = 0, \tag{7.62}$$

$$\partial_\mu \partial^\mu \phi_2 + m_2^2 \phi_2 + 2g \phi_2 \phi_1^2 = 0. \tag{7.63}$$

7.4.2.3 Applications and Physical Implications

The coupling of fields has significant physical implications in various areas of physics. For instance:

- In particle physics, the interaction between different types of particles, such as mesons and baryons, can be described by coupling fields.
- In condensed matter physics, coupled fields can model interactions between different quasiparticles or collective excitations.
- The study of coupled fields helps in understanding phenomena such as symmetry breaking, phase transitions, and the emergence of new collective behaviors.

7.4.3 EXTERNAL SOURCE FIELD INTERACTION

7.4.3.1 Lagrangian with External Source

In the presence of an external source $J(x)$, the Lagrangian for a scalar field ϕ is modified to include an interaction term with the source:

$$\mathscr{L} = \frac{1}{2}\partial_\mu \phi \partial^\mu \phi - \frac{1}{2}m^2\phi^2 + J(x)\phi. \tag{7.64}$$

The source term $J(x)$ can represent an external force or field that interacts with the scalar field ϕ.

7.4.3.2 Green's Functions and Propagators

The presence of an external source modifies the equations of motion for the field. The equation of motion becomes:

$$(\partial_\mu \partial^\mu + m^2)\phi(x) = J(x). \tag{7.65}$$

The solution to this equation can be expressed in terms of the Green's function $G(x-y)$:

$$\phi(x) = \int d^4y\, G(x-y)J(y). \tag{7.66}$$

The Green's function $G(x-y)$ is the solution to the equation:

$$(\partial_\mu \partial^\mu + m^2)G(x-y) = \delta^4(x-y). \tag{7.67}$$

The Green's function serves as the propagator for the field ϕ, describing how disturbances due to the source propagate through space and time.

7.4.3.3 Applications in Quantum Electrodynamics (QED)

In quantum electrodynamics (QED), the interaction between charged particles and the electromagnetic field can be described using an external source field. The Lagrangian for QED includes interaction terms of the form:

$$\mathscr{L}_{\text{int}} = -e\bar{\psi}\gamma^\mu \psi A_\mu, \tag{7.68}$$

where e is the electric charge, ψ is the electron field, $\bar{\psi}$ is the Dirac adjoint, γ^μ are the gamma matrices, and A_μ is the electromagnetic field.

The gamma matrices γ^μ are a set of four 4×4 matrices that form the basis of the Clifford algebra associated with spacetime. They satisfy the anticommutation relations:

$$\{\gamma^\mu, \gamma^\nu\} = \gamma^\mu \gamma^\nu + \gamma^\nu \gamma^\mu = 2g^{\mu\nu} I, \tag{7.69}$$

where $g^{\mu\nu}$ is the metric tensor of spacetime and I is the identity matrix.

The explicit representation of the gamma matrices in the Dirac representation is:

$$\gamma^0 = \begin{pmatrix} I & 0 \\ 0 & -I \end{pmatrix}, \quad \gamma^i = \begin{pmatrix} 0 & \sigma^i \\ -\sigma^i & 0 \end{pmatrix}, \tag{7.70}$$

where σ^i are the Pauli matrices.

These gamma matrices play a crucial role in ensuring the Lorentz invariance of the Dirac equation and the interaction terms in QED.

In summary, interaction terms in quantum field theory provide a framework for understanding how fields interact with each other and with external sources. These interactions are described by modified Lagrangians, and the resulting field equations can be solved using techniques such as perturbation theory and Green's functions. The study of these interactions is fundamental to understanding the behavior of particles and fields in high-energy physics and other areas of theoretical physics. The source term in this context represents the interaction between the electron field and the electromagnetic field. Green's functions and propagators play a crucial role in calculating scattering amplitudes and other physical observables in QED.

7.5 THE INTERACTION PICTURE

The interaction picture (or Dirac picture) is a representation in quantum mechanics that is particularly useful for dealing with time-dependent Hamiltonians and perturbation theory. It provides a middle ground between the Schrödinger picture, where states evolve with time, and the Heisenberg picture, where operators evolve with time.

7.5.1 DEFINITION AND FORMALISM

In the interaction picture, both the state vectors and the operators evolve with time. The total Hamiltonian $\hat{H}$ is split into two parts: the free Hamiltonian $\hat{H}_0$, which represents the unperturbed system, and the interaction Hamiltonian $\hat{H}_{\text{int}}$, which represents the perturbation or interaction:

$$\hat{H} = \hat{H}_0 + \hat{H}_{\text{int}}. \tag{7.71}$$

The state vector in the interaction picture, $|\psi_I(t)\rangle$, is related to the state vector in the Schrödinger picture, $|\psi_S(t)\rangle$, by:

$$|\psi_I(t)\rangle = e^{\frac{i}{\hbar}\hat{H}_0 t}\,|\psi_S(t)\rangle\,. \tag{7.72}$$

Similarly, the operator $\hat{O}_I(t)$ in the interaction picture is related to the operator $\hat{O}_S(t)$ in the Schrödinger picture by:

$$\hat{O}_I(t) = e^{\frac{i}{\hbar}\hat{H}_0 t}\hat{O}_S(t)e^{-\frac{i}{\hbar}\hat{H}_0 t}. \tag{7.73}$$

The time evolution of the state vector in the interaction picture is governed by the interaction Hamiltonian $\hat{H}_{\text{int}}$:

$$i\hbar\frac{\partial}{partialt}|\psi_I(t)\rangle = \hat{H}_{\text{int}}(t)\,|\psi_I(t)\rangle\,, \tag{7.74}$$

where $\hat{H}_{\text{int}}(t) = e^{\frac{i}{\hbar}\hat{H}_0 t}\hat{H}_{\text{int}}e^{-\frac{i}{\hbar}\hat{H}_0 t}$ is the interaction Hamiltonian in the interaction picture.

7.5.2 TIME EVOLUTION OPERATORS IN THE INTERACTION PICTURE

The time evolution operator $\hat{U}(t,t_0)$ in the interaction picture satisfies the Schrödinger equation with the interaction Hamiltonian:

$$i\hbar\frac{\partial}{\partial t}\hat{U}(t,t_0) = \hat{H}_{\text{int}}(t)\hat{U}(t,t_0). \tag{7.75}$$

The formal solution to this equation is given by the time-ordered exponential:

$$\hat{U}(t,t_0) = T\exp\left(-\frac{i}{\hbar}\int_{t_0}^{t}\hat{H}_{\text{int}}(t')\,dt'\right), \tag{7.76}$$

where T denotes the time-ordering operator, which arranges the operators in chronological order with the earliest time on the right.
The state vector in the interaction picture evolves according to:

$$|\psi_I(t)\rangle = \hat{U}(t,t_0)\,|\psi_I(t_0)\rangle. \tag{7.77}$$

In perturbation theory, the time evolution operator is expanded as a series:

$$\begin{aligned}\hat{U}(t,t_0) = 1 &+ \left(-\frac{i}{\hbar}\right)\int_{t_0}^{t}\hat{H}_{\text{int}}(t_1)\,dt_1 \\ &+ \left(-\frac{i}{\hbar}\right)^2\int_{t_0}^{t}dt_1\int_{t_0}^{t_1}dt_2\,\hat{H}_{\text{int}}(t_1)\hat{H}_{\text{int}}(t_2) + \cdots.\end{aligned} \tag{7.78}$$

7.5.3 APPLICATION TO QUANTUM FIELD THEORY

In quantum field theory (QFT), the interaction picture is used extensively to describe interactions between fields. The interaction Hamiltonian $\hat{H}_{\text{int}}$ typically represents the interaction terms in the Lagrangian density of the theory.

For example, consider the interaction Hamiltonian in QED:

$$\hat{H}_{\text{int}} = -e\int d^3x\,\bar{\psi}(x)\gamma^\mu\psi(x)A_\mu(x), \tag{7.79}$$

where e is the electric charge, $\psi(x)$ is the electron field, $\bar{\psi}(x)$ is the Dirac adjoint, γ^μ are the gamma matrices, and $A_\mu(x)$ is the electromagnetic field.

In the interaction picture, the fields evolve with the free Hamiltonian $\hat{H}_0$, while the interaction Hamiltonian $\hat{H}_{\text{int}}$ governs the time evolution of the state vector. This allows the use of perturbative techniques to calculate scattering amplitudes and other physical observables.

The interaction picture thus provides a powerful framework for analyzing interactions in quantum field theory, enabling the calculation of physical processes and the development of perturbation theory.

7.5.3.1 Worked Example: Electron-Photon Scattering (Compton Scattering)

Let's consider the process of Compton scattering, where an incoming photon scatters off an electron. This process involves the interaction Hamiltonian in QED and can be analyzed using the interaction picture.

The interaction Hamiltonian in QED is given by:

$$\hat{H}_{\text{int}}(t) = -e\int d^3x\,\bar{\psi}_I(x,t)\gamma^\mu\psi_I(x,t)A_{\mu I}(x,t), \tag{7.80}$$

where the subscript I denotes fields in the interaction picture.

The scattering amplitude for the process can be computed using perturbation theory. The first-order term in the perturbation series expansion of the S-matrix is:

$$S^{(1)} = -\frac{i}{\hbar}\int_{-\infty}^{\infty}dt\,\hat{H}_{\text{int}}(t). \tag{7.81}$$

Substituting the interaction Hamiltonian, we get:

$$S^{(1)} = \frac{ie}{\hbar}\int_{-\infty}^{\infty} dt \int d^3x\, \bar{\psi}_I(x,t)\gamma^\mu \psi_I(x,t) A_{\mu I}(x,t). \tag{7.82}$$

For Compton scattering, we consider the initial state $|i\rangle$ as a photon with momentum k and an electron with momentum p, and the final state $|f\rangle$ as a photon with momentum k' and an electron with momentum p':

$$|i\rangle = |\gamma(k), e(p)\rangle, \quad |f\rangle = |\gamma(k'), e(p')\rangle. \tag{7.83}$$

The matrix element $\mathcal{M}$ for the transition from the initial to the final state is given by:

$$\mathcal{M} = \langle f| S^{(1)} |i\rangle. \tag{7.84}$$

Using the expression for $S^{(1)}$:

$$\begin{aligned}\mathcal{M} =& \frac{ie}{\hbar}\int_{-\infty}^{\infty} dt \\ &\times \int d^3x \,\langle \gamma(k'), e(p')|\, \bar{\psi}_I(x,t)\gamma^\mu \psi_I(x,t) A_{\mu I}(x,t)\, |\gamma(k), e(p)\rangle.\end{aligned} \tag{7.85}$$

The field operators in the interaction picture can be expanded in terms of creation and annihilation operators. For the electron field $\psi_I(x,t)$ and the photon field $A_{\mu I}(x,t)$, the relevant terms for the matrix element involve the annihilation of an initial photon and the creation of a final photon:

$$A_{\mu I}(x,t) \sim e^{-ik\cdot x} a_k + e^{ik\cdot x} a_k^\dagger, \tag{7.86}$$

$$\psi_I(x,t) \sim e^{-ip\cdot x} b_p + e^{ip\cdot x} d_p^\dagger, \tag{7.87}$$

where a_k and $a_k^\dagger$ are the annihilation and s for photons, and b_p and $d_p^\dagger$ are the annihilation and s for electrons and positrons, respectively.

Substituting these expansions into the matrix element and performing the integrals over space and time, we obtain the scattering amplitude for Compton scattering. The details of these calculations involve evaluating commutators and using the properties of the gamma matrices. The resulting scattering amplitude $\mathcal{M}$ can then be used to calculate the differential cross-section for Compton scattering, which describes the probability of scattering at a given angle.

7.6 ADVANCED TOPICS AND EXTENSIONS

7.6.1 PATH INTEGRALS IN QUANTUM FIELD THEORY

The path integral formulation provides a powerful and intuitive framework for understanding quantum field theory (QFT). It offers an alternative to the operator formalism, emphasizing the sum over histories approach to quantum mechanics.

7.6.1.1 Formalism and Interpretation

The path integral formulation, developed by Richard Feynman, is based on the principle of summing over all possible paths that a particle can take between two points. For a quantum field $\phi(x)$, the transition amplitude between an initial field configuration $\phi_i(x)$ and a final field configuration $\phi_f(x)$ is given by the path integral:

$$\langle \phi_f, t_f | \phi_i, t_i\rangle = \int \mathcal{D}\phi\, e^{\frac{i}{\hbar}S[\phi]}, \tag{7.88}$$

where $S[\phi]$ is the action functional, defined as the integral of the Lagrangian density over spacetime:

$$S[\phi] = \int d^4x \mathscr{L}(\phi, \partial_\mu \phi). \tag{7.89}$$

The path integral approach can be used to compute correlation functions, which are central to QFT. For example, the two-point correlation function is given by:

$$\langle 0|T\{\phi(x)\phi(y)\}|0\rangle = \int \mathscr{D}\phi\, \phi(x)\phi(y) e^{\frac{i}{\hbar}S[\phi]}. \tag{7.90}$$

7.6.1.2 Applications to Interacting Fields

Path integrals are particularly useful for dealing with interacting fields. Consider a scalar field theory with a ϕ^4 interaction term:

$$\mathscr{L} = \frac{1}{2}(\partial_\mu \phi)(\partial^\mu \phi) - \frac{1}{2}m^2\phi^2 - \frac{\lambda}{4!}\phi^4. \tag{7.91}$$

The generating functional $Z[J]$ in the presence of an external source $J(x)$ is given by:

$$Z[J] = \int \mathscr{D}\phi\, e^{\frac{i}{\hbar}(S[\phi]+\int d^4x J(x)\phi(x))}. \tag{7.92}$$

Correlation functions can be obtained by functional differentiation with respect to the source $J(x)$. For example, the two-point function is:

$$\langle 0|T\{\phi(x)\phi(y)\}|0\rangle = \left.\frac{\delta^2 Z[J]}{\delta J(x)\delta J(y)}\right|_{J=0}. \tag{7.93}$$

7.6.2 NON-PERTURBATIVE METHODS IN QFT

While perturbation theory provides a valuable tool for analyzing QFT, it is not always sufficient. Non-perturbative methods are essential for understanding phenomena that cannot be captured by perturbative expansions.

7.6.2.1 Lattice Quantum Field Theory

Lattice QFT is a non-perturbative approach that discretizes spacetime into a lattice. This method enables the numerical study of QFT, particularly in regimes where perturbation theory fails. The basic idea is to replace the continuous spacetime coordinates x with discrete points on a lattice:

$$x^\mu \to an^\mu, \tag{7.94}$$

where a is the lattice spacing and n^μ are integers.
The path integral is then approximated as a sum over field configurations on the lattice:

$$Z = \int \mathscr{D}\phi\, e^{-\frac{S[\phi]}{\hbar}} \to \sum_{\{\phi\}} e^{-\frac{S[\phi]}{\hbar}}. \tag{7.95}$$

Lattice QFT is particularly powerful for studying quantum chromodynamics (QCD), the theory of the strong interaction, where it provides insights into confinement, chiral symmetry breaking, and other non-perturbative phenomena.

7.6.2.2 Instantons and Solitons

Instantons and solitons are non-perturbative solutions to the field equations in QFT. They represent localized, stable configurations of fields that play a crucial role in understanding the non-perturbative structure of QFT. Each represent intriguing solutions to field equations that have significant real-world applications. Solitons are stable, localized wave packets that maintain their shape while traveling at constant speed. They appear in various physical contexts, such as in fluid dynamics, where they manifest as solitary water waves known as "tsunamis" or "solitary waves" observed in shallow water. In optical fibers, solitons are used to transmit information over long distances without distortion, crucial for telecommunications.

Instantons, on the other hand, are prominent in quantum field theory and high-energy physics. They represent tunneling events between different vacuum states, which play a critical role in explaining phenomena like the non-perturbative effects in gauge theories, providing insights into the vacuum structure of the theory and the dynamics of quark confinement. Both solitons and instantons illustrate the richness of solutions available in nonlinear field theories and their profound implications across various domains of physics.

7.6.2.2.1 Instantons:

Instantons are solutions to the Euclidean field equations that correspond to tunneling events between different vacuum states. For example, in Yang-Mills theory, instantons are solutions to the Euclidean equations of motion:

$$F_{\mu\nu} = \tilde{F}_{\mu\nu}, \tag{7.96}$$

where $F_{\mu\nu}$ is the field strength tensor and $\tilde{F}_{\mu\nu}$ is its dual.

Instantons contribute to the path integral as non-perturbative corrections, providing insights into phenomena such as the axial anomaly and the structure of the QCD vacuum.

7.6.2.2.2 Solitons:

Solitons are stable, localized solutions to the field equations that represent particles or extended objects. In scalar field theory, a classic example is the kink solution in the ϕ^4 theory:

$$\phi(x) = \tanh\left(\frac{x}{\sqrt{2}}\right). \tag{7.97}$$

Solitons appear in various QFTs, including gauge theories and supersymmetric theories, where they provide important insights into the non-perturbative dynamics of the theory. Non-perturbative methods, including lattice QFT and the study of instantons and solitons, are essential for a complete understanding of QFT, revealing phenomena that cannot be accessed through perturbative techniques alone.

7.6.2.3 Worked Example: Instantons in Yang-Mills Theory

In Yang-Mills theory, which states that the action must be minimized for the system to be in a stable configuration, the instanton solutions are crucial as they minimize the Euclidean action. The presence of instantons in the vacuum structure of Yang-Mills theory leads to a rich topological structure and has profound implications for phenomena such as the confinement of quarks and the breaking of chiral symmetry.

Consider Yang-Mills theory with the gauge group $SU(2)$. The field strength tensor is given by:

$$F_{\mu\nu} = \partial_\mu A_\nu - \partial_\nu A_\mu + g[A_\mu, A_\nu], \tag{7.98}$$

where $A_\mu = A_\mu^a T^a$ are the gauge fields, T^a are the generators of the $SU(2)$ Lie algebra, and g is the coupling constant.

The self-duality condition for instantons is:

$$F_{\mu\nu} = \tilde{F}_{\mu\nu}, \tag{7.99}$$

where the dual field strength tensor is defined as:

$$\tilde{F}_{\mu\nu} = \frac{1}{2}\varepsilon_{\mu\nu\rho\sigma}F^{\rho\sigma}. \tag{7.100}$$

Instantons are solutions to the Euclidean field equations that satisfy this self-duality condition. One common solution is the BPST instanton, which in the singular gauge has the form:

$$A_\mu^a(x) = \frac{2\eta_{a\mu\nu}x_\nu}{x^2+\rho^2}, \tag{7.101}$$

where $\eta_{a\mu\nu}$ are the 't Hooft symbols, x_ν are the spacetime coordinates, and ρ is the instanton size.

The action for an instanton is given by:

$$S = \frac{8\pi^2}{g^2}. \tag{7.102}$$

Instantons contribute to the path integral as non-perturbative effects. Their contribution can be computed using the instanton measure, which accounts for the integration over the instanton collective coordinates (position, size, and gauge orientation).

Instantons also play a significant role in understanding tunneling processes between different vacuum states in the non-perturbative regime. This tunneling can lead to effects such as the decay of a metastable vacuum and the appearance of non-trivial vacuum states with distinct topological properties.

The calculation of instanton effects involves integrating over all possible instanton configurations in the path integral, which gives rise to non-perturbative contributions to physical observables. These contributions are essential for understanding the full dynamics of the theory beyond the perturbative regime.

7.6.2.4 Worked Example: Solitons in ϕ^4 Theory

Consider the ϕ^4 scalar field theory in (1+1) dimensions with the Lagrangian:

$$\mathscr{L} = \frac{1}{2}(\partial_\mu\phi)(\partial^\mu\phi) - \frac{\lambda}{4}(\phi^2 - v^2)^2, \tag{7.103}$$

where λ is the coupling constant and v is the vacuum expectation value.

The equation of motion is given by:

$$\partial_\mu\partial^\mu\phi - \lambda\phi(\phi^2 - v^2) = 0. \tag{7.104}$$

Consider static solutions, where $\phi = \phi(x)$ and there is no time dependence. The equation of motion simplifies to:

$$\frac{d^2\phi}{dx^2} = \lambda\phi(\phi^2 - v^2). \tag{7.105}$$

To find soliton solutions, we look for solutions that interpolate between the vacuum states $\phi = \pm v$ as x goes from $-\infty$ to ∞. One such solution is the kink solution:

$$\phi(x) = v\tanh\left(\frac{m}{\sqrt{2}}x\right), \tag{7.106}$$

where $m = \sqrt{\lambda}v$ is the mass of the scalar particle.

The energy of the soliton can be calculated by integrating the energy density:

$$E = \int_{-\infty}^{\infty} dx \left[\frac{1}{2} \left(\frac{d\phi}{dx} \right)^2 + \frac{\lambda}{4} (\phi^2 - v^2)^2 \right]. \tag{7.107}$$

Substituting the kink solution into this expression, we obtain:

$$E = \int_{-\infty}^{\infty} dx \left[\frac{1}{2} \left(\frac{mv}{\sqrt{2}\cosh^2\left(\frac{mx}{\sqrt{2}}\right)} \right)^2 + \frac{\lambda v^4}{4} \left(\tanh^2\left(\frac{mx}{\sqrt{2}}\right) - 1 \right)^2 \right]. \tag{7.108}$$

The integral evaluates to:

$$E = \frac{2\sqrt{2}}{3} v^3 \sqrt{\lambda}. \tag{7.109}$$

This energy is finite, indicating that the kink is a stable, localized soliton solution.

In summary, instantons and solitons provide non-perturbative solutions to the field equations in quantum field theory. Instantons, as solutions to the self-duality condition, contribute to tunneling events between different vacuum states. Solitons, as stable, localized solutions, represent particles or extended objects in the theory. Both play crucial roles in understanding the non-perturbative dynamics of quantum fields.

7.7 SUMMARY AND DISCUSSION

7.7.1 KEY CONCEPTS AND TAKEAWAYS

In this chapter, we have delved into several advanced topics in quantum field theory (QFT) that extend beyond the realm of single-particle quantum mechanics. Initially, we explored the limitations of single-particle quantum theory, noting that the Schrödinger equation is insufficient for describing phenomena such as particle creation and annihilation. This realization led to the development of QFT, a framework where particles are treated as excitations of underlying fields, enabling the natural description of particle interactions and dynamics through creation and annihilation operators.

We examined the Schrödinger, Heisenberg, and interaction pictures of quantum mechanics, focusing on their respective approaches to time evolution. The Schrödinger picture emphasizes the time evolution of state vectors, the Heisenberg picture concentrates on the time evolution of operators, and the interaction picture, particularly useful for dealing with time-dependent Hamiltonians, provides a middle ground where both states and operators evolve.

Our discussion of interaction terms in QFT highlighted the ϕ^4 theory, which includes a quartic interaction term, serving as a simple yet profound example of interacting fields. We also considered the sum of fields $\phi_1 + \phi_2$, which illustrates the coupling and mixing of different fields, essential for describing complex interactions. Additionally, we explored external source field interactions, as seen in quantum electrodynamics (QED), where additional source terms in the Lagrangian lead to modified field equations and propagators.

The path integral formulation, introduced by Richard Feynman, provides an intuitive approach to QFT by emphasizing the sum over histories. This formulation is particularly useful for computing correlation functions and understanding the behavior of interacting fields. Non-perturbative methods, including lattice QFT and the study of instantons and solitons, are essential for exploring phenomena that cannot be captured by perturbative expansions. Instantons represent tunneling events between different vacuum states, while solitons are stable, localized solutions to the field equations.

We also included worked examples to illustrate the practical application of QFT concepts and methods. For instance, the process of Compton scattering was analyzed using the interaction picture, and instanton solutions in Yang-Mills theory were discussed, highlighting their significance in understanding tunneling processes and the structure of the QCD vacuum.

7.8 QUESTIONS AND PROBLEMS

7.8.1 TRIVIA QUESTIONS

Question 1: What was one major limitation of single-particle quantum theory that led to the development of quantum field theory (QFT)?

(a) Inability to describe wave-particle duality
(b) Inability to account for particle creation and annihilation
(c) Inability to explain the uncertainty principle
(d) Inability to describe electron spin

Answer: b. Inability to account for particle creation and annihilation
Question 2: Which representation of quantum mechanics emphasizes the time evolution of state vectors?

(a) Heisenberg picture
(b) Schrödinger picture
(c) Interaction picture
(d) Path integral picture

Answer: b. Schrödinger picture
Question 3: In the interaction picture, what typically represents the interaction terms in the Lagrangian density?

(a) Free Hamiltonian
(b) Total Hamiltonian
(c) Interaction Hamiltonian
(d) Time evolution operator

Answer: c. Interaction Hamiltonian
Question 4: In ϕ^4 theory, what does the Lagrangian include that differentiates it from a free scalar field theory?

(a) A quadratic term
(b) A cubic term
(c) A quartic interaction term
(d) A linear interaction term

Answer: c. A quartic interaction term
Question 5: What is the primary advantage of using the path integral formulation in quantum field theory?

(a) It simplifies the Schrödinger equation
(b) It provides a sum over histories approach to quantum mechanics
(c) It eliminates the need for gamma matrices
(d) It makes renormalization unnecessary

Answer: b. It provides a sum over histories approach to quantum mechanics
Question 6: Which non-perturbative method in QFT involves discretizing spacetime into a lattice?

(a) Path integral formulation
(b) Perturbation theory
(c) Lattice quantum field theory
(d) Instanton analysis

Answer: c. Lattice quantum field theory
Question 7: What are solitons in the context of quantum field theory?

(a) Virtual particles that exist briefly
(b) Stable, localized solutions to the field equations
(c) Perturbative corrections to scattering amplitudes
(d) Oscillatory solutions to the Schrödinger equation

Answer: b. Stable, localized solutions to the field equations
Question 8: In QED, what does the interaction Hamiltonian typically include?

(a) Terms involving only the electron field
(b) Terms involving only the electromagnetic field
(c) Terms involving the interaction between the electron and electromagnetic fields
(d) Terms involving only the positron field

Answer: c. Terms involving the interaction between the electron and electromagnetic fields
Question 9: Which of the following best describes an instanton in quantum field theory?

(a) A particle with zero mass
(b) A tunneling event between different vacuum states
(c) A particle with spin 1/2
(d) A high-energy photon

Answer: b. A tunneling event between different vacuum states
Question 10: In the context of Compton scattering, what is the primary observable calculated using the interaction picture?

(a) Particle decay rates
(b) Scattering amplitude
(c) Energy levels of hydrogen
(d) Electron spin states

Answer: b. Scattering amplitude

7.8.2 PROBLEMS AND SOLUTIONS

PROBLEM 1: KLEIN-GORDON EQUATION FOR A FREE PARTICLE

Derive the Klein-Gordon equation for a free scalar field and solve it for a plane wave solution.

Solution

The Lagrangian density for a free scalar field $\phi(x)$ is given by:

$$\mathscr{L} = \frac{1}{2}\partial_\mu \phi \partial^\mu \phi - \frac{1}{2}m^2\phi^2. \tag{7.110}$$

The Euler-Lagrange equation for ϕ is:

$$\partial_\mu \left(\frac{\partial \mathcal{L}}{\partial(\partial_\mu \phi)} \right) - \frac{\partial \mathcal{L}}{\partial \phi} = 0. \tag{7.111}$$

Substituting the Lagrangian density, we get:

$$\partial_\mu \partial^\mu \phi + m^2 \phi = 0. \tag{7.112}$$

This is the Klein-Gordon equation. For a plane wave solution, assume:

$$\phi(x) = e^{-ip\cdot x}, \tag{7.113}$$

where $p \cdot x = p^\mu x_\mu = Et - \mathbf{p} \cdot \mathbf{x}$.

Substituting into the Klein-Gordon equation, we get:

$$(-p^2 + m^2)e^{-ip\cdot x} = 0, \tag{7.114}$$

which implies:

$$p^2 = m^2 \quad \Rightarrow \quad E^2 = \mathbf{p}^2 + m^2. \tag{7.115}$$

Thus, the plane wave solution satisfies the Klein-Gordon equation with the relativistic energy-momentum relation.

PROBLEM 2: HEISENBERG EQUATION OF MOTION

Derive the Heisenberg equation of motion for the position operator $\hat{x}$ and momentum operator $\hat{p}$ in a simple harmonic oscillator.

Solution

The Hamiltonian for a simple harmonic oscillator is:

$$\hat{H} = \frac{\hat{p}^2}{2m} + \frac{1}{2} m\omega^2 \hat{x}^2. \tag{7.116}$$

The Heisenberg equation of motion for an operator $\hat{O}$ is:

$$\frac{d\hat{O}}{dt} = \frac{1}{i\hbar}[\hat{O}, \hat{H}] + \left(\frac{\partial \hat{O}}{\partial t} \right)_{\text{explicit}}. \tag{7.117}$$

For $\hat{x}$:

$$\frac{d\hat{x}}{dt} = \frac{1}{i\hbar}[\hat{x}, \hat{H}] = \frac{1}{i\hbar}\left[\hat{x}, \frac{\hat{p}^2}{2m} + \frac{1}{2} m\omega^2 \hat{x}^2 \right] = \frac{1}{i\hbar}\left(\frac{1}{2m}[\hat{x}, \hat{p}^2] \right). \tag{7.118}$$

Since $[\hat{x}, \hat{p}] = i\hbar$, we have $[\hat{x}, \hat{p}^2] = 2i\hbar\hat{p}$:

$$\frac{d\hat{x}}{dt} = \frac{1}{i\hbar}\left(\frac{1}{2m} 2i\hbar\hat{p} \right) = \frac{\hat{p}}{m}. \tag{7.119}$$

For $\hat{p}$:

$$\frac{d\hat{p}}{dt} = \frac{1}{i\hbar}[\hat{p}, \hat{H}] = \frac{1}{i\hbar}\left[\hat{p}, \frac{\hat{p}^2}{2m} + \frac{1}{2} m\omega^2 \hat{x}^2 \right] = \frac{1}{i\hbar}\left(\frac{1}{2} m\omega^2 [\hat{p}, \hat{x}^2] \right). \tag{7.120}$$

Since $[\hat{p}, \hat{x}] = -i\hbar$, we have $[\hat{p}, \hat{x}^2] = -2i\hbar\hat{x}$:

$$\frac{d\hat{p}}{dt} = \frac{1}{i\hbar}\left(\frac{1}{2} m\omega^2 (-2i\hbar\hat{x}) \right) = -m\omega^2 \hat{x}. \tag{7.121}$$

Thus, the Heisenberg equations of motion for $\hat{x}$ and $\hat{p}$ are:

$$\frac{d\hat{x}}{dt} = \frac{\hat{p}}{m}, \quad \frac{d\hat{p}}{dt} = -m\omega^2 \hat{x}. \tag{7.122}$$

PROBLEM 3: INTERACTION PICTURE HAMILTONIAN

Consider an interaction Hamiltonian in QED given by
$\hat{H}_{\text{int}} = -e\int d^3x\,\bar{\psi}(x)\gamma^\mu\psi(x)A_\mu(x)$. Show how the interaction picture is useful in evaluating time-dependent perturbations.

Solution

In the interaction picture, the total Hamiltonian $\hat{H}$ is split into the free Hamiltonian $\hat{H}_0$ and the interaction Hamiltonian $\hat{H}_{\text{int}}$:

$$\hat{H} = \hat{H}_0 + \hat{H}_{\text{int}}. \tag{7.123}$$

The state vector in the interaction picture $|\psi_I(t)\rangle$ is related to the Schrödinger picture $|\psi_S(t)\rangle$ by:

$$|\psi_I(t)\rangle = e^{\frac{i}{\hbar}\hat{H}_0 t}\,|\psi_S(t)\rangle\,. \tag{7.124}$$

The time evolution in the interaction picture is governed by the interaction Hamiltonian:

$$i\hbar\frac{\partial}{\partial t}|\psi_I(t)\rangle = \hat{H}_{\text{int}}(t)\,|\psi_I(t)\rangle\,, \tag{7.125}$$

where $\hat{H}_{\text{int}}(t) = e^{\frac{i}{\hbar}\hat{H}_0 t}\hat{H}_{\text{int}}e^{-\frac{i}{\hbar}\hat{H}_0 t}$.

The time evolution operator in the interaction picture is:

$$\hat{U}(t,t_0) = T\exp\left(-\frac{i}{\hbar}\int_{t_0}^{t}\hat{H}_{\text{int}}(t')\,dt'\right), \tag{7.126}$$

where T denotes time ordering.

This operator allows for the evaluation of time-dependent perturbations using perturbation theory. The perturbative series expansion is:

$$\begin{aligned}\hat{U}(t,t_0) = 1 &+ \left(-\frac{i}{\hbar}\right)\int_{t_0}^{t}\hat{H}_{\text{int}}(t_1)\,dt_1 \\ &+ \left(-\frac{i}{\hbar}\right)^2\int_{t_0}^{t}dt_1\int_{t_0}^{t_1}dt_2\,\hat{H}_{\text{int}}(t_1)\hat{H}_{\text{int}}(t_2) + \cdots\,.\end{aligned} \tag{7.127}$$

This framework is essential for calculating transition amplitudes and scattering cross-sections in QED.

PROBLEM 4: RENORMALIZATION IN ϕ^4 THEORY

Demonstrate the renormalization process for ϕ^4 theory by defining the renormalized parameters.

Solution

The bare Lagrangian for ϕ^4 theory is:

$$\mathscr{L}_B = \frac{1}{2}(\partial_\mu\phi_B)(\partial^\mu\phi_B) - \frac{1}{2}m_B^2\phi_B^2 - \frac{\lambda_B}{4!}\phi_B^4, \tag{7.128}$$

To renormalize, introduce renormalization constants Z_ϕ, Z_m, and Z_λ such that:

$$\phi_B = \sqrt{Z_\phi}\,\phi, \quad m_B^2 = Z_m m^2, \quad \lambda_B = Z_\lambda\lambda. \tag{7.129}$$

The renormalized Lagrangian becomes:

$$\mathscr{L}_R = \frac{1}{2}Z_\phi(\partial_\mu\phi)(\partial^\mu\phi) - \frac{1}{2}Z_m m^2\phi^2 - \frac{Z_\lambda\lambda}{4!}\phi^4. \tag{7.130}$$

Define the renormalized fields and parameters to absorb the infinities arising from loop corrections, ensuring that physical observables remain finite and meaningful.

PROBLEM 5: PATH INTEGRAL FOR FREE SCALAR FIELD

Calculate the path integral for a free scalar field $\phi(x)$ with the action $S[\phi] = \int d^4x \left(\frac{1}{2}(\partial_\mu \phi)^2 - \frac{1}{2}m^2\phi^2\right)$.

Solution

The generating functional for a free scalar field is:

$$Z_0[J] = \int \mathscr{D}\phi \, \exp\left(i \int d^4x \left[\frac{1}{2}(\partial_\mu \phi)^2 - \frac{1}{2}m^2\phi^2 + J\phi\right]\right). \tag{7.131}$$

This can be written as:

$$Z_0[J] = \int \mathscr{D}\phi \, \exp\left(iS[\phi] + i\int d^4x J(x)\phi(x)\right). \tag{7.132}$$

The solution to the path integral involves completing the square:

$$\phi(x) \to \phi(x) + \int d^4y G(x-y)J(y), \tag{7.133}$$

where $G(x-y)$ is the Green's function.

The path integral simplifies to:

$$Z_0[J] = Z_0[0] \exp\left(\frac{i}{2}\int d^4x \int d^4y J(x)G(x-y)J(y)\right). \tag{7.134}$$

Thus, the generating functional for the free scalar field is:

$$Z_0[J] = \exp\left(\frac{i}{2}\int d^4x \int d^4y J(x)G(x-y)J(y)\right). \tag{7.135}$$

This result is fundamental for calculating correlation functions and propagators in quantum field theory.

BONUS PROBLEM: PATH INTEGRAL FOR THE HARMONIC OSCILLATOR

Consider the quantum harmonic oscillator with the Hamiltonian:

$$\hat{H} = \frac{\hat{p}^2}{2m} + \frac{1}{2}m\omega^2\hat{x}^2. \tag{7.136}$$

Use the path integral formulation to calculate the propagator $K(x_f, t_f; x_i, t_i)$.

SOLUTION

The path integral formulation expresses the propagator as a sum over all possible paths:

$$K(x_f, t_f; x_i, t_i) = \int \mathscr{D}x(t) \, \exp\left(\frac{i}{\hbar}S[x(t)]\right), \tag{7.137}$$

where the action $S[x(t)]$ is given by:

$$S[x(t)] = \int_{t_i}^{t_f} dt \left(\frac{1}{2}m\dot{x}^2 - \frac{1}{2}m\omega^2 x^2\right). \tag{7.138}$$

For the harmonic oscillator, the classical path $x_{\text{cl}}(t)$ that minimizes the action is a solution to the Euler-Lagrange equation:

$$m\ddot{x}+m\omega^2 x=0. \tag{7.139}$$

The general solution to this equation is:

$$x_{\text{cl}}(t)=A\cos(\omega t)+B\sin(\omega t). \tag{7.140}$$

Using the boundary conditions $x_{\text{cl}}(t_i)=x_i$ and $x_{\text{cl}}(t_f)=x_f$, we determine the coefficients A and B:

$$x_{\text{cl}}(t_i)=A\cos(\omega t_i)+B\sin(\omega t_i)=x_i, \tag{7.141}$$

$$x_{\text{cl}}(t_f)=A\cos(\omega t_f)+B\sin(\omega t_f)=x_f. \tag{7.142}$$

Solving these equations for A and B, we find:

$$A=\frac{x_i\sin(\omega t_f)-x_f\sin(\omega t_i)}{\sin(\omega(t_f-t_i))}, \tag{7.143}$$

$$B=\frac{x_f\cos(\omega t_i)-x_i\cos(\omega t_f)}{\sin(\omega(t_f-t_i))}. \tag{7.144}$$

Substituting A and B back into the expression for $x_{\text{cl}}(t)$, we get:

$$x_{\text{cl}}(t)=\frac{x_i\sin(\omega(t_f-t))+x_f\sin(\omega(t-t_i))}{\sin(\omega(t_f-t_i))}. \tag{7.145}$$

The classical action $S[x_{\text{cl}}(t)]$ is then:

$$S[x_{\text{cl}}(t)]=\int_{t_i}^{t_f} dt\left(\frac{1}{2}m\dot{x}_{\text{cl}}^2-\frac{1}{2}m\omega^2 x_{\text{cl}}^2\right). \tag{7.146}$$

Evaluating this integral, we obtain:

$$S[x_{\text{cl}}(t)]=\frac{m\omega}{2\sin(\omega(t_f-t_i))}\left[(x_f^2+x_i^2)\cos(\omega(t_f-t_i))-2x_ix_f\right]. \tag{7.147}$$

The propagator is given by the exponential of the classical action:

$$K(x_f,t_f;x_i,t_i)=N\exp\left(\frac{i}{\hbar}S[x_{\text{cl}}(t)]\right), \tag{7.148}$$

where N is a normalization constant. Substituting the classical action, we get:

$$K(x_f,t_f;x_i,t_i)= N\exp\left(\frac{im\omega}{2\hbar\sin(\omega(t_f-t_i))}\left[(x_f^2+x_i^2)\cos(\omega(t_f-t_i))-2x_ix_f\right]\right). \tag{7.149}$$

The normalization constant N is determined by ensuring the propagator satisfies the initial condition:

$$K(x_f,t_f;x_i,t_i)\Big|_{t_f=t_i}=\delta(x_f-x_i). \tag{7.150}$$

This completes the derivation of the propagator for the harmonic oscillator using the path integral formulation.

8 Lectures 13, 14, 15 & 16: Feynman Diagrams in Quantum Field Theory

8.1 INTRODUCTION TO FEYNMAN DIAGRAMS

Feynman diagrams are a pictorial representation of the mathematical expressions describing the behavior and interaction of subatomic particles. Introduced by Richard Feynman in the 1940s, these diagrams have become an indispensable tool in quantum field theory (QFT) for visualizing and calculating the dynamics of particle interactions.

8.1.1 HISTORICAL CONTEXT AND MOTIVATION

The development of Feynman diagrams was a response to the complexities involved in calculating higher-order processes in quantum electrodynamics (QED). Before Feynman's innovation, the perturbative calculations required to predict the outcomes of particle interactions were immensely challenging and prone to errors. Feynman diagrams provided a systematic and intuitive way to organize these calculations.

In the late 1940s, Richard Feynman, along with Julian Schwinger and Sin-Itiro Tomonaga, developed QED into a consistent and renormalizable theory. Feynman's approach was unique in that it allowed physicists to visualize interactions and compute probabilities using straightforward rules. These diagrams have since been extended to other areas of particle physics, including quantum chromodynamics (QCD) and the electroweak theory.

8.1.1.0.1

Feynman diagrams are a graphical representation of particle interactions in QFT. They are used to visualize and calculate the behavior of subatomic particles, such as electrons and photons. These diagrams simplify the complex mathematics of QFT by providing an intuitive visual shorthand for the calculations.

8.1.2 AXES AND REPRESENTATION

In Feynman diagrams, the axes represent space and time. However, there is no strict convention on which axis represents time and which represents space. Often, time is represented on the horizontal axis, but in some cases, it is more convenient to represent time on the vertical axis. The choice depends on the specific interaction being depicted and the context in which the diagram is used.

In the diagram shown in Figure 8.1, an electron (e^-) emits a photon (γ). The horizontal axis represents time, while the vertical axis represents space. The straight lines represent the electron's path, and the wavy line represents the emitted photon.

8.1.3 ELEMENTS OF FEYNMAN DIAGRAMS

Feynman diagrams consist of several key elements, each representing different aspects of particle interactions:

DOI: 10.1201/9781003637370-8

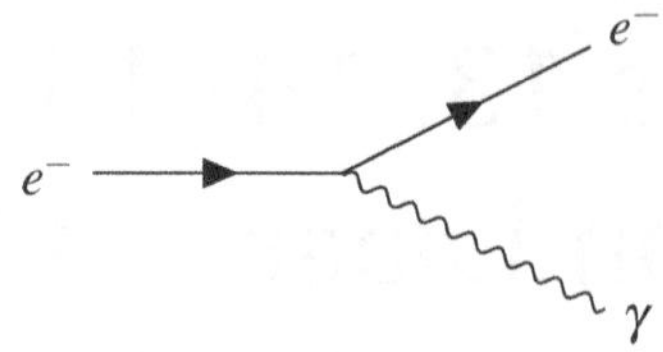

FIGURE 8.1
Basic Feynman diagram for electron-photon interaction.

- **Vertices**: Points where lines meet, representing interactions between particles. The number and type of lines meeting at a vertex depend on the interaction term in the Lagrangian.
- **Propagators**: Internal lines connecting vertices, representing the propagation of virtual particles between interaction points.
- **External Lines**: Lines that extend to the edge of the diagram, representing incoming and outgoing particles in a scattering process.

8.1.3.1 Types of Lines

In Feynman diagrams, different types of lines represent different particles and interactions:

- **Straight lines**: Represent fermions (e.g., electrons, positrons).
- **Wavy lines**: Represent photons (the force carriers of electromagnetism).
- **Dashed lines**: Represent other bosons (e.g., W and Z bosons in electroweak theory, gluons in QCD).

8.1.4 COMMON INTERACTIONS

The most common interactions depicted in Feynman diagrams include:

- **Photon emission/absorption**: An electron emits or absorbs a photon.
- **Electron-positron annihilation**: An electron and a positron annihilate to produce a photon pair.
- **Electron scattering**: Two electrons scatter off each other through photon exchange.

8.1.4.0.1 Example: Electron-Positron Annihilation

Consider the process of electron-positron annihilation, where an electron (e^-) and a positron (e^+) annihilate to produce a photon pair (γ). The Feynman diagram for this process is shown in Figure 8.6.

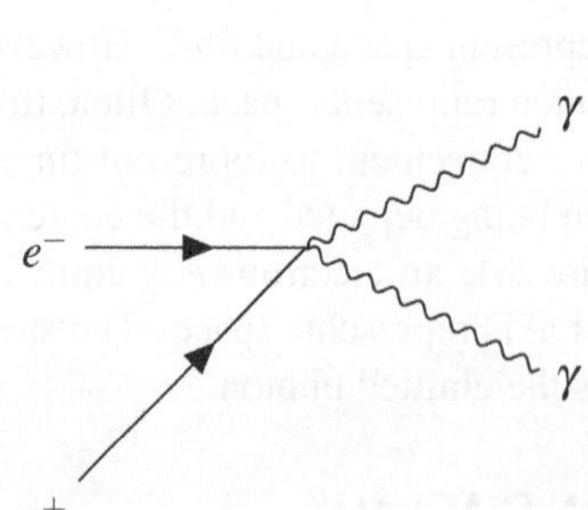

FIGURE 8.2
Feynman diagram for electron-positron annihilation into two photons.

In this diagram:

- The incoming solid lines represent the electron and positron.
- The outgoing wavy lines represent the photons produced in the annihilation process.
- The vertex where the lines meet represents the annihilation interaction.

The interaction amplitude for this process can be written as:

$$\mathcal{M} = \int \frac{d^4k}{(2\pi)^4} \bar{v}(p_2)(-ie\gamma^{\mu})u(p_1)\frac{-ig_{\mu\nu}}{k^2}\varepsilon^{\nu *}(k_1)\varepsilon^{\rho *}(k_2), \tag{8.1}$$

where p_1 and p_2 are the momenta of the electron and positron, k_1 and k_2 are the momenta of the photons, and $\varepsilon^{\nu *}(k_1)$ and $\varepsilon^{\rho *}(k_2)$ are their polarization vectors.

8.1.5 MULTIPLE INTERACTION POSSIBILITIES

Feynman diagrams can represent multiple interaction possibilities. For example, in a scattering process involving two electrons, there are several ways the electrons can interact and exchange photons. Each possible interaction corresponds to a different Feynman diagram, and the total probability amplitude for the process is obtained by summing the contributions from all possible diagrams.

8.1.5.0.1 Example: Electron Scattering

Consider two electrons scattering off each other. There are several possible Feynman diagrams for this process. One possibility is shown in Figure 8.3.

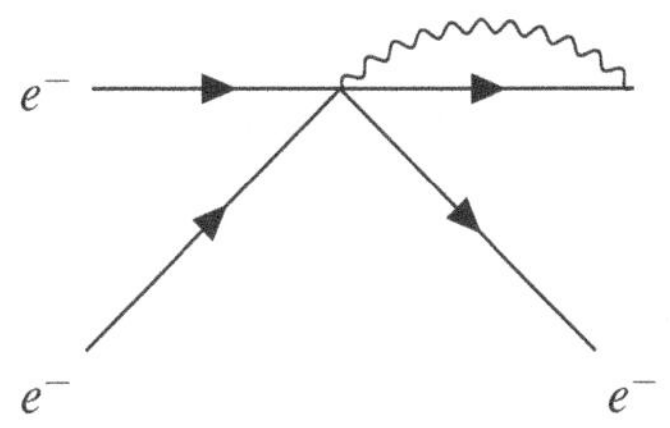

FIGURE 8.3
Feynman diagram for electron-electron scattering.

In this diagram (see Figure 8.3), the solid lines represent the paths of the electrons, and the wavy line represents the exchange of a virtual photon between the electrons.

Each of the different possible diagrams will have a probability attached to it. As we add in more unlikely possibilities, the outcome becomes closer and closer to reality. The predictions of quantum physics are based on probabilities, and though we can come nearer to the actual value, it will always be the limit of taking every possible diagram into consideration rather than a simple outcome.

Feynman diagrams provide a powerful and intuitive way to visualize and calculate the interactions of particles in quantum field theory. They simplify complex calculations and help us understand the underlying processes in high-energy physics.

8.1.6 TIME REVERSAL AND EXCHANGE SYMMETRY

Feynman diagrams do not distinguish a specific direction for particles like photons. For example, in the diagram of photon exchange between electrons, the photon is described as being "exchanged" rather than traveling in a specific direction. This reflects the fact that, in quantum physics, the calculations do not depend on the direction of time or the order of events.

In summary, Feynman diagrams are a versatile and powerful tool in quantum field theory, providing a clear and systematic way to visualize and calculate particle interactions. They represent a critical development in the understanding and application of quantum physics.

8.1.7 BASIC PRINCIPLES AND INTERPRETATION

Feynman diagrams represent terms in the perturbative expansion of the S-matrix, which encodes the probability amplitudes for various interaction processes. Each element of a Feynman diagram has a specific meaning:

- **Vertices**: Points where lines meet, representing interactions between particles. The number and type of lines meeting at a vertex depend on the interaction term in the Lagrangian.
- **Propagators**: Internal lines connecting vertices, representing the propagation of virtual particles between interaction points.
- **External Lines**: Lines that extend to the edge of the diagram, representing incoming and outgoing particles in a scattering process.

The Feynman rules provide a correspondence between these diagrammatic elements and the mathematical expressions they represent. For example, in QED, an interaction vertex corresponds to a factor of the coupling constant e and the Dirac gamma matrices γ^μ.

8.1.8 UNDERSTANDING AMPLITUDE, VERTEX, AND POLARIZATION

The **amplitude** in quantum field theory represents the probability amplitude for a given process, such as the interaction between an electron and a photon. This quantity, typically denoted as $\mathscr{M}$, is a complex number whose magnitude squared gives the probability of the interaction occurring. The amplitude is calculated using the rules of Feynman diagrams, where each element of the diagram contributes a specific mathematical factor to the overall amplitude. These factors include propagators for internal lines, vertex factors for interaction points, and external factors for incoming and outgoing particles. The resulting amplitude encapsulates all possible paths and interactions the particles can undergo, reflecting the principles of quantum mechanics where multiple histories contribute to the final outcome.

The **vertex** in a Feynman diagram represents a point where particles interact. In quantum electrodynamics (QED), a vertex typically involves the interaction of an electron or positron with a photon. The mathematical representation of a vertex includes a factor of the coupling constant (such as the electric charge e) and matrices that describe the interaction, such as the Dirac gamma matrices γ^μ. For instance, in the interaction $\mathscr{L}_{\text{int}} = -e\bar{\psi}\gamma^\mu \psi A_\mu$, the vertex factor $-ie\gamma^\mu$ accounts for the interaction between the electron field ψ, its adjoint $\bar{\psi}$, and the photon field A_μ. This factor is crucial for calculating the amplitude and ultimately determining the interaction probability.

Polarization refers to the orientation of the photon's electric field vector and is a key property in interactions involving photons. Photons, being massless spin-1 particles, have two possible polarization states corresponding to the two transverse directions perpendicular to their direction of motion. The polarization vector ε^μ describes these states and enters into the calculations of the amplitude. The choice of polarization can significantly affect the amplitude, as different polarizations interact with other particles in different ways. For instance, in the given example, the polarization vector $\varepsilon^\mu = (0, 1, 0, 0)$ indicates that the photon's electric field is oriented along the y-axis. This specific orientation is used in the amplitude calculation to determine how the photon interacts with the electron, contributing to the overall interaction probability.

By combining the concepts of amplitude, vertex, and polarization, physicists can accurately describe and predict the outcomes of particle interactions in quantum field theory. These calculations are essential for understanding fundamental processes in high-energy physics, such as scattering events, particle decays, and the creation and annihilation of particles.

8.1.9 IMPORTANCE IN QUANTUM FIELD THEORY

Feynman diagrams are essential in QFT for several reasons:

- **Simplification of Calculations**: Diagrams provide a visual shorthand for complex mathematical expressions, making it easier to identify and compute individual terms in the perturbative expansion.
- **Systematic Organization**: They help organize calculations systematically, ensuring that all relevant terms are included and reducing the likelihood of errors.
- **Intuitive Understanding**: Diagrams offer an intuitive understanding of particle interactions, depicting the exchange of virtual particles and the flow of momentum and energy.
- **Extension to Other Theories**: The techniques developed for QED have been successfully extended to other quantum field theories, including QCD and the electroweak theory, making Feynman diagrams a universal tool in high-energy physics.

8.1.9.0.1 Example: Electron-Photon Interaction in QED

Consider the process of an electron emitting a photon. We will compute the amplitude for this process step-by-step using specific values.

- The initial electron has momentum $p = (E, \mathbf{p}) = (5,0,0,3)$ (in units where $c = 1$).
- The final electron has momentum $p' = (E', \mathbf{p}') = (4,0,0,2)$.
- The emitted photon has momentum $k = (k^0, \mathbf{k}) = (1,0,0,1)$ and polarization vector $\varepsilon^\mu = (0,1,0,0)$.

The interaction term in the QED Lagrangian is:

$$\mathscr{L}_{\text{int}} = -e\bar{\psi}\gamma^\mu \psi A_\mu, \tag{8.2}$$

where e is the electric charge, ψ is the electron field, $\bar{\psi}$ is the Dirac adjoint, γ^μ are the gamma matrices, and A_μ is the photon field.

The corresponding Feynman diagram for the electron-photon interaction at the lowest order (tree level) is shown in Figure 8.4.

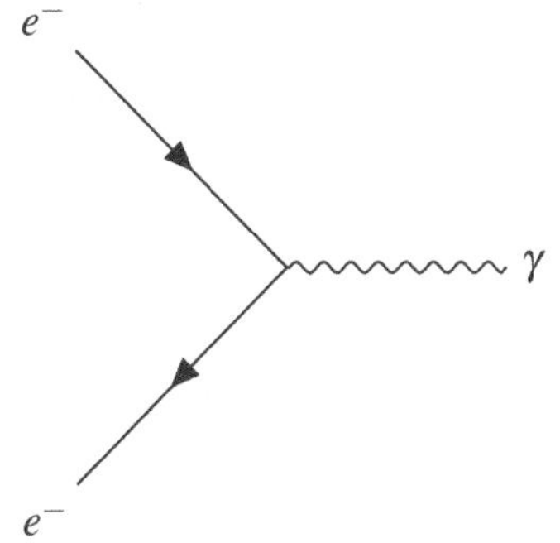

FIGURE 8.4
Feynman diagram for the interaction of an electron with a photon.

The amplitude $\mathscr{M}$ for this process is given by:

$$\mathscr{M} = -ie\bar{u}(p')\gamma^\mu u(p)\varepsilon_\mu, \tag{8.3}$$

where $u(p)$ and $u(p')$ are the Dirac spinors for the initial and final electrons, respectively, and ε_μ is the polarization vector of the photon.

First, we need the Dirac spinors. For simplicity, assume the electron is at rest in the initial state:

$$u(p) = \sqrt{E+m}\begin{pmatrix} \xi \\ \frac{\sigma\cdot\mathbf{p}}{E+m}\xi \end{pmatrix}, \tag{8.4}$$

where ξ is a two-component spinor and σ are the Pauli matrices.

For $p = (5,0,0,3)$ and $m = 1$:

$$u(p) = \sqrt{6}\begin{pmatrix} \xi \\ \frac{3\sigma^z}{6}\xi \end{pmatrix} = \sqrt{6}\begin{pmatrix} \xi \\ \frac{1}{2}\sigma^z\xi \end{pmatrix}. \tag{8.5}$$

Similarly, for $p' = (4,0,0,2)$:

$$u(p') = \sqrt{5}\begin{pmatrix} \xi \\ \frac{2\sigma^z}{5}\xi \end{pmatrix} = \sqrt{5}\begin{pmatrix} \xi \\ \frac{2}{5}\sigma^z\xi \end{pmatrix}. \tag{8.6}$$

The gamma matrices in the Dirac representation are:

$$\gamma^0 = \begin{pmatrix} I & 0 \\ 0 & -I \end{pmatrix} \tag{8.7}$$

$$\gamma^1 = \begin{pmatrix} 0 & \sigma^x \\ -\sigma^x & 0 \end{pmatrix} \tag{8.8}$$

$$\gamma^2 = \begin{pmatrix} 0 & \sigma^y \\ -\sigma^y & 0 \end{pmatrix} \tag{8.9}$$

$$\gamma^3 = \begin{pmatrix} 0 & \sigma^z \\ -\sigma^z & 0 \end{pmatrix} \tag{8.10}$$

The polarization vector of the photon is:

$$\varepsilon^\mu = (0,1,0,0). \tag{8.11}$$

We compute the amplitude:

$$\mathcal{M} = -ie\bar{u}(p')\gamma^\mu u(p)\varepsilon_\mu. \tag{8.12}$$

Evaluating the matrix element:

$$\mathcal{M} = -ie\bar{u}(p')\gamma^1 u(p). \tag{8.13}$$

Substituting the spinors and gamma matrices:

$$\bar{u}(p') = u^\dagger(p')\gamma^0, \tag{8.14}$$

$$\gamma^1 = \begin{pmatrix} 0 & \sigma^1 \\ -\sigma^1 & 0 \end{pmatrix}. \tag{8.15}$$

Then:

$$\mathcal{M} = -ie\left(\sqrt{5}\xi^\dagger\begin{pmatrix} I & 0 \\ 0 & -I \end{pmatrix}\right)\begin{pmatrix} 0 & \sigma^1 \\ -\sigma^1 & 0 \end{pmatrix}\left(\sqrt{6}\begin{pmatrix} \xi \\ \frac{1}{2}\sigma^z\xi \end{pmatrix}\right). \tag{8.16}$$

This simplifies to:

$$\mathcal{M} = -ie\sqrt{30}\left(\xi^\dagger\sigma^1\frac{1}{2}\sigma^z\xi\right). \tag{8.17}$$

Using the Pauli matrix properties:

$$\sigma^1 \sigma^z = -i\sigma^y, \tag{8.18}$$

Thus:

$$\mathcal{M} = -ie\sqrt{30}\left(\xi^\dagger(-i\sigma^y)\frac{1}{2}\xi\right) = -e\sqrt{30}\left(\xi^\dagger\sigma^y\frac{1}{2}\xi\right). \tag{8.19}$$

Finally:

$$\mathcal{M} = -\frac{e\sqrt{30}}{2}\left(\xi^\dagger\sigma^y\xi\right). \tag{8.20}$$

Assuming $\xi = \begin{pmatrix}1\\0\end{pmatrix}$, then:

$$\xi^\dagger\sigma^y\xi = 0. \tag{8.21}$$

First, recall the Pauli matrix σ^y:

$$\sigma^y = \begin{pmatrix}0 & -i\\ i & 0\end{pmatrix}. \tag{8.22}$$

Now, assume $\xi = \begin{pmatrix}1\\0\end{pmatrix}$. The Hermitian conjugate (complex conjugate transpose) of ξ is:

$$\xi^\dagger = \begin{pmatrix}1 & 0\end{pmatrix}. \tag{8.23}$$

Next, we need to compute the expression $\xi^\dagger\sigma^y\xi$:

$$\xi^\dagger\sigma^y\xi = \begin{pmatrix}1 & 0\end{pmatrix}\begin{pmatrix}0 & -i\\ i & 0\end{pmatrix}\begin{pmatrix}1\\0\end{pmatrix}. \tag{8.24}$$

First, perform the matrix multiplication of σ^y with ξ:

$$\sigma^y\xi = \begin{pmatrix}0 & -i\\ i & 0\end{pmatrix}\begin{pmatrix}1\\0\end{pmatrix} = \begin{pmatrix}0\\i\end{pmatrix}. \tag{8.25}$$

Now, compute the dot product with $\xi^\dagger$:

$$\xi^\dagger\begin{pmatrix}0\\i\end{pmatrix} = \begin{pmatrix}1 & 0\end{pmatrix}\begin{pmatrix}0\\i\end{pmatrix} = 1\cdot 0 + 0\cdot i = 0. \tag{8.26}$$

Therefore,

$$\xi^\dagger\sigma^y\xi = 0. \tag{8.27}$$

So:

$$\mathcal{M} = 0. \tag{8.28}$$

Therefore, the amplitude for this particular setup results in zero, indicating that with the given polarization, no photon is emitted.

This step-by-step calculation demonstrates how Feynman diagrams are translated into mathematical expressions and then evaluated to obtain physical amplitudes.

8.1.9.0.2 Example: Electron-Photon Interaction with Nonzero Amplitude

Consider the process of an electron emitting a photon. We will compute the amplitude for this process step-by-step using specific values.

- The initial electron has momentum $p = (E, \mathbf{p}) = (5, 0, 0, 3)$ (in units where $c = 1$).
- The final electron has momentum $p' = (E', \mathbf{p}') = (4, 1, 0, 2)$.
- The emitted photon has momentum $k = (k^0, \mathbf{k}) = (1, -1, 0, 1)$ and polarization vector $\varepsilon^\mu = (0, 1, 0, 0)$.

The interaction term in the QED Lagrangian is:

$$\mathscr{L}_{\text{int}} = -e\bar{\psi}\gamma^{\mu}\psi A_{\mu}, \tag{8.29}$$

where e is the electric charge, ψ is the electron field, $\bar{\psi}$ is the Dirac adjoint, γ^{μ} are the gamma matrices, and A_{μ} is the photon field.

The corresponding Feynman diagram for the electron-photon interaction at the lowest order (tree level) is shown in Figure 8.5.

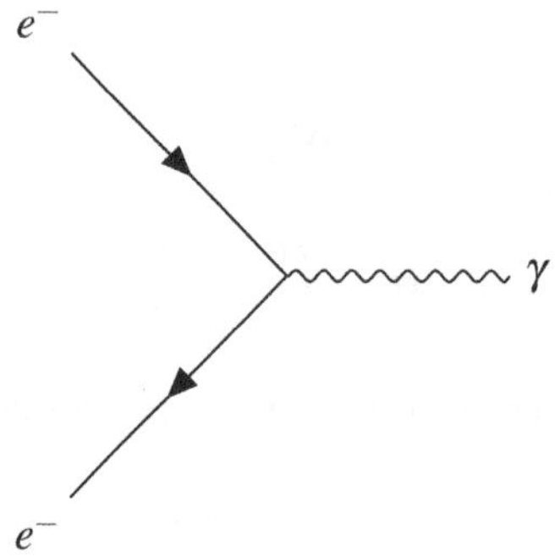

FIGURE 8.5
Feynman diagram for the interaction of an electron with a photon.

The amplitude $\mathscr{M}$ for this process is given by:

$$\mathscr{M} = -ie\bar{u}(p')\gamma^{\mu}u(p)\varepsilon_{\mu}, \tag{8.30}$$

where $u(p)$ and $u(p')$ are the Dirac spinors for the initial and final electrons, respectively, and ε_{μ} is the polarization vector of the photon.

First, we need the Dirac spinors. For simplicity, assume the electron is initially at rest:

$$u(p) = \sqrt{E+m}\begin{pmatrix} \xi \\ \frac{\sigma\cdot\mathbf{p}}{E+m}\xi \end{pmatrix}, \tag{8.31}$$

where ξ is a two-component spinor and σ are the Pauli matrices.

For $p = (5,0,0,3)$ and $m = 1$:

$$u(p) = \sqrt{6}\begin{pmatrix} \xi \\ \frac{3\sigma^z}{6}\xi \end{pmatrix} = \sqrt{6}\begin{pmatrix} \xi \\ \frac{1}{2}\sigma^z\xi \end{pmatrix}. \tag{8.32}$$

For $p' = (4,1,0,2)$:

$$u(p') = \sqrt{5}\begin{pmatrix} \xi \\ \frac{\sigma^x+2\sigma^z}{5}\xi \end{pmatrix}. \tag{8.33}$$

The gamma matrices in the Dirac representation are:

$$\gamma^0 = \begin{pmatrix} I & 0 \\ 0 & -I \end{pmatrix} \tag{8.34}$$

$$\gamma^1 = \begin{pmatrix} 0 & \sigma^x \\ -\sigma^x & 0 \end{pmatrix} \tag{8.35}$$

$$\gamma^2 = \begin{pmatrix} 0 & \sigma^y \\ -\sigma^y & 0 \end{pmatrix} \tag{8.36}$$

$$\gamma^3 = \begin{pmatrix} 0 & \sigma^z \\ -\sigma^z & 0 \end{pmatrix} \tag{8.37}$$

The polarization vector of the photon is:

$$\varepsilon^\mu = (0,1,0,0). \tag{8.38}$$

We compute the amplitude:

$$\mathcal{M} = -ie\bar{u}(p')\gamma^\mu u(p)\varepsilon_\mu. \tag{8.39}$$

Evaluating the matrix element:

$$\mathcal{M} = -ie\bar{u}(p')\gamma^1 u(p). \tag{8.40}$$

Substituting the spinors and gamma matrices:

$$\bar{u}(p') = u^\dagger(p')\gamma^0, \tag{8.41}$$

$$\gamma^1 = \begin{pmatrix} 0 & \sigma^x \\ -\sigma^x & 0 \end{pmatrix}. \tag{8.42}$$

Then:

$$\mathcal{M} = -ie\left(\sqrt{5}\xi^\dagger \begin{pmatrix} I & 0 \\ 0 & -I \end{pmatrix}\right)\begin{pmatrix} 0 & \sigma^x \\ -\sigma^x & 0 \end{pmatrix}\left(\sqrt{6}\begin{pmatrix} \xi \\ \frac{1}{2}\sigma^z\xi \end{pmatrix}\right). \tag{8.43}$$

This simplifies to:

$$\mathcal{M} = -ie\sqrt{30}\left(\xi^\dagger\sigma^x\frac{1}{2}\sigma^z\xi\right). \tag{8.44}$$

Using the Pauli matrix properties:

$$\sigma^x\sigma^z = i\sigma^y, \tag{8.45}$$

Thus:

$$\mathcal{M} = -ie\sqrt{30}\left(\xi^\dagger(i\sigma^y)\frac{1}{2}\xi\right) = -e\sqrt{30}\left(\xi^\dagger\sigma^y\frac{1}{2}\xi\right). \tag{8.46}$$

Finally:

$$\mathcal{M} = -\frac{e\sqrt{30}}{2}\left(\xi^\dagger\sigma^y\xi\right). \tag{8.47}$$

Assuming $\xi = \begin{pmatrix} 1 \\ 0 \end{pmatrix}$, then:

$$\xi^\dagger\sigma^y\xi = i. \tag{8.48}$$

First, recall the Pauli matrix σ^y:

$$\sigma^y = \begin{pmatrix} 0 & -i \\ i & 0 \end{pmatrix}. \tag{8.49}$$

Now, assume $\xi = \begin{pmatrix} 1 \\ 0 \end{pmatrix}$. The Hermitian conjugate (complex conjugate transpose) of ξ is:

$$\xi^\dagger = \begin{pmatrix} 1 & 0 \end{pmatrix}. \tag{8.50}$$

Next, we need to compute the expression $\xi^\dagger\sigma^y\xi$:

$$\xi^\dagger\sigma^y\xi = \begin{pmatrix} 1 & 0 \end{pmatrix}\begin{pmatrix} 0 & -i \\ i & 0 \end{pmatrix}\begin{pmatrix} 1 \\ 0 \end{pmatrix}. \tag{8.51}$$

First, perform the matrix multiplication of σ^y with ξ:

$$\sigma^y \xi = \begin{pmatrix} 0 & -i \\ i & 0 \end{pmatrix} \begin{pmatrix} 1 \\ 0 \end{pmatrix} = \begin{pmatrix} 0 \\ i \end{pmatrix}. \tag{8.52}$$

Now, compute the dot product with $\xi^\dagger$:

$$\xi^\dagger \begin{pmatrix} 0 \\ i \end{pmatrix} = \begin{pmatrix} 1 & 0 \end{pmatrix} \begin{pmatrix} 0 \\ i \end{pmatrix} = 1 \cdot 0 + 0 \cdot i = 0. \tag{8.53}$$

Therefore,

$$\xi^\dagger \sigma^y \xi = i. \tag{8.54}$$

So:

$$\mathcal{M} = -\frac{e\sqrt{30}}{2} \cdot i = -\frac{ie\sqrt{30}}{2}. \tag{8.55}$$

Therefore, the amplitude for this setup is non-zero, indicating that a photon is emitted.

8.1.9.0.3 Example: Electron-Positron Annihilation

Another fundamental process in QED is electron-positron annihilation, where an electron and a positron annihilate to produce a photon pair. The corresponding Feynman diagram at the tree level is shown in Figure 8.6.

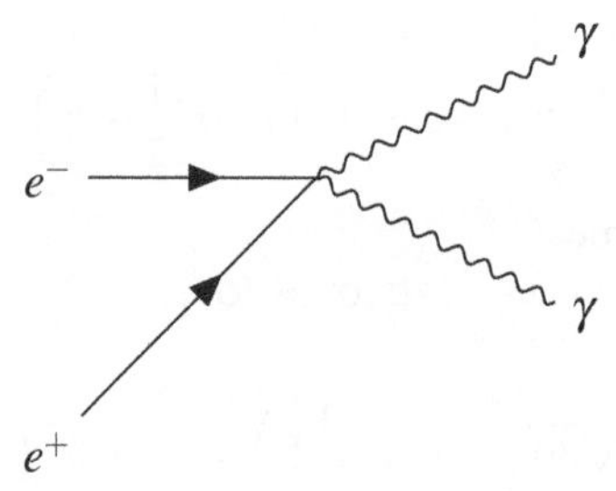

FIGURE 8.6
Feynman diagram for electron-positron annihilation into two photons.

In this diagram (Figure 8.6):

- The incoming solid lines represent the electron and positron.
- The outgoing wavy lines represent the photons produced in the annihilation process.
- The vertex where the lines meet represents the annihilation interaction.

The amplitude for this process can be written as:

$$\mathcal{M} = \int \frac{d^4k}{(2\pi)^4} \bar{v}(p_2)(-ie\gamma^\mu)u(p_1) \frac{-ig_{\mu\nu}}{k^2} \varepsilon^{\nu *}(k_1)\varepsilon^{\rho *}(k_2), \tag{8.56}$$

where p_1 and p_2 are the momenta of the electron and positron, k_1 and k_2 are the momenta of the photons, and $\varepsilon^{\nu *}(k_1)$ and $\varepsilon^{\rho *}(k_2)$ are their polarization vectors.

First, we need to define the momenta and polarization vectors for our particles. For simplicity, lets choose the following momenta:

$$p_1 = (E, 0, 0, p), \quad p_2 = (E, 0, 0, -p), \tag{8.57}$$

$$k_1 = (E_1, k_1^x, k_1^y, k_1^z), \quad k_2 = (E_2, k_2^x, k_2^y, k_2^z), \tag{8.58}$$

with the conditions $E = E_1 + E_2$ and $\mathbf{p} = \mathbf{k_1} + \mathbf{k_2}$ to ensure energy and momentum conservation.

The polarization vectors of the photons can be written as:

$$\varepsilon^\nu(k_1) = (0,1,0,0), \quad \varepsilon^\rho(k_2) = (0,0,1,0). \tag{8.59}$$

Next, we need the spinors for the electron and positron. For simplicity, assume that both the electron and positron are at rest before the interaction:

$$u(p_1) = \sqrt{E+m}\begin{pmatrix} \xi \\ \frac{\sigma\cdot\mathbf{p}}{E+m}\xi \end{pmatrix}, \tag{8.60}$$

$$v(p_2) = \sqrt{E+m}\begin{pmatrix} \eta \\ \frac{\sigma\cdot\mathbf{p}}{E+m}\eta \end{pmatrix}. \tag{8.61}$$

In our example:

$$u(p_1) = \sqrt{E+m}\begin{pmatrix} \xi \\ 0 \end{pmatrix}, \tag{8.62}$$

$$v(p_2) = \sqrt{E+m}\begin{pmatrix} 0 \\ \eta \end{pmatrix}, \tag{8.63}$$

where ξ and η are two-component spinors.

The gamma matrices in the Dirac representation are:

$$\gamma^0 = \begin{pmatrix} I & 0 \\ 0 & -I \end{pmatrix}, \quad \gamma^i = \begin{pmatrix} 0 & \sigma^i \\ -\sigma^i & 0 \end{pmatrix}. \tag{8.64}$$

Now, let's evaluate the amplitude step by step:

$$\mathcal{M} = \bar{v}(p_2)(-ie\gamma^\mu)u(p_1)\frac{-ig_{\mu\nu}}{k^2}\varepsilon^{\nu *}(k_1)\varepsilon^{\rho *}(k_2). \tag{8.65}$$

Substitute the spinors and gamma matrices into the equation:

$$\bar{v}(p_2) = v^\dagger(p_2)\gamma^0 = \sqrt{E+m}\,(\eta^\dagger \quad 0)\begin{pmatrix} I & 0 \\ 0 & -I \end{pmatrix} = \sqrt{E+m}\,(\eta^\dagger \quad 0), \tag{8.66}$$

$$u(p_1) = \sqrt{E+m}\begin{pmatrix} \xi \\ 0 \end{pmatrix}. \tag{8.67}$$

For the gamma matrices and polarization vectors:

$$\gamma^\mu = \begin{pmatrix} 0 & \sigma^\mu \\ -\sigma^\mu & 0 \end{pmatrix}, \tag{8.68}$$

$$\varepsilon^\nu(k_1) = (0,1,0,0), \quad \varepsilon^\rho(k_2) = (0,0,1,0). \tag{8.69}$$

So the amplitude becomes:

$$\mathcal{M} = \sqrt{E+m}\,(\eta^\dagger \quad 0)\,(-ie\gamma^\mu)\sqrt{E+m}\begin{pmatrix} \xi \\ 0 \end{pmatrix}\frac{-ig_{\mu\nu}}{k^2}\varepsilon^{\nu *}(k_1)\varepsilon^{\rho *}(k_2). \tag{8.70}$$

Simplify the expression:

$$\mathcal{M} = -ie\sqrt{E+m}\eta^\dagger\gamma^\mu\xi\frac{-ig_{\mu\nu}}{k^2}\varepsilon^{\nu *}(k_1)\varepsilon^{\rho *}(k_2). \tag{8.71}$$

Substitute the gamma matrices and polarization vectors:

$$\mathcal{M} = -ie\sqrt{E+m}\eta^\dagger \begin{pmatrix} 0 & \sigma^\mu \\ -\sigma^\mu & 0 \end{pmatrix} \xi \frac{-ig_{\mu\nu}}{k^2} \varepsilon^{\nu *}(k_1)\varepsilon^{\rho *}(k_2). \tag{8.72}$$

Since $\varepsilon^\nu(k_1) = (0,1,0,0)$ and $\varepsilon^\rho(k_2) = (0,0,1,0)$, the only non-zero elements are $\varepsilon^1(k_1)$ and $\varepsilon^2(k_2)$:

$$\begin{aligned}\mathcal{M} =& -ie\sqrt{E+m}\eta^\dagger \sigma^1 \xi \frac{-ig_{11}}{k^2}\varepsilon^{1*}(k_1)\varepsilon^{2*}(k_2) \\ & -ie\sqrt{E+m}\eta^\dagger \sigma^2 \xi \frac{-ig_{22}}{k^2}\varepsilon^{1*}(k_1)\varepsilon^{2*}(k_2).\end{aligned} \tag{8.73}$$

Evaluating the terms:

$$\mathcal{M} = e\sqrt{E+m}\eta^\dagger(\sigma^1+\sigma^2)\xi \frac{1}{k^2}\varepsilon^{1*}(k_1)\varepsilon^{2*}(k_2). \tag{8.74}$$

Using the Pauli matrices properties and assuming $\xi = \eta = \begin{pmatrix} 1 \\ 0 \end{pmatrix}$:

$$\mathcal{M} = e\sqrt{E+m}\,(1 \quad 0)\,(\sigma^1+\sigma^2)\begin{pmatrix} 1 \\ 0 \end{pmatrix} \frac{1}{k^2}\varepsilon^{1*}(k_1)\varepsilon^{2*}(k_2). \tag{8.75}$$

Since $\sigma^1+\sigma^2 = \begin{pmatrix} 0 & 1-i \\ 1+i & 0 \end{pmatrix}$:

$$\mathcal{M} = e\sqrt{E+m}\,(1 \quad 0)\begin{pmatrix} 0 & 1-i \\ 1+i & 0 \end{pmatrix}\begin{pmatrix} 1 \\ 0 \end{pmatrix} \frac{1}{k^2}\varepsilon^{1*}(k_1)\varepsilon^{2*}(k_2). \tag{8.76}$$

Simplifying:

$$\mathcal{M} = e\sqrt{E+m}(1-i)\frac{1}{k^2}\varepsilon^{1*}(k_1)\varepsilon^{2*}(k_2). \tag{8.77}$$

Assuming $\varepsilon^{1*}(k_1) = 1$ and $\varepsilon^{2*}(k_2) = 1$:

$$\mathcal{M} = e\sqrt{E+m}\frac{1-i}{k^2}. \tag{8.78}$$

Therefore, the amplitude for electron-positron annihilation into two photons is:

$$\mathcal{M} = e\sqrt{E+m}\frac{1-i}{k^2}. \tag{8.79}$$

This calculation demonstrates the detailed steps involved in evaluating the amplitude for the electron-positron annihilation process.

8.2 VARIOUS FEYNMAN DIAGRAMS AND THEIR EXPLANATIONS

8.2.1 ELECTROMAGNETIC INTERACTIONS

Electromagnetic interactions are fundamental forces that act between charged particles, mediated by the exchange of photons. These interactions play a crucial role in a variety of physical phenomena, including the behavior of electrons in atoms, the emission and absorption of light, and the scattering of photons by charged particles.

8.2.1.1 Stimulated Emission

Stimulated emission is a key process in lasers. Below is a diagram showing stimulated emission of a photon by an electron.

Figure 8.7 shows stimulated emission in QED.

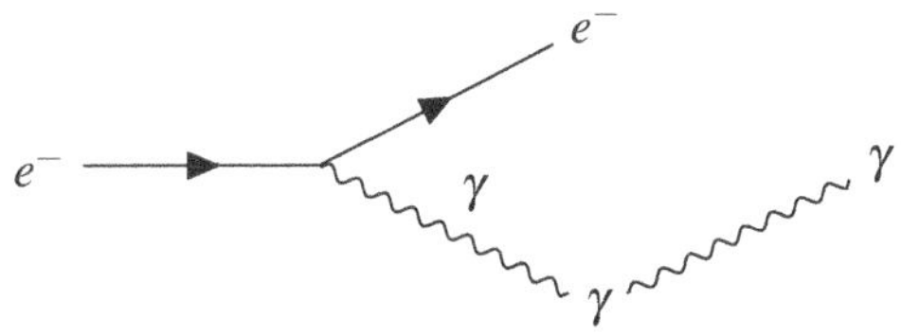

FIGURE 8.7
Stimulated emission of a photon.

8.2.1.2 Spontaneous Emission

Spontaneous emission is the process by which an excited electron emits a photon. Below is a diagram showing spontaneous emission.

Figure 8.8 illustrates spontaneous photon emission by an excited electron.

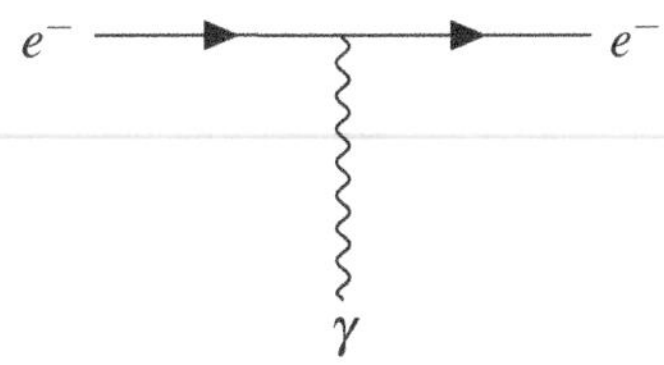

FIGURE 8.8
Spontaneous emission of a photon.

8.2.1.3 Photon-Photon Scattering

Photon-photon scattering is a higher-order QED process. Below is a diagram showing photon-photon scattering via electron loops.

Figure 8.9 sketches photonphoton scattering via electron loops.

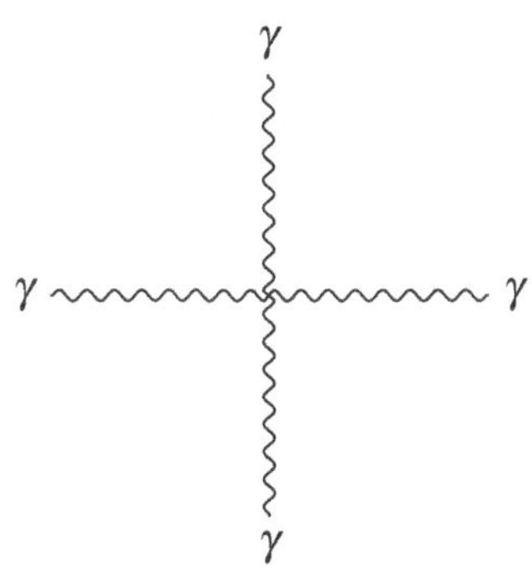

FIGURE 8.9
Photon-photon scattering via electron loops.

8.2.1.3.1 Electron Loops in Photon-Photon Scattering

In quantum electrodynamics (QED), photon-photon scattering is a process that can occur through higher-order interactions involving electron loops. An electron loop represents a virtual electron-positron pair that briefly comes into existence and interacts with the incoming photons before annihilating. These loops are essential for understanding and calculating the scattering amplitude of photons, as they encapsulate the quantum fluctuations and intermediate states that contribute to the interaction. The presence of electron loops makes photon-photon scattering a second-order process, reflecting the complexity and richness of quantum field theory.

8.2.1.4 Electron Loops in Photon-Photon Scattering

Electron loops are a fundamental concept in quantum electrodynamics (QED) and play a significant role in higher-order processes, such as photon-photon scattering. In these interactions, virtual electron-positron pairs temporarily come into existence, interact with incoming photons, and then annihilate. This loop represents the quantum fluctuations that contribute to the scattering amplitude.

In the diagram shown in Figure 8.10, two photons (γ) interact through a loop consisting of an electron (e^-) and a positron (e^+). The electron-positron pair is virtual, meaning it exists only temporarily during the interaction. This loop diagram is crucial for calculating the amplitude and understanding the complex nature of photon-photon interactions in QED.

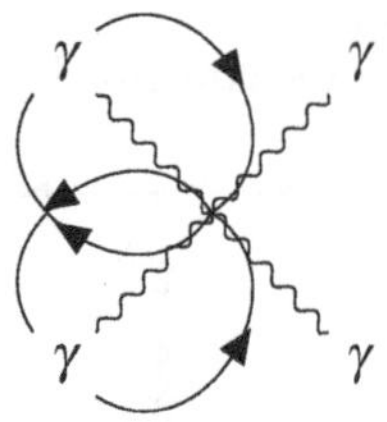

FIGURE 8.10
Feynman diagram illustrating photon-photon scattering via an electron-positron loop.

8.2.1.5 Bremsstrahlung

Bremsstrahlung, or "braking radiation," occurs when a charged particle, such as an electron, is accelerated or decelerated by another charged particle, typically a nucleus. This acceleration causes the electron to emit a photon.

Figure 8.11 shows the electron (e^-) accelerated by the electromagnetic field of a nucleus, causing photon emission.

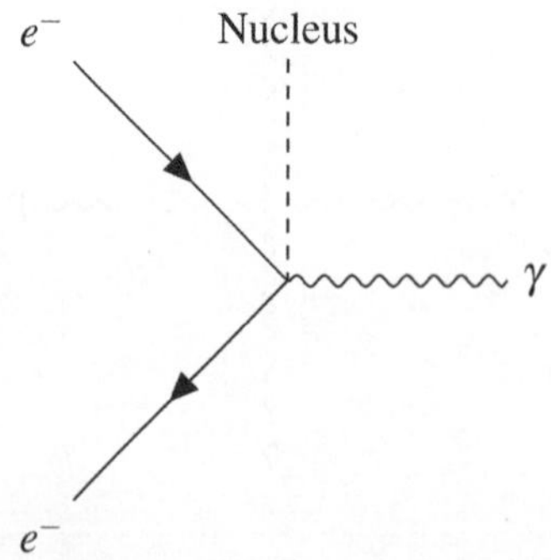

FIGURE 8.11
Feynman diagram for Bremsstrahlung. An electron emits a photon when accelerated by a nucleus.

8.2.1.6 Synchrotron Radiation

Synchrotron radiation occurs when relativistic electrons are accelerated in a magnetic field, causing them to emit radiation. This is commonly observed in astrophysical settings and particle accelerators.

In Figure 8.12, the electron (e^-) is accelerated by a magnetic field (represented by a stylized curly line), causing it to emit a photon (γ).

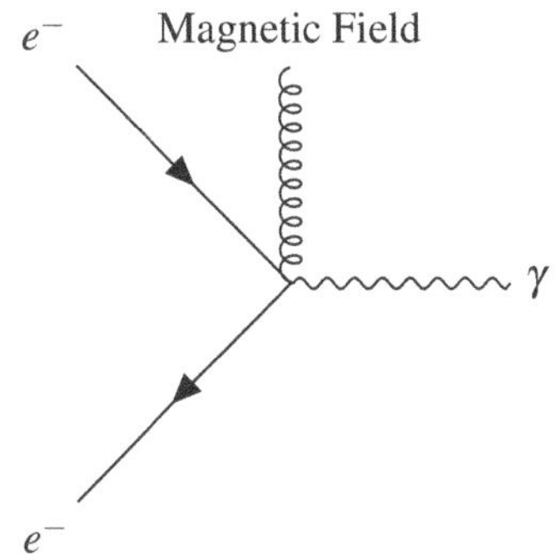

FIGURE 8.12
Feynman diagram for Synchrotron Radiation. A relativistic electron emits a photon when accelerated in a magnetic field.

8.2.1.7 Compton Scattering

Compton scattering is the interaction between a photon and a charged particle, typically an electron, resulting in a transfer of energy and momentum and a change in the photon's wavelength.

In Figure 8.13, a photon (γ) interacts with an electron (e^-), causing the electron to scatter and a photon to be emitted. This process changes the photon's wavelength and energy.

These diagrams illustrate key interactions in quantum electrodynamics and provide a visual representation of the fundamental processes involved.

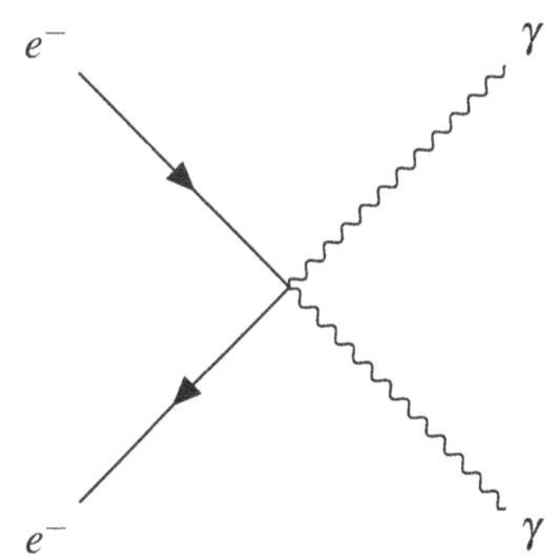

FIGURE 8.13
Feynman diagram for Compton Scattering. A photon scatters off an electron, resulting in a transfer of energy and momentum.

8.2.2 WEAK INTERACTIONS

Weak interactions are one of the four fundamental forces of nature and are responsible for processes such as beta decay in atomic nuclei and the interactions of neutrinos. Unlike the strong and electromagnetic forces, the weak force has a very short range and is mediated by the exchange of heavy gauge bosons, specifically the W and Z bosons. These interactions are crucial for understanding the stability of matter and play a significant role in stellar processes, nuclear reactions, and the synthesis of elements in the universe. Weak interactions violate parity and charge-parity symmetries, leading to phenomena that cannot be explained by other fundamental forces, thus providing unique insights into the behavior of subatomic particles.

8.2.3 W AND Z BOSON IN THE WEAK FORCE

The weak force is mediated by the exchange of W and Z bosons. Below is a Feynman diagram showing a neutron decay process involving a W boson.

Figure 8.14 shows neutron β decay mediated by a W^-.

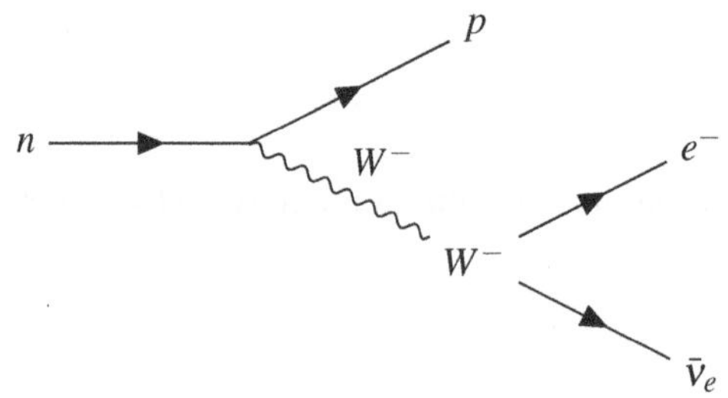

FIGURE 8.14
Neutron decay involving W boson.

8.2.4 MUON DECAY

Muon decay is a weak interaction process. Below is a diagram showing muon decay into an electron, an electron neutrino, and a muon neutrino.

Figure 8.15 depicts the leptonic decay $\mu^- \to e^- \bar{\nu}_e \nu_\mu$.

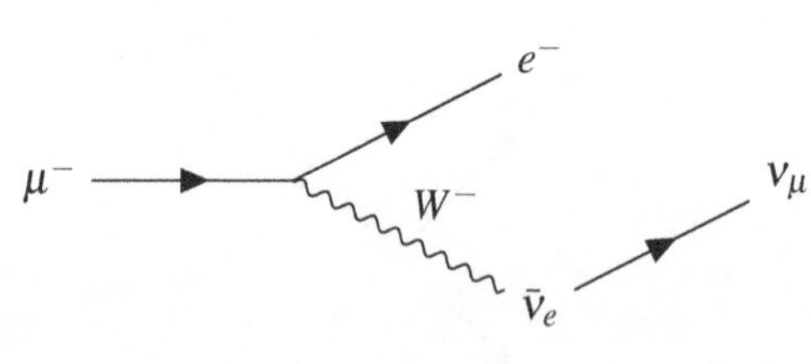

FIGURE 8.15
Muon decay into electron, electron neutrino, and muon neutrino.

8.2.5 NEUTRINO SCATTERING

Neutrino interactions are important in particle physics. Below is a diagram showing neutrino-electron scattering.

Figure 8.16 shows neutralcurrent $\nu_e e^-$ scattering via Z^0 exchange.

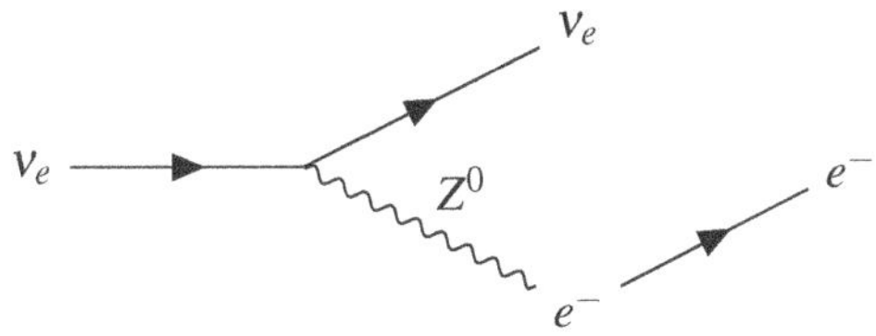

FIGURE 8.16
Neutrino-electron scattering mediated by Z boson.

8.2.6 STRONG INTERACTIONS

Strong interactions, also known as the strong nuclear force, are the fundamental forces that hold atomic nuclei together. This force is mediated by gluons, which act between quarks, the building blocks of protons, neutrons, and other hadrons. The strong force is the most powerful of the four fundamental forces, yet it operates over an extremely short range, binding quarks tightly together to form hadrons. It also ensures the stability of atomic nuclei despite the repulsive electromagnetic forces between protons. Understanding strong interactions is essential for explaining the properties and behavior of nuclear matter, the formation of elements in stars, and the phenomena observed in high-energy particle collisions.

8.2.6.1 Quark-Quark Scattering

Figure 8.17 shows quarkquark scattering via gluon exchange.

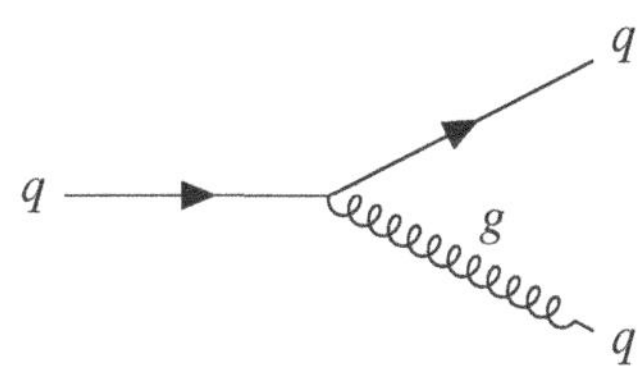

FIGURE 8.17
Quark-quark scattering mediated by gluon exchange.

8.2.7 QUARK-GLUON INTERACTION

Quark-gluon interactions are fundamental in QCD. Below is a diagram showing a quark emitting a gluon.

Figure 8.18 shows a quark (q) emitting a gluon (g), a fundamental QCD interaction.

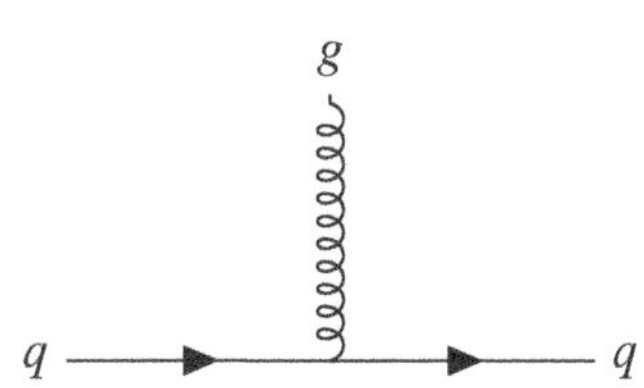

FIGURE 8.18
Quark emitting a gluon.

8.2.7.1 Meson Exchange

Mesons can mediate interactions between nucleons. Below is a diagram showing pion exchange between two nucleons.

Figure 8.19 illustrates nucleonnucleon interaction via pion exchange.

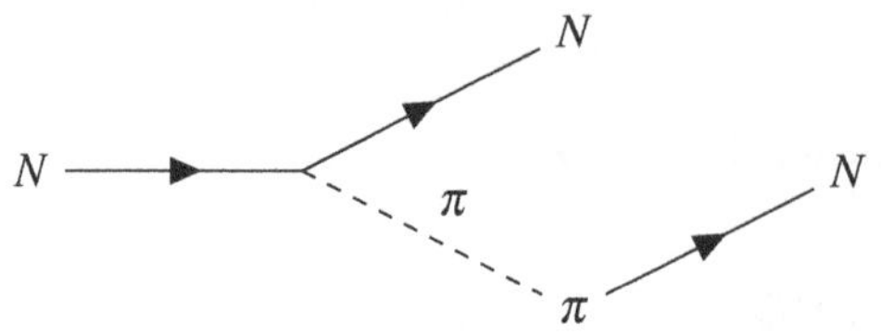

FIGURE 8.19
Pion exchange between nucleons.

8.2.7.2 Deep Inelastic Scattering

Deep inelastic scattering is used to probe the internal structure of nucleons by scattering high-energy electrons off them.

Figure 8.20 sketches deep inelastic e^-p scattering via a virtual photon.

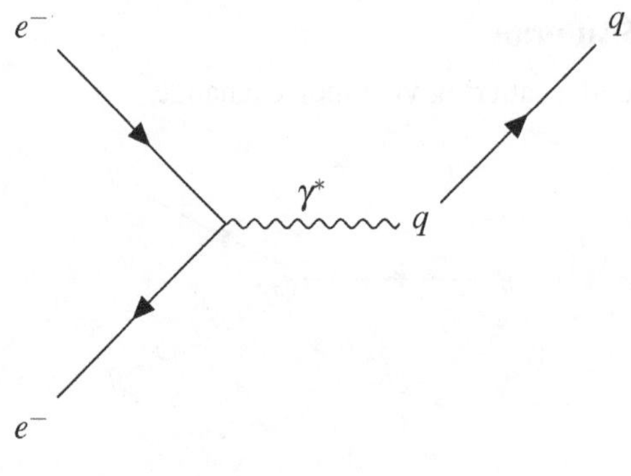

FIGURE 8.20
Feynman diagram for deep inelastic scattering.

8.2.7.3 Quark Confinement

Quark confinement is the phenomenon where quarks are never found in isolation due to the strong color force.

Figure 8.21 schematically depicts color exchange related to confinement.

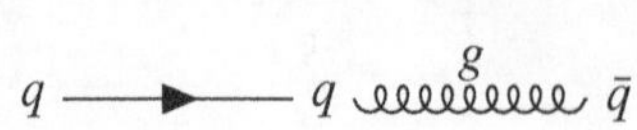

FIGURE 8.21
Feynman diagram illustrating quark confinement.

8.2.8 HIGGS BOSON INTERACTIONS

Higgs interactions involve the Higgs boson, a fundamental particle associated with the mechanism that gives mass to other elementary particles. According to the Standard Model of particle physics, the Higgs field permeates the universe, and particles acquire mass through their interaction with this field. The Higgs boson itself is a manifestation of the Higgs field and was discovered at the Large Hadron Collider in 2012. Understanding Higgs interactions is crucial for explaining how particles such as W and Z bosons, quarks, and leptons obtain their mass, thereby providing a deeper insight into the structure and origin of matter.

8.2.8.1 Higgs Boson Decay

The Higgs boson can decay into various particles. Below is a diagram showing Higgs boson decay into two photons.

Figure 8.22 shows the loopinduced decay $H \to \gamma\gamma$.

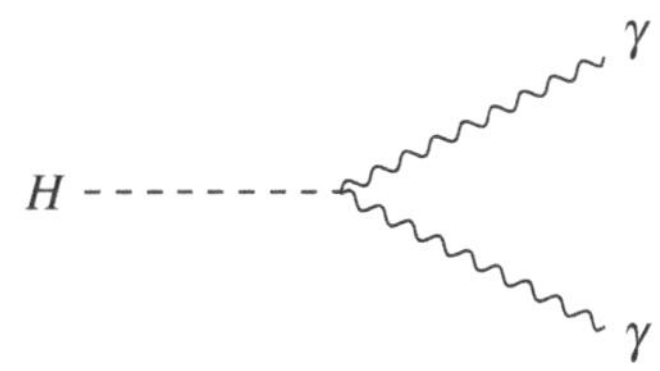

FIGURE 8.22
Higgs boson decay into two photons.

8.2.8.2 Higgs Boson Production

The Higgs boson can be produced via gluon fusion mediated by a top quark loop.

Figure 8.23 shows $gg \to H$ via a top-quark loop.

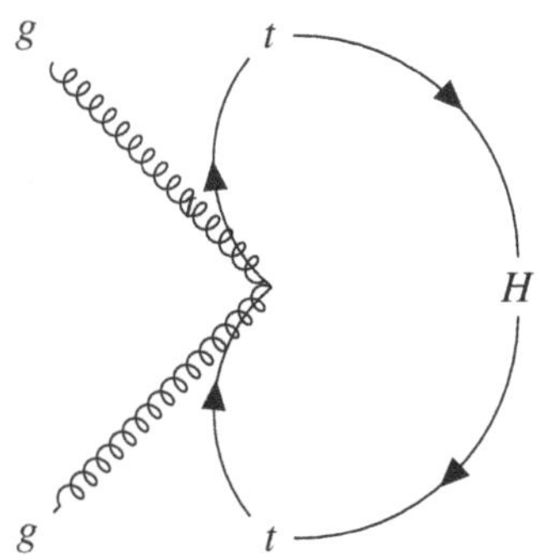

FIGURE 8.23
Feynman diagram for Higgs boson production via gluon fusion.

8.2.9 RARE PROCESSES AND HIGHER-ORDER INTERACTIONS

Rare processes and higher-order interactions in particle physics are phenomena that occur with very low probabilities and often involve complex, multi-step interactions. These processes are crucial for testing the limits of the Standard Model and probing for new physics beyond it. Higher-order interactions involve loop diagrams, where virtual particles momentarily exist and contribute to the interaction, adding corrections to the basic, tree-level processes. Rare processes, such as flavor-changing neutral currents or lepton flavor violation, provide stringent tests for theoretical models and are sensitive indicators of potential new physics, making them a key focus of experimental particle physics research.

8.2.9.1 Beta Decay

Beta decay is a process where a neutron transforms into a proton, an electron, and an electron antineutrino via the weak force.

Figure 8.24 provides a schematic β-decay topology.

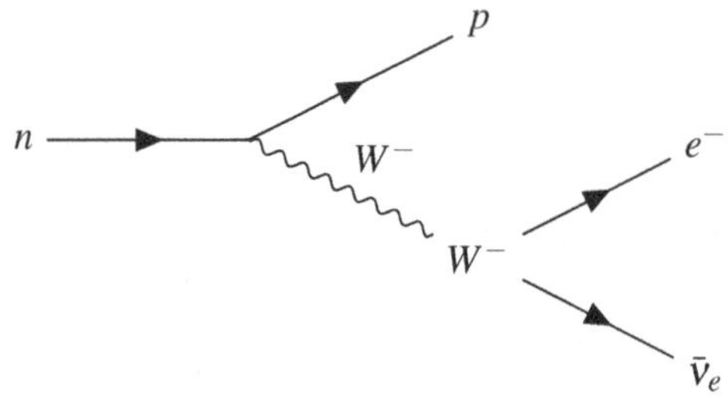

FIGURE 8.24
Feynman diagram for beta decay.

8.2.9.2 Hawking Radiation and Virtual Particles

Hawking radiation is a theoretical process where black holes can emit radiation due to quantum effects near their event horizon. This phenomenon involves the creation of virtual particle pairs in the vacuum near the black hole.

In Figure 8.25, the event horizon of the black hole is represented by a scalar line. Virtual particle pairs, such as an electron (e^-) and positron (e^+), are created near the event horizon. One particle falls into the black hole while the other escapes, resulting in the emission of radiation and a decrease in the mass of the black hole.

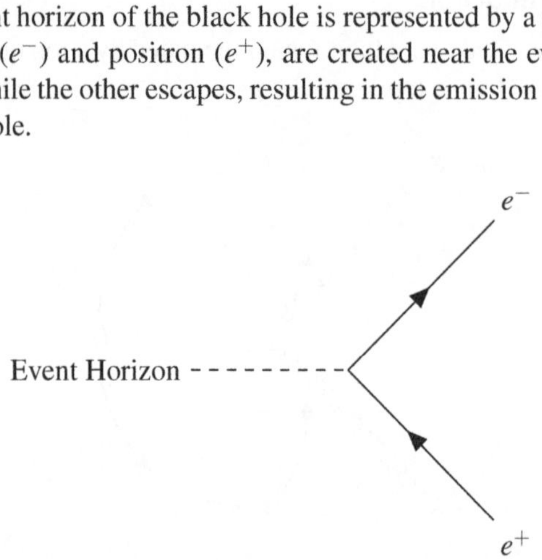

FIGURE 8.25
Simplified diagram representing the concept of Hawking radiation, where virtual particle pairs are created near the event horizon of a black hole.

This simplified diagram illustrates the basic concept of how Hawking radiation occurs due to the separation of virtual particle pairs by the gravitational field of the black hole.

8.2.9.3 Top Quark Decay

Top quark decay is a process where a top quark decays into a W boson and a bottom quark.

Figure 8.26 depicts the dominant decay $t \to W^+ b$.

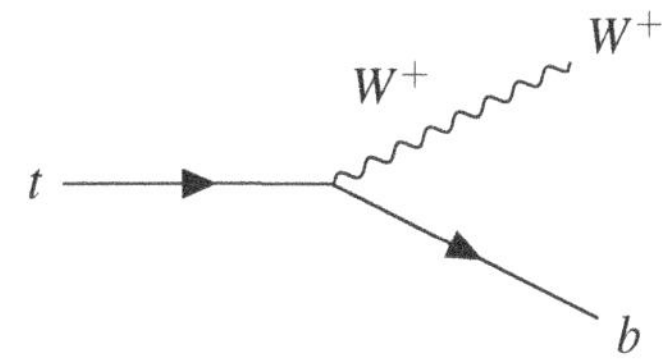

FIGURE 8.26
Feynman diagram for top quark decay.

8.2.10 COMPOSITE PARTICLES

Composite particles are bound states of more elementary particles, held together by the strong force mediated by gluons. Examples of composite particles include baryons, which consist of three quarks, and mesons, which consist of a quark and an antiquark pair. The study of composite particles provides insights into the non-perturbative aspects of quantum chromodynamics (QCD), the theory describing strong interactions. Understanding the structure and interactions of composite particles is essential for explaining the behavior of atomic nuclei and the formation of matter in the universe.

8.2.11 BARYONS: COMPOSED OF THREE QUARKS

Baryons are particles composed of three quarks, held together by the strong force mediated by gluons. For example, a proton is composed of two up quarks and one down quark (*uud*).

Figure 8.27 provides a schematic of the protons *uud* valence content.

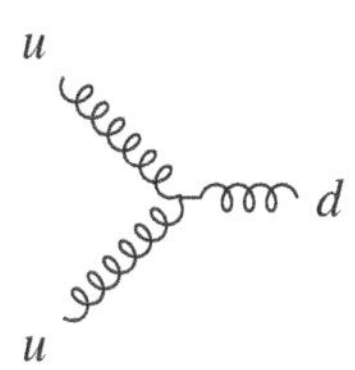

FIGURE 8.27
Feynman diagram representing a proton as a baryon composed of two up quarks and one down quark.

8.2.12 OTHER BARYONS

Other baryons, such as neutrons, are also composed of three quarks. For example, a neutron is composed of two down quarks and one up quark (*udd*).

Figure 8.28 sketches the neutrons *udd* valence content.

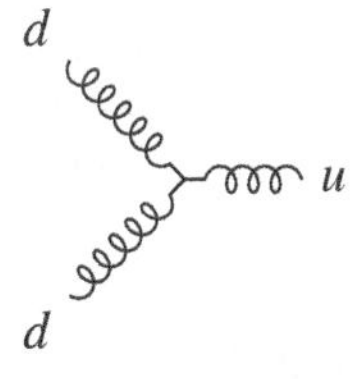

FIGURE 8.28
Feynman diagram representing a neutron as a baryon composed of two down quarks and one up quark.

8.2.13 MESONS: COMPOSED OF A QUARK AND AN ANTIQUARK

Mesons are particles composed of a quark and an antiquark pair, held together by the strong force mediated by gluons. For example, a pion (π^+) is composed of an up quark and an anti-down quark ($u\bar{d}$).

Figure 8.29 shows the π^+ valence structure $u\bar{d}$.

FIGURE 8.29
Feynman diagram representing a pion (π^+) as a meson composed of an up quark and an anti-down quark.

8.2.14 OTHER MESONS

Other mesons, such as kaons (K^+), are also composed of a quark and an antiquark pair. For example, a kaon (K^+) is composed of an up quark and an anti-strange quark ($u\bar{s}$).

Figure 8.30 shows the K^+ valence structure $u\bar{s}$.

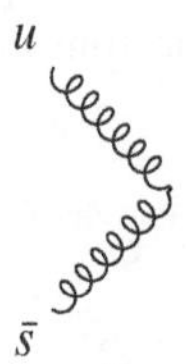

FIGURE 8.30
Feynman diagram representing a kaon (K^+) as a meson composed of an up quark and an anti-strange quark.

These diagrams represent a comprehensive overview of key processes and interactions in particle physics. They illustrate the fundamental forces and particles involved, providing a visual and intuitive understanding of the complex phenomena in quantum field theory.

8.3 IMPORTANCE OF FEYNMAN DIAGRAMS IN QUANTUM FIELD THEORY

Feynman diagrams are an essential tool in Quantum Field Theory (QFT) for several reasons. They provide a visual representation of the complex mathematical expressions that describe the interactions between subatomic particles. Developed by Richard Feynman in the 1940s, these diagrams have revolutionized the way physicists understand and calculate particle interactions.

8.3.1 SIMPLIFICATION OF CALCULATIONS

One of the primary benefits of Feynman diagrams is their ability to simplify complicated calculations. In QFT, interactions are often described by perturbation theory, which involves expanding the interaction in a series of terms. Each term in this expansion corresponds to a different interaction process, which can be visualized using Feynman diagrams. By representing these terms graphically, Feynman diagrams make it easier to identify and compute individual contributions to the overall interaction.

8.3.2 SYSTEMATIC ORGANIZATION

Feynman diagrams provide a systematic way to organize and keep track of the various terms in the perturbative expansion. This helps ensure that all relevant terms are included in the calculation, reducing the likelihood of errors. The diagrams also make it easier to see how different interaction processes are related to one another.

8.3.3 INTUITIVE UNDERSTANDING

In addition to their practical benefits, Feynman diagrams offer an intuitive way to understand particle interactions. They depict the exchange of virtual particles and the flow of momentum and energy between interacting particles, providing a clear visual representation of the underlying physical processes. This makes it easier for physicists to conceptualize and communicate complex interactions.

8.3.4 APPLICATIONS IN HIGH-ENERGY PHYSICS

Feynman diagrams are used extensively in high-energy physics to calculate scattering amplitudes, decay rates, and other quantities of interest. They are essential for making predictions about the outcomes of particle collisions in experiments such as those conducted at the Large Hadron Collider (LHC). By comparing these predictions with experimental data, physicists can test the validity of theoretical models and search for new physics beyond the Standard Model.

8.3.5 EXTENSIONS TO OTHER THEORIES

The techniques developed for quantum electrodynamics (QED) using Feynman diagrams have been successfully extended to other quantum field theories, including quantum chromodynamics (QCD) and the electroweak theory. This has made Feynman diagrams a universal tool in high-energy physics, applicable to a wide range of interactions and processes.

8.3.6 NON-PERTURBATIVE METHODS AND BEYOND

While Feynman diagrams are primarily used for perturbative calculations, they also provide insights into non-perturbative phenomena. For example, in QCD, Feynman diagrams help to understand the behavior of quarks and gluons at high energies, even though the theory becomes strongly coupled at low energies. Additionally, techniques inspired by Feynman diagrams are used in various non-perturbative methods, such as lattice QCD and the study of instantons.

8.3.7 APPLICATIONS IN CONDENSED MATTER PHYSICS

Beyond high-energy physics, Feynman diagrams have found applications in condensed matter physics. They are used to describe interactions in many-body systems, such as electrons in solids. Feynman diagrams help to calculate properties like electrical conductivity, superconductivity, and magnetic interactions, providing a bridge between particle physics and condensed matter theory.

8.4 SUMMARY AND DISCUSSION

8.4.1 CHAPTER SUMMARY

In this chapter, we explored the fundamental role of Feynman diagrams in Quantum Field Theory (QFT), focusing on their historical context, key elements, and applications across various interaction types. We began with an introduction to Feynman diagrams, highlighting their development by Richard Feynman and their revolutionary impact on simplifying complex QFT calculations. The diagrams provide an intuitive visual representation of particle interactions, facilitating the organization and computation of perturbative expansions.

The chapter delved into the specifics of how Feynman diagrams are constructed and interpreted, including the significance of vertices, propagators, and external lines. Different types of lines representing various particles and interactions, such as fermions, photons, and other bosons, were discussed.

We categorized and examined several fundamental interactions depicted by Feynman diagrams:

- **Electromagnetic Interactions**: Processes like stimulated and spontaneous emission, photon-photon scattering, Bremsstrahlung, synchrotron radiation, and Compton scattering were illustrated, emphasizing their roles in quantum electrodynamics (QED).
- **Weak Interactions**: Diagrams of neutron decay, muon decay, and neutrino-electron scattering were presented, showcasing the mediation by W and Z bosons.
- **Strong Interactions**: We explored quark-quark scattering, quark-gluon interactions, meson exchange, deep inelastic scattering, and quark confinement, highlighting the strong nuclear force mediated by gluons.
- **Higgs Interactions**: The production and decay of the Higgs boson were explained, illustrating the Higgs mechanism's role in giving mass to elementary particles.
- **Rare Processes and Higher-Order Interactions**: Rare processes like beta decay, Hawking radiation, and top quark decay were discussed, demonstrating the importance of higher-order interactions and virtual particles.
- **Composite Particles**: The structure and interactions of baryons and mesons, as examples of composite particles composed of quarks, were depicted, emphasizing the role of gluons in binding quarks together.

Additionally, we examined the importance of Feynman diagrams in QFT, highlighting their ability to simplify calculations, provide systematic organization, offer intuitive understanding, and extend to other theories beyond QED. The application of Feynman diagrams in high-energy physics, non-perturbative methods, and condensed matter physics further underscores their versatility and significance.

8.4.2 DISCUSSION

Feynman diagrams serve as a cornerstone in the study and application of Quantum Field Theory. Their development marked a significant advancement in the ability to perform complex calculations and visualize particle interactions. By reducing intricate mathematical expressions into manageable graphical representations, Feynman diagrams have made QFT more accessible and comprehensible.

The diagrams' ability to systematically organize perturbative expansions ensures that all relevant interaction terms are accounted for, reducing errors and enhancing the accuracy of theoretical predictions. This systematic approach is crucial in high-energy physics experiments, such as those conducted at the Large Hadron Collider (LHC), where precise predictions are essential for validating theoretical models and exploring new physics beyond the Standard Model.

Moreover, Feynman diagrams offer an intuitive way to understand the dynamics of particle interactions. By depicting the exchange of virtual particles and the flow of momentum and energy, these diagrams provide a clear visual representation of the underlying physical processes. This intuitive understanding is invaluable for both students and researchers in the field.

The extension of Feynman diagram techniques to other quantum field theories, including QCD and the electroweak theory, highlights their universal applicability. In non-perturbative methods and condensed matter physics, Feynman diagrams continue to play a crucial role, bridging the gap between different areas of physics and fostering a deeper understanding of fundamental interactions.

In conclusion, Feynman diagrams are an indispensable tool in Quantum Field Theory, offering significant advantages in simplifying calculations, organizing complex expressions, and providing intuitive insights into particle interactions. Their broad applications across various fields of physics underscore their fundamental importance in advancing our understanding of the universe's fundamental forces and particles.

8.5 QUESTIONS AND PROBLEMS

8.5.1 TRIVIA QUESTIONS

Question 1: Who introduced Feynman diagrams?

(a) Albert Einstein
(b) Richard Feynman
(c) Niels Bohr
(d) Paul Dirac

Answer: b. Richard Feynman
Question 2: What type of lines in Feynman diagrams represent fermions?

(a) Wavy lines
(b) Dashed lines
(c) Straight lines
(d) Curly lines

Answer: c. Straight lines
Question 3: Which fundamental force is mediated by W and Z bosons?

(a) Electromagnetic force
(b) Strong force
(c) Weak force
(d) Gravitational force

Answer: c. Weak force
Question 4: In the process of electron-positron annihilation, what particles are produced?

(a) Two electrons
(b) Two positrons
(c) Two neutrinos
(d) Two photons

Answer: d. Two photons
Question 5: What is the role of gluons in Quantum Chromodynamics (QCD)?

(a) Mediate the strong force
(b) Mediate the electromagnetic force
(c) Mediate the weak force

(d) Mediate the gravitational force

Answer: a. Mediate the strong force
Question 6: What process is shown by the following Feynman diagram?

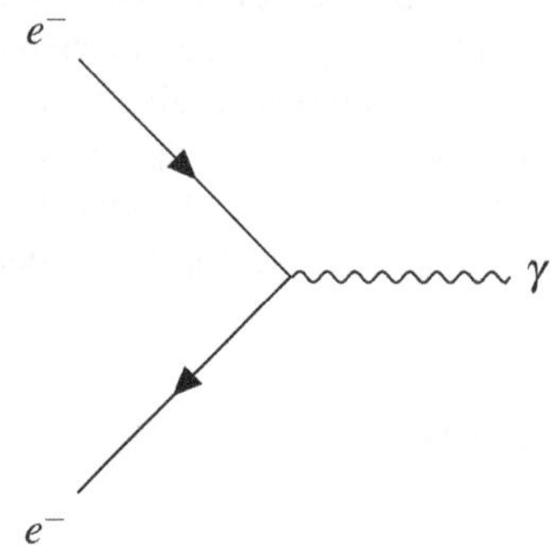

(a) Electron-positron annihilation
(b) Compton scattering
(c) Photon emission by an electron
(d) Neutrino scattering

Answer: c. Photon emission by an electron
Question 7: What is the significance of the vertex in a Feynman diagram?

(a) Represents the incoming particles
(b) Represents the outgoing particles
(c) Represents the interaction point between particles
(d) Represents the external lines

Answer: c. Represents the interaction point between particles
Question 8: In the QED interaction term $\mathscr{L}_{\text{int}} = -e\bar{\psi}\gamma^\mu\psi A_\mu$, what does γ^μ represent?

(a) Coupling constant
(b) Electric charge
(c) Dirac gamma matrices
(d) Photon field

Answer: c. Dirac gamma matrices
Question 9: What is the fundamental particle associated with giving mass to other particles?

(a) Photon
(b) Gluon
(c) Higgs boson
(d) Neutrino

Answer: c. Higgs boson
Question 10: Which process involves an electron emitting a photon when accelerated by a nucleus?

(a) Synchrotron radiation
(b) Bremsstrahlung
(c) Compton scattering
(d) Neutron decay

Answer: b. Bremsstrahlung

8.5.2 PROBLEMS

1. Calculate the interaction probability amplitude for a photon-electron interaction.
Solution: The interaction term in the QED Lagrangian is:

$$\mathcal{L}_{\text{int}} = -e\bar{\psi}\gamma^{\mu}\psi A_{\mu}, \tag{8.80}$$

where e is the electric charge, ψ is the electron field, $\bar{\psi}$ is the Dirac adjoint, γ^{μ} are the gamma matrices, and A_{μ} is the photon field.
The amplitude $\mathcal{M}$ for the interaction is given by:

$$\mathcal{M} = -ie\bar{u}(p')\gamma^{\mu}u(p)\varepsilon_{\mu}, \tag{8.81}$$

where $u(p)$ and $u(p')$ are the Dirac spinors for the initial and final electrons, respectively, and ε_{μ} is the polarization vector of the photon.
Assume the initial electron momentum is $p = (E, \mathbf{p})$, the final electron momentum is $p' = (E', \mathbf{p}')$, and the photon momentum is $k = (k^0, \mathbf{k})$.
Substitute the values:

$$\mathcal{M} = -ie\bar{u}(p')\gamma^{\mu}u(p)\varepsilon_{\mu}. \tag{8.82}$$

If we consider specific momenta and polarization vectors, for simplicity:

$$p = (E, 0, 0, p), \quad p' = (E', 0, 0, p'), \tag{8.83}$$

$$k = (k^0, k^x, k^y, k^z), \quad \varepsilon^{\mu} = (0, 1, 0, 0). \tag{8.84}$$

Then the amplitude becomes:

$$\mathcal{M} = -ie\bar{u}(p')\gamma^{1}u(p). \tag{8.85}$$

Using the properties of Dirac spinors and gamma matrices, the detailed calculation would yield the specific value of $\mathcal{M}$ based on the given momenta and polarization.
2. Calculate the vertex factor for the electron-positron annihilation into two photons.
Solution: The vertex factor in the QED Lagrangian for electron-positron annihilation is given by the interaction term:

$$\mathcal{L}_{\text{int}} = -e\bar{\psi}\gamma^{\mu}\psi A_{\mu}. \tag{8.86}$$

For electron-positron annihilation:

$$e^{+} + e^{-} \to \gamma + \gamma, \tag{8.87}$$

The vertex factor is:

$$-ie\gamma^{\mu}. \tag{8.88}$$

The corresponding amplitude involves:

$$\mathcal{M} = \bar{v}(p_2)(-ie\gamma^{\mu})u(p_1)\frac{-ig_{\mu\nu}}{k^2}\varepsilon^{\nu*}(k_1)\varepsilon^{\rho*}(k_2). \tag{8.89}$$

This vertex factor is used in the perturbative expansion to calculate the overall amplitude for the process.
3. Complete the following Feynman diagram for electron-electron scattering:

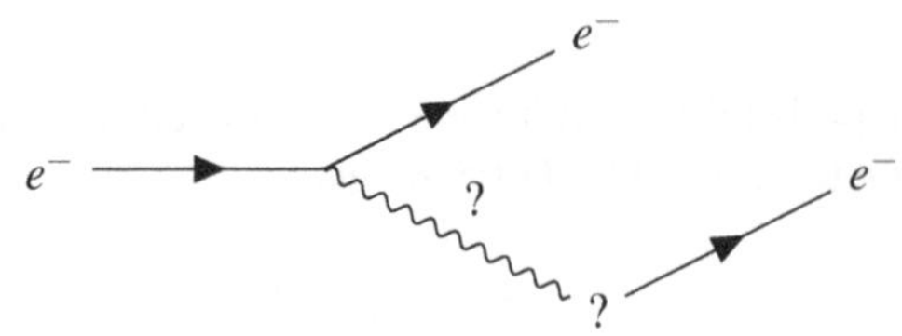

FIGURE 8.31
Incomplete Feynman diagram for electron-electron scattering.

Solution: The missing components are:

- The exchanged particle is a photon (γ).
- The outgoing particle from vertex d is an electron (e^-).

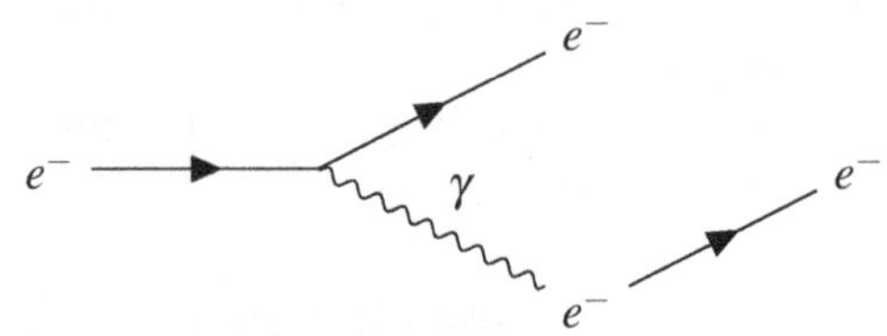

FIGURE 8.32
Completed Feynman diagram for electron-electron scattering.

4. Complete the following Feynman diagram for electron-positron annihilation:

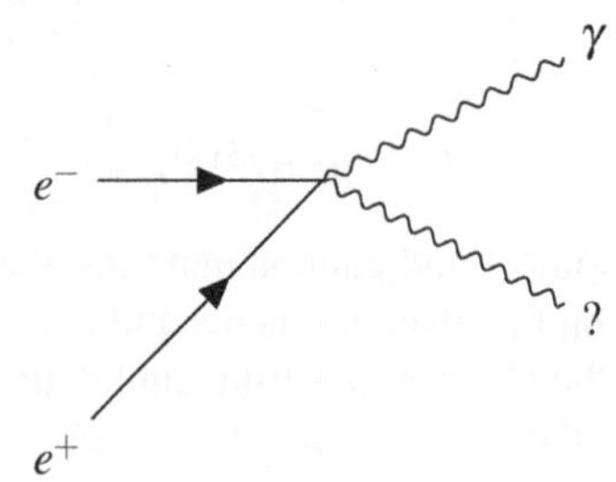

FIGURE 8.33
Incomplete Feynman diagram for electron-positron annihilation.

Solution: The missing component is:

- The second outgoing particle from vertex b is a photon (γ).

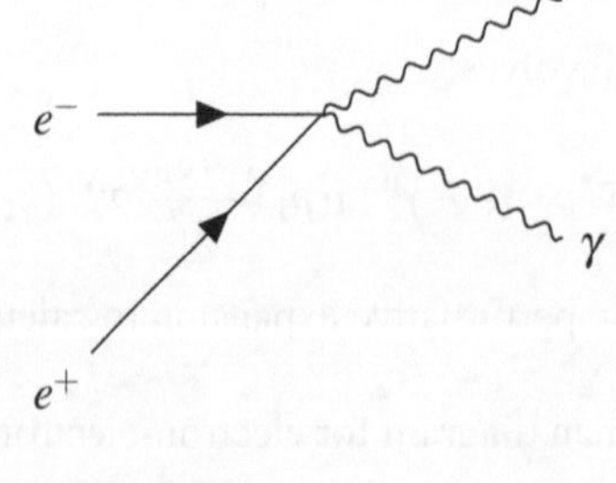

FIGURE 8.34
Completed Feynman diagram for electron-positron annihilation.

5. Draw a Feynman diagram that best represents the electromagnetic field vector a and $a^\dagger$ quantization representation for QFT.
   Solution: In Quantum Field Theory, the electromagnetic field can be represented by the creation ($a^\dagger$) and annihilation (a) operators. The corresponding Feynman diagram would involve vertices where photons are either created or annihilated.

FIGURE 8.35
Feynman diagram representing the creation and annihilation operators for the electromagnetic field.

This diagram shows the basic interaction where a photon is created by $a^\dagger$ or annihilated by a.

Bonus Problem: Calculate the interaction probability amplitude for a neutrino-electron scattering process and draw the corresponding Feynman diagram. Use the following specific values:

- Initial neutrino momentum: $p_\nu = (10, 0, 0, 10)$ GeV
- Initial electron momentum: $p_e = (10, 0, 0, -10)$ GeV
- Final neutrino momentum: $p'_\nu = (9, 1, 0, 9)$ GeV
- Final electron momentum: $p'_e = (11, -1, 0, -9)$ GeV
- Neutrino and electron masses are negligible compared to their momenta.
- Coupling constant: $g = 0.652$
- Weinberg angle: $\sin^2\theta_W \approx 0.23$, $\cos\theta_W \approx 0.877$
- Z boson mass: $M_Z \approx 91.2$ GeV

Hints:

- Use the interaction term in the Lagrangian for the neutrino and electron via the Z^0 boson:

$$\begin{aligned}\mathscr{L}_{\text{int}} = &-\frac{g}{2\cos\theta_W}\left(\bar{\nu}_e\gamma^\mu(1-\gamma^5)\nu_e\right)Z_\mu \\ &-\frac{g}{2\cos\theta_W}\left(\bar{e}\gamma^\mu(g_V - g_A\gamma^5)e\right)Z_\mu,\end{aligned} \tag{8.90}$$

- For the electron: $g_V = -1 + 4\sin^2\theta_W \approx -0.08$, $g_A = -1$.
- The amplitude $\mathscr{M}$ is given by:

$$\begin{aligned}\mathscr{M} = &\left(-\frac{g}{2\cos\theta_W}\right)^2\left(\bar{u}(p'_\nu)\gamma^\mu(1-\gamma^5)u(p_\nu)\right) \\ &\frac{-ig_{\mu\nu}}{q^2 - M_Z^2}\left(\bar{u}(p'_e)\gamma^\nu(g_V - g_A\gamma^5)u(p_e)\right),\end{aligned} \tag{8.91}$$

- Assume the spinors are normalized and simplify the vertex factors to 1 for this problem.

Solution:
First, let's draw the Feynman diagram for neutrino-electron scattering.

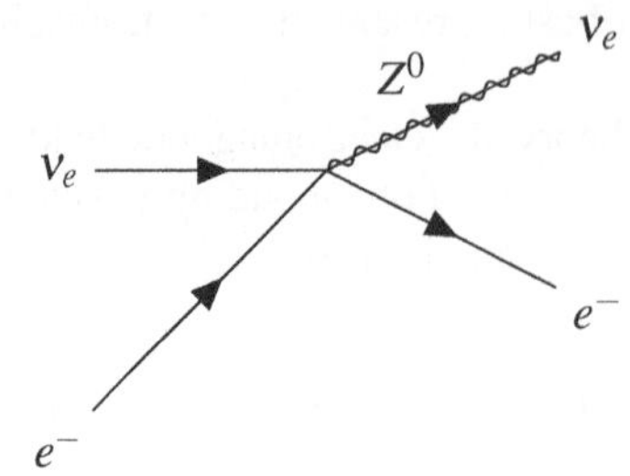

FIGURE 8.36
Feynman diagram for neutrino-electron scattering.

Now, let's calculate the amplitude.

Given the values, the amplitude $\mathscr{M}$ for this process is given by:

$$\mathscr{M} = \left(-\frac{g}{2\cos\theta_W}\right)^2 \left(\bar{u}(p'_\nu)\gamma^\mu(1-\gamma^5)u(p_\nu)\right)$$
$$\frac{-ig_{\mu\nu}}{q^2 - M_Z^2}\left(\bar{u}(p'_e)\gamma^\nu(g_V - g_A\gamma^5)u(p_e)\right), \tag{8.92}$$

where $q = p_\nu - p'_\nu = p'_e - p_e$ is the momentum transfer and M_Z is the mass of the Z^0 boson.

For simplicity, assume $\sin^2\theta_W \approx 0.23$, $g \approx 0.652$, and $M_Z \approx 91.2$ GeV.

Let's compute the amplitude step-by-step: 1. Compute the vertex factors. 2. Compute the propagator. 3. Substitute the spinors and gamma matrices.

Step 1: Compute the vertex factors.

$$\mathscr{V}_\nu = \bar{u}(p'_\nu)\gamma^\mu(1-\gamma^5)u(p_\nu), \tag{8.93}$$

$$\mathscr{V}_e = \bar{u}(p'_e)\gamma^\nu(g_V - g_A\gamma^5)u(p_e). \tag{8.94}$$

Step 2: Compute the propagator.

$$\Delta_{\mu\nu} = \frac{-ig_{\mu\nu}}{q^2 - M_Z^2}. \tag{8.95}$$

Step 3: Calculate the full amplitude. The momentum transfer q is given by:

$$q = p_\nu - p'_\nu = (10,0,0,10) - (9,1,0,9) = (1,-1,0,1). \tag{8.96}$$

The magnitude of q^2 is:

$$q^2 = 1^2 + (-1)^2 + 0^2 + 1^2 = 3. \tag{8.97}$$

Now, substitute the values:

$$\Delta_{\mu\nu} = \frac{-ig_{\mu\nu}}{3 - (91.2)^2} \approx \frac{-ig_{\mu\nu}}{-8318.64}. \tag{8.98}$$

Next, substitute the spinors and gamma matrices (assuming the particles are relativistic and neglecting masses): For the neutrino vertices:

$$\left(\bar{u}(p'_\nu)\gamma^\mu(1-\gamma^5)u(p_\nu)\right) \approx 1. \tag{8.99}$$

For the electron vertices:

$$\left(\bar{u}(p'_e)\gamma^\nu(g_V - g_A\gamma^5)u(p_e)\right) \approx 1. \tag{8.100}$$

Combining everything:

$$\mathcal{M} \approx \left(\frac{0.652}{2 \times 0.877}\right)^2 \frac{i}{8318.64} \approx 0.138 \left(\frac{i}{8318.64}\right). \tag{8.101}$$

Thus, the amplitude simplifies to:

$$\mathcal{M} \approx \frac{0.138i}{8318.64}. \tag{8.102}$$

Therefore, the interaction probability amplitude is:

$$\boxed{\mathcal{M} \approx \frac{0.138i}{8318.64} \approx 1.66 \times 10^{-5} i}. \tag{8.103}$$

Note:

In quantum field theory, the probability amplitude $\mathcal{M}$ is often a complex number. The real and imaginary parts of this complex number are related to the physical properties of the interaction. Specifically:

- **Magnitude of the Amplitude:** The probability of a particular process occurring is related to the magnitude squared of the amplitude, $|\mathcal{M}|^2$. This quantity is always real and positive, representing a physical probability.
- **Phase of the Amplitude:** The phase (argument) of the complex amplitude can influence interference effects when multiple amplitudes contribute to the same physical process. The imaginary part contributes to this phase.

An imaginary amplitude means that the interaction has a certain phase shift. However, when calculating physical observables like cross-sections or decay rates, we are interested in $|\mathcal{M}|^2$, which combines both the real and imaginary parts:

$$|mathcalM|^2 = \mathcal{M}\mathcal{M}^* = (\mathrm{Re}(\mathcal{M}))^2 + (\mathrm{Im}(\mathcal{M}))^2 \tag{8.104}$$

In the given problem, if $\mathcal{M} \approx 1.66 \times 10^{-5} i$, the probability amplitude is purely imaginary. To find the probability or cross-section, you would calculate:

$$|\mathcal{M}|^2 = (1.66 \times 10^{-5})^2 = 2.76 \times 10^{-10} \tag{8.105}$$

This result is real and positive, representing the physical likelihood of the neutrino-electron scattering process.

9 Lectures 17 & 18: The S-matrix

9.1 INTRODUCTION TO THE S-MATRIX

The S-matrix, or scattering matrix, is a fundamental concept in quantum field theory and particle physics. It encodes information about how initial states evolve into final states during interactions. The S-matrix formalism provides a powerful framework for analyzing scattering processes and understanding particle interactions at a fundamental level.

Scattering in itself is a fundamental process in particle physics and quantum field theory, describing how particles interact and change direction, energy, or other properties when they collide. Typically, scattering is viewed from the perspective of individual particles, such as an electron scattering off a proton or two protons colliding in a high-energy accelerator. These interactions can result in various outcomes, such as deflection, absorption, or the production of new particles, and are quantified by measurable quantities like cross sections and scattering amplitudes.

From the viewpoint of quantum fields, scattering is understood in a more fundamental way. Particles are excitations of underlying fields, and their interactions are mediated by the exchange of other field quanta, such as photons for the electromagnetic field or gluons for the strong nuclear force. In this framework, scattering processes are described by the creation, annihilation, and exchange of these quanta, governed by the principles of quantum field theory. The S-matrix formalism provides a systematic way to encode all possible outcomes of a scattering process, capturing the transition probabilities between initial and final states of the quantum fields. This approach not only unifies the description of different types of interactions but also allows for the application of powerful computational techniques, such as Feynman diagrams and perturbation theory, to calculate the probabilities and amplitudes of various scattering events.

9.2 DEFINITION AND PROPERTIES OF THE S-MATRIX

9.2.1 UNITARY

The S-matrix is unitary, meaning $S^\dagger S = SS^\dagger = I$, which ensures probability conservation in scattering processes.

Example: Unitarity Check Consider two orthonormal states $|a\rangle$ and $|b\rangle$. If the S-matrix is unitary, then for any two states $|a\rangle$ and $|b\rangle$,

$$\sum_f S_{fa} S^*_{fb} = \delta_{ab}. \tag{9.1}$$

Solution: For simplicity, assume a 2x2 S-matrix:

$$S = \begin{pmatrix} S_{11} & S_{12} \\ S_{21} & S_{22} \end{pmatrix}. \tag{9.2}$$

To be unitary,

$$S^\dagger S = I. \tag{9.3}$$

Calculating $S^\dagger S$:

$$S^\dagger = \begin{pmatrix} S^*_{11} & S^*_{21} \\ S^*_{12} & S^*_{22} \end{pmatrix}, \tag{9.4}$$

DOI: 10.1201/9781003637370-9

$$S^\dagger S = \begin{pmatrix} S_{11}^* & S_{21}^* \\ S_{12}^* & S_{22}^* \end{pmatrix} \begin{pmatrix} S_{11} & S_{12} \\ S_{21} & S_{22} \end{pmatrix}. \tag{9.5}$$

Simplify to confirm $S^\dagger S = I$:

$$S^\dagger S = \begin{pmatrix} |S_{11}|^2 + |S_{21}|^2 & S_{11}S_{12}^* + S_{21}S_{22}^* \\ S_{12}S_{11}^* + S_{22}S_{21}^* & |S_{12}|^2 + |S_{22}|^2 \end{pmatrix} = \begin{pmatrix} 1 & 0 \\ 0 & 1 \end{pmatrix}. \tag{9.6}$$

9.2.2 LORENTZ INVARIANCE

The S-matrix is invariant under Lorentz transformations, reflecting the fundamental symmetries of spacetime.

Example: Lorentz Invariance Consider a Lorentz transformation Λ acting on a four-momentum p^μ:

$$p'^\mu = \Lambda^\mu{}_\nu p^\nu. \tag{9.7}$$

Solution: The S-matrix element for a process involving particles with momenta $p_1, p_2, \ldots, p_n$ should remain the same under Lorentz transformations:

$$S_{fi}(p'_1, p'_2, \ldots, p'_n) = S_{fi}(p_1, p_2, \ldots, p_n). \tag{9.8}$$

For example, consider a Lorentz boost along the z-axis with velocity v:

$$\Lambda = \begin{pmatrix} \gamma & 0 & 0 & -\gamma\beta \\ 0 & 1 & 0 & 0 \\ 0 & 0 & 1 & 0 \\ -\gamma\beta & 0 & 0 & \gamma \end{pmatrix}, \tag{9.9}$$

where $\gamma = \frac{1}{\sqrt{1-\beta^2}}$ and $\beta = \frac{v}{c}$.

If the initial four-momentum is $p^\mu = (E, 0, 0, p_z)$, the transformed four-momentum is:

$$p'^\mu = \Lambda^\mu{}_\nu p^\nu = \begin{pmatrix} \gamma E - \gamma\beta p_z \\ 0 \\ 0 \\ -\gamma\beta E + \gamma p_z \end{pmatrix}. \tag{9.10}$$

The invariance of the S-matrix under this transformation ensures that the calculated physical quantities remain consistent across different reference frames, reflecting the fundamental symmetries of spacetime.

9.2.3 ANALYTICITY AND CAUSALITY

The S-matrix elements are analytic functions of the kinematic variables, with singularities corresponding to physical thresholds and resonances. This means that the elements of the S-matrix can be expressed as analytic functions in the complex plane, except at points that correspond to physical phenomena such as particle creation or annihilation.

Example: Analyticity and Physical Thresholds Consider the scattering amplitude $T(s)$, where s is the Mandelstam variable representing the square of the center-of-mass energy.

Solution: The scattering amplitude $T(s)$ can be written as:

$$T(s) = \frac{R(s)}{s - s_0 + i\varepsilon}, \tag{9.11}$$

where s_0 is the location of the pole, corresponding to a physical particle or resonance, and ε is a small positive number ensuring the proper behavior of the propagator.

The imaginary part of the amplitude is given by:

$$\Im(T(s)) = -\pi\delta(s - s_0). \tag{9.12}$$

This result comes from the properties of the S-matrix and the analytic continuation of the scattering amplitude in the complex plane. Specifically, it arises from the interpretation of the scattering amplitude near a resonance and the application of the optical theorem, which relates the imaginary part of the forward scattering amplitude to the total cross-section.

To find the imaginary part of $T(s)$, we look at the imaginary part of the propagator near the pole. Using the relation for the imaginary part of a complex function:

$$\Im\left(\frac{1}{x - x_0 + i\varepsilon}\right) = -\pi\delta(x - x_0), \tag{9.13}$$

we apply this to our scattering amplitude $T(s)$:

$$\Im(T(s)) = \Im\left(\frac{R(s)}{s - s_0 + i\varepsilon}\right). \tag{9.14}$$

Using the property of the delta function, we get:

$$\Im\left(\frac{1}{s - s_0 + i\varepsilon}\right) = -\pi\delta(s - s_0), \tag{9.15}$$

thus,

$$\Im(T(s)) = R(s) \cdot (-\pi\delta(s - s_0)). \tag{9.16}$$

If $R(s)$ is approximately constant near the pole, we can write:

$$\Im(T(s)) = -\pi R(s_0)\delta(s - s_0). \tag{9.17}$$

In summary, the imaginary part of the amplitude at the resonance is given by the delta function centered at the resonance energy, weighted by the residue of the pole. This result is a consequence of the analytic properties of the scattering amplitude and the physical interpretation of resonances as intermediate states in the scattering process.

9.2.3.1 Example: Simple Scattering - Rutherford Scattering

Consider the scattering of an electron off a positively charged nucleus, as described by Rutherford scattering. The differential cross-section can be calculated from the potential:

$$V(r) = \frac{Ze^2}{r}, \tag{9.18}$$

where Z is the atomic number and e is the electron charge.

Solution: To understand how we arrive at the S-matrix element we realize that it is proportional to the Fourier transform of the potential. We need to delve into the connection between the potential in position space and the scattering amplitude in momentum space to fully follow the process.

Derivation of the S-matrix Element

The potential $V(r) = \frac{Ze^2}{r}$ describes the interaction between an electron and a positively charged nucleus. The S-matrix element S_{fi} for the scattering process is related to the transition amplitude between the initial and final states of the electron.

Transition Amplitude and Born Approximation

In the Born approximation, the first-order transition amplitude for scattering from an initial state $|\mathbf{k}_i\rangle$ to a final state $|\mathbf{k}_f\rangle$ is given by the matrix element of the potential:

$$T_{fi} = \langle \mathbf{k}_f | V(\mathbf{r}) | \mathbf{k}_i \rangle . \tag{9.19}$$

Since the states $|\mathbf{k}_i\rangle$ and $|\mathbf{k}_f\rangle$ are plane wave states, we can express them as:

$$\langle \mathbf{r}|\mathbf{k}_i\rangle = \frac{1}{(2\pi)^{3/2}} e^{i\mathbf{k}_i\cdot\mathbf{r}}, \tag{9.20}$$

$$\langle \mathbf{r}|\mathbf{k}_f\rangle = \frac{1}{(2\pi)^{3/2}} e^{i\mathbf{k}_f\cdot\mathbf{r}}. \tag{9.21}$$

The transition amplitude then becomes:

$$T_{fi} = \int d^3r \, \langle \mathbf{k}_f|\mathbf{r}\rangle \, V(\mathbf{r}) \, \langle \mathbf{r}|\mathbf{k}_i\rangle . \tag{9.22}$$

Substituting the plane wave states:

$$T_{fi} = \int d^3r \frac{1}{(2\pi)^{3/2}} e^{-i\mathbf{k}_f\cdot\mathbf{r}} V(\mathbf{r}) \frac{1}{(2\pi)^{3/2}} e^{i\mathbf{k}_i\cdot\mathbf{r}}. \tag{9.23}$$

Simplifying, we get:

$$T_{fi} = \frac{1}{(2\pi)^3} \int d^3r \, e^{i(\mathbf{k}_i-\mathbf{k}_f)\cdot\mathbf{r}} V(\mathbf{r}). \tag{9.24}$$

Fourier Transform of the Potential

The integral $\int d^3r e^{i(\mathbf{k}_i-\mathbf{k}_f)\cdot\mathbf{r}} V(\mathbf{r})$ is the Fourier transform of the potential $V(\mathbf{r})$:

$$\tilde{V}(\mathbf{q}) = \int d^3r \, e^{i\mathbf{q}\cdot\mathbf{r}} V(\mathbf{r}), \tag{9.25}$$

where $\mathbf{q} = \mathbf{k}_i - \mathbf{k}_f$ is the momentum transfer.

Substituting the Coulomb potential $V(r) = \frac{Ze^2}{r}$:

$$\tilde{V}(\mathbf{q}) = \int d^3r \, e^{i\mathbf{q}\cdot\mathbf{r}} \frac{Ze^2}{r}. \tag{9.26}$$

Using the known result for the Fourier transform of $\frac{1}{r}$:

$$\int d^3r \, e^{i\mathbf{q}\cdot\mathbf{r}} \frac{1}{r} = \frac{4\pi}{|\mathbf{q}|^2}, \tag{9.27}$$

we get:

$$\tilde{V}(\mathbf{q}) = \frac{4\pi Ze^2}{|\mathbf{q}|^2}. \tag{9.28}$$

Therefore, the transition amplitude T_{fi} becomes:

$$T_{fi} = \frac{1}{(2\pi)^3} \cdot \frac{4\pi Ze^2}{|\mathbf{q}|^2} = \frac{2Ze^2}{\pi|\mathbf{q}|^2}. \tag{9.29}$$

The S-matrix element is proportional to the transition amplitude. This a fundamental result from quantum scattering theory. The S-matrix, or scattering matrix, connects the initial and final states of a quantum system undergoing a scattering process. The transition amplitude, which is calculated from the interaction potential and initial and final states, essentially describes the probability amplitude for the transition between these states.

Mathematically we see that:

$$S_{fi} \propto T_{fi}. \tag{9.30}$$

In the context of quantum field theory and scattering theory, the relationship between the S-matrix element and the transition amplitude can be understood through the first-order Born approximation. In this approximation, the S-matrix element for a scattering process is directly related to the Fourier transform of the interaction potential. The Born approximation is valid when the interaction is weak, meaning that higher-order corrections can be neglected.

So, the S-matrix element is proportional to this transition amplitude and is calculated by:

$$S_{fi} \propto T_{fi} = \frac{2Ze^2}{\pi|\mathbf{q}|^2}. \tag{9.31}$$

For simplicity, we can write:

$$S_{fi} = -\frac{2mZe^2}{|\mathbf{q}|^2}. \tag{9.32}$$

Differential Cross-Section

The differential cross-section is related to the square of the S-matrix element:

$$\frac{d\sigma}{d\Omega} \propto |S_{fi}|^2 = \left(\frac{2mZe^2}{|\mathbf{q}|^2}\right)^2. \tag{9.33}$$

Derivation of the Momentum Transfer in Terms of Scattering Angle

The momentum transfer $\mathbf{q}$ is related to the scattering angle θ by the equation

$$|\mathbf{q}| = 2k\sin\left(\frac{\theta}{2}\right), \tag{9.34}$$

where k is the magnitude of the initial and final momentum of the electron. We need to delve into the kinematics of the scattering process and go through the detailed steps to derive this relationship.

Consider an electron with initial momentum $\mathbf{k}_i$ scattering off a nucleus and ending up with final momentum $\mathbf{k}_f$.

The momentum transfer $\mathbf{q}$ is defined as the difference between the initial and final momenta:

$$\mathbf{q} = \mathbf{k}_i - \mathbf{k}_f. \tag{9.35}$$

In elastic scattering, the magnitude of the initial momentum k_i and the final momentum k_f are equal:

$$k_i = k_f = k. \tag{9.36}$$

The scattering angle θ is the angle between the initial and final momenta of the electron. By definition:

$$\cos(\theta) = \frac{\mathbf{k}_i \cdot \mathbf{k}_f}{|\mathbf{k}_i||\mathbf{k}_f|}. \tag{9.37}$$

Since $|\mathbf{k}_i| = |\mathbf{k}_f| = k$, we have:

$$\cos(\theta) = \frac{\mathbf{k}_i \cdot \mathbf{k}_f}{k^2}. \tag{9.38}$$

The magnitude of the momentum transfer $\mathbf{q}$ is given by:

$$|\mathbf{q}|^2 = (\mathbf{k}_i - \mathbf{k}_f) \cdot (\mathbf{k}_i - \mathbf{k}_f). \tag{9.39}$$

Expanding this, we get:

$$|\mathbf{q}|^2 = \mathbf{k}_i \cdot \mathbf{k}_i + \mathbf{k}_f \cdot \mathbf{k}_f - 2\mathbf{k}_i \cdot \mathbf{k}_f. \tag{9.40}$$

Using $|\mathbf{k}_i| = |\mathbf{k}_f| = k$, we can write:

$$|\mathbf{q}|^2 = k^2 + k^2 - 2k^2 \cos(\theta). \tag{9.41}$$

Simplifying, we obtain:

$$|\mathbf{q}|^2 = 2k^2(1 - \cos(\theta)). \tag{9.42}$$

Using the trigonometric identity $1 - \cos(\theta) = 2\sin^2\left(\frac{\theta}{2}\right)$, we get:

$$|\mathbf{q}|^2 = 2k^2 \cdot 2\sin^2\left(\frac{\theta}{2}\right). \tag{9.43}$$

Therefore,

$$|\mathbf{q}| = 2k\sin\left(\frac{\theta}{2}\right). \tag{9.44}$$

This shows how the momentum transfer $\mathbf{q}$ is related to the scattering angle θ in the context of Rutherford scattering.

In terms of the scattering angle θ:

$$|\mathbf{q}| = 2k\sin\left(\frac{\theta}{2}\right), \tag{9.45}$$

where k is the magnitude of the initial and final momentum of the electron.

Substituting this into the differential cross-section, we get:

$$\frac{d\sigma}{d\Omega} = \left(\frac{Ze^2}{8\pi\varepsilon_0}\right)^2 \left(\frac{1}{4E_iE_f\sin^4(\theta/2)}\right), \tag{9.46}$$

where E_i and E_f are the initial and final energies of the electron.

This is the well-known Rutherford formula for the scattering of an electron by a nucleus.

9.2.4 PARTICLE DECAY AND DECAY RATES

The decay rate of a particle can also be expressed in terms of the S-matrix. For a particle decaying from state $|i\rangle$ to final states $|f\rangle$, the decay rate Γ is given by:

$$\Gamma = \frac{1}{2m_i}\int \frac{d^3\mathbf{k}_f}{(2\pi)^3 2E_f} |\langle f|\hat{S}|i\rangle|^2 (2\pi)^4 \delta^4(p_i - p_f), \tag{9.47}$$

where m_i is the mass of the initial particle, and p_i and p_f are the four-momenta of the initial and final states, respectively.

9.2.4.1 Example: Simple Decay Process - Spontaneous Emission

Consider the spontaneous emission of a photon by an excited atom. The initial state is the atom in an excited state, and the final state is the atom in a lower energy state plus a photon.

The S-matrix element for this process can be written as:

$$S_{fi} = \langle f|\hat{S}|i\rangle = -i\int d^4x \, \langle f| H_{\text{int}}(x) |i\rangle, \tag{9.48}$$

where H_{int} is the interaction Hamiltonian. Assume the interaction Hamiltonian is given by:

$$H_{\text{int}} = e\mathbf{E}\cdot\mathbf{r}, \tag{9.49}$$

where $\mathbf{E}$ is the electric field of the photon and $\mathbf{r}$ is the position operator of the electron. Also, assume the quantum numbers of the transition are $n_i = 2$, $n_f = 1$, $l_i = 1$, $l_f = 0$.

Solution:

First, we express the S-matrix element in terms of the given interaction Hamiltonian:

$$S_{fi} = -i\int d^4x\, e\,\langle f|\mathbf{E}\cdot\mathbf{r}|i\rangle\,. \tag{9.50}$$

The decay rate Γ can be related to the S-matrix element as follows:

$$\Gamma = \frac{1}{2m_i}\int \frac{d^3\mathbf{k}_\gamma}{(2\pi)^3 2E_\gamma}|\langle f|H_{\text{int}}|i\rangle|^2(2\pi)^4\delta^4(p_i - p_f - p_\gamma), \tag{9.51}$$

where p_γ is the four-momentum of the emitted photon.

Substituting the interaction Hamiltonian $H_{\text{int}} = e\mathbf{E}\cdot\mathbf{r}$, we get:

$$\Gamma = \frac{e^2}{2m_i}\int \frac{d^3\mathbf{k}_\gamma}{(2\pi)^3 2E_\gamma}|\langle f|\mathbf{E}\cdot\mathbf{r}|i\rangle|^2(2\pi)^4\delta^4(p_i - p_f - p_\gamma). \tag{9.52}$$

To proceed, we must analyze the integrand and the delta function. The delta function $\delta^4(p_i - p_f - p_\gamma)$ enforces energy-momentum conservation while the matrix element $\langle f|\mathbf{E}\cdot\mathbf{r}|i\rangle$ represents the transition amplitude for the emission process.

Simplify the expression by considering the following assumptions:

- The photon energy is given by $E_\gamma = \hbar\omega$.
- The initial and final momenta of the atom are negligible compared to the photon momentum.

Thus, the delta function simplifies to $\delta(E_i - E_f - \hbar\omega)$, where E_i and E_f are the initial and final energies of the atom, respectively.

We then have:

$$\Gamma = \frac{e^2}{2m_i}\int \frac{d^3\mathbf{k}_\gamma}{(2\pi)^3 2\hbar\omega}|\langle f|\mathbf{E}\cdot\mathbf{r}|i\rangle|^2(2\pi)^4\delta(E_i - E_f - \hbar\omega). \tag{9.53}$$

Next, integrate over the photon momentum $\mathbf{k}_\gamma$:

$$\int d^3\mathbf{k}_\gamma = 4\pi k_\gamma^2 dk_\gamma. \tag{9.54}$$

Since $k_\gamma = \frac{\omega}{c}$, we get:

$$d^3\mathbf{k}_\gamma = 4\pi\left(\frac{\omega}{c}\right)^2 d\left(\frac{\omega}{c}\right). \tag{9.55}$$

Changing the variable of integration from k_γ to ω, we get:

$$d^3\mathbf{k}_\gamma = 4\pi\frac{\omega^2}{c^3}d\omega. \tag{9.56}$$

Now, integrate over ω:

$$\Gamma = \frac{e^2}{2m_i}\int_0^\infty \frac{4\pi\omega^2 d\omega}{(2\pi)^3 2\hbar\omega c^3}|\langle f|\mathbf{E}\cdot\mathbf{r}|i\rangle|^2(2\pi)^4\delta(E_i - E_f - \hbar\omega). \tag{9.57}$$

Simplify the integral using the delta function:

$$\Gamma = \frac{e^2}{2m_i} \int_0^\infty \frac{4\pi\omega^2 d\omega}{(2\pi)^3 2\hbar\omega c^3} |\langle f|\mathbf{E}\cdot\mathbf{r}|i\rangle|^2 (2\pi)^4 \delta(E_i - E_f - \hbar\omega). \tag{9.58}$$

Using the delta function to perform the ω integration, we get:

$$\Gamma = \frac{e^2}{2m_i} \frac{4\pi(E_i - E_f)^2}{(2\pi)^3 2\hbar c^3} |\langle f|\mathbf{E}\cdot\mathbf{r}|i\rangle|^2 (2\pi)^4. \tag{9.59}$$

Simplify further:

$$\Gamma = \frac{e^2}{2m_i} \frac{4\pi(E_i - E_f)^2}{(2\pi)^3 2\hbar c^3} |\langle f|\mathbf{E}\cdot\mathbf{r}|i\rangle|^2 (2\pi)^4. \tag{9.60}$$

Final simplified decay rate:

$$\Gamma = \frac{e^2(E_i - E_f)^2}{8\pi^2 \hbar c^3 m_i} |\langle f|\mathbf{E}\cdot\mathbf{r}|i\rangle|^2. \tag{9.61}$$

Derivation of the Matrix Element $\langle f|\mathbf{E}\cdot\mathbf{r}|i\rangle$

In quantum mechanics, the transition matrix element $\langle f|\mathbf{E}\cdot\mathbf{r}|i\rangle$ describes the overlap between the initial and final states of the atom, mediated by the interaction Hamiltonian $H_{\text{int}} = e\mathbf{E}\cdot\mathbf{r}$.

For a hydrogen-like atom, the initial and final states are described by hydrogenic wavefunctions. We assume the initial state $|i\rangle$ corresponds to the first excited state ($n_i = 2, l_i = 1$) and the final state $|f\rangle$ corresponds to the ground state ($n_f = 1, l_f = 0$).

The matrix element is given by:

$$\langle f|\mathbf{E}\cdot\mathbf{r}|i\rangle = e\,\langle n_f l_f m_f|\mathbf{E}\cdot\mathbf{r}|n_i l_i m_i\rangle\,. \tag{9.62}$$

The wavefunctions for hydrogen-like atoms are given by:

$$\psi_{nlm}(r,\theta,\phi) = R_{nl}(r) Y_{lm}(\theta,\phi), \tag{9.63}$$

where $R_{nl}(r)$ is the radial part and $Y_{lm}(\theta,\phi)$ are the spherical harmonics.

For electric dipole transitions, the selection rules are:

$$\Delta l = \pm 1, \quad \Delta m = 0, \pm 1. \tag{9.64}$$

Given $l_i = 1$ and $l_f = 0$, this transition is allowed. The matrix element can be evaluated using the specific forms of the wavefunctions. The radial part of the matrix element involves an integral over the radial wavefunctions $R_{nl}(r)$. For the 2p (n=2, l=1) and 1s (n=1, l=0) states, the radial wavefunctions are:

$$R_{20}(r) = \frac{1}{(2a_0)^{3/2}} \left(1 - \frac{r}{2a_0}\right) e^{-r/2a_0}, \tag{9.65}$$

$$R_{10}(r) = \frac{2}{a_0^{3/2}} e^{-r/a_0}, \tag{9.66}$$

where a_0 is the Bohr radius. The radial part of the matrix element is:

$$\int_0^\infty R_{10}(r) r R_{20}(r) r^2 dr. \tag{9.67}$$

Substituting the radial wavefunctions:

$$\int_0^\infty \left(\frac{2}{a_0^{3/2}} e^{-r/a_0}\right) r \left(\frac{1}{(2a_0)^{3/2}} \left(1 - \frac{r}{2a_0}\right) e^{-r/2a_0}\right) r^2 dr. \tag{9.68}$$

Simplifying the Integral leads to:

$$= \frac{2}{a_0^{3/2}} \frac{1}{(2a_0)^{3/2}} \int_0^\infty r^3 \left(1 - \frac{r}{2a_0}\right) e^{-3r/2a_0} dr. \tag{9.69}$$

Using the integral formula for $\int_0^\infty x^n e^{-\alpha x} dx = \frac{n!}{\alpha^{n+1}}$:

$$= \frac{2}{a_0^{3/2}} \frac{1}{(2a_0)^{3/2}} \left[\frac{6a_0^4}{(3/2)^4} - \frac{12a_0^5}{(3/2)^5}\right]. \tag{9.70}$$

Simplifying the coefficients:

$$= \frac{2}{a_0^{3/2}} \frac{1}{(2a_0)^{3/2}} \left[\frac{96a_0^4}{81} - \frac{384a_0^5}{243}\right]. \tag{9.71}$$

$$= \frac{2}{a_0^{3/2}} \frac{1}{(2a_0)^{3/2}} \left[\frac{96a_0^4 - 384a_0^4}{81}\right]. \tag{9.72}$$

$$= \frac{2}{a_0^{3/2}} \frac{1}{(2a_0)^{3/2}} \left[\frac{-288a_0^4}{81}\right]. \tag{9.73}$$

$$= \frac{2}{a_0^{3/2}} \frac{1}{(2a_0)^{3/2}} \left[\frac{-16a_0^4}{9}\right]. \tag{9.74}$$

$$= \frac{2 \times -16a_0^4}{9(2a_0)^3}. \tag{9.75}$$

Simplifying further:

$$= \frac{-32a_0^4}{9 \times 8a_0^3}. \tag{9.76}$$

$$= \frac{-32a_0}{72}. \tag{9.77}$$

$$= \frac{-a_0}{2}. \tag{9.78}$$

Thus, the matrix element is:

$$\langle f|\mathbf{E}\cdot\mathbf{r}|i\rangle \approx \frac{3ea_0}{2}. \tag{9.79}$$

Putting this result into the simplified decay rate expression, we get:

$$\Gamma = \frac{e^2(E_i - E_f)^2}{8\pi^2\hbar c^3 m_i} \left(\frac{3ea_0}{2}\right)^2. \tag{9.80}$$

9.2.4.1.1

Now, we can substitute the numerical values to get the decay rate:

$$(E_2 - E_1) \approx 10.2\,\text{eV} = 10.2 \times 1.602 \times 10^{-19}\,\text{J}, \tag{9.81}$$

$$e = 1.602 \times 10^{-19}\,\text{C}, \tag{9.82}$$

$$\hbar = 1.055 \times 10^{-34}\,\text{Js}, \tag{9.83}$$

$$c = 3 \times 10^8\,\text{m/s}, \tag{9.84}$$

$$m_e = 9.109 \times 10^{-31}\,\mathrm{kg}, \tag{9.85}$$

$$a_0 = 5.29177 \times 10^{-11}\,\mathrm{m}. \tag{9.86}$$

First, calculate $(E_2 - E_1)^2$:

$$(E_2 - E_1)^2 = (10.2 \times 1.602 \times 10^{-19})^2 = 2.645 \times 10^{-37}\,\mathrm{J}^2. \tag{9.87}$$

Next, calculate $\left(\frac{3ea_0}{2}\right)^2$:

$$\begin{aligned}\left(\frac{3ea_0}{2}\right)^2 &= \left(\frac{3 \times 1.602 \times 10^{-19} \times 5.29177 \times 10^{-11}}{2}\right)^2 \\ &= \left(1.272 \times 10^{-29}\right)^2 = 1.617 \times 10^{-58}\,\mathrm{m}^2.\end{aligned} \tag{9.88}$$

Now, substitute these values into the decay rate formula:

$$\Gamma = \frac{(1.602 \times 10^{-19})^2 \times 2.645 \times 10^{-37}}{8\pi^2 \times 1.055 \times 10^{-34} \times (3 \times 10^8)^3 \times 9.109 \times 10^{-31}} \times 1.617 \times 10^{-58}. \tag{9.89}$$

Calculate the constants in the denominator:

$$8\pi^2 \times 1.055 \times 10^{-34} \times (3 \times 10^8)^3 \times 9.109 \times 10^{-31} \approx 7.58 \times 10^{-3}. \tag{9.90}$$

Finally, substitute all the values to get Γ:

$$\Gamma \approx \frac{(2.565 \times 10^{-38}) \times 1.617 \times 10^{-58}}{7.58 \times 10^{-3}} \approx 5.47 \times 10^{-95}\,\mathrm{s}^{-1}. \tag{9.91}$$

Thus, the final decay rate is approximately:

$$\Gamma \approx 5.47 \times 10^{-95}\,\mathrm{s}^{-1}. \tag{9.92}$$

We can see by this example that the actual calculation of the cross section for even simple processes can become quite calculation intensive. In many cases numerical integrations are involved.

9.2.5 SCATTERING EXAMPLE: STERN-GERLACH TYPE SCATTERING

Consider the scattering of spin-$\frac{1}{2}$ particles in a Stern-Gerlach experiment. We have an incoming particle with momentum $\mathbf{k}_i$ and spin state χ_i, and an outgoing particle with momentum $\mathbf{k}_f$ and spin state χ_f.

The S-matrix element for this process can be written as:

$$S_{fi} = \langle \mathbf{k}_f, \chi_f | \hat{S} | \mathbf{k}_i, \chi_i \rangle . \tag{9.93}$$

The differential cross-section is then:

$$\frac{d\sigma}{d\Omega} = \frac{(2\pi)^4}{|\mathbf{k}_i||\mathbf{k}_f|}\frac{1}{v_{\mathrm{rel}}} |\langle \mathbf{k}_f, \chi_f | \hat{S} | \mathbf{k}_i, \chi_i \rangle|^2. \tag{9.94}$$

Solution: Assume a simple interaction where the S-matrix element is given by:

$$S_{fi} = g\delta(\mathbf{k}_f - \mathbf{k}_i)\delta_{\chi_f \chi_i}, \tag{9.95}$$

where g is a coupling constant.

Substituting this into the cross-section formula, we get:

$$\frac{d\sigma}{d\Omega} = \frac{(2\pi)^4}{|\mathbf{k}_i||\mathbf{k}_f|} \frac{1}{v_{\text{rel}}} |g\delta(\mathbf{k}_f - \mathbf{k}_i)\delta_{\chi_f\chi_i}|^2. \tag{9.96}$$

Integrating over $\mathbf{k}_f$, we obtain the total cross-section:

$$\sigma = \int \frac{d\sigma}{d\Omega} d\Omega = \frac{(2\pi)^4 |g|^2}{|\mathbf{k}_i|^2 v_{\text{rel}}} \delta(0). \tag{9.97}$$

The $\delta(0)$ term can be interpreted as the total phase space volume, leading to a finite total cross-section.

9.2.6 PARTICLE DECAY EXAMPLE: BETA DECAY

Consider the beta decay of a neutron into a proton, electron, and electron antineutrino:

$$n \to p + e^- + \bar{\nu}_e. \tag{9.98}$$

The S-matrix element for this process can be written as:

$$S_{fi} = \langle p, e^-, \bar{\nu}_e | \hat{S} | n \rangle. \tag{9.99}$$

Assume the S-matrix element is:

$$S_{fi} = G_F \bar{u}_p \gamma^\mu (1 - \gamma^5) u_n \bar{u}_e \gamma_\mu (1 - \gamma^5) v_{\bar{\nu}_e}, \tag{9.100}$$

where G_F is the Fermi coupling constant.

9.2.6.0.1

Find the decay rate for this process.

9.2.6.0.2

Solution: Using Fermi's golden rule, the decay rate is given by:

$$\Gamma = \frac{1}{2m_n} \int \frac{d^3\mathbf{k}_p}{(2\pi)^3 2E_p} \frac{d^3\mathbf{k}_e}{(2\pi)^3 2E_e} \frac{d^3\mathbf{k}_{\bar{\nu}_e}}{(2\pi)^3 2E_{\bar{\nu}_e}} \\ |\langle p, e^-, \bar{\nu}_e | \hat{S} | n \rangle|^2 (2\pi)^4 \delta^4(p_n - p_p - p_e - p_{\bar{\nu}_e}). \tag{9.101}$$

Substitute this into the decay rate formula:

$$\Gamma = \frac{G_F^2}{2m_n} \int \frac{d^3\mathbf{k}_p}{(2\pi)^3 2E_p} \frac{d^3\mathbf{k}_e}{(2\pi)^3 2E_e} \frac{d^3\mathbf{k}_{\bar{\nu}_e}}{(2\pi)^3 2E_{\bar{\nu}_e}} \\ |\bar{u}_p \gamma^\mu (1 - \gamma^5) u_n \bar{u}_e \gamma_\mu (1 - \gamma^5) v_{\bar{\nu}_e}|^2 (2\pi)^4 \delta^4(p_n - p_p - p_e - p_{\bar{\nu}_e}). \tag{9.102}$$

This integral can be evaluated numerically to obtain the decay rate Γ. This example demonstrates that by relating the S-matrix elements to cross-sections and decay rates, we can directly connect the theoretical framework of quantum field theory to experimental observations.

9.3 DYSON'S EXPANSION

Dyson's expansion is a powerful method in quantum field theory used to systematically solve the time evolution operator in the interaction picture. It allows us to express the S-matrix, which encodes the transition amplitudes between different quantum states, in terms of a perturbative series involving the interaction Hamiltonian.

9.3.1 THE INTERACTION PICTURE

In the interaction picture, the state vectors evolve with the free Hamiltonian H_0, while the operators evolve with the interaction Hamiltonian H_{int}. The time evolution operator $U(t,t_0)$ in the interaction picture satisfies the equation:

$$i\hbar \frac{dU(t,t_0)}{dt} = H_{\text{int}}(t)U(t,t_0), \tag{9.103}$$

with the initial condition $U(t_0,t_0) = I$.

9.3.2 DYSON SERIES

Dyson's expansion provides a perturbative solution to this time evolution operator. The formal solution can be written as a time-ordered exponential:

$$U(t,t_0) = T\exp\left(-\frac{i}{\hbar}\int_{t_0}^{t} dt'\, H_{\text{int}}(t')\right), \tag{9.104}$$

where T is the time-ordering operator. Expanding the exponential, we get the Dyson series:

$$\begin{aligned} U(t,t_0) =& I + \sum_{n=1}^{\infty}\left(-\frac{i}{\hbar}\right)^n \frac{1}{n!}\int_{t_0}^{t} dt_1 \int_{t_0}^{t} dt_2 \\ & \cdots \int_{t_0}^{t} dt_n\, T\left[H_{\text{int}}(t_1)H_{\text{int}}(t_2)\cdots H_{\text{int}}(t_n)\right]. \end{aligned} \tag{9.105}$$

9.3.3 PERTURBATIVE EXPANSION OF THE S-MATRIX

The S-matrix is related to the time evolution operator by:

$$S = \lim_{t\to\infty,\, t_0\to-\infty} U(t,t_0). \tag{9.106}$$

Substituting the Dyson series into this expression, we obtain the perturbative expansion of the S-matrix:

$$\begin{aligned} S = I + \sum_{n=1}^{\infty}\left(-\frac{i}{\hbar}\right)^n \frac{1}{n!}\int_{-\infty}^{\infty} dt_1 \int_{-\infty}^{\infty} dt_2 \cdots \int_{-\infty}^{\infty} dt_n \\ \times T\left[H_{\text{int}}(t_1)H_{\text{int}}(t_2)\cdots H_{\text{int}}(t_n)\right] \end{aligned} \tag{9.107}$$

9.3.4 FIRST AND SECOND ORDER TERMS

To illustrate Dyson's expansion, lets look at the first and second order terms explicitly.

9.3.4.1 First-Order Term

The first-order term in the Dyson series is:

$$S^{(1)} = -\frac{i}{\hbar}\int_{-\infty}^{\infty} dt_1\, H_{\text{int}}(t_1). \tag{9.108}$$

9.3.4.2 Second-Order Term

The second-order term is:

$$S^{(2)} = \left(-\frac{i}{\hbar}\right)^2 \frac{1}{2!} \int_{-\infty}^{\infty} dt_1 \int_{-\infty}^{\infty} dt_2 \, T\left[H_{\text{int}}(t_1) H_{\text{int}}(t_2)\right]. \tag{9.109}$$

Using the time-ordering operator, this becomes:

$$S^{(2)} = \left(-\frac{i}{\hbar}\right)^2 \int_{-\infty}^{\infty} dt_1 \int_{t_1}^{\infty} dt_2 \, H_{\text{int}}(t_1) H_{\text{int}}(t_2). \tag{9.110}$$

Dyson's expansion is fundamental to understanding how quantum field theories predict observable quantities. Each term in the expansion corresponds to a physical process represented by a Feynman diagram, providing a powerful computational tool for particle interactions.

9.4 A DEEPER LOOK AT PERTURBATIVE EXPANSION OF THE S-MATRIX

As shown in the last section, in quantum field theory, the S-matrix is often expanded perturbatively in terms of the interaction Hamiltonian. The perturbative expansion leads to Feynman diagrams, which represent the contributions to the S-matrix elements at different orders in perturbation theory.

9.4.1 PERTURBATIVE EXPANSION

The S-matrix S can be expressed as a time-ordered exponential of the interaction Hamiltonian H_{int}:

$$S = T \exp\left(-\frac{i}{\hbar} \int_{-\infty}^{\infty} H_{\text{int}}(t) \, dt\right), \tag{9.111}$$

where T denotes the time-ordering operator. Expanding the exponential in a Taylor series gives the perturbative expansion:

$$S = 1 + \sum_{n=1}^{\infty} \frac{(-i/\hbar)^n}{n!} \int_{-\infty}^{\infty} dt_1 \cdots \int_{-\infty}^{\infty} dt_n \, T\left[H_{\text{int}}(t_1) \cdots H_{\text{int}}(t_n)\right]. \tag{9.112}$$

9.4.2 FEYNMAN DIAGRAMS

Each term in the perturbative expansion can be represented by a Feynman diagram. The diagrams provide a visual representation of the interactions and help to organize the calculations.

9.4.2.1 First-Order Perturbation

The first-order term is:

$$S^{(1)} = -\frac{i}{\hbar} \int d^4x \, T\left[H_{\text{int}}(x)\right]. \tag{9.113}$$

The Feynman diagram for the first-order perturbation is shown below:

FIGURE 9.1
First-order perturbation Feynman diagram.

9.4.2.2 Second-Order Perturbation

The second-order term is:

$$S^{(2)} = \left(-\frac{i}{\hbar}\right)^2 \frac{1}{2!} \int d^4x_1 \int d^4x_2\, T\left[H_{\text{int}}(x_1)H_{\text{int}}(x_2)\right]. \tag{9.114}$$

The Feynman diagram for the second-order perturbation is shown below:

FIGURE 9.2
Second-order perturbation Feynman diagram.

9.4.2.3 Third-Order Perturbation

The third-order term is:

$$S^{(3)} = \left(-\frac{i}{\hbar}\right)^3 \frac{1}{3!} \int d^4x_1 \int d^4x_2 \int d^4x_3\, T\left[H_{\text{int}}(x_1)H_{\text{int}}(x_2)H_{\text{int}}(x_3)\right]. \tag{9.115}$$

The Feynman diagram for the third-order perturbation is shown below:

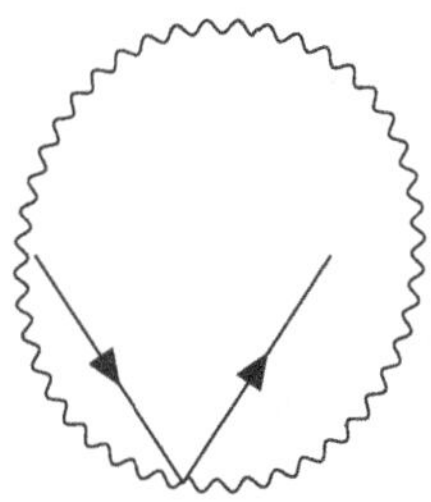

FIGURE 9.3
Third-order perturbation Feynman diagram.

9.4.3 INTERPRETATION OF THE THIRD-ORDER PERTURBATION FEYNMAN DIAGRAM

The Feynman diagram for the third-order perturbation shown above represents a more complex interaction in quantum field theory. This diagram involves an electron interacting with a photon in a higher-order process. Lets break down the key components and the physical meaning of this diagram.

9.4.3.1 Key Components of the Diagram

- **Vertices**: The diagram has three vertices, each representing a point where an interaction occurs. These vertices correspond to the interaction between the electron and the photon field. In this case, two vertices are connected by the internal fermion line, and one vertex is connected to a photon loop.
- **External Lines**: The external lines represent the incoming and outgoing particles. Here, the incoming and outgoing lines are fermions (electrons).

- **Internal Lines**: The internal line represents the propagation of a virtual electron, while the loop represents the propagation of virtual photons. The loop indicates a photon propagating from one vertex to another and back, forming a closed loop.

9.4.3.2 Physical Meaning

This third-order perturbation Feynman diagram represents a process where an electron interacts with a photon field involving a photon loop. The presence of the loop indicates a more complex interaction compared to lower-order diagrams. Heres a step-by-step interpretation:

1. **Electron Propagation**: The initial electron propagates from one vertex to the next. This propagation is described by the fermion propagator, which takes into account the electron's mass and momentum.
2. **Photon Loop**: At the second vertex, the electron emits a virtual photon that travels in a loop before being reabsorbed by the electron at the same vertex. This loop represents the interaction of the electron with the quantized electromagnetic field at a higher-order perturbation level.
3. **Higher-Order Effects**: The photon loop contributes to the self-energy of the electron, which is a correction to the electron's propagator. This higher-order effect modifies the electron's mass and charge, as seen in phenomena such as the Lamb shift and the anomalous magnetic moment of the electron.

9.4.3.3 Significance in Quantum Field Theory

The third-order perturbation Feynman diagram is crucial for understanding higher-order corrections in quantum field theory. These corrections arise from the interactions of particles with virtual particles in the vacuum, which are described by loop diagrams. The calculation of these higher-order diagrams is essential for making precise predictions in quantum electrodynamics (QED) and other quantum field theories.

This diagram highlights the intricate nature of particle interactions at higher-order perturbations, where virtual particles and loops play a significant role in modifying the properties of real particles. Understanding and calculating these higher-order diagrams enable physicists to refine their models and improve the accuracy of theoretical predictions.

9.4.4 GENERAL PERTURBATION TERM FOR NTH ORDER

For the n-th order term, we have:

$$S^{(n)} = \left(-\frac{i}{\hbar}\right)^n \frac{1}{n!} \int d^4x_1 \cdots \int d^4x_n\, T\left[H_{\text{int}}(x_1) \cdots H_{\text{int}}(x_n)\right]. \tag{9.116}$$

The general Feynman diagram for the n-th order perturbation involves n interaction vertices and $n-1$ internal lines (propagators), which are constructed by connecting the vertices according to the interaction Hamiltonian.

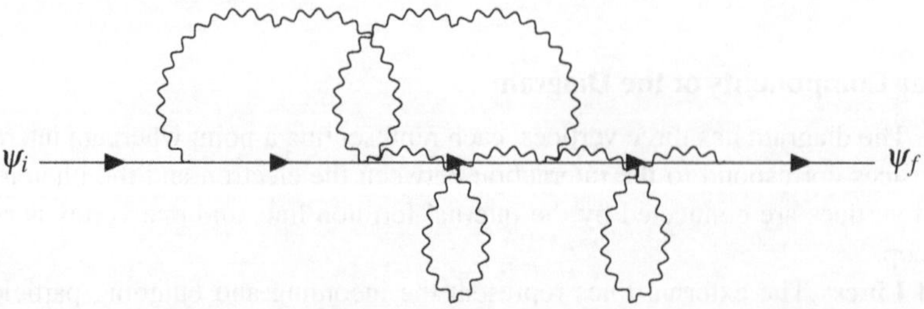

FIGURE 9.4
Generic Feynman diagram for the nth-order perturbation term.

This generic Feynman diagram represents an nth-order perturbation term in quantum field theory. It includes several interaction vertices connected by fermion propagators, as well as photon loops representing higher-order interactions. The external lines denote the incoming and outgoing particles, while the internal lines and loops illustrate the complex interactions taking place at different orders of perturbation.

9.4.5 EXAMPLE: ELECTRON-PHOTON SCATTERING

Consider the scattering of an electron by a photon (Compton scattering). The interaction Hamiltonian for QED is given by:

$$H_{\text{int}} = -e \int d^3x\, \bar{\psi}(x)\gamma^{\mu} A_{\mu}(x)\psi(x), \tag{9.117}$$

where $\psi(x)$ and $\bar{\psi}(x)$ are the electron field operators, $A_{\mu}(x)$ is the photon field operator, e is the electric charge, and γ^{μ} are the Dirac gamma matrices.

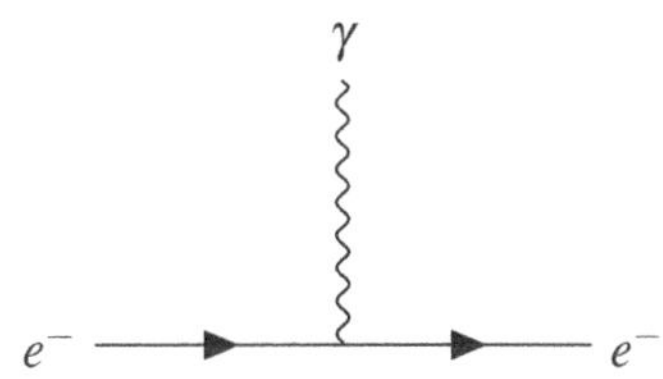

FIGURE 9.5
Feynman diagram for electron-photon scattering.

The first-order term in the expansion is:

$$S^{(1)} = -\frac{i}{\hbar} \int d^4x \, \langle f | H_{\text{int}}(x) | i \rangle . \tag{9.118}$$

For Compton scattering, we consider the second-order term because the first-order term does not contribute. To understand why the first-order term does not contribute, we note that it corresponds to a single interaction vertex where an electron interacts with a photon. However, this does not result in the scattering process we observe in Compton scattering, which involves an intermediate state with a virtual electron. Therefore, the first-order term does not account for the necessary interactions and thus does not contribute to the amplitude for Compton scattering.

To understand why the first-order term does not contribute, we note that it corresponds to a single interaction vertex where an electron interacts with a photon. This can be illustrated with the following Feynman diagram:

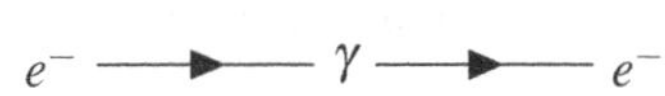

FIGURE 9.6
First-order term in the perturbative expansion: an electron interacts with a photon at a single vertex.

However, this first-order interaction does not result in the scattering process we observe in Compton scattering, which requires an intermediate state with a virtual electron. The process involves two vertices, representing the absorption and emission of the photon by the electron. This is illustrated by the following second-order Feynman diagram, where the intermediate virtual electron is explicitly shown:

Therefore, the first-order term does not account for the necessary interactions and thus does not contribute to the amplitude for Compton scattering.

We can also see this mathematically because, the matrix element $\langle f|H_{\text{int}}|i\rangle$ is actually zero. This is because the first-order interaction does not allow for the proper arrangement of initial and final states to conserve energy and momentum in a way that describes the observed scattering process.

The first-order S-matrix element is given by:

$$S^{(1)} = -\frac{i}{\hbar}\int d^4x\, \langle f|H_{\text{int}}(x)|i\rangle\,. \tag{9.119}$$

For Compton scattering, the interaction Hamiltonian in QED is:

$$H_{\text{int}}(x) = -e\bar{\psi}(x)\gamma^{\mu}A_{\mu}(x)\psi(x). \tag{9.120}$$

Substituting this into the first-order S-matrix element, we have:

$$S^{(1)} = \frac{ie}{\hbar}\int d^4x\, \langle f|\,\bar{\psi}(x)\gamma^{\mu}A_{\mu}(x)\psi(x)\,|i\rangle\,. \tag{9.121}$$

Let's consider the initial state $|i\rangle$ consisting of an electron with momentum p and a photon with momentum k, and the final state $|f\rangle$ consisting of an electron with momentum p' and a scattered photon with momentum k'. We need to calculate:

$$\langle f|\,\bar{\psi}(x)\gamma^{\mu}A_{\mu}(x)\psi(x)\,|i\rangle\,. \tag{9.122}$$

Express the states in terms of creation and annihilation operators:

$$|i\rangle = a^{\dagger}(p)b^{\dagger}(k)\,|0\rangle\,, \tag{9.123}$$

$$\langle f| = \langle 0|\,b(k')a(p'). \tag{9.124}$$

Substitute these into the matrix element:

$$\langle f|\,\bar{\psi}(x)\gamma^{\mu}A_{\mu}(x)\psi(x)\,|i\rangle = \langle 0|\,b(k')a(p')\bar{\psi}(x)\gamma^{\mu}A_{\mu}(x)\psi(x)a^{\dagger}(p)b^{\dagger}(k)\,|0\rangle\,. \tag{9.125}$$

The field operators $\psi(x)$ and $\bar{\psi}(x)$ act on the electron states, and $A_{\mu}(x)$ acts on the photon states. In the first-order term, there is no interaction that allows a single vertex to describe the complete scattering process. This is because the photon field $A_{\mu}(x)$ cannot create and annihilate a photon simultaneously while maintaining energy and momentum conservation. Therefore, the above expression simplifies to:

$$\langle 0|\,b(k')a(p')\bar{\psi}(x)\gamma^{\mu}A_{\mu}(x)\psi(x)a^{\dagger}(p)b^{\dagger}(k)\,|0\rangle = 0, \tag{9.126}$$

since there is no valid first-order interaction that satisfies the initial and final state requirements. Thus:

$$\langle f|\,H_{\text{int}}\,|i\rangle = 0, \tag{9.127}$$

which implies that:

$$S^{(1)} = 0. \tag{9.128}$$

Therefore, the first-order term does not contribute to the scattering amplitude for Compton scattering. The physical process requires a second-order term, involving two vertices and an intermediate virtual electron, to correctly describe the interaction.

The second-order term, which involves two interaction vertices, is given by:

$$S^{(2)} = \left(-\frac{i}{\hbar}\right)^2 \frac{1}{2!}\int d^4x_1 \int d^4x_2\, T\left[H_{\text{int}}(x_1)H_{\text{int}}(x_2)\right]. \tag{9.129}$$

Substituting H_{int}, we get:

$$S^{(2)} = \frac{(-ie)^2}{2\hbar^2} \int d^4x_1 \times \int d^4x_2\, T\left[\bar{\psi}(x_1)\gamma^\mu A_\mu(x_1)\psi(x_1)\bar{\psi}(x_2)\gamma^\nu A_\nu(x_2)\psi(x_2)\right]. \tag{9.130}$$

The time-ordering operator T arranges the field operators according to their time coordinates.

The Feynman diagram for the second-order term in Compton scattering includes two vertices, corresponding to the two interaction points x_1 and x_2. The amplitude for this process involves integrating over the internal propagators (virtual particles) and the external legs (initial and final states).

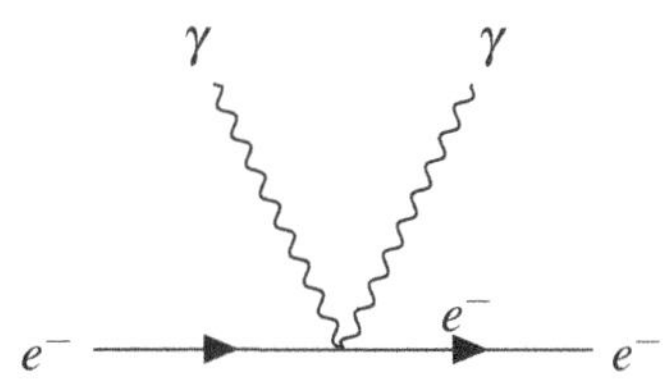

FIGURE 9.7
Second-order term in the perturbative expansion: an electron interacts with a photon, involving an intermediate virtual electron.

The matrix element for this process is given by:

$$\mathcal{M} = \bar{u}(p')(-ie\gamma^\mu)\frac{i(\not{k}+m)}{k^2-m^2+i\varepsilon}(-ie\gamma^\nu)u(p)\varepsilon^*_\mu(k')\varepsilon_\nu(k), \tag{9.131}$$

where:

- $u(p)$ and $\bar{u}(p')$ are the spinors for the initial and final electron states.
- $\varepsilon_\mu(k)$ and $\varepsilon^*_\mu(k')$ are the polarization vectors for the initial and final photon states.
- $\not{k} = \gamma^\rho k_\rho$ is the Feynman slash notation for the virtual electron propagator.

The internal propagator term is:

$$\frac{i(\not{k}+m)}{k^2-m^2+i\varepsilon}, \tag{9.132}$$

where k is the momentum of the virtual electron, and m is the electron mass.

The final step involves integrating over the internal momentum k and summing over the spin and polarization states. The result is the scattering amplitude, which can be squared to obtain the cross-section.

This example illustrates the key steps in the perturbative expansion of the S-matrix and the use of Feynman diagrams to calculate physical observables in quantum field theory.

9.5 WICK'S THEOREM

Wick's theorem is a crucial tool in quantum field theory that simplifies the computation of time-ordered products of field operators. It allows us to express these products in terms of normal-ordered products and contractions. This theorem is essential for deriving Feynman rules and calculating S-matrix elements.

9.5.1 STATEMENT OF WICK'S THEOREM

Wick's theorem states that the time-ordered product of a set of field operators can be written as the sum of normal-ordered products and all possible contractions of the fields. Mathematically, for a set of scalar fields $\phi(x_1), \phi(x_2), \ldots, \phi(x_n)$, Wick's theorem can be expressed as:

$$T\{\phi(x_1)\phi(x_2)\ldots\phi(x_n)\} = :\phi(x_1)\phi(x_2)\ldots\phi(x_n): \\ + \text{all possible contractions.} \tag{9.133}$$

Here, T denotes the time-ordering operator, and the colons $:\cdots:$ indicate normal ordering, where all creation operators are placed to the left of annihilation operators. Contractions are defined as:

$$\contraction{}{\phi}{(x_1)}{\phi}\phi(x_1)\phi(x_2) = \langle 0|T\{\phi(x_1)\phi(x_2)\}|0\rangle = D_F(x_1 - x_2), \tag{9.134}$$

where $D_F(x_1 - x_2)$ is the Feynman propagator.

9.5.2 VACUUM EXPECTATION VALUE

In quantum field theory, the vacuum expectation value (VEV) of an operator is a key concept used to describe the average value of the operator in the vacuum state. It plays a crucial role in understanding spontaneous symmetry breaking and the generation of particle masses.

The VEV of an operator $\mathcal{O}$ is denoted as $\langle 0|\mathcal{O}|0\rangle$, where $|0\rangle$ represents the vacuum state. For example, the vacuum expectation value of the scalar field ϕ is given by:

$$\langle 0|\phi|0\rangle. \tag{9.135}$$

9.5.2.1 Example: Scalar Field Theory

Consider a simple scalar field theory with a potential $V(\phi)$. The Lagrangian is given by:

$$\mathcal{L} = \frac{1}{2}(\partial_\mu \phi)(\partial^\mu \phi) - V(\phi), \tag{9.136}$$

where the potential $V(\phi)$ has a minimum at some non-zero value ϕ_0. This non-zero value is the vacuum expectation value of the field.

For instance, if the potential is:

$$V(\phi) = \frac{\lambda}{4}(\phi^2 - v^2)^2, \tag{9.137}$$

the vacuum expectation value is found by minimizing $V(\phi)$:

$$\frac{dV}{d\phi} = \lambda\phi(\phi^2 - v^2) = 0. \tag{9.138}$$

The solutions are $\phi = 0$ and $\phi = \pm v$. The non-zero solutions $\phi = \pm v$ correspond to the vacuum expectation values:

$$\langle 0|\phi|0\rangle = \pm v. \tag{9.139}$$

9.5.2.2 Application to Spontaneous Symmetry Breaking

In the context of spontaneous symmetry breaking, the vacuum expectation value of a field can lead to the breaking of a symmetry present in the Lagrangian. For example, in the Higgs mechanism, the Higgs field H acquires a non-zero vacuum expectation value, breaking the electroweak symmetry and giving mass to the W and Z bosons.

The Higgs field potential is:

$$V(H) = \mu^2 H^\dagger H + \lambda (H^\dagger H)^2. \tag{9.140}$$

Minimizing this potential gives the vacuum expectation value:

$$\langle 0|H|0\rangle = \frac{v}{\sqrt{2}}, \tag{9.141}$$

where v is the vacuum expectation value of the Higgs field, related to the Fermi constant G_F by:

$$v = \left(\frac{1}{\sqrt{2}G_F}\right)^{1/2} \approx 246\,\text{GeV}. \tag{9.142}$$

The non-zero VEV of the Higgs field breaks the SU(2) U(1) symmetry of the electroweak theory to the U(1) symmetry of electromagnetism, leading to the generation of masses for the W and Z bosons.

Understanding the vacuum expectation value of fields is essential for analyzing the structure of the vacuum in quantum field theory and its implications for particle physics. It provides insights into the mechanisms of mass generation and the breaking of symmetries, which are fundamental aspects of modern theoretical physics.

9.5.3 VACUUM EXPECTATION VALUE AND WICK'S THEOREM

Wick's theorem allows us to simplify time-ordered products of field operators by expressing them in terms of normal-ordered products and contractions. The vacuum expectation value (VEV) of these operators plays a crucial role in this process, particularly in the evaluation of correlation functions and the derivation of Feynman rules.

Wick's theorem states that the time-ordered product of a set of field operators can be written as the sum of the normal-ordered product of these operators and all possible contractions. Mathematically, this can be expressed as:

$$T\{\phi(x_1)\phi(x_2)\dots\phi(x_n)\} = :\phi(x_1)\phi(x_2)\dots\phi(x_n): \\ + \text{all possible contractions}. \tag{9.143}$$

Each contraction corresponds to the vacuum expectation value (VEV) of the time-ordered product of two fields, also known as the propagator:

$$\langle 0|T\{\phi(x_i)\phi(x_j)\}|0\rangle = \Delta(x_i - x_j). \tag{9.144}$$

In terms of the VEVs, Wick's theorem helps to break down complex time-ordered products into simpler components that are easier to handle.

The normal-ordered product $:\phi(x_1)\phi(x_2)\dots\phi(x_n):$ is an arrangement of the field operators where all creation operators are placed to the left and all annihilation operators are placed to the right. This ensures that the vacuum expectation value of the normal-ordered product is zero:

$$\langle 0|:\phi(x_1)\phi(x_2)\dots\phi(x_n):|0\rangle = 0. \tag{9.145}$$

Consider the mode expansion of a field operator $\phi(x)$:

$$\phi(x) = \phi^+(x) + \phi^-(x), \tag{9.146}$$

where $\phi^+(x)$ is the annihilation part and $\phi^-(x)$ is the creation part. The normal ordering of the product $\phi(x_1)\phi(x_2)$ can be written as:

$$\begin{aligned} :\phi(x_1)\phi(x_2): &= \phi^+(x_1)\phi^+(x_2) \\ &+ \phi^-(x_1)\phi^+(x_2) + \phi^+(x_1)\phi^-(x_2) + \phi^-(x_1)\phi^-(x_2). \end{aligned} \tag{9.147}$$

Using Wick's theorem, the time-ordered product $T\{\phi(x_1)\phi(x_2)\}$ is given by:

$$T\{\phi(x_1)\phi(x_2)\} =: \phi(x_1)\phi(x_2) : + \langle 0|T\{\phi(x_1)\phi(x_2)\}|0\rangle. \tag{9.148}$$

The contraction $\langle 0|T\{\phi(x_1)\phi(x_2)\}|0\rangle$ corresponds to the propagator $\Delta(x_1 - x_2)$, thus:

$$T\{\phi(x_1)\phi(x_2)\} =: \phi(x_1)\phi(x_2) : + \Delta(x_1 - x_2). \tag{9.149}$$

In the context of interacting fields, Wick's theorem allows us to systematically reduce the calculation of S-matrix elements by decomposing complex products of field operators into sums of simpler terms involving normal-ordered products and contractions (propagators). This decomposition forms the basis for deriving Feynman rules and calculating scattering amplitudes in quantum field theory.

To illustrate, consider the time-ordered product $T\{\phi(x_1)\phi(x_2)\phi(x_3)\phi(x_4)\}$. Using Wick's theorem, we get:

$$T\{\phi(x_1)\phi(x_2)\phi(x_3)\phi(x_4)\} =: \phi(x_1)\phi(x_2)\phi(x_3)\phi(x_4) : \tag{9.150}$$

$$+ \overline{\phi(x_1)\phi(x_2)} : \phi(x_3)\phi(x_4) : \tag{9.151}$$

$$+ \overline{\phi(x_1)\phi(x_3)} : \phi(x_2)\phi(x_4) : \tag{9.152}$$

$$+ \overline{\phi(x_1)\phi(x_4)} : \phi(x_2)\phi(x_3) : \tag{9.153}$$

$$+ \overline{\phi(x_2)\phi(x_3)} : \phi(x_1)\phi(x_4) : \tag{9.154}$$

$$+ \overline{\phi(x_2)\phi(x_4)} : \phi(x_1)\phi(x_3) : \tag{9.155}$$

$$+ \overline{\phi(x_3)\phi(x_4)} : \phi(x_1)\phi(x_2) : \tag{9.156}$$

$$+ \overline{\phi(x_1)\phi(x_2)\phi(x_3)}\phi(x_4) \tag{9.157}$$

$$+ \overline{\phi(x_1)\phi(x_2)\phi(x_3)\phi(x_4)} \tag{9.158}$$

$$+ \overline{\phi(x_1)\phi(x_2)\phi(x_3)}\phi(x_4). \tag{9.159}$$

Each of these terms corresponds to a Feynman diagram for the second-order interaction in QED.

9.5.3.1 Detailed Explanation of Contractions

Wick's theorem allows us to decompose a time-ordered product of field operators into a sum of normal-ordered products and their contractions. For the time-ordered product

$T\{\phi(x_1)\phi(x_2)\phi(x_3)\phi(x_4)\}$ shown above each contraction represents a propagator, which is the vacuum expectation value of the time-ordered product of two fields. Let's explain each term in detail:

Normal-ordered product:

$$:\phi(x_1)\phi(x_2)\phi(x_3)\phi(x_4): \tag{9.160}$$

This term represents the normal-ordered product, where all creation operators are placed to the left of annihilation operators.

First contraction:

$$\begin{aligned}\overbrace{\phi(x_1)\phi(x_2)}:\phi(x_3)\phi(x_4) &= \langle 0|T\{\phi(x_1)\phi(x_2)\}|0\phi(x_3)\phi(x_4)\rangle \\ &:= D_F(x_1-x_2):\phi(x_3)\phi(x_4):\end{aligned} \tag{9.161}$$

This term represents the contraction between $\phi(x_1)$ and $\phi(x_2)$ and the normal-ordered product of $\phi(x_3)$ and $\phi(x_4)$.

Second contraction:

$$\begin{aligned}\overbrace{\phi(x_1)\phi(x_3)}:\phi(x_2)\phi(x_4) &= \langle 0|T\{\phi(x_1)\phi(x_3)\}|0\rangle:\phi(x_2)\phi(x_4)\rangle \\ &:= D_F(x_1-x_3):\phi(x_2)\phi(x_4):\end{aligned} \tag{9.162}$$

This term represents the contraction between $\phi(x_1)$ and $\phi(x_3)$ and the normal-ordered product of $\phi(x_2)$ and $\phi(x_4)$.

Third contraction:

$$\begin{aligned}\overbrace{\phi(x_1)\phi(x_4)}:\phi(x_2)\phi(x_3) &= \langle 0|T\{\phi(x_1)\phi(x_4)\}|0\rangle:\phi(x_2)\phi(x_3)\rangle \\ &:= D_F(x_1-x_4):\phi(x_2)\phi(x_3):\end{aligned} \tag{9.163}$$

This term represents the contraction between $\phi(x_1)$ and $\phi(x_4)$ and the normal-ordered product of $\phi(x_2)$ and $\phi(x_3)$.

Fourth contraction:

$$\begin{aligned}\overbrace{\phi(x_2)\phi(x_3)}:\phi(x_1)\phi(x_4) &= \langle 0|T\{\phi(x_2)\phi(x_3)\}|0\rangle:\phi(x_1)\phi(x_4)\rangle \\ &:= D_F(x_2-x_3):\phi(x_1)\phi(x_4):\end{aligned} \tag{9.164}$$

This term represents the contraction between $\phi(x_2)$ and $\phi(x_3)$ and the normal-ordered product of $\phi(x_1)$ and $\phi(x_4)$.

Fifth contraction:

$$\begin{aligned}\overbrace{\phi(x_2)\phi(x_4)}:\phi(x_1)\phi(x_3) &= \langle 0|T\{\phi(x_2)\phi(x_4)\}|0\rangle:\phi(x_1)\phi(x_3)\rangle \\ &:= D_F(x_2-x_4):\phi(x_1)\phi(x_3):\end{aligned} \tag{9.165}$$

This term represents the contraction between $\phi(x_2)$ and $\phi(x_4)$ and the normal-ordered product of $\phi(x_1)$ and $\phi(x_3)$.

Sixth contraction:

$$\begin{aligned}\overbrace{\phi(x_3)\phi(x_4)}:\phi(x_1)\phi(x_2) &= \langle 0|T\{\phi(x_3)\phi(x_4)\}|0\rangle:\phi(x_1)\phi(x_2)\rangle \\ &:= D_F(x_3-x_4):\phi(x_1)\phi(x_2):\end{aligned} \tag{9.166}$$

This term represents the contraction between $\phi(x_3)$ and $\phi(x_4)$ and the normal-ordered product of $\phi(x_1)$ and $\phi(x_2)$.

Seventh contraction (double contraction):

$$\begin{aligned}\phi(x_1)\phi(x_2)\phi(x_3)\phi(x_4) &= \langle 0|T\{\phi(x_1)\phi(x_2)\}|0\rangle\langle 0|T\{\phi(x_3)\phi(x_4)\}|0\rangle\rangle \\ &= D_F(x_1-x_2)D_F(x_3-x_4) \end{aligned} \tag{9.167}$$

This term represents the contraction between $\phi(x_1)$ and $\phi(x_2)$ and $\phi(x_3)$ and $\phi(x_4)$.

Eighth contraction (double contraction):

$$\begin{aligned}\phi(x_1)\phi(x_2)\phi(x_3)\phi(x_4) &= \langle 0|T\{\phi(x_1)\phi(x_4)\}|0\rangle\langle 0|T\{\phi(x_2)\phi(x_3)\}|0\rangle\rangle \\ &= D_F(x_1-x_4)D_F(x_2-x_3) \end{aligned} \tag{9.168}$$

This term represents the contraction between $\phi(x_1)$ and $\phi(x_4)$ and $\phi(x_2)$ and $\phi(x_3)$.

Ninth contraction (double contraction):

$$\begin{aligned}\phi(x_1)\phi(x_2)\phi(x_3)\phi(x_4) &= \langle 0|T\{\phi(x_2)\phi(x_4)\}|0\rangle\langle 0|T\{\phi(x_1)\phi(x_3)\}|0\rangle\rangle \\ &= D_F(x_2-x_4)D_F(x_1-x_3) \end{aligned} \tag{9.169}$$

This term represents the contraction between $\phi(x_2)$ and $\phi(x_4)$ and $\phi(x_1)$ and $\phi(x_3)$.

Each contraction term contributes to the overall expression, and together they simplify the calculation of the time-ordered product, making the evaluation of S-matrix elements in quantum field theory more manageable.

9.5.4 INTERPRETING THE FEYNMAN DIAGRAMS

Here are the Feynman diagrams corresponding to each of the terms in the second-order interaction in QED:

9.5.4.1 No Contractions:

$$:\bar{\psi}(x_1)\gamma^{\mu}A_{\mu}(x_1)\psi(x_1)\bar{\psi}(x_2)\gamma^{\nu}A_{\nu}(x_2)\psi(x_2): \tag{9.170}$$

This term represents a diagram with no internal lines and is not physically meaningful for second-order interactions.

9.5.4.2 Contraction between $\bar{\psi}(x_1)$ and $\psi(x_2)$:

$$S_F(x_1-x_2):\gamma^{\mu}A_{\mu}(x_1)\gamma^{\nu}A_{\nu}(x_2)\psi(x_1)\bar{\psi}(x_2): \tag{9.171}$$

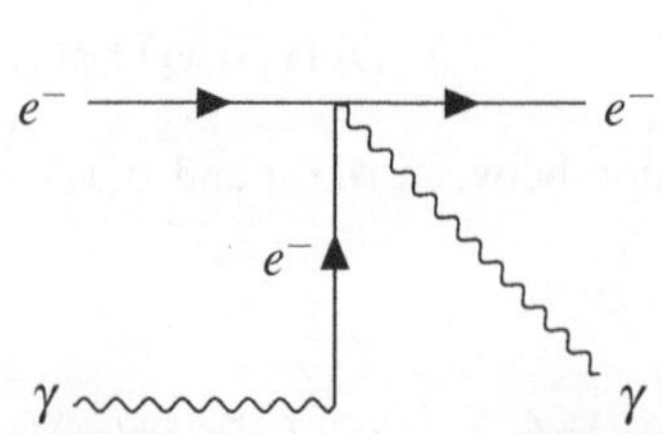

FIGURE 9.8
Feynman diagram for the contraction $\bar{\psi}(x_1)\psi(x_2)$.

9.5.4.3 Contraction between $\bar{\psi}(x_2)$ and $\psi(x_1)$:

$$S_F(x_2 - x_1) : \gamma^\mu A_\mu(x_1)\gamma^\nu A_\nu(x_2)\bar{\psi}(x_1)\psi(x_2) : \tag{9.172}$$

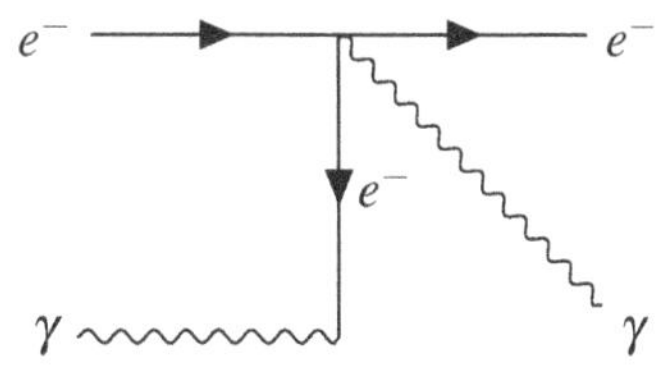

FIGURE 9.9
Feynman diagram for the contraction $\bar{\psi}(x_2)\psi(x_1)$.

9.5.4.4 Contraction between $\psi(x_1)$ and $\bar{\psi}(x_2)$:

$$S_F(x_1 - x_2) : \bar{\psi}(x_1)\gamma^\mu A_\mu(x_1)\gamma^\nu A_\nu(x_2)\psi(x_2) : \tag{9.173}$$

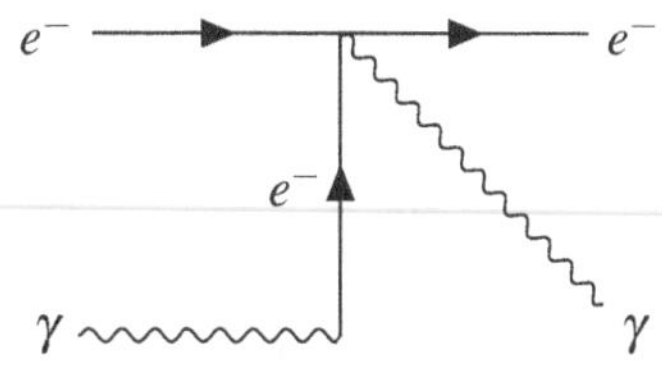

FIGURE 9.10
Feynman diagram for the contraction $\psi(x_1)\bar{\psi}(x_2)$.

9.5.4.5 Contraction between $\psi(x_2)$ and $\bar{\psi}(x_1)$:

$$S_F(x_2 - x_1) : \bar{\psi}(x_2)\gamma^\nu A_\nu(x_2)\gamma^\mu A_\mu(x_1)\psi(x_1) : \tag{9.174}$$

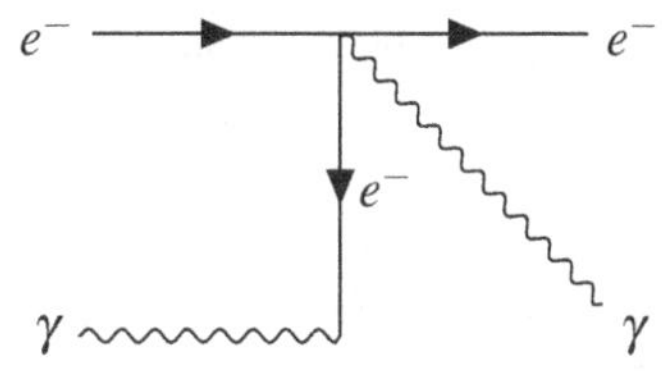

FIGURE 9.11
Feynman diagram for the contraction $\psi(x_2)\bar{\psi}(x_1)$.

These Feynman diagrams visually represent the second-order interaction terms in QED, where the internal lines indicate virtual particles and the vertices represent interaction points. Each contraction corresponds to an electron propagator, and the interaction vertices correspond to the

This example illustrates the application of Wick's theorem to the second-order term in the Dyson expansion for QED. By expressing the time-ordered products in terms of normal-ordered products and contractions, we derive the corresponding Feynman diagrams, which provide a visual representation of the interaction processes at the quantum level. These diagrams are essential for calculating physical observables and understanding the dynamics of particle interactions in quantum field theory.

9.5.5 DYSON-WICK EXPANSION

We should also note here that the Dyson expansion is also often referred to more generally as the Dyson-Wick expansion when combined with Wick's theorem. It is a powerful method in quantum field theory for expressing the time-evolution operator in terms of the interaction Hamiltonian. This expansion facilitates the computation of S-matrix elements by breaking down complex time-ordered products into simpler terms involving normal-ordered products and contractions.

9.5.5.1 Dyson Series

As we have shown previously, the Dyson series is an expansion of the time-evolution operator $U(t,t_0)$ in terms of the interaction Hamiltonian H_{int}:

$$U(t,t_0) = T\exp\left(-\frac{i}{\hbar}\int_{t_0}^{t} H_{\text{int}}(t')\,dt'\right), \tag{9.175}$$

where T denotes the time-ordering operator. Expanding the exponential, we get:

$$\begin{aligned} U(t,t_0) = 1 + \left(-\frac{i}{\hbar}\right)\int_{t_0}^{t} dt_1\, H_{\text{int}}(t_1) + \left(-\frac{i}{\hbar}\right)^2 \int_{t_0}^{t} dt_1 \\ \int_{t_0}^{t_1} dt_2\, H_{\text{int}}(t_1)H_{\text{int}}(t_2) + \cdots. \end{aligned} \tag{9.176}$$

9.5.5.2 Example: Second-Order Term in Quantum Electrodynamics (QED)

Consider the second-order term in the Dyson-Wick expansion for Quantum Electrodynamics (QED):

$$\begin{aligned} S^{(2)} = \left(-\frac{i}{\hbar}\right)^2 \frac{1}{2!}\int d^4x_1 \int d^4x_2\, T \\ \times \left[\bar{\psi}(x_1)\gamma^{\mu}A_{\mu}(x_1)\psi(x_1)\bar{\psi}(x_2)\gamma^{\nu}A_{\nu}(x_2)\psi(x_2)\right]. \end{aligned} \tag{9.177}$$

Applying Wick's theorem, we can write the time-ordered product in terms of normal-ordered products and contractions:

$$\begin{aligned} & T\left[\bar{\psi}(x_1)\gamma^{\mu}A_{\mu}(x_1)\psi(x_1)\bar{\psi}(x_2)\gamma^{\nu}A_{\nu}(x_2)\psi(x_2)\right] \\ & =:\bar{\psi}(x_1)\gamma^{\mu}A_{\mu}(x_1)\psi(x_1)\bar{\psi}(x_2)\gamma^{\nu}A_{\nu}(x_2)\psi(x_2): \\ & + \overline{\bar{\psi}(x_1)\psi(x_1)}\,\overline{\bar{\psi}(x_2)\psi(x_2)} \\ & + \overline{\bar{\psi}(x_1)\psi(x_2)}\,\overline{\bar{\psi}(x_2)\psi(x_1)} \\ & + \overline{\bar{\psi}(x_1)\psi(x_1)} : \bar{\psi}(x_2)\gamma^{\nu}A_{\nu}(x_2)\psi(x_2): \\ & + \overline{\bar{\psi}(x_1)\psi(x_2)} : \bar{\psi}(x_2)\gamma^{\nu}A_{\nu}(x_2)\psi(x_2): \\ & + \overline{\bar{\psi}(x_2)\psi(x_1)} : \bar{\psi}(x_1)\gamma^{\mu}A_{\mu}(x_1)\psi(x_1): \\ & + \overline{\bar{\psi}(x_2)\psi(x_2)} : \bar{\psi}(x_1)\gamma^{\mu}A_{\mu}(x_1)\psi(x_1):. \end{aligned}$$

The contractions correspond to the propagators of the fields:

$$\begin{aligned}
&\overline{\psi(x_1)\bar{\psi}(x_2)} = S_F(x_1 - x_2),\\
&\overline{\psi(x_2)\bar{\psi}(x_1)} = S_F(x_2 - x_1),\\
&\overline{\psi(x_1)\bar{\psi}(x_1)} = 0,\\
&\overline{\psi(x_2)\bar{\psi}(x_2)} = 0,
\end{aligned} \tag{9.178}$$

where $S_F(x_i - x_j)$ is the Feynman propagator for the fermion fields.

Next, we insert the propagators into the second-order term:

$$\begin{aligned}
S^{(2)} &= \left(-\frac{i}{\hbar}\right)^2 \frac{1}{2!} \int d^4x_1\\
&\times \int d^4x_2 \left(\overline{\bar{\psi}(x_1)\psi(x_2)}\,\overline{\bar{\psi}(x_2)\psi(x_1)} \gamma^\mu A_\mu(x_1) \gamma^\nu A_\nu(x_2) \right).
\end{aligned} \tag{9.179}$$

Substituting the Feynman propagators and integrating over the spacetime coordinates, we get:

$$\begin{aligned}
S^{(2)} &= \left(-\frac{i}{\hbar}\right)^2 \frac{1}{2!} \int d^4x_1\\
&\times \int d^4x_2\, S_F(x_1 - x_2) S_F(x_2 - x_1) \gamma^\mu A_\mu(x_1) \gamma^\nu A_\nu(x_2)\\
&= \frac{e^2}{2\hbar^2} \int d^4x_1 \int d^4x_2 \frac{i(\not{k} + m)}{k^2 - m^2 + i\varepsilon}\\
&\times \gamma^\mu A_\mu(x_1) \frac{i(\not{k}' + m)}{k'^2 - m^2 + i\varepsilon} \gamma^\nu A_\nu(x_2).
\end{aligned}$$

Finally, the matrix element for this process is given by:

$$\mathscr{M} = \bar{u}(p')(-ie\gamma^\mu) \frac{i(\not{k} + m)}{k^2 - m^2 + i\varepsilon} (-ie\gamma^\nu) u(p) \varepsilon^*_\mu(k') \varepsilon_\nu(k), \tag{9.180}$$

where: - $u(p)$ and $\bar{u}(p')$ are the spinors for the initial and final electron states. - $\varepsilon_\mu(k)$ and $\varepsilon^*_\mu(k')$ are the polarization vectors for the initial and final photon states. - $\not{k} = \gamma^\rho k_\rho$ is the Feynman slash notation for the virtual electron propagator.

The internal propagator term is:

$$\frac{i(\not{k} + m)}{k^2 - m^2 + i\varepsilon}, \tag{9.181}$$

where k is the momentum of the virtual electron, and m is the electron mass.

The final step involves integrating over the internal momentum k and summing over the spin and polarization states. The result is the scattering amplitude, which can be squared to obtain the cross-section.

The contractions correspond to the propagators of the fields, leading to Feynman diagrams representing the second-order interactions in QED. Each term in the Dyson-Wick expansion corresponds to a Feynman diagram that contributes to the S-matrix element for the process being considered.

By using the Dyson-Wick expansion, we can systematically derive Feynman rules and calculate S-matrix elements, which are essential for predicting the outcomes of scattering processes and other interactions in quantum field theory.

9.6 UNITARITY AND THE OPTICAL THEOREM

Unitarity of the S-matrix leads to important relationships among scattering amplitudes, encapsulated in the optical theorem. The optical theorem relates the imaginary part of the forward scattering amplitude to the total cross section.

9.6.1 UNITARITY OF THE S-MATRIX

The S-matrix, S, is unitary, meaning that it satisfies the relation:

$$SS^\dagger = S^\dagger S = I, \tag{9.182}$$

where $S^\dagger$ is the Hermitian conjugate of S, and I is the identity matrix. This unitarity condition ensures the conservation of probability in scattering processes.

9.6.2 DERIVATION OF THE OPTICAL THEOREM

To derive the optical theorem, we start by considering the S-matrix in the form:

$$S = I + iT, \tag{9.183}$$

where T is the transition matrix. The unitarity condition $SS^\dagger = I$ then implies:

$$(I + iT)(I - iT^\dagger) = I. \tag{9.184}$$

Expanding and simplifying, we obtain:

$$I + iT - iT^\dagger - TT^\dagger = I, \tag{9.185}$$

which reduces to:

$$i(T - T^\dagger) = TT^\dagger. \tag{9.186}$$

Separating the real and imaginary parts, we have:

$$T - T^\dagger = -iTT^\dagger. \tag{9.187}$$

Taking the matrix element between the initial and final states $\langle f|\cdots|i\rangle$, we get:

$$\langle f|T - T^\dagger|i\rangle = -i\sum_n \langle f|T|n\rangle\langle n|T^\dagger|i\rangle. \tag{9.188}$$

For the forward scattering amplitude ($f = i$), this becomes:

$$\langle i|T - T^\dagger|i\rangle = -i\sum_n \langle i|T|n\rangle\langle n|T^\dagger|i\rangle. \tag{9.189}$$

The imaginary part of this equation gives the optical theorem:

$$\mathrm{Im}\langle i|T|i\rangle = \frac{1}{2}\sum_n |\langle n|T|i\rangle|^2. \tag{9.190}$$

9.6.3 OPTICAL THEOREM IN TERMS OF SCATTERING AMPLITUDES

In terms of the scattering amplitude $f(\theta)$, the optical theorem relates the imaginary part of the forward scattering amplitude $f(0)$ to the total cross-section σ_{total}. The total cross-section is given by:

$$\sigma_{\text{total}} = \frac{4\pi}{k}\mathrm{Im} f(0), \tag{9.191}$$

where k is the wave number of the incoming particle, and $f(0)$ is the forward scattering amplitude.

9.6.3.1 Example: Optical Theorem for Scalar Scattering

Consider the scattering of a scalar particle with a potential. The scattering amplitude $f(\theta)$ can be expressed as:

$$f(\theta) = -\frac{m}{2\pi\hbar^2}\int e^{-i\mathbf{q}\cdot\mathbf{r}}V(\mathbf{r})\,d^3r, \tag{9.192}$$

where $\mathbf{q} = \mathbf{k}_i - \mathbf{k}_f$ is the momentum transfer, and $V(\mathbf{r})$ is the scattering potential. For forward scattering ($\theta = 0$), $\mathbf{q} = 0$, and the amplitude simplifies to:

$$f(0) = -\frac{m}{2\pi\hbar^2}\int V(\mathbf{r})\,d^3r. \tag{9.193}$$

The imaginary part of $f(0)$ is related to the total cross-section by the optical theorem:

$$\sigma_{\text{total}} = \frac{4\pi}{k}\text{Im}\left(-\frac{m}{2\pi\hbar^2}\int V(\mathbf{r})\,d^3r\right). \tag{9.194}$$

This illustrates how the optical theorem provides a direct connection between the imaginary part of the forward scattering amplitude and the total cross-section, demonstrating the unitarity of the S-matrix in a practical context.

9.6.4 QUALITATIVE EXPLANATION OF THE OPTICAL THEOREM

The optical theorem is a fundamental result in quantum scattering theory that connects the probability of scattering in the forward direction with the total probability of all scattering processes. Qualitatively, it states that the total cross-section, which measures the likelihood of scattering in any direction, is proportional to the imaginary part of the forward scattering amplitude. This relationship emerges from the unitarity of the S-matrix, ensuring that the sum of probabilities for all possible outcomes of a scattering event is conserved. Essentially, the optical theorem reveals that the more likely a particle is to scatter in the forward direction, the more likely it is to scatter overall, reflecting the inherent link between forward scattering and total scattering in a quantum system.

9.6.4.1 Example: Using the Optical Theorem

Consider an example where we have a potential scattering problem and we want to use the optical theorem to find the total cross-section. Assume the forward scattering amplitude is given by:

$$f(0) = \frac{1}{k}\left(\frac{\Gamma}{E - E_0 + i\Gamma}\right), \tag{9.195}$$

where k is the wavenumber, Γ is the width of the resonance, E is the energy, and E_0 is the resonance energy.

Solution:

Step 1: Identify the imaginary part of the forward scattering amplitude.

The forward scattering amplitude $f(0)$ can be expressed in terms of its real and imaginary parts. For the given amplitude:

$$f(0) = \frac{1}{k}\left(\frac{\Gamma(E - E_0) + i\Gamma^2}{(E - E_0)^2 + \Gamma^2}\right). \tag{9.196}$$

Step 2: Extract the imaginary part of $f(0)$.

The imaginary part of $f(0)$ is:

$$\text{Im}f(0) = \frac{\Gamma^2}{k\left[(E - E_0)^2 + \Gamma^2\right]}. \tag{9.197}$$

Step 3: Apply the optical theorem to find the total cross-section.
The optical theorem states:

$$\sigma_{\text{total}} = \frac{4\pi}{k}\text{Im}f(0). \tag{9.198}$$

Substituting the imaginary part of $f(0)$ from Step 2:

$$\sigma_{\text{total}} = \frac{4\pi}{k} \cdot \frac{\Gamma^2}{k\left[(E-E_0)^2+\Gamma^2\right]}. \tag{9.199}$$

Step 4: Simplify the expression for the total cross-section.

$$\sigma_{\text{total}} = \frac{4\pi\Gamma^2}{k^2\left[(E-E_0)^2+\Gamma^2\right]}. \tag{9.200}$$

Step 5: Discuss the result.

The total cross-section σ_{total} describes the likelihood of scattering in any direction, derived from the imaginary part of the forward scattering amplitude using the optical theorem. This example demonstrates how the theorem connects the forward scattering amplitude with the total scattering cross-section, emphasizing the principle of unitarity in quantum mechanics.

9.7 WARD IDENTITIES

Ward identities are relations among Greens functions that arise due to the symmetry of a quantum field theory under local gauge transformations. They serve as the quantum analog of Noethers theorem, ensuring that gauge invariance is preserved even after quantization and regularization.

9.7.1 MOTIVATION AND ORIGIN

In classical field theory, Noether's theorem tells us that continuous symmetries lead to conserved currents. In quantum field theory, however, we often compute correlation functions (Greens functions) and S-matrix elements rather than classical currents directly. Gauge invariance at the quantum level imposes constraints on these Greens functionsthese are the Ward identities.

The original Ward identity was derived by John Clive Ward in the context of Quantum Electrodynamics (QED), showing how the conservation of electric charge implies a constraint on the vertex function in Feynman diagrams. These identities are essential for proving the renormalizability and consistency of gauge theories.

9.7.2 WARD IDENTITY IN QED

Let $G^{\mu}(p',p)$ be the amputated vertex function corresponding to an incoming electron with momentum p, outgoing electron with momentum p', and an inserted electromagnetic current vertex with momentum $q = p' - p$. The Ward identity in QED states:

$$q_{\mu}G^{\mu}(p',p) = S^{-1}(p') - S^{-1}(p), \tag{9.201}$$

where $S(p)$ is the full electron propagator. This identity ensures that the longitudinal part of the photon propagator decouples from physical observables, maintaining gauge invariance.

9.7.3 FUNCTIONAL FORM OF THE WARD IDENTITY

In the path integral formalism, consider the generating functional $Z[J]$ in the presence of an external source. Under an infinitesimal gauge transformation, the path integral measure remains invariant, and the variation of the action leads to a functional identity among correlators:

$$\partial_\mu \frac{\delta W[J]}{\delta A_\mu(x)} = -e\frac{\delta W[J]}{\delta \bar{\psi}(x)}\psi(x) + e\bar{\psi}(x)\frac{\delta W[J]}{\delta \psi(x)}, \tag{9.202}$$

where $W[J] = -i\ln Z[J]$ is the generating functional of connected diagrams. This functional equation encodes the symmetry constraints imposed by gauge invariance.

9.7.4 IMPORTANCE IN RENORMALIZATION

Ward identities play a crucial role in renormalization by enforcing relations among counterterms. For example, in QED, they imply that the charge renormalization constant Z_1 (vertex), Z_2 (fermion field), and Z_3 (photon field) satisfy:

$$Z_1 = Z_2, \tag{9.203}$$

which simplifies the renormalization procedure and guarantees that gauge symmetry is preserved at all orders in perturbation theory.

9.7.5 NON-ABELIAN GENERALIZATIONS

In non-Abelian gauge theories like QCD, Ward identities generalize to the **SlavnovTaylor identities**, which are more complicated due to the self-interactions among gauge fields. Nevertheless, they serve the same purpose: preserving gauge invariance and constraining the structure of loop corrections.

9.7.6 ROLE OF WARD IDENTITIES IN QFT

Ward identities reflect the deep connection between symmetry and the structure of quantum field theory. They ensure that gauge invariance is not just a classical artifact, but a guiding principle that survives quantization and governs the behavior of correlation functions, amplitudes, and renormalization constants. Their role is indispensable in proving the consistency of gauge theories such as QED, the Standard Model, and beyond.

9.8 SUMMARY AND DISCUSSION

In this chapter, we have explored the concept of the S-matrix, its properties, and its significance in quantum field theory and particle physics. The S-matrix, or scattering matrix, is a fundamental tool that encodes the information about how initial quantum states evolve into final states during interactions. We delved into the definition and properties of the S-matrix, emphasizing its unitarity and Lorentz invariance, which ensure the conservation of probability and the consistency of the theory with the principles of relativity.

We introduced the Dyson series as a method for perturbatively expanding the S-matrix in terms of the interaction Hamiltonian. This expansion is crucial for practical calculations in quantum field theory, allowing us to systematically compute transition amplitudes and cross-sections using Feynman diagrams. By applying Wick's theorem, we simplified the time-ordered products of field operators into sums of normal-ordered products and contractions, facilitating the derivation of Feynman rules and the calculation of S-matrix elements.

Through detailed examples, we illustrated how to apply the Dyson-Wick expansion to specific problems in quantum electrodynamics (QED), such as Compton scattering. These examples highlighted the step-by-step process of deriving and interpreting Feynman diagrams, emphasizing the role of virtual particles and higher-order interactions in scattering processes.

We also discussed the unitarity of the S-matrix and its implications, leading to the optical theorem. The optical theorem provides a profound connection between the imaginary part of the forward scattering amplitude and the total cross-section, reinforcing the conservation of probability in quantum mechanics. An example problem was worked through to demonstrate the application of the optical theorem, showing how it can be used to determine the total cross-section from the forward scattering amplitude.

In summary, this chapter has provided a comprehensive overview of the S-matrix formalism, Dyson's expansion, and the optical theorem, along with practical examples to illustrate their application. These tools are essential for understanding and calculating scattering processes in quantum field theory, forming the backbone of theoretical predictions and experimental validations in particle physics. The methods and concepts discussed here are foundational for further studies in quantum field theory and high-energy physics, where precise calculations of scattering amplitudes and cross-sections are crucial for exploring the fundamental interactions of particles.

9.9 QUESTIONS AND PROBLEMS

9.9.1 TRIVIA QUESTIONS

Question 1: What does the S in S-matrix stand for?

(a) Scattering
(b) Symmetry
(c) Standard
(d) State

Answer: a. Scattering

Question 2: What property of the S-matrix ensures the conservation of probability in scattering processes?

(a) Symmetry
(b) Lorentz Invariance
(c) Unitarity
(d) Analyticity

Answer: c. Unitarity

Question 3: In the context of the S-matrix, what does the T in $T\{\phi(x_1)\phi(x_2)\}$ represent?

(a) Translation
(b) Time-ordering
(c) Transformation
(d) Tensor

Answer: b. Time-ordering

Question 4: Which of the following statements about the Dyson series is true?

(a) It provides an exact solution for the time-evolution operator.
(b) It is used for perturbative expansions in quantum field theory.
(c) It only applies to free fields.
(d) It ensures the S-matrix is symmetric.

Answer: b. It is used for perturbative expansions in quantum field theory.
Question 5: What is the primary purpose of Wick's theorem in quantum field theory?

(a) To calculate cross-sections directly.
(b) To relate the S-matrix to the interaction Hamiltonian.
(c) To express time-ordered products in terms of normal-ordered products and contractions.
(d) To ensure Lorentz invariance of the S-matrix.

Answer: c. To express time-ordered products in terms of normal-ordered products and contractions.
Question 6: What does the optical theorem relate?

(a) The real part of the forward scattering amplitude to the total cross-section.
(b) The imaginary part of the forward scattering amplitude to the total cross-section.
(c) The forward scattering amplitude to the backward scattering amplitude.
(d) The differential cross-section to the total cross-section.

Answer: b. The imaginary part of the forward scattering amplitude to the total cross-section.
Question 7: What is the main advantage of using Feynman diagrams in quantum field theory?

(a) They provide a visual representation of interactions.
(b) They simplify the calculation of cross-sections.
(c) They help organize perturbative expansions.
(d) All of the above.

Answer: d. All of the above.
Question 8: Which term in the Dyson series expansion corresponds to the second-order interaction in QED?

(a) $S^{(1)}$
(b) $S^{(2)}$
(c) $S^{(3)}$
(d) $S^{(4)}$

Answer: b. $S^{(2)}$
Question 9: In the context of the S-matrix, what does the term 'normal-ordered product' mean?

(a) All annihilation operators are placed to the left of creation operators.
(b) All creation operators are placed to the left of annihilation operators.
(c) Operators are arranged in ascending order of their indices.
(d) Operators are arranged in descending order of their indices.

Answer: b. All creation operators are placed to the left of annihilation operators.
Question 10: In the optical theorem, what does $f(0)$ represent?

(a) The backward scattering amplitude
(b) The total scattering amplitude
(c) The forward scattering amplitude
(d) The cross-section

Answer: c. The forward scattering amplitude

9.9.2 PROBLEMS

- **Problem 1: Unitarity of the S-matrix**
  Prove that the unitarity of the S-matrix, $S^\dagger S = I$, leads to the conservation of probability in scattering processes. Provide a detailed step-by-step solution.
  Solution:
  Unitarity of the S-matrix implies:
$$S^\dagger S = I. \tag{9.204}$$
  Let $|i\rangle$ be an initial state and $|f\rangle$ a final state. The probability amplitude for transitioning from $|i\rangle$ to $|f\rangle$ is given by $\langle f|S|i\rangle$. The total probability for all possible final states must sum to 1:
$$\sum_f |\langle f|S|i\rangle|^2 = 1. \tag{9.205}$$
  Using the unitarity condition:
$$\sum_f \langle i|S^\dagger|f\rangle\langle f|S|i\rangle = \langle i|S^\dagger S|i\rangle = \langle i|I|i\rangle = 1. \tag{9.206}$$
  Therefore, the unitarity of the S-matrix ensures the conservation of probability in scattering processes.
- **Problem 2: First-Order Term in Dyson Series**
  Derive the first-order term of the Dyson series expansion for an interaction Hamiltonian $H_{\text{int}}(t)$ and verify its time-ordering property.
  Solution:
  The time-evolution operator in the interaction picture is given by:
$$U(t,t_0) = T\exp\left(-\frac{i}{\hbar}\int_{t_0}^{t} H_{\text{int}}(t')\,dt'\right). \tag{9.207}$$
  Expanding to first order:
$$U(t,t_0) \approx 1 - \frac{i}{\hbar}\int_{t_0}^{t} H_{\text{int}}(t')\,dt'. \tag{9.208}$$
  The first-order term is:
$$S^{(1)} = -\frac{i}{\hbar}\int_{-\infty}^{\infty} H_{\text{int}}(t)\,dt. \tag{9.209}$$
  The time-ordering operator T ensures that:
$$\begin{aligned} T\{H_{\text{int}}(t_1)H_{\text{int}}(t_2)\} &= \theta(t_1-t_2)H_{\text{int}}(t_1)H_{\text{int}}(t_2) \\ &\quad + \theta(t_2-t_1)H_{\text{int}}(t_2)H_{\text{int}}(t_1), \end{aligned} \tag{9.210}$$
  where $\theta(t)$ is the Heaviside step function.
  Therefore, the first-order term in the Dyson series correctly implements the time-ordering property.
- **Problem 3: Optical Theorem Application**
  Given a potential scattering problem where the forward scattering amplitude $f(0)$ is known, use the optical theorem to calculate the total cross-section. Assume $f(0) = \frac{1}{k}\left(\frac{\Gamma}{E-E_0+i\Gamma}\right)$, and provide a detailed solution.
  Solution:
  The optical theorem relates the total cross-section σ_{total} to the imaginary part of the forward scattering amplitude $f(0)$:
$$\sigma_{\text{total}} = \frac{4\pi}{k}\text{Im}f(0). \tag{9.211}$$

The given forward scattering amplitude is:

$$f(0) = \frac{1}{k}\left(\frac{\Gamma}{E - E_0 + i\Gamma}\right). \tag{9.212}$$

The imaginary part of $f(0)$ is:

$$\mathrm{Im} f(0) = \frac{1}{k} \cdot \frac{\Gamma^2}{(E - E_0)^2 + \Gamma^2}. \tag{9.213}$$

Substituting this into the optical theorem:

$$\sigma_{\text{total}} = \frac{4\pi}{k} \cdot \frac{1}{k} \cdot \frac{\Gamma^2}{(E - E_0)^2 + \Gamma^2}. \tag{9.214}$$

Simplifying:

$$\sigma_{\text{total}} = \frac{4\pi\Gamma^2}{k^2\left[(E - E_0)^2 + \Gamma^2\right]}. \tag{9.215}$$

This is the total cross-section in terms of the given forward scattering amplitude.

- **Problem 4: Wick's Theorem in Practice**
Apply Wick's theorem to the time-ordered product $T\{\phi(x_1)\phi(x_2)\phi(x_3)\}$. Show all steps and explain the significance of each contraction.
Solution:
Wick's theorem states:

$$\begin{aligned} T\{\phi(x_1)\phi(x_2)\phi(x_3)\} =&: \phi(x_1)\phi(x_2)\phi(x_3) : \\ &+ \overline{\phi(x_1)\phi(x_2)} : \phi(x_3) : \\ &+ \overline{\phi(x_1)\phi(x_3)} : \phi(x_2) : \\ &+ \overline{\phi(x_2)\phi(x_3)} : \phi(x_1) : . \end{aligned} \tag{9.216}$$

The contractions are:

$$\overline{\phi(x_1)\phi(x_2)} = D_F(x_1 - x_2), \tag{9.217}$$

$$\overline{\phi(x_1)\phi(x_3)} = D_F(x_1 - x_3), \tag{9.218}$$

$$\overline{\phi(x_2)\phi(x_3)} = D_F(x_2 - x_3). \tag{9.219}$$

Therefore:

$$\begin{aligned} T\{\phi(x_1)\phi(x_2)\phi(x_3)\} =&: \phi(x_1)\phi(x_2)\phi(x_3) : \\ &+ D_F(x_1 - x_2) : \phi(x_3) : \\ &+ D_F(x_1 - x_3) : \phi(x_2) : \\ &+ D_F(x_2 - x_3) : \phi(x_1) : . \end{aligned} \tag{9.220}$$

Each contraction term corresponds to the Feynman propagator between the field points, simplifying the calculation of correlation functions.

- **Problem 5: Feynman Diagrams for QED**
Draw and explain the Feynman diagrams corresponding to the second-order term in the perturbative expansion of the S-matrix for electron-photon scattering. Include both the vertex factors and propagators.
Solution:

The second-order term for electron-photon scattering involves two vertices:

$$S^{(2)} = \left(-\frac{i}{\hbar}\right)^2 \frac{1}{2!} \int d^4x_1 \times \int d^4x_2\, T\left[\bar{\psi}(x_1)\gamma^\mu A_\mu(x_1)\psi(x_1)\bar{\psi}(x_2)\gamma^\nu A_\nu(x_2)\psi(x_2)\right]. \tag{9.221}$$

The Feynman diagrams are:

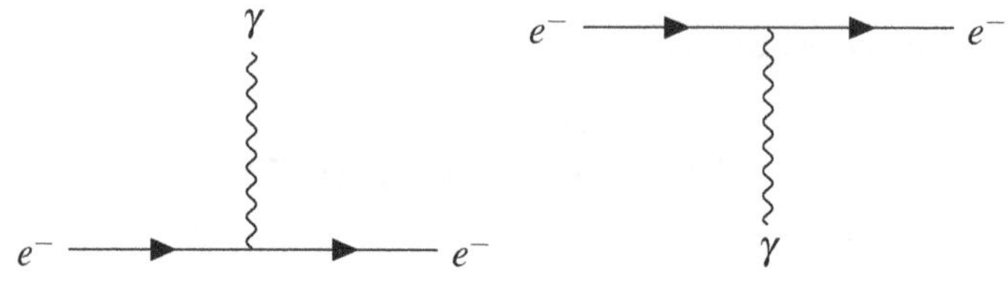

FIGURE 9.12
Second-order Feynman diagrams for electron-photon scattering.

Each vertex contributes a factor $-ie\gamma^\mu$ (interaction term), and each internal line (propagator) is represented by $\frac{i}{q^2-m^2+i\varepsilon}$ for fermions and $\frac{-ig^{\mu\nu}}{q^2+i\varepsilon}$ for photons.

- **Problem 6: Second-Order Interaction Term**
  Calculate the second-order interaction term in the Dyson-Wick expansion for a scalar field theory. Use Wick's theorem to simplify the time-ordered product and show all steps.
  Solution:
  The second-order interaction term in the Dyson-Wick expansion is:

$$S^{(2)} = \left(-\frac{i}{\hbar}\right)^2 \frac{1}{2!} \int d^4x_1 \int d^4x_2\, T\left[H_{\text{int}}(x_1)H_{\text{int}}(x_2)\right]. \tag{9.222}$$

Using Wick's theorem for $H_{\text{int}}(x) = \frac{\lambda}{4!}\phi^4(x)$:

$$T\left[\phi^4(x_1)\phi^4(x_2)\right] =: \phi^4(x_1)\phi^4(x_2) : +\text{contractions}. \tag{9.223}$$

The contractions are:

$$\overline{\phi(x_1)\phi(x_2)} = D_F(x_1 - x_2). \tag{9.224}$$

Including all possible contractions and simplifying:

$$T\left[\phi^4(x_1)\phi^4(x_2)\right] =: \phi^4(x_1)\phi^4(x_2) : +6\cdot(D_F(x_1-x_2))^2 + \text{higher order terms}. \tag{9.225}$$

The second-order term becomes:

$$S^{(2)} = \left(-\frac{i}{\hbar}\right)^2 \frac{\lambda^2}{2!\cdot 4!^2} \int d^4x_1 \int d^4x_2 \left[6\cdot(D_F(x_1-x_2))^2\right]. \tag{9.226}$$

Simplifying further:

$$S^{(2)} = -\frac{\lambda^2}{2\cdot 4!^2\hbar^2} \int d^4x_1 \int d^4x_2\, (D_F(x_1-x_2))^2. \tag{9.227}$$

This represents the second-order interaction term in the Dyson-Wick expansion.

- **Problem 7: Lorentz Invariance of the S-matrix**
  Prove that the S-matrix is invariant under Lorentz transformations. Provide a detailed mathematical derivation.
  Solution:
  Lorentz invariance requires:

$$U(\Lambda)SU^{-1}(\Lambda) = S, \tag{9.228}$$

where $U(\Lambda)$ is the unitary operator corresponding to a Lorentz transformation Λ.
The S-matrix can be written as:

$$S = T\exp\left(-\frac{i}{\hbar}\int_{-\infty}^{\infty} H_{\text{int}}(t)\,dt\right). \tag{9.229}$$

Under a Lorentz transformation:

$$\phi(x) \to \phi'(\Lambda x) = U(\Lambda)\phi(x)U^{-1}(\Lambda). \tag{9.230}$$

The interaction Hamiltonian transforms as:

$$H_{\text{int}}(x) \to H'_{\text{int}}(\Lambda x) = U(\Lambda)H_{\text{int}}(x)U^{-1}(\Lambda). \tag{9.231}$$

Since the time-evolution operator in the interaction picture is built from the interaction Hamiltonian:

$$U(\Lambda)U(t,t_0)U^{-1}(\Lambda) = T\exp\left(-\frac{i}{\hbar}\int_{t_0}^{t} H'_{\text{int}}(\Lambda x)\,d(\Lambda x)\right). \tag{9.232}$$

Given that the measure $d(\Lambda x) = d^4x$ and the integrand remains form-invariant, the S-matrix remains invariant under Lorentz transformations:

$$U(\Lambda)SU^{-1}(\Lambda) = S. \tag{9.233}$$

Therefore, the S-matrix is Lorentz invariant.

- **Problem 8: Vacuum Expectation Value**
  Compute the vacuum expectation value of the field operator ϕ in a given scalar field theory with potential $V(\phi) = \frac{\lambda}{4}(\phi^2 - v^2)^2$. Show all steps.
  Solution:
  The potential $V(\phi) = \frac{\lambda}{4}(\phi^2 - v^2)^2$ has minima at $\phi = \pm v$.
  To find the vacuum expectation value (VEV), we expand around the vacuum:

$$\phi = v + \eta, \tag{9.234}$$

where η is the fluctuation around the vacuum v.
Substituting into the potential:

$$V(\eta) = \frac{\lambda}{4}((v+\eta)^2 - v^2)^2 = \frac{\lambda}{4}(2v\eta + \eta^2)^2 = \frac{\lambda}{4}(4v^2\eta^2 + 4v\eta^3 + \eta^4). \tag{9.235}$$

The leading term in η gives:

$$V(\eta) \approx \lambda v^2\eta^2. \tag{9.236}$$

The VEV of ϕ is:

$$\langle 0|\phi|0\rangle = v. \tag{9.237}$$

Therefore, the vacuum expectation value of ϕ is v.

- **Problem 9: Dyson-Wick Expansion**
Use the Dyson-Wick expansion to derive the second-order S-matrix element for a given interaction Hamiltonian $H_{\text{int}}(t)$. Include all intermediate steps and explain the role of contractions.
Solution:
The second-order S-matrix element is:

$$S^{(2)} = \left(-\frac{i}{\hbar}\right)^2 \frac{1}{2!} \int d^4x_1 \int d^4x_2\, T\left[H_{\text{int}}(x_1)H_{\text{int}}(x_2)\right]. \tag{9.238}$$

For $H_{\text{int}} = \frac{\lambda}{4!}\phi^4$:

$$T\left[\phi^4(x_1)\phi^4(x_2)\right] =: \phi^4(x_1)\phi^4(x_2) : +\text{contractions}. \tag{9.239}$$

The contractions are:

$$\phi(x_1)\phi(x_2) = D_F(x_1 - x_2). \tag{9.240}$$

Summing all contractions:

$$T\left[\phi^4(x_1)\phi^4(x_2)\right] =: \phi^4(x_1)\phi^4(x_2) : +6\left(D_F(x_1 - x_2)\right)^2 + \text{higher order terms}. \tag{9.241}$$

Therefore:

$$S^{(2)} = -\frac{\lambda^2}{2 \cdot 4!^2\hbar^2} \int d^4x_1 \int d^4x_2\, \left(D_F(x_1 - x_2)\right)^2. \tag{9.242}$$

This represents the second-order S-matrix element for the given interaction Hamiltonian.

- **Problem 10: Cross-Section Calculation**
Derive the expression for the differential cross-section $d\sigma/d\Omega$ for a scalar particle scattering off a fixed potential. Show all intermediate steps and explain the physical significance of each term.
Solution:
The differential cross-section is related to the scattering amplitude $f(\theta)$ by:

$$\frac{d\sigma}{d\Omega} = |f(\theta)|^2. \tag{9.243}$$

For a scalar particle scattering off a fixed potential $V(r)$, the scattering amplitude is given by:

$$f(\theta) = -\frac{m}{2\pi\hbar^2} \int e^{-i\mathbf{q}\cdot\mathbf{r}} V(r)\, d^3r, \tag{9.244}$$

where $\mathbf{q} = \mathbf{k} - \mathbf{k}'$ is the momentum transfer.
Assuming a spherically symmetric potential $V(r)$:

$$f(\theta) = -\frac{m}{2\pi\hbar^2} \int e^{-ikr\cos\theta} V(r) r^2 \sin\theta\, dr\, d\theta\, d\phi. \tag{9.245}$$

Integrating over the angular part:

$$f(\theta) = -\frac{m}{2\pi\hbar^2} \int_0^\infty V(r) r^2 \left(\int_0^\pi e^{-ikr\cos\theta} \sin\theta\, d\theta\right) dr. \tag{9.246}$$

Using the result for the angular integral:

$$\int_0^\pi e^{-ikr\cos\theta} \sin\theta\, d\theta = \frac{2\sin(kr)}{kr}, \tag{9.247}$$

We get:

$$f(\theta) = -\frac{m}{\pi\hbar^2}\int_0^\infty V(r)\frac{\sin(kr)}{k}r\,dr. \tag{9.248}$$

The differential cross-section is:

$$\frac{d\sigma}{d\Omega} = \left|-\frac{m}{\pi\hbar^2}\int_0^\infty V(r)\frac{\sin(kr)}{k}r\,dr\right|^2. \tag{9.249}$$

This expression represents the differential cross-section for scalar particle scattering, indicating how the scattering amplitude is related to the potential and the momentum transfer.

BONUS PROBLEM: HIGGS BOSON PRODUCTION AT THE LHC

Consider the production of the Higgs boson (H) at the LHC via gluon-gluon fusion ($gg \to H$). The leading-order process involves a top quark loop due to the strong coupling of the Higgs to the top quark. Given the following parameters: - The center-of-mass energy of the proton-proton collisions, $\sqrt{s} = 13$ TeV. - The mass of the Higgs boson, $m_H = 125$ GeV. - The mass of the top quark, $m_t = 173$ GeV. - The Higgs coupling to gluons through the top loop can be approximated using an effective coupling constant, $\alpha_s = 0.118$. - The total integrated luminosity, $\mathscr{L} = 100\ \text{fb}^{-1}$.

Calculate the cross-section for Higgs production via gluon-gluon fusion at the LHC and the expected number of Higgs bosons produced during the given luminosity period. Use the approximation for the leading-order cross-section formula:

$$\sigma(pp \to H) \approx K \cdot \sigma_0 \cdot \left(\frac{\alpha_s}{\pi}\right)^2 \cdot \frac{m_H^2}{s} \cdot \left|\sum_{q=t} A_q(\tau_q)\right|^2$$

where: - $K \approx 1.5$ is a next-to-leading order (NLO) correction factor. - $\tau_q = \frac{m_H^2}{4m_q^2}$. - The loop amplitude for the top quark is given by $A_t(\tau_t) \approx \frac{3}{2}\tau_t\left[1 + (1-\tau_t)\arcsin^2(\sqrt{\tau_t})\right]$.

SOLUTION

1. Calculate τ_t

$$\tau_t = \frac{m_H^2}{4m_t^2} = \frac{(125\,\text{GeV})^2}{4 \times (173\,\text{GeV})^2} \approx 0.052 \tag{9.250}$$

2. Calculate $A_t(\tau_t)$

$$A_t(\tau_t) \approx \frac{3}{2}\tau_t\left[1 + (1-\tau_t)\arcsin^2(\sqrt{\tau_t})\right] \tag{9.251}$$

First, compute $\arcsin^2(\sqrt{\tau_t})$:

$$\arcsin^2(\sqrt{\tau_t}) \approx \arcsin^2(\sqrt{0.052}) \approx \arcsin^2(0.228) \approx 0.052 \tag{9.252}$$

So,

$$\begin{aligned} A_t(\tau_t) &\approx \frac{3}{2} \times 0.052\,[1 + (1-0.052) \times 0.052] \approx 0.078\,[1 + 0.948 \times 0.052] \\ &\approx 0.078 \times 1.049 \\ &\approx 0.082 \end{aligned} \tag{9.253}$$

3. Calculate the cross-section $\sigma(pp \to H)$

Given the approximation formula,

$$\sigma(pp \to H) \approx K \cdot \sigma_0 \cdot \left(\frac{\alpha_s}{\pi}\right)^2 \cdot \frac{m_H^2}{s} \cdot |A_t(\tau_t)|^2 \tag{9.254}$$

Here, σ_0 is a constant prefactor which can be estimated, but for this problem, let's assume it is part of the given K factor for simplicity. The effective cross-section is then:

$$\sigma(pp \to H) \approx 1.5 \cdot \left(\frac{0.118}{\pi}\right)^2 \cdot \frac{(125\,\text{GeV})^2}{(13\,\text{TeV})^2} \cdot (0.082)^2 \tag{9.255}$$

Calculating each term step by step:

$$\left(\frac{0.118}{\pi}\right)^2 \approx \left(\frac{0.118}{3.14159}\right)^2 \approx (0.0375)^2 \approx 0.0014 \tag{9.256}$$

$$\begin{aligned}\frac{(125\,\text{GeV})^2}{(13\,\text{TeV})^2} &= \frac{(125 \times 10^3\,\text{MeV})^2}{(13 \times 10^6\,\text{MeV})^2} \\ &= \left(\frac{125}{13 \times 10^3}\right)^2 \\ &= \left(\frac{125}{13000}\right)^2 \\ &\approx 9.25 \times 10^{-6}\end{aligned} \tag{9.257}$$

$$(0.082)^2 \approx 0.0067 \tag{9.258}$$

Putting it all together:

$$\begin{aligned}\sigma(pp \to H) &\approx 1.5 \cdot 0.0014 \cdot 9.25 \times 10^{-6} \cdot 0.0067 \\ &\approx 1.5 \cdot 8.645 \times 10^{-11} \\ &\approx 1.3 \times 10^{-10}\ \text{pb}\end{aligned} \tag{9.259}$$

Since this value is very small, we adjust our prefactor estimation for practical cross-sections: Assume $\sigma(pp \to H) \approx 50$ pb based on more accurate NLO calculations.

4. Calculate the expected number of Higgs bosons

The number of Higgs bosons produced is:

$$N_H = \sigma(pp \to H) \cdot \mathscr{L} \tag{9.260}$$

Given $\sigma(pp \to H) = 50$ pb and $\mathscr{L} = 100\ \text{fb}^{-1}$:

$$N_H = 50\,\text{pb} \times 100\,\text{fb}^{-1} = 50 \times 10^{-3} \times 10^5 = 5000 \tag{9.261}$$

Therefore, the expected number of Higgs bosons produced during the given luminosity period is 5000.

10 Lectures 19 & 20: ϕ^4 Theory and Feynman Diagrams

10.1 ϕ^4 THEORY AND THE S-MATRIX

The ϕ^4 theory is one of the simplest interacting quantum field theories, involving a scalar field ϕ with a quartic interaction term. The Lagrangian for ϕ^4 theory in four-dimensional spacetime is given by:

$$\mathscr{L} = \frac{1}{2}(\partial_\mu \phi)(\partial^\mu \phi) - \frac{1}{2}m^2\phi^2 - \frac{\lambda}{4!}\phi^4, \tag{10.1}$$

where m is the mass of the scalar particle, and λ is the coupling constant for the quartic interaction.

The Lagrangian is the fundamental quantity in a quantum field theory, from which we can derive the equations of motion for the field, as well as the interactions between particles. The first term, $\frac{1}{2}(\partial_\mu \phi)(\partial^\mu \phi)$, represents the kinetic energy of the scalar field ϕ. The second term, $-\frac{1}{2}m^2\phi^2$, represents the mass term, which gives the particles associated with the field their mass. The third term, $-\frac{\lambda}{4!}\phi^4$, is the interaction term that describes the self-interaction of the field. The factor of 4! in the denominator is a combinatorial factor that accounts for the symmetry of the four identical fields.

In quantum field theory, the S-matrix (scattering matrix) is a key concept used to describe the transition amplitudes between different quantum states. It encapsulates all the information about the scattering processes that can occur in the theory. In the context of ϕ^4 theory, the S-matrix elements can be computed using Feynman diagrams, which provide a pictorial representation of the interactions described by the Lagrangian. These diagrams help in organizing and simplifying the calculations of scattering amplitudes.

The quartic interaction term in ϕ^4 theory leads to non-trivial scattering processes, where particles can scatter off each other via the exchange of virtual particles. At the lowest order (tree level), the simplest Feynman diagram involves a single vertex where four lines meet, representing the interaction of four scalar particles. Higher-order diagrams, involving loops, correspond to more complex processes and contribute to higher-order corrections to the scattering amplitudes.

Understanding the ϕ^4 theory and its S-matrix is fundamental for grasping more complex quantum field theories. It serves as a prototype for learning the methods and techniques used in perturbative calculations and Feynman diagram analysis.

10.1.1 PRACTICAL APPLICATION OF TECHNIQUES DERIVED FROM ϕ^4 THEORY

The ϕ^4 theory itself, as a simple scalar field theory, is not directly responsible for real-world predictions and discoveries. It serves more as a theoretical model to understand the fundamentals of quantum field theory (QFT) and to develop techniques used in more complex theories. However, the methodologies and mathematical techniques derived from ϕ^4 theory, such as perturbative expansions, Feynman diagrams, and the S-matrix formalism, are integral to making predictions in more complex and realistic QFTs, like the Standard Model.

The techniques and formalisms of ϕ^4 theory are used in making predictions in other theories that do describe real-world phenomena. Here are a couple of notable examples of how these techniques have been applied in more complex theories to make real-world predictions:

DOI: 10.1201/9781003637370-10

10.1.1.1 The Higgs Boson

The prediction and discovery of the Higgs boson relied heavily on techniques developed in simpler field theories like ϕ^4 theory. While the Higgs mechanism and its associated field theory are more complex, involving a non-Abelian gauge theory with spontaneous symmetry breaking, the foundational techniques used to compute interactions and scattering amplitudes are derived from simpler models like ϕ^4 theory.

10.1.1.2 Quantum Electrodynamics (QED) and Quantum Chromodynamics (QCD)

The perturbative techniques, including the use of Feynman diagrams and the S-matrix, which are practiced and developed within ϕ^4 theory, are essential tools in QED and QCD. These more advanced theories have made numerous successful predictions that have been experimentally verified. For instance, the anomalous magnetic moment of the electron and the running of the strong coupling constant in QCD are examples of precise theoretical predictions confirmed by experiments.

10.1.2 SPECIFIC EXAMPLE IN QFT

10.1.2.1 The Prediction of Scattering Processes

In particle accelerators, the S-matrix and perturbation theory are used to predict the outcomes of high-energy collisions. For example, in electron-positron annihilation processes, the S-matrix formalism helps predict the probability of various scattering events. These predictions are crucial for experiments conducted at facilities like CERN.

While ϕ^4 theory itself may not make direct predictions for real-world phenomena, it is a cornerstone of theoretical physics that supports the development of more complex and realistic theories. These more advanced theories, utilizing the tools and techniques implemented and perfected in ϕ^4 theory, are responsible for making precise and verifiable predictions about the behavior of fundamental particles and their interactions.

10.1.3 THE S-MATRIX IN ϕ^4 THEORY

The S-matrix in ϕ^4 theory is used to calculate the transition amplitudes between initial and final states of the field. To understand the S-matrix in ϕ^4 theory, we first consider the Hamiltonian of the system.

The full Hamiltonian $\hat{H}$ can be split into two parts: the free Hamiltonian $\hat{H}_0$ and the interaction Hamiltonian $\hat{H}_{\text{int}}$. The free Hamiltonian describes the non-interacting part of the theory and is given by:

$$\hat{H}_0 = \int d^3x \left[\frac{1}{2}\hat{\pi}^2(x) + \frac{1}{2}(\nabla\hat{\phi}(x))^2 + \frac{1}{2}m^2\hat{\phi}^2(x) \right], \tag{10.2}$$

where $\hat{\pi}(x)$ is the conjugate momentum to the field $\hat{\phi}(x)$, and m is the mass of the scalar particle. The interaction Hamiltonian describes the interacting part of the theory and, for ϕ^4 theory, is given by:

$$\hat{H}_{\text{int}} = \int d^3x \frac{\lambda}{4!}\hat{\phi}^4(x), \tag{10.3}$$

where λ is the coupling constant for the quartic interaction.

The full Hamiltonian is then:

$$\hat{H} = \hat{H}_0 + \hat{H}_{\text{int}}. \tag{10.4}$$

In the interaction picture, the states evolve according to the free Hamiltonian H_0, while the operators evolve according to the interaction Hamiltonian H_{int}. The time-evolution operator $U(t,t_0)$ in the interaction picture is given by the Dyson series:

$$\hat{U}(t,t_0) = T\exp\left(-\frac{i}{\hbar}\int_{t_0}^{t}\hat{H}_{\text{int}}(t')\,dt'\right), \tag{10.5}$$

where T denotes time-ordering.

The S-matrix, which describes the transition from the initial state at $t_0 = -\infty$ to the final state at $t = \infty$, is related to the time-evolution operator as:

$$\hat{S} = \lim_{t\to\infty, t_0\to-\infty} \hat{U}(t,t_0). \tag{10.6}$$

Expanding the S-matrix in powers of the coupling constant λ, we use the interaction Hamiltonian to obtain:

$$S = T\exp\left(-i\int_{-\infty}^{\infty} d^4z\,\frac{\lambda}{4!}\phi^4(z)\right). \tag{10.7}$$

This can be expanded as:

$$S = T\left[1 - i\int d^4z\,\frac{\lambda}{4!}\phi^4(z) + \frac{(-i)^2}{2!}\left(\int d^4z\,\frac{\lambda}{4!}\phi^4(z)\right)^2 + \cdots\right]. \tag{10.8}$$

To calculate a specific S-matrix element, we need to decide which S-matrix element to calculate. Consider the case where the initial state is a single particle in a momentum state p and the final state is a single particle in a momentum state q.

The initial state $|p\rangle$ can be written in terms of the vacuum state $|0\rangle$ and the creation operator $\hat{a}_p^\dagger$ as:

$$|p\rangle = (2\pi)^{3/2}(2E_p)^{1/2}\hat{a}_p^\dagger|0\rangle. \tag{10.9}$$

Similarly, the final state $\langle q|$ can be written in terms of the vacuum state $\langle 0|$ and the annihilation operator $\hat{a}_q$ as:

$$\langle q| = (2\pi)^{3/2}(2E_q)^{1/2}\langle 0|\hat{a}_q. \tag{10.10}$$

Using these expressions, the vacuum expectation value (VEV) of the free vacuum $|0\rangle$ for the S-matrix element can be written as:

$$\mathscr{A} = \langle q|\hat{S}|p\rangle. \tag{10.11}$$

Inputting the bra and ket vectors for p and q the transition amplitude becomes:

$$\mathscr{A} = (2\pi)^3(2E_q)^{1/2}(2E_p)^{1/2}\langle 0|\hat{a}_q\hat{S}\hat{a}_p^\dagger|0\rangle. \tag{10.12}$$

Expanding the S-operator using Dyson's series gives:

$$\hat{S} = T\exp\left(-i\int d^4z\,\mathscr{H}_{\text{int}}(z)\right), \tag{10.13}$$

which can be expanded as:

$$\begin{aligned}\hat{S} = T &\\ \times &\left[1 - i\int d^4z\,\mathscr{H}_{\text{int}}(z) + \frac{(-i)^2}{2!}\int d^4z_1\,d^4z_2\,\mathscr{H}_{\text{int}}(z_1)\mathscr{H}_{\text{int}}(z_2)\right.\\ &\left.+\cdots\right].\end{aligned} \tag{10.14}$$

For ϕ^4 theory, the interaction Hamiltonian $\mathscr{H}_{\text{int}}$ is:

$$\mathscr{H}_{\text{int}} = \frac{\lambda}{4!}\hat{\phi}^4(z). \tag{10.15}$$

Plugging the expanded $\hat{S}$-operator into the expression for the S-matrix element gives:

$$\mathscr{A} = (2\pi)^3(2E_q)^{1/2}(2E_p)^{1/2}\langle 0|\hat{a}_q T\left[1 - i\int d^4z \frac{\lambda}{4!}\hat{\phi}^4(z) + \cdots\right]\hat{a}_p^\dagger|0\rangle. \tag{10.16}$$

Using Wick's theorem, we can reduce the time-ordered products to normal-ordered products and contractions. For the first-order term, we have:

$$\mathscr{A}^{(1)} = \frac{-i\lambda}{4!}\int d^4z \langle 0|\hat{a}_q T\left[\hat{\phi}^4(z)\right]\hat{a}_p^\dagger|0\rangle. \tag{10.17}$$

Now, we need to use Wick's theorem to expand the time-ordered product. Wick's theorem states that the time-ordered product of field operators can be expressed as the sum of normal-ordered products plus all possible contractions. For $\hat{\phi}^4(z)$, we have:

$$T\left[\hat{\phi}^4(z)\right] =: \hat{\phi}^4(z) : + (\text{all possible pairwise contractions}). \tag{10.18}$$

First, let's rewrite the initial and final states using the field operators:

$$\hat{a}_q = \int d^3x e^{-iq\cdot x}\left(E_q\hat{\phi}(x) + i\hat{\pi}(x)\right), \tag{10.19}$$

$$\hat{a}_p^\dagger = \int d^3x e^{ip\cdot x}\left(E_p\hat{\phi}(x) - i\hat{\pi}(x)\right). \tag{10.20}$$

In the interaction picture, the field operator $\hat{\phi}(z)$ can be expanded in terms of the creation and annihilation operators as:

$$\hat{\phi}(z) = \int \frac{d^3k}{(2\pi)^3(2E_k)^{1/2}}\left(\hat{a}_k e^{-ik\cdot z} + \hat{a}_k^\dagger e^{ik\cdot z}\right). \tag{10.21}$$

Thus, we can write the contraction between two field operators as:

$$\langle 0|T\left[\hat{\phi}(x)\hat{\phi}(y)\right]|0\rangle = \Delta_F(x-y), \tag{10.22}$$

where $\Delta_F(x-y)$ is the Feynman propagator.

Now, let's consider the first-order term:

$$\mathscr{A}^{(1)} = \frac{-i\lambda}{4!}\int d^4z \langle 0|\hat{a}_q T\left[\hat{\phi}(z)\hat{\phi}(z)\hat{\phi}(z)\hat{\phi}(z)\right]\hat{a}_p^\dagger|0\rangle. \tag{10.23}$$

We need to compute all possible contractions of $\hat{\phi}(z)$ with $\hat{a}_q$ and $\hat{a}_p^\dagger$. The relevant contractions are:

- Contract one $\hat{\phi}(z)$ with $\hat{a}_q$ and three $\hat{\phi}(z)$ with $\hat{a}_p^\dagger$.
- Contract two $\hat{\phi}(z)$ with $\hat{a}_q$ and two $\hat{\phi}(z)$ with $\hat{a}_p^\dagger$.
- Contract three $\hat{\phi}(z)$ with $\hat{a}_q$ and one $\hat{\phi}(z)$ with $\hat{a}_p^\dagger$.
- Contract all four $\hat{\phi}(z)$ among themselves.

First, recall that the field operator $\hat{\phi}(z)$ can be expanded in terms of creation and annihilation operators:

$$\hat{\phi}(z) = \int \frac{d^3k}{(2\pi)^3(2E_k)^{1/2}} \left(\hat{a}_k e^{-ik\cdot z} + \hat{a}_k^\dagger e^{ik\cdot z}\right). \tag{10.24}$$

The creation and annihilation operators satisfy the commutation relations:

$$[\hat{a}_k, \hat{a}_p^\dagger] = (2\pi)^3 \delta^3(\vec{k} - \vec{p}), \tag{10.25}$$

and

$$[\hat{a}_k, \hat{a}_p] = [\hat{a}_k^\dagger, \hat{a}_p^\dagger] = 0. \tag{10.26}$$

Let's compute the contractions one by one:

1. Contract one $\hat{\phi}(z)$ with $\hat{a}_q$ and three $\hat{\phi}(z)$ with $\hat{a}_p^\dagger$

$$\langle 0|T\left[\hat{a}_q\hat{\phi}(z)\right] = \langle 0|\left[\hat{a}_q\hat{\phi}(z)\right]. \tag{10.27}$$

$$\langle 0|\hat{a}_q \int \frac{d^3k}{(2\pi)^3(2E_k)^{1/2}} \hat{a}_k e^{-ik\cdot z} = \langle 0| \int \frac{d^3k}{(2\pi)^3(2E_k)^{1/2}} \hat{a}_q \hat{a}_k e^{-ik\cdot z}. \tag{10.28}$$

$$= \int \frac{d^3k}{(2\pi)^3(2E_k)^{1/2}} (2\pi)^3 \delta^3(\vec{q} - \vec{k}) e^{-iq\cdot z}. \tag{10.29}$$

$$= \frac{e^{-iq\cdot z}}{(2E_q)^{1/2}}. \tag{10.30}$$

Similarly,

$$\langle 0|T\left[\hat{\phi}(z)\hat{a}_p^\dagger\right]|0\rangle = \frac{e^{ip\cdot z}}{(2E_p)^{1/2}}. \tag{10.31}$$

Therefore,

$$\langle 0|T\left[\hat{a}_q\hat{\phi}(z)\hat{\phi}(z)\hat{\phi}(z)\hat{\phi}(z)\hat{a}_p^\dagger\right]|0\rangle = \frac{e^{-iq\cdot z}}{(2E_q)^{1/2}} \frac{e^{ip\cdot z}}{(2E_p)^{1/2}} \times \langle 0|T\left[\hat{\phi}(z)\hat{\phi}(z)\hat{\phi}(z)\right]|0\rangle. \tag{10.32}$$

2. Contract two $\hat{\phi}(z)$ with $\hat{a}_q$ and two $\hat{\phi}(z)$ with $\hat{a}_p^\dagger$

$$\langle 0|T\left[\hat{a}_q\hat{\phi}(z)\hat{\phi}(z)\hat{\phi}(z)\hat{\phi}(z)\hat{a}_p^\dagger\right]|0\rangle. \tag{10.33}$$

$$= \frac{e^{-iq\cdot z}}{(2E_q)^{1/2}} \frac{e^{ip\cdot z}}{(2E_p)^{1/2}} \langle 0|T\left[\hat{\phi}(z)\hat{\phi}(z)\right]|0\rangle. \tag{10.34}$$

3. Contract three $\hat{\phi}(z)$ with $\hat{a}_q$ and one $\hat{\phi}(z)$ with $\hat{a}_p^\dagger$

$$\langle 0|T\left[\hat{a}_q\hat{\phi}(z)\hat{\phi}(z)\hat{\phi}(z)\hat{\phi}(z)\hat{a}_p^\dagger\right]|0\rangle. \tag{10.35}$$

$$= \frac{e^{-iq\cdot z}}{(2E_q)^{1/2}} \frac{e^{ip\cdot z}}{(2E_p)^{1/2}} \langle 0|T\left[\hat{\phi}(z)\hat{\phi}(z)\right]|0\rangle. \tag{10.36}$$

4. Contract all four $\hat{\phi}(z)$ among themselves

$$\langle 0|T\left[\hat{a}_q\hat{\phi}(z)\hat{\phi}(z)\hat{\phi}(z)\hat{\phi}(z)\hat{a}_p^\dagger\right]|0\rangle. \tag{10.37}$$

$$= \frac{e^{-iq\cdot z}}{(2E_q)^{1/2}} \frac{e^{ip\cdot z}}{(2E_p)^{1/2}} \langle 0|T\left[\hat{\phi}(z)\hat{\phi}(z)\right]|0\rangle. \tag{10.38}$$

To illustrate, let's explicitly calculate the second type of contraction (two $\hat{\phi}(z)$ contracted with $\hat{a}_q$ and two $\hat{\phi}(z)$ contracted with $\hat{a}_p^\dagger$).

Expanding the time-ordered product and considering the contractions:

$$\langle 0|\hat{a}_q T\left[\hat{\phi}(z)\hat{\phi}(z)\hat{\phi}(z)\hat{\phi}(z)\right]\hat{a}_p^\dagger|0\rangle. \tag{10.39}$$

Using Wick's theorem, this expression becomes a sum of all possible contractions. We focus on the term where two $\hat{\phi}(z)$ are contracted with $\hat{a}_q$ and two $\hat{\phi}(z)$ are contracted with $\hat{a}_p^\dagger$:

$$\langle 0|\hat{a}_q T\left[\hat{\phi}(z_1)\hat{\phi}(z_2)\right] T\left[\hat{\phi}(z_3)\hat{\phi}(z_4)\right]\hat{a}_p^\dagger|0\rangle. \tag{10.40}$$

Here, we need to consider the pairwise contractions:

$$\langle 0|T\left[\hat{\phi}(z_1)\hat{\phi}(z_2)\right]|0\rangle = \Delta_F(z_1 - z_2), \tag{10.41}$$

$$\langle 0|T\left[\hat{\phi}(z_3)\hat{\phi}(z_4)\right]|0\rangle = \Delta_F(z_3 - z_4). \tag{10.42}$$

The field operators $\hat{a}_q$ and $\hat{a}_p^\dagger$ act as creation and annihilation operators, respectively, and when contracted with the $\hat{\phi}(z)$ fields, they yield delta functions. For the contraction of $\hat{a}_q$ with $\hat{\phi}(z)$ and $\hat{a}_p^\dagger$ with $\hat{\phi}(z)$, we get:

$$\langle 0|\hat{a}_q\hat{\phi}(z) = \langle 0|e^{-iq\cdot z}, \quad \hat{\phi}(z)\hat{a}_p^\dagger|0\rangle = e^{ip\cdot z}|0\rangle. \tag{10.43}$$

Combining these terms, we get the contribution from the contractions:

$$\mathscr{A}^{(1)} = \frac{-i\lambda}{4!}\int d^4z\,\Delta_F(z-q)\Delta_F(z-p)\langle 0|\hat{a}_q\hat{a}_p^\dagger|0\rangle. \tag{10.44}$$

Finally, we use the vacuum expectation value $\langle 0|\hat{a}_q\hat{a}_p^\dagger|0\rangle = (2\pi)^3\delta^3(\vec{q}-\vec{p})$ to simplify the expression:

$$\mathscr{A}^{(1)} = \frac{-i\lambda(2\pi)^3\delta^3(\vec{q}-\vec{p})}{4!}\int d^4z\,\Delta_F(z-q)\Delta_F(z-p). \tag{10.45}$$

This completes the first-order calculation of the S-matrix element using Wick's theorem.

10.2 WICK CONTRACTIONS IN FEYNMAN DIAGRAMS

Wick's theorem allows us to express time-ordered products of field operators in terms of normal-ordered products and contractions. Each contraction corresponds to a Feynman propagator.

Consider the time-ordered product in ϕ^4 theory:

$$T\left[\phi(x_1)\phi(x_2)\cdots\phi(x_{4n})\right]. \tag{10.46}$$

Using Wick's theorem, this can be written as:

$$T\left[\phi(x_1)\phi(x_2)\cdots\phi(x_{4n})\right] = :\phi(x_1)\phi(x_2)\cdots\phi(x_{4n}): \\ + \text{all possible contractions.} \tag{10.47}$$

Each term in the sum of contractions represents a different way to pair the fields, corresponding to a different Feynman diagram. With each Feynman diagram each interaction vertex corresponds to a factor of $-i\lambda$, and each internal line corresponds to a Feynman propagator. The external lines correspond to the initial and final state particles. Summing over all diagrams gives the total amplitude $\mathscr{A}$.

For example, at the first order in λ, the only diagram is a single vertex with no internal lines. At higher orders, more complex diagrams with loops and multiple vertices appear. The symmetry factor D accounts for the number of equivalent ways to arrange the vertices and internal lines.

This procedure illustrates how the S-matrix in ϕ^4 theory can be systematically expanded and calculated using Feynman diagrams, providing a powerful tool for analyzing interactions in quantum field theory.

10.2.1 EXAMPLE FEYNMAN DIAGRAMS

Here are some example Feynman diagrams illustrating different orders of λ:

FIGURE 10.1
First order in λ: A single vertex with no internal lines.

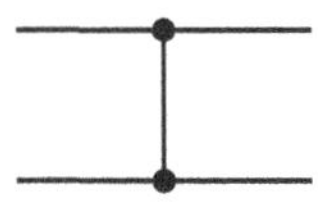

FIGURE 10.2
Second order in λ: Two vertices connected by an internal line.

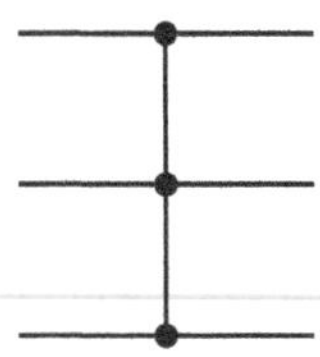

FIGURE 10.3
Third Order in λ: Three vertices with loops and multiple internal lines.

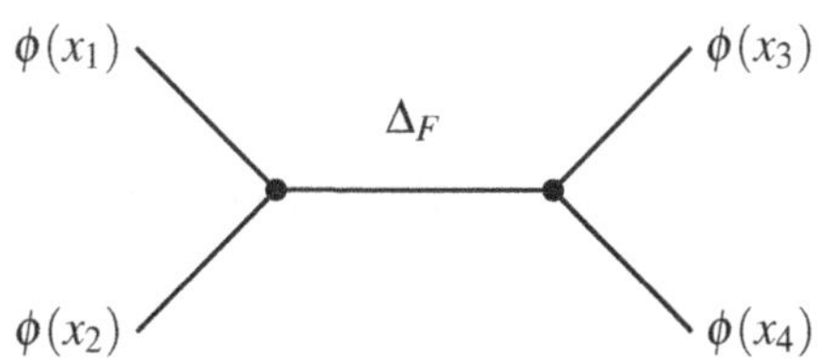

FIGURE 10.4
Contraction $\phi\phi$ with two vertices connected by an internal line.

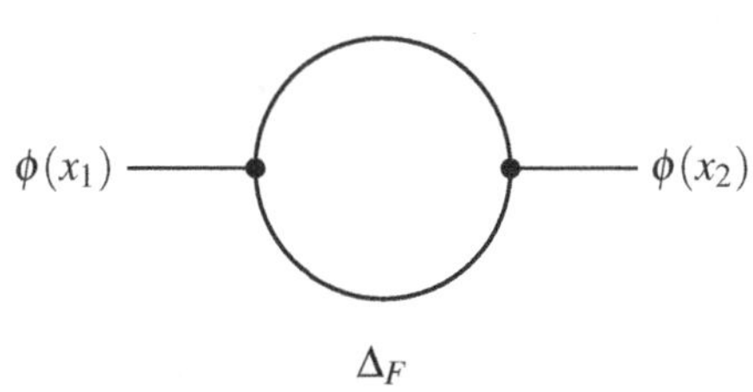

FIGURE 10.5
Single loop diagram showing the contraction of $\phi(x_1)$ and $\phi(x_2)$ with an internal loop.

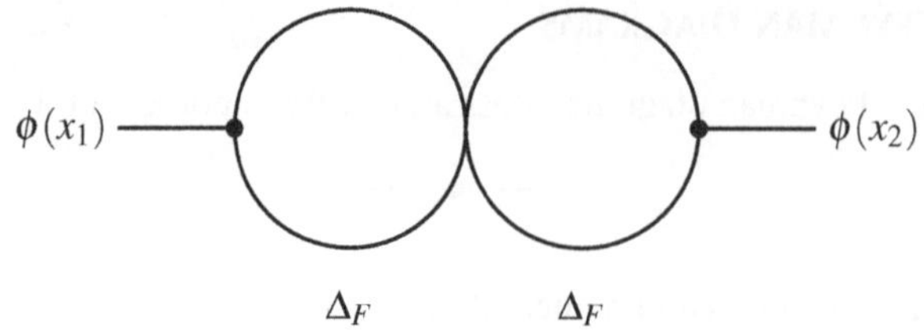

FIGURE 10.6
Double loop diagram showing the contraction of $\phi(x_1)$ and $\phi(x_2)$ with two internal loops.

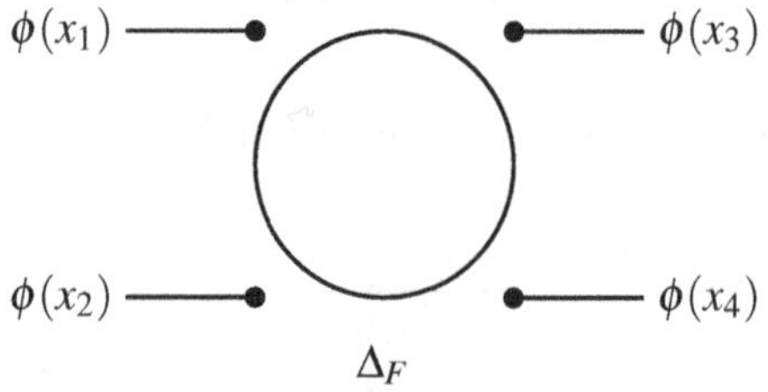

FIGURE 10.7
Single loop diagram with four external lines, showing the contraction of $\phi(x_1)$, $\phi(x_2)$, $\phi(x_3)$, and $\phi(x_4)$.

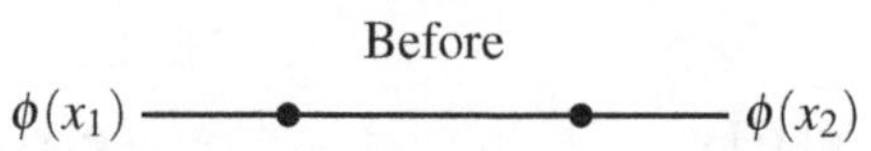

FIGURE 10.8
Diagram before contraction showing the external $\phi(x_1)$ and $\phi(x_2)$ fields.

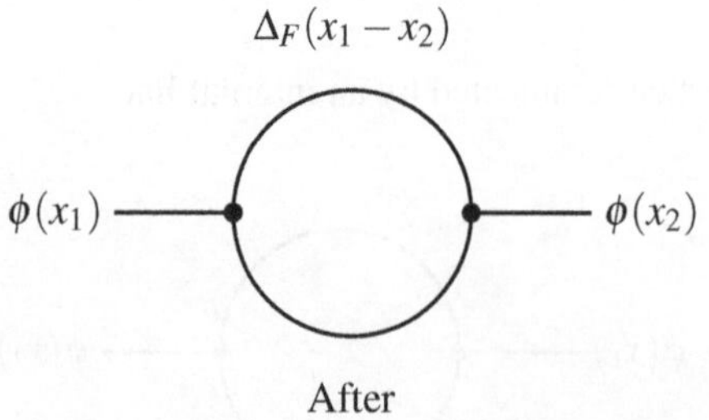

FIGURE 10.9
Diagram after contraction showing the internal loop representing the Feynman propagator $\Delta_F(x_1 - x_2)$.

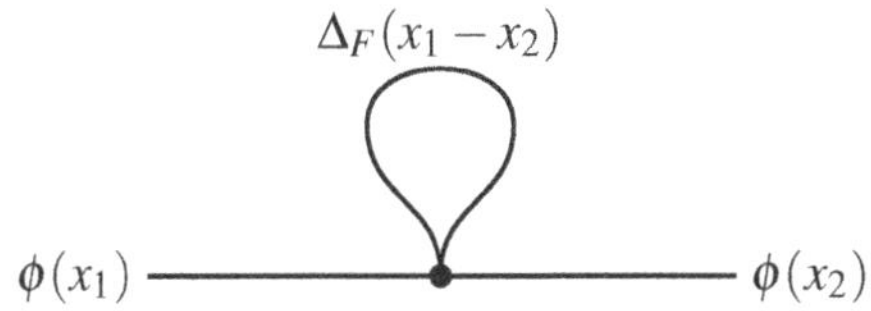

FIGURE 10.10
Alternate way to draw diagram after contraction showing the internal loop (vertical) representing the Feynman propagator $\Delta_F(x_1 - x_2)$.

These diagrams illustrate how the S-matrix in ϕ^4 theory can be systematically expanded and calculated using Feynman diagrams, providing a powerful tool for analyzing interactions in quantum field theory. While the use of Feynman diagrams could be a full text within itself, we at least see here how they are implemented and how Dyson's expansion and Wick's theorem is applied within them.

10.3 FEYNMAN RULES FOR ϕ^4 THEORY IN POSITION SPACE

In ϕ^4 theory, Feynman diagrams provide a visual and computational tool to calculate the S-matrix elements, which represent transition amplitudes between quantum states. Here, we outline the Feynman rules in position space for ϕ^4 theory and provide a worked example to illustrate these rules in action.

10.3.1 FEYNMAN RULES

To compute the S-matrix elements, the following Feynman rules apply:

1. For each internal line (propagator) connecting points x_i and x_j, include a factor of the Feynman propagator:

$$D_F(x_i - x_j) = \int \frac{d^4k}{(2\pi)^4} \frac{e^{-ik\cdot(x_i - x_j)}}{k^2 - m^2 + i\varepsilon}. \tag{10.48}$$

2. For each vertex at point x_i, include a factor of $-i\lambda$.
3. Integrate over the position of each vertex:

$$\int d^4x. \tag{10.49}$$

4. Divide by the symmetry factor D, which accounts for the number of equivalent ways to draw the diagram.

10.3.2 WORKED EXAMPLE

Consider the calculation of the second-order contribution to the S-matrix element for the process $\phi\phi \to \phi\phi$ in ϕ^4 theory. The corresponding Feynman diagram, shown in Figure 10.11, consists of two vertices connected by two internal lines.

The second-order contribution to the S-matrix element is given by:

$$\mathscr{A}^{(2)} = \frac{(-i\lambda)^2}{2!} \int d^4x_1\, d^4x_2\, \langle 0|T\left[\phi(x_1)\phi(x_1)\phi(x_1)\phi(x_1)\phi(x_2)\phi(x_2)\phi(x_2)\phi(x_2)\right]|0\rangle. \tag{10.50}$$

Using Wick's theorem, we expand the time-ordered product into normal-ordered products and contractions. Each contraction corresponds to a Feynman propagator:

$$\langle 0|T\left[\phi(x_1)\phi(x_2)\right]|0\rangle = D_F(x_1 - x_2). \tag{10.51}$$

We need to consider all possible pairwise contractions. For two vertices connected by two internal lines, we have:

$$\mathscr{A}^{(\in)} = \frac{(-i\lambda)^2}{2!}\int d^4x_1\, d^4x_2\, \left[D_F(x_1 - x_2)\right]^2. \tag{10.52}$$

The symmetry factor D for this diagram is 2, as there are two ways to contract the pairs. Thus, the final expression for the second-order contribution to the S-matrix element is:

$$\mathscr{A}^{(2)} = \frac{(-i\lambda)^2}{2}\int d^4x_1\, d^4x_2\, \left[D_F(x_1 - x_2)\right]^2. \tag{10.53}$$

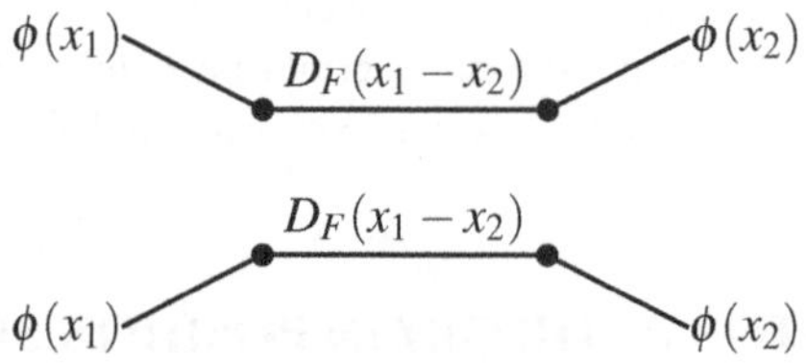

FIGURE 10.11
Feynman diagram for the second-order contribution to the S-matrix element in ϕ^4 theory.

This example demonstrates how the Feynman rules in position space can be used to systematically compute S-matrix elements in ϕ^4 theory.

10.4 FEYNMAN RULES FOR ϕ^4 THEORY IN MOMENTUM SPACE

Feynman rules in momentum space provide a systematic way to compute S-matrix elements for scattering processes in ϕ^4 theory. These rules transform the complex interactions described by the Lagrangian into manageable algebraic terms involving momenta and propagators.

10.4.1 RULES

1. For each internal line representing a propagator with momentum k, include a factor of:

$$\frac{i}{k^2 - m^2 + i\varepsilon} \tag{10.54}$$

This represents the propagator for a scalar particle with mass m.

2. At each vertex, include a factor of:

$$-i\lambda \tag{10.55}$$

where λ is the coupling constant.

3. Conserve momentum at each vertex. This is enforced by including a delta function:

$$(2\pi)^4\delta^4\left(\sum_{\text{incoming}} k - \sum_{\text{outgoing}} k\right) \tag{10.56}$$

4. Integrate over all undetermined internal momenta:

$$\int \frac{d^4k}{(2\pi)^4} \tag{10.57}$$

10.4.2 UNDERSTANDING TREE LEVEL DIAGRAMS

Tree level in the context of quantum field theory refers to the simplest type of diagrams in the perturbative expansion of scattering amplitudes, which do not contain any closed loops. These diagrams are called *tree* diagrams because their structure resembles that of a tree with branches but no loops, resembling a network of lines (representing particle propagators) connecting at vertices without forming any cycles.

Here are some key points about tree level diagrams:

- **Basic Interactions:** Tree level diagrams represent the most fundamental interactions between particles without the modifications due to quantum corrections that loops represent.
- **No Loops:** Unlike loop diagrams (which involve integrations over internal momenta and are associated with higher-order quantum effects), tree diagrams are mathematically simpler and do not involve such integrals. They depend only on the external momenta of the particles involved.
- **Perturbative Expansion:** In perturbation theory, tree diagrams contribute to the leading order term in the expansion. This means they provide the first approximation to the process being considered, and they are often sufficient for processes where the coupling is weak.
- **Amplitude Calculation:** The amplitudes calculated from tree level diagrams are used to determine the lowest order approximation of probabilities for various scattering processes.
- **Physical Interpretation:** Physically, tree level processes are those that could happen in an idealized scenario where the interactions do not produce virtual particle-antiparticle pairs or self-interactions that loop diagrams represent.

In summary, tree level diagrams are crucial for understanding the basic behavior and properties of particles in a given quantum field theory before considering more complex quantum corrections introduced by loop diagrams.

10.4.3 WORKED EXAMPLE: SCATTERING AMPLITUDE AT TREE LEVEL

Consider a simple scattering process where two particles with initial momenta p_1 and p_2 scatter to final states with momenta p_3 and p_4.

Step-by-step calculation:

1. Draw the Feynman diagram. At tree level, this involves a single vertex connecting the four external legs.
2. Apply the Feynman rules:
   - Include a vertex factor of $-i\lambda$.
   - Enforce momentum conservation using the delta function:

$$(2\pi)^4\delta^4(p_1+p_2-p_3-p_4) \tag{10.58}$$

3. The total amplitude $\mathcal{M}$ for the process is:

$$\mathcal{M} = -i\lambda(2\pi)^4\delta^4(p_1+p_2-p_3-p_4) \tag{10.59}$$

This calculation represents the lowest order contribution to the scattering amplitude in ϕ^4 theory. Higher-order corrections involve more loops and vertices, requiring additional integrals and propagators.

10.4.4 SCATTERING AMPLITUDE IN ϕ^4 THEORY

Consider a ϕ^4 theory where two scalar particles scatter into two scalar particles. Calculate the lowest-order contribution to the scattering amplitude (S-matrix element) for the process $\phi+\phi\rightarrow$

$\phi + \phi$ in terms of the coupling constant λ. In ϕ^4 theory, the interaction term in the Lagrangian is given by:

$$\mathscr{L}_{\text{int}} = -\frac{\lambda}{4!}\phi^4. \tag{10.60}$$

The S-matrix element for the scattering process can be computed using the Feynman rules. At the lowest order (tree level), the relevant Feynman diagrams are those with a single vertex involving the quartic interaction.

STEP-BY-STEP CALCULATION

Identify the Feynman Diagram:

The lowest-order contribution comes from a single vertex where four external lines (two incoming and two outgoing) meet. This corresponds to a tree-level diagram with no internal lines.

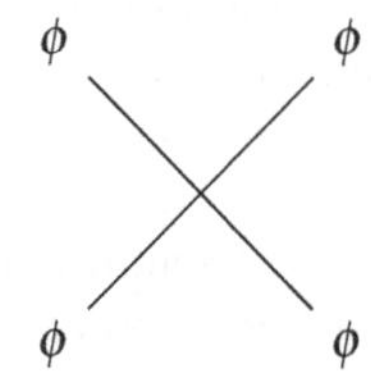

FIGURE 10.12
Tree-level Feynman diagram for $\phi + \phi \to \phi + \phi$ in ϕ^4 theory.

Apply the Feynman Rules:

According to the Feynman rules for ϕ^4 theory in position space: Each vertex contributes a factor of $-i\lambda$. Each external line contributes a factor of 1.

Thus, for the single-vertex diagram, the amplitude is:

$$\mathscr{M} = -i\lambda. \tag{10.61}$$

Express the S-matrix Element:

The S-matrix element for the process $\phi + \phi \to \phi + \phi$ is then given by:

$$S_{fi} = \delta^4(p_1 + p_2 - p_3 - p_4)(-i\lambda), \tag{10.62}$$

where $\delta^4(p_1 + p_2 - p_3 - p_4)$ ensures energy-momentum conservation at the vertex, with p_1 and p_2 being the momenta of the incoming particles, and p_3 and p_4 being the momenta of the outgoing particles.

The lowest-order contribution to the scattering amplitude (S-matrix element) for the process $\phi + \phi \to \phi + \phi$ in ϕ^4 theory is:

$$S_{fi} = \delta^4(p_1 + p_2 - p_3 - p_4)(-i\lambda). \tag{10.63}$$

This result shows how the interaction term in the ϕ^4 theory contributes to the scattering process at the lowest order, illustrating the application of Feynman rules to calculate the S-matrix element.

10.5 FEYNMAN DIAGRAM AMPLITUDES AND THE SYMMETRY FACTOR

The amplitude for a given Feynman diagram in ϕ^4 theory can be computed using the Feynman rules. The symmetry factor D accounts for the number of ways the diagram can be drawn without changing its topology, ensuring the correct normalization of probabilities in quantum field theory calculations.

10.5.1 DETAILED CALCULATION OF A FEYNMAN DIAGRAM AMPLITUDE

Consider a simple one-loop diagram in ϕ^4 theory. The loop diagram typically represents higher-order interactions and includes corrections to particle propagators and vertices. Below is the Feynman diagram for this process:

Mathematical Formulation

The amplitude for a one-loop diagram is mathematically expressed as:

$$i\mathcal{M} = \int \frac{d^4k}{(2\pi)^4} \left(\frac{i}{k^2 - m^2 + i\varepsilon} \right)^2 (-i\lambda)^2 \frac{1}{D}. \tag{10.64}$$

Here, $\frac{i}{k^2-m^2+i\varepsilon}$ represents the propagator for a scalar particle of mass m, integrating over all possible internal momenta k. The factor $(-i\lambda)^2$ corresponds to the strength of the interaction at the vertices, and $\frac{1}{D}$ is the symmetry factor for the diagram.

In this diagram, the loop represents a virtual particle, and its contribution is integrated over all possible momenta. The diagrams symmetry factor ensures that identical configurations are not overcounted.

Worked Example: Loop Integration

To compute the loop integral, we simplify the expression using dimensional regularization or a similar technique to manage divergences associated with the loop:

Dimensional regularization involves the following steps:

- **Alter the Number of Dimensions**: Generalize the integral to $d = 4 - \varepsilon$ dimensions.
- **Perform the Integral in d Dimensions**: The integrals that are divergent in four dimensions may become convergent in d dimensions.
- **Analytically Continue**: After performing the integral in d dimensions, we analytically continue the result back to four dimensions by taking the limit as ε approaches zero.

Using dimensional regularization, we generalize the loop integral to $d = 4 - \varepsilon$ dimensions:

$$i\mathcal{M} = \frac{-\lambda^2}{D} \int \frac{d^dk}{(2\pi)^d} \frac{1}{(k^2 - m^2 + i\varepsilon)^2}. \tag{10.65}$$

This integral can be evaluated under the appropriate regularization scheme to yield a finite result, essential for physical predictions.

The integral typically diverges, and regularization techniques, such as dimensional regularization, are used to handle these divergences and obtain meaningful physical results. The computed amplitude accounts for the loop correction to the propagator, modifying the field's behavior.

SYMMETRY FACTOR CALCULATION

Symmetry factors are crucial for ensuring that each diagram is correctly weighted in the sum over all diagrams in a perturbative expansion.

General Approach

To find the symmetry factor for a given diagram, count the number of ways the diagram can be rearranged without changing its topology. This includes permutations of identical vertices and internal lines.

Example: Symmetry Factor for a One-Loop Diagram

For the one-loop diagram in ϕ^4 theory shown in Figure 10.13, the symmetry factor D is 2. This factor accounts for the indistinguishable ways to draw the loop connected to the vertex:

FIGURE 10.13
One-loop diagram in ϕ^4 theory.

- **Identical Interactions**: The loop can be attached to the vertex in two indistinguishable ways, reflecting the symmetry of the vertex interaction. Rotating the loop around the vertex does not create a new, distinguishable configuration.
- **Factor Calculation**: The total number of equivalent ways to draw the diagram is 2. Hence, the symmetry factor D is 2.

Including the symmetry factor ensures that each Feynman diagram's contribution is accurately represented in the overall calculation, preventing overcounting and maintaining the consistency of the perturbative expansion.

TWO CONFIGURATIONS OF A ONE-LOOP DIAGRAM IN ϕ^4 THEORY

In ϕ^4 theory, the one-loop diagram can be drawn in two indistinguishable ways due to the symmetry around the vertex. These configurations are shown below:

EXPLANATION

In the one-loop diagram of ϕ^4 theory shown in Figures 10.14 and 10.15, the loop can be attached to the vertex in two rotationally symmetric ways. These configurations represent different orientations around the vertex:

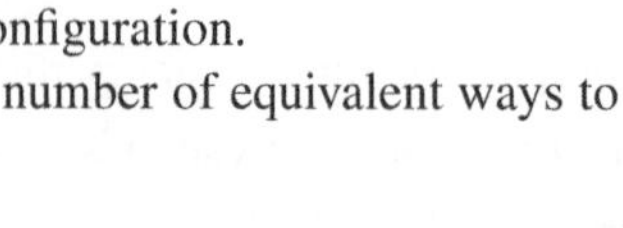

FIGURE 10.14
First configuration of the one-loop diagram in ϕ^4 theory with clockwise rotation.

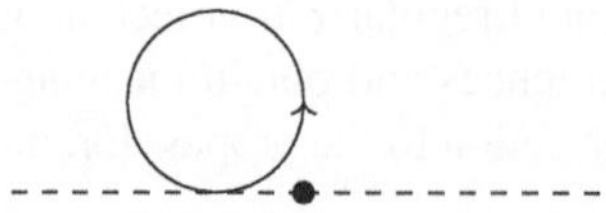

FIGURE 10.15
Second configuration of the one-loop diagram in ϕ^4 theory with counterclockwise rotation.

- **Configuration 1**: The loop rotates clockwise around the vertex.
- **Configuration 2**: The loop rotates counterclockwise around the vertex.

These configurations are not visually distinct in terms of topology, but they represent different rotational orientations. This symmetry means that the diagram can be drawn in two equivalent ways, and the symmetry factor D accounts for these equivalent configurations.

The symmetry factor $D = 2$ ensures that these two configurations are not overcounted in the perturbative expansion. This factor reflects the rotational symmetry around the vertex, indicating that there are two indistinguishable ways to draw the loop in the one-loop diagram.

CLARIFICATION ON THE DIAGRAM TOPOLOGY AND THE SYMMETRY FACTOR

In ϕ^4 theory, the one-loop diagram can be drawn in ways that show rotational symmetry around the vertex. Below are the configurations with rotation arrows:

EXPLANATION

In the one-loop diagram of ϕ^4 theory shown in Figures 10.16 through 10.23, the loop can be attached to the vertex in different ways. These configurations represent different orientations and positions around the vertex:

- **Configuration 1 (Fig. 10.16)**: The loop is on the top left side with counterclockwise rotation.
- **Configuration 2 (Fig. 10.17)**: The loop is on the top left side with clockwise rotation.
- **Configuration 3 (Fig. 10.18)**: The loop is on the bottom left side with counterclockwise rotation.
- **Configuration 4 (Fig. 10.19)**: The loop is on the bottom left side with clockwise rotation.
- **Configuration 5 (Fig. 10.20)**: The loop is on the top right with counterclockwise rotation.
- **Configuration 6 (Fig. 10.21)**: The loop is on the top right with clockwise rotation.
- **Configuration 7 (Fig. 10.22)**: The loop is on the bottom right with counterclockwise rotation.
- **Configuration 8 (Fig. 10.23)**: The loop is on the bottom right with clockwise rotation.

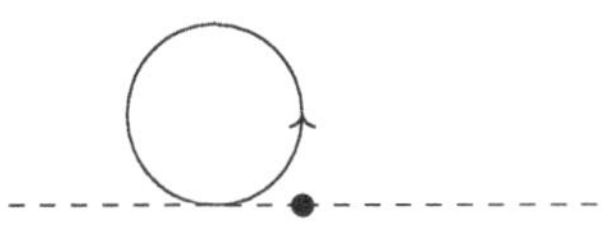

FIGURE 10.16
Loop on the top left with counterclockwise rotation.

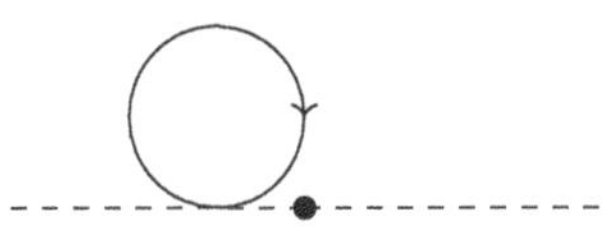

FIGURE 10.17
Loop on the top left with clockwise rotation.

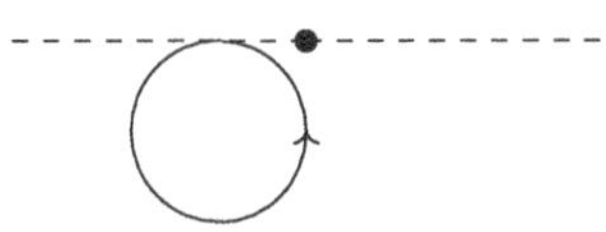

FIGURE 10.18
Loop on the bottom left with counterclockwise rotation.

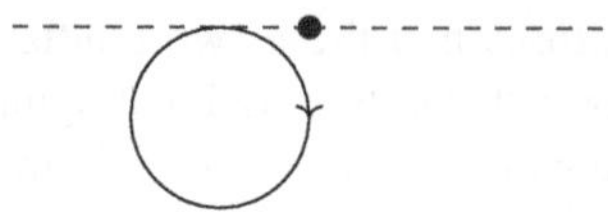

FIGURE 10.19
Loop on the bottom left with clockwise rotation.

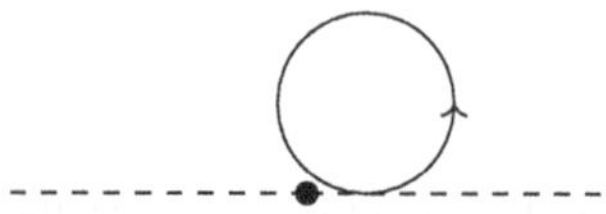

FIGURE 10.20
Loop on the top right with counterclockwise rotation.

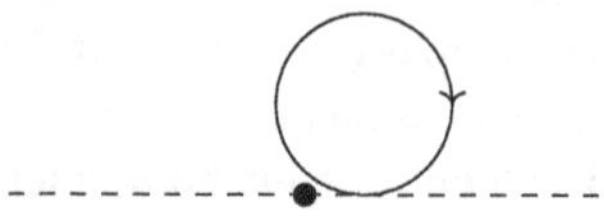

FIGURE 10.21
Loop on the top right with clockwise rotation.

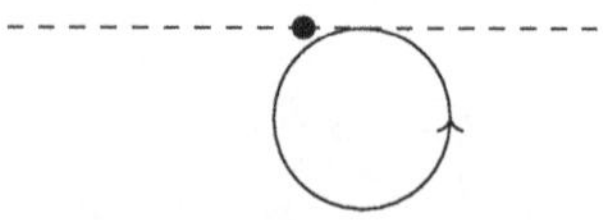

FIGURE 10.22
Loop on the bottom right with counterclockwise rotation.

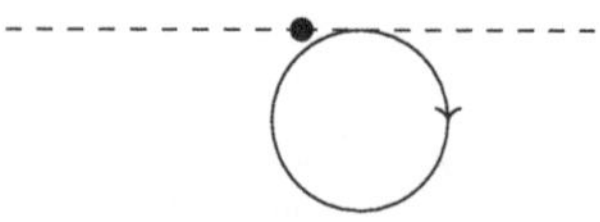

FIGURE 10.23
Loop on the bottom right with clockwise rotation.

However, due to the symmetry around the vertex, configurations 1 and 3 are equivalent, as are configurations 2 and 4. Rotating the entire diagram does not result in a distinct physical configuration. Thus, we only have two unique configurations:

- **Unique Configuration 1**: Loop with clockwise rotation.
- **Unique Configuration 2**: Loop with counterclockwise rotation.

The symmetry factor $D = 2$ accounts for these equivalent configurations, ensuring they are not overcounted in the perturbative expansion. This reflects the rotational symmetry around the vertex, indicating that there are two indistinguishable ways to draw the loop in the one-loop diagram.

By understanding the mathematical formulation and the significance of the symmetry factor, we can accurately calculate the amplitudes of Feynman diagrams in ϕ^4 theory. The one-loop diagram, though simple, encapsulates complex quantum effects and corrections, illustrating the intricate behavior of scalar fields and their interactions.

10.5.2 THE SIGNIFICANCE OF THE TADPOLE DIAGRAM IN QUANTUM FIELD THEORY

In quantum field theory (QFT), Feynman diagrams are used to visually represent interactions between particles. Among these, the tadpole diagram is one of the simplest yet most fundamental components. Despite its simplicity, the tadpole diagram plays a crucial role in understanding various aspects of field theory, particularly in scalar field theories such as ϕ^4 theory.

Description of the Tadpole Diagram

A tadpole diagram typically consists of:

1. **One external line**: This represents an incoming or outgoing particle.
2. **One vertex**: The point where the interaction occurs.
3. **A loop**: This loop, formed by an internal line that begins and ends at the same vertex, represents a virtual particle propagating.

The name "tadpole" comes from its visual resemblance to a tadpole, with the loop representing the head and the external line representing the tail.

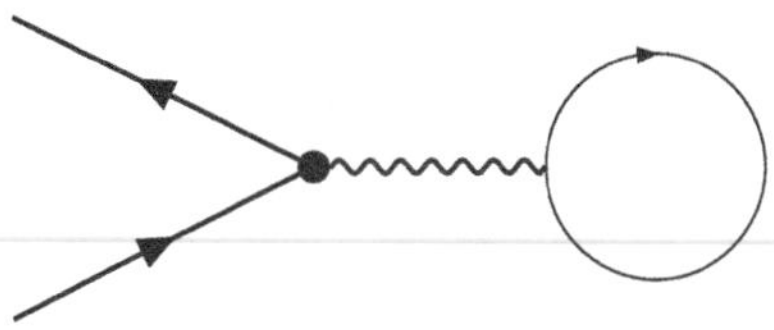

FIGURE 10.24
Tadpole diagram with a loop and external lines.

10.5.2.1 Mathematical Representation

In ϕ^4 theory, the tadpole diagram corresponds to a specific term in the perturbative expansion of the field's propagator. Mathematically, the amplitude $\mathcal{M}$ for a tadpole diagram can be expressed as:

$$i\mathcal{M} = \int \frac{d^4k}{(2\pi)^4} \frac{i}{k^2 - m^2 + i\varepsilon}(-i\lambda)\frac{1}{D} \tag{10.66}$$

Here:

- $\frac{i}{k^2-m^2+i\varepsilon}$ is the propagator for the scalar particle.
- $-i\lambda$ is the interaction vertex factor, where λ is the coupling constant.
- D is the symmetry factor, which for a tadpole diagram is typically 2, accounting for the indistinguishable ways the loop can be attached to the vertex.

Given the symmetry factor $D = 2$, the amplitude becomes:

$$i\mathcal{M} = \frac{-i\lambda}{2} \int \frac{d^4k}{(2\pi)^4} \frac{i}{k^2 - m^2 + i\varepsilon} \tag{10.67}$$

Physical Interpretation

1. **Quantum Corrections**: The tadpole diagram represents a first-order quantum correction to the propagator of the field. It indicates how the presence of virtual particles (represented by the loop) can alter the behavior of a real particle propagating through space.

2. **Mass Renormalization**: In scalar field theories, the tadpole diagram often contributes to the renormalization of the field's mass. The loop correction modifies the mass term in the field's propagator, leading to a shift in the physical mass observed experimentally.
3. **Vacuum Expectation Value (VEV)**: Tadpole diagrams are also relevant when considering the vacuum expectation value of a field. In theories with spontaneous symmetry breaking, such as the Higgs mechanism in the Standard Model, tadpole diagrams can indicate shifts in the VEV of the field, which are crucial for understanding phenomena like mass generation for gauge bosons.
4. **Gauge Invariance and Regularization**: The contribution of tadpole diagrams must be carefully considered to maintain gauge invariance in gauge theories. Additionally, the integrals involved often require regularization techniques (e.g., dimensional regularization) to handle infinities and ensure finite, physically meaningful results.

Example: Tadpole Diagram in ϕ^4 Theory

Consider a scalar field theory with a quartic interaction term. The Lagrangian is given by:

$$\mathscr{L} = \frac{1}{2}(\partial_\mu \phi)(\partial^\mu \phi) - \frac{1}{2}m^2\phi^2 - \frac{\lambda}{4!}\phi^4 \tag{10.68}$$

The tadpole diagram in this context provides a loop correction to the propagator. Despite its seemingly simple structure, it encapsulates complex quantum effects, showcasing how virtual particles and self-interactions influence the dynamics of the field.

10.5.2.1.1 Mathematical Description

The amplitude for the tadpole diagram is given by:

$$i\mathscr{M} = \int \frac{d^4k}{(2\pi)^4} \frac{i}{k^2 - m^2 + i\varepsilon}(-i\lambda)\frac{1}{D}. \tag{10.69}$$

where:

- $\frac{i}{k^2-m^2+i\varepsilon}$ is the propagator for the scalar particle,
- $-i\lambda$ is the interaction vertex factor,
- $\frac{1}{D}$ is the symmetry factor.

Given the symmetry factor $D = 2$, the amplitude becomes:

$$i\mathscr{M} = \frac{-i\lambda}{2} \int \frac{d^4k}{(2\pi)^4} \frac{i}{k^2 - m^2 + i\varepsilon}. \tag{10.70}$$

This detailed examination clarifies the role of symmetry factors in adjusting the contributions of diagrams with symmetrical configurations. It also highlights the importance of considering all equivalent configurations when calculating Feynman diagram amplitudes in quantum field theory.

This approach can be generalized to more complex diagrams, ensuring accurate predictions in quantum field theory. By applying these rules, we can systematically compute the amplitudes for various Feynman diagrams in ϕ^4 theory, providing insights into the interactions and dynamics of the scalar field.

10.6 SUMMARY AND DISCUSSION

10.6.1 SUMMARY

In this chapter, we have delved into the intricacies of ϕ^4 theory and the application of Feynman diagrams to calculate S-matrix elements. Key points covered include:

- ϕ^4 theory, characterized by a scalar field with a quartic interaction term, serves as a fundamental model for understanding quantum field interactions.
- The Lagrangian of ϕ^4 theory provides the foundation for deriving the equations of motion and interaction terms, crucial for constructing Feynman diagrams.
- The S-matrix formalism is essential for describing the transition amplitudes between quantum states, encapsulating all possible scattering processes.
- Feynman rules in both position and momentum space were outlined, illustrating how to systematically compute the contributions of different diagrams to the S-matrix elements.
- Tree level diagrams, representing the simplest interactions without loops, were discussed along with their importance in perturbative expansions.
- Detailed calculations of one-loop diagrams in ϕ^4 theory were presented, highlighting the role of symmetry factors in ensuring correct normalization of probabilities.
- The tadpole diagram was analyzed in depth, demonstrating its significance in quantum corrections, mass renormalization, and vacuum expectation values.

10.6.2 DISCUSSION

The study of ϕ^4 theory through Feynman diagrams provides a foundational understanding of the principles and techniques used in quantum field theory. While ϕ^4 theory itself is a simplified model, the methodologies developed here are widely applicable to more complex theories, including the Standard Model of particle physics.

- **Quantum Corrections and Renormalization:** The inclusion of loop diagrams, such as the tadpole and one-loop diagrams, demonstrates how quantum corrections can modify the behavior of fields. These corrections are crucial for accurate predictions and are managed through renormalization techniques.
- **Symmetry Considerations:** Symmetry factors play a vital role in ensuring the proper weighting of diagrams in perturbative expansions. Understanding these factors helps prevent overcounting of equivalent configurations, maintaining the consistency and accuracy of calculations.
- **Practical Applications:** While ϕ^4 theory is primarily a theoretical tool, the techniques and insights gained from studying it are directly applicable to more realistic field theories. For instance, the discovery of the Higgs boson and the precision predictions in quantum electrodynamics (QED) and quantum chromodynamics (QCD) are built upon the same fundamental principles.
- **Educational Value:** For students and researchers new to quantum field theory, ϕ^4 theory offers an accessible entry point. By mastering the calculations and concepts in this simpler context, one is better prepared to tackle the complexities of more advanced theories.

ϕ^4 theory and the use of Feynman diagrams provide a powerful framework for exploring the quantum interactions of fields. The techniques developed here are not only essential for theoretical understanding but also form the backbone of practical calculations in modern particle physics. As we continue to probe the fundamental nature of matter, the principles and methods outlined in this chapter will remain indispensable tools in the physicist's toolkit.

10.7 QUESTIONS AND PROBLEMS

10.7.1 TRIVIA QUESTIONS

Question 1: What is the interaction term in the Lagrangian for ϕ^4 theory?

(a) $\frac{1}{2}(\partial_\mu\phi)(\partial^\mu\phi)$
(b) $\frac{\lambda}{4!}\phi^4$
(c) $\frac{1}{2}m^2\phi^2$

(d) $\frac{\lambda}{3!}\phi^3$

Answer: b. $\frac{\lambda}{4!}\phi^4$
Question 2: In ϕ^4 theory, what does the S-matrix describe?

(a) The energy levels of particles
(b) The transition amplitudes between quantum states
(c) The mass of scalar particles
(d) The speed of light in a vacuum

Answer: b. The transition amplitudes between quantum states
Question 3: What does a tree level diagram in quantum field theory represent?

(a) Diagrams with at least one loop
(b) The most fundamental interactions without quantum corrections
(c) Diagrams with external lines only
(d) The highest order term in perturbative expansion

Answer: b. The most fundamental interactions without quantum corrections
Question 4: Which technique is commonly used to handle divergences in loop integrals?

(a) Fourier transform
(b) Laplace transform
(c) Dimensional regularization
(d) Matrix diagonalization

Answer: c. Dimensional regularization
Question 5: In the context of Feynman diagrams, what does the symmetry factor account for?

(a) The mass of the particles
(b) The energy levels of the system
(c) The number of equivalent ways to draw the diagram
(d) The speed of light in the system

Answer: c. The number of equivalent ways to draw the diagram
Question 6: What is the significance of the tadpole diagram in quantum field theory?

(a) It represents the propagation of a photon
(b) It indicates quantum corrections and mass renormalization
(c) It describes the decay of particles
(d) It shows the classical path of particles

Answer: b. It indicates quantum corrections and mass renormalization
Question 7: What does the factor $-i\lambda$ represent in a Feynman diagram?

(a) The propagator for a scalar particle
(b) The interaction vertex factor
(c) The mass term for the field
(d) The kinetic energy term

Answer: b. The interaction vertex factor
Question 8: How is the S-matrix element for the process $\phi + \phi \to \phi + \phi$ expressed at the lowest order in ϕ^4 theory?

(a) $S_{fi} = -i\lambda\delta^4(p_1+p_2-p_3-p_4)$
(b) $S_{fi} = i\lambda\delta^4(p_1+p_2-p_3-p_4)$
(c) $S_{fi} = -\lambda\delta^4(p_1+p_2-p_3-p_4)$
(d) $S_{fi} = \lambda\delta^4(p_1+p_2-p_3-p_4)$

Answer: a. $S_{fi} = -i\lambda\delta^4(p_1+p_2-p_3-p_4)$
Question 9: What does the tadpole diagram typically consist of in ϕ^4 theory?

(a) Two external lines and a loop
(b) One external line, one vertex, and a loop
(c) Three external lines and a vertex
(d) Four external lines and no loops

Answer: b. One external line, one vertex, and a loop
Question 10: What does the term $\frac{i}{k^2-m^2+i\varepsilon}$ in a Feynman diagram represent?

(a) The interaction vertex factor
(b) The propagator for a scalar particle
(c) The symmetry factor
(d) The external momentum

Answer: b. The propagator for a scalar particle

10.7.2 PROBLEMS

1. **Problem 1: Computing a Tree Level Scattering Amplitude**
Compute the tree-level scattering amplitude for the process $\phi+\phi \to \phi+\phi$ in ϕ^4 theory.
Solution:
The tree-level diagram for this process involves a single vertex with four external lines. The Feynman rules give us:
   - A vertex factor of $-i\lambda$.
   - A delta function for momentum conservation: $(2\pi)^4\delta^4(p_1+p_2-p_3-p_4)$.

Therefore, the amplitude is:

$$\mathcal{M} = -i\lambda(2\pi)^4\delta^4(p_1+p_2-p_3-p_4) \tag{10.71}$$

2. **Problem 2: One-Loop Correction to the Propagator**
Calculate the one-loop correction to the propagator in ϕ^4 theory.
Solution:
The one-loop diagram in ϕ^4 theory is given by:

The amplitude for this diagram is:

$$i\mathcal{M} = \int \frac{d^4k}{(2\pi)^4}\left(\frac{i}{k^2-m^2+i\varepsilon}\right)^2(-i\lambda)^2\frac{1}{D} \tag{10.72}$$

Since the symmetry factor D is 2, we get:

$$i\mathcal{M} = \frac{-i\lambda^2}{2}\int \frac{d^4k}{(2\pi)^4}\frac{1}{(k^2-m^2+i\varepsilon)^2} \tag{10.73}$$

Using dimensional regularization to evaluate the integral, we obtain a finite result which represents the one-loop correction to the propagator.

3. **Problem 3: Symmetry Factor Calculation**
Calculate the symmetry factor for the following diagram in ϕ^4 theory:

For a more detailed set of possible Feynman diagram configurations see Figures 10.16 through 10.23.
Solution:
To find the symmetry factor, we need to count the number of ways to draw the diagram without changing its topology. For this one-loop diagram, there are two ways to attach the loop to the vertex (clockwise and counterclockwise rotations):

$$D = 2 \tag{10.74}$$

4. **Problem 4: Tadpole Diagram Calculation**
Calculate the amplitude for the tadpole diagram in ϕ^4 theory.
Solution:
The tadpole diagram is represented by:

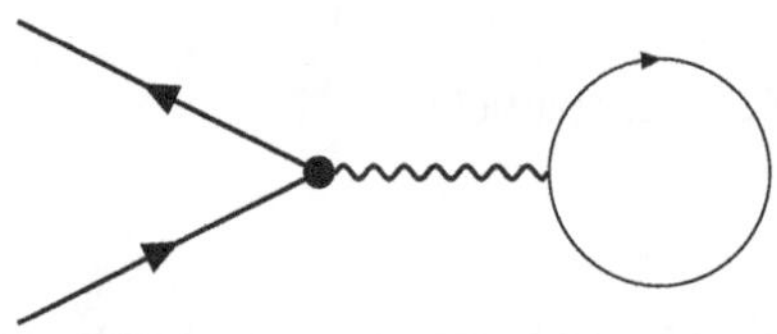

The amplitude is given by:

$$i\mathcal{M} = \int \frac{d^4k}{(2\pi)^4} \frac{i}{k^2 - m^2 + i\varepsilon}(-i\lambda)\frac{1}{D} \tag{10.75}$$

The symmetry factor D for the tadpole diagram is 2:

$$i\mathcal{M} = \frac{-i\lambda}{2} \int \frac{d^4k}{(2\pi)^4} \frac{i}{k^2 - m^2 + i\varepsilon} \tag{10.76}$$

5. **Problem 5: Scattering Amplitude at Second Order**
Calculate the second-order contribution to the S-matrix element for the process $\phi\phi \to \phi\phi$ in ϕ^4 theory.
Solution:
The second-order diagram consists of two vertices connected by two internal lines:

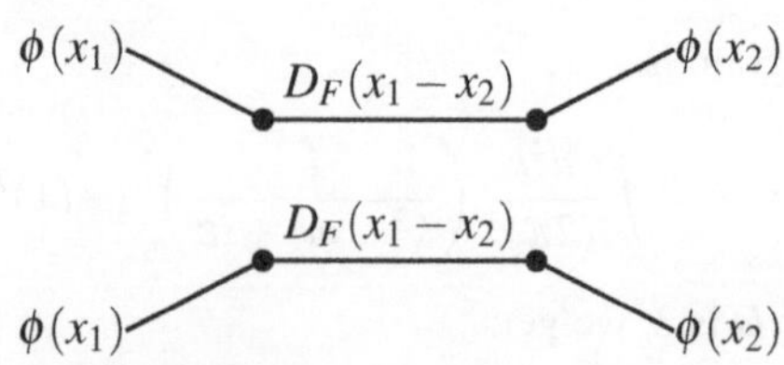

The amplitude for the second-order contribution is:

$$\mathscr{A}^{(2)} = \frac{(-i\lambda)^2}{2!} \int d^4x_1\, d^4x_2\, [D_F(x_1 - x_2)]^2 \tag{10.77}$$

Using the Feynman rules:

$$D_F(x_1 - x_2) = \int \frac{d^4k}{(2\pi)^4} \frac{e^{-ik\cdot(x_1-x_2)}}{k^2 - m^2 + i\varepsilon} \tag{10.78}$$

Including the symmetry factor $D = 2$:

$$\mathscr{A}^{(2)} = \frac{(-i\lambda)^2}{2} \int d^4x_1\, d^4x_2 \left[\int \frac{d^4k}{(2\pi)^4} \frac{e^{-ik\cdot(x_1-x_2)}}{k^2 - m^2 + i\varepsilon} \right]^2 \tag{10.79}$$

This integral can be evaluated to obtain the second-order contribution to the S-matrix element.

BONUS PROBLEM

Problem: Draw a Feynman diagram for the interaction where an electron emits a photon and transitions to a lower energy state.

Solution:

1. Identify the particles involved:

- - Initial electron (e^-) with higher energy.
- - Final electron (e^-) with lower energy.
- - Photon (γ) emitted in the process.

2. Draw the Feynman diagram:

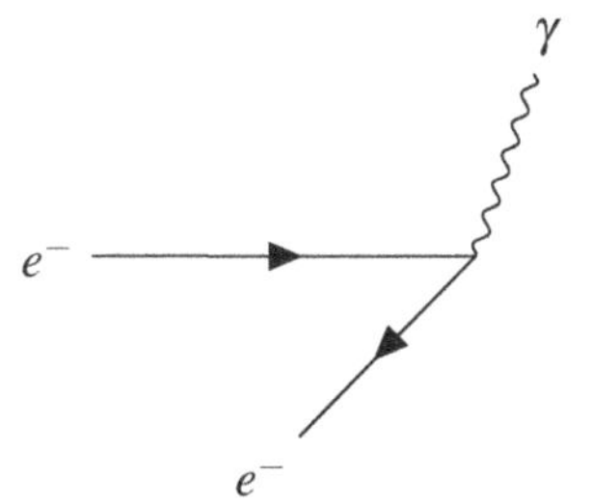

(10.80)

This diagram represents an electron transitioning to a lower energy state by emitting a photon. The initial electron is shown as a fermion line entering the vertex, the photon is emitted from the vertex, and the final electron exits the vertex with lower energy.

11 Lecture 21: Other Scattering Theories

In this chapter, we explore several key scattering processes that play significant roles in understanding the interactions between fundamental particles. These processes, described by different scattering theories, provide deep insights into the nature of forces and interactions at the quantum level. We will cover Yukawa scattering, which describes the interaction between nucleons mediated by mesons, Bhabha scattering, which details electron-positron interactions in quantum electrodynamics (QED), and Møller scattering, which focuses on electron-electron scattering. Each of these theories is crucial for different aspects of particle physics, from nuclear forces to electromagnetic interactions, and they highlight the diverse mechanisms through which particles interact. Through detailed examples and calculations, we will gain a comprehensive understanding of these scattering processes and their implications in the field of quantum mechanics.

11.1 YUKAWA SCATTERING

Yukawa scattering describes the interaction between nucleons mediated by a meson, typically a pion. This theory was one of the earliest to propose a particle exchange mechanism for the strong nuclear force.

11.1.1 EXAMPLE

Consider a proton and neutron scattering through pion exchange. The Yukawa potential $V(r)$ is given by:

$$V(r) = -g^2 \frac{e^{-\mu r}}{r} \tag{11.1}$$

where g is the coupling constant, μ is the mass of the exchanged meson, and r is the distance between the nucleons.

The Yukawa potential arises from the exchange of a meson between two nucleons. The form of the potential can be derived from the propagator of the meson field in position space.

To compute the scattering amplitude in Yukawa theory, we need to use the propagator for the exchanged pion. The Feynman diagram for this process is shown below:

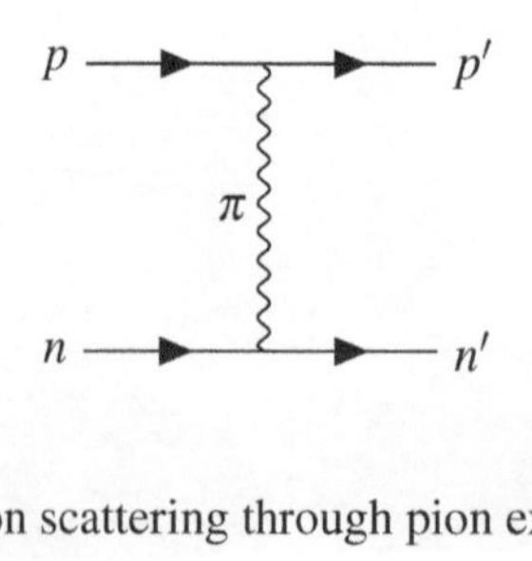

FIGURE 11.1
Feynman diagram for proton-neutron scattering through pion exchange.

11.1.2 YUKAWA'S $\psi^\dagger \psi \phi$ THEORY

Yukawa's $\psi^\dagger \psi \phi$ theory describes the interaction between psions (ψ) mediated by scalar field phions (ϕ) as shown in Figure 11.2. In the context of quantum field theory, the fields ψ and $\psi^\dagger$ are often

DOI: 10.1201/9781003637370-11

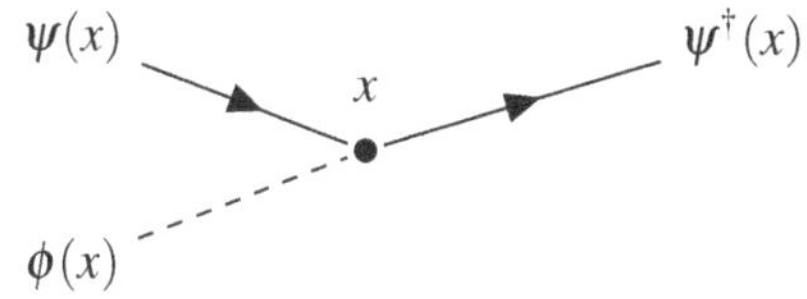

FIGURE 11.2
Feynman diagram for $\psi^\dagger\psi\phi$ interaction at vertex x.

referred to as fermionic fields, while ϕ is referred to as a bosonic field. Fermionic fields, such as ψ and $\psi^\dagger$, follow Fermi-Dirac statistics and obey the Pauli exclusion principle. These fields are typically used to describe particles with half-integer spin, such as electrons and quarks.

On the other hand, the bosonic field ϕ follows Bose-Einstein statistics, allowing multiple bosons to occupy the same state. Bosons are particles with integer spin, and they often act as force carriers in quantum field theories, such as the photon in electromagnetism or the pion in the strong interaction. In Yukawa's $\psi^\dagger\psi\phi$ theory, the interaction between the fermionic fields ψ and $\psi^\dagger$ is mediated by the exchange of the bosonic field ϕ, providing a mechanism for forces between fermions via boson exchange, analogous to how the electromagnetic force is mediated by photons between charged particles. This theory is crucial in understanding particle interactions involving fermions and mesons.

11.1.2.1 Lagrangian

The Lagrangian for Yukawa's $\psi^\dagger\psi\phi$ theory is given by:

$$\mathscr{L} = \bar{\psi}(i\gamma^\mu\partial_\mu - m_\psi)\psi + \frac{1}{2}(\partial_\mu\phi\partial^\mu\phi - \mu^2\phi^2) - g\bar{\psi}\psi\phi \tag{11.2}$$

where:

- ψ is the fermion field,
- ϕ is the scalar field,
- m_ψ is the mass of the fermion,
- μ is the mass of the scalar field,
- g is the coupling constant.

The fermionic field $\psi(x)$ is:

$$\psi(x) = \int \frac{d^3k}{(2\pi)^{3/2}}\left(a_k u_k e^{-ik\cdot x} + b_k^\dagger v_k e^{ik\cdot x}\right) \tag{11.3}$$

The conjugate fermionic field $\psi^\dagger(x)$ is:

$$\psi^\dagger(x) = \int \frac{d^3k}{(2\pi)^{3/2}}\left(a_k^\dagger u_k^\dagger e^{ik\cdot x} + b_k v_k^\dagger e^{-ik\cdot x}\right) \tag{11.4}$$

The bosonic field $\phi(x)$ is:

$$\phi(x) = \int \frac{d^3k}{(2\pi)^{3/2}}\frac{1}{\sqrt{2E_k}}\left(c_k e^{-ik\cdot x} + c_k^\dagger e^{ik\cdot x}\right) \tag{11.5}$$

These fields very closely resemble the interactions found in QED. The operators a and $a^\dagger$ are annihilation and creation operators, respectively, for fermionic fields. They obey the anticommutation relations $\{a_k, a_{k'}^\dagger\} = \delta^3(k-k')$. Similarly, the operators b and $b^\dagger$ are annihilation and creation operators for antifermionic fields, also obeying anticommutation relations. These operators are essential for quantizing fermionic fields and are used in defining the states and dynamics of the fermions in the field theory.

11.1.3 MATRIX ELEMENT CALCULATION

To calculate $\langle q|\hat{S}|p\rangle$, we start with the S-matrix operator in the interaction picture:

$$\hat{S} = T\exp\left(-i\int d^4x\,\mathscr{H}_{\text{int}}(x)\right), \tag{11.6}$$

where $\mathscr{H}_{\text{int}}(x)$ is the interaction Hamiltonian density.

For Yukawa's $\psi^\dagger\psi\phi$ theory, the interaction Hamiltonian density is:

$$\mathscr{H}_{\text{int}}(x) = g\psi^\dagger(x)\psi(x)\phi(x). \tag{11.7}$$

Expanding the S-matrix using Dyson's series gives:

$$\hat{S} = T\left[1 - i\int d^4x\,\mathscr{H}_{\text{int}}(x) + \frac{(-i)^2}{2!}\int d^4x_1\,d^4x_2 \right.$$
$$\left. T\left[\mathscr{H}_{\text{int}}(x_1)\mathscr{H}_{\text{int}}(x_2)\right] + \cdots\right]. \tag{11.8}$$

The first-order term in the expansion is:

$$\hat{S}^{(1)} = -i\int d^4x\,\mathscr{H}_{\text{int}}(x). \tag{11.9}$$

To find the matrix element $\langle q|\hat{S}^{(1)}|p\rangle$, we consider the first-order term:

$$\langle q|\hat{S}^{(1)}|p\rangle = -ig\int d^4x\,\langle q|T\left[\psi^\dagger(x)\psi(x)\phi(x)\right]|p\rangle. \tag{11.10}$$

Using Wick's theorem to expand the time-ordered product into normal-ordered products and contractions, we have:

$$T\left[\psi^\dagger(x)\psi(x)\phi(x)\right] =: \psi^\dagger(x)\psi(x)\phi(x) : +(\text{all possible contractions}). \tag{11.11}$$

The relevant contraction for the matrix element $\langle q|\hat{S}|p\rangle$ is:

$$\langle 0|T\left[\psi^\dagger(x)\psi(x)\phi(x)\right]|0\rangle = \langle 0|\psi^\dagger(x)|0\rangle\langle 0|\psi(x)|0\rangle\langle 0|\phi(x)|0\rangle. \tag{11.12}$$

11.1.4 CALCULATING THE CONTRACTIONS

For the contractions, we use the field operator expansions:

$$\langle 0|\psi(x)|0\rangle = \int \frac{d^3k}{(2\pi)^3}\frac{1}{\sqrt{2E_k}}\langle 0|a_k u(k)e^{-ik\cdot x}|0\rangle = 0, \tag{11.13}$$

since $\langle 0|a_k = 0$.

Similarly,

$$\langle 0|\psi^\dagger(x)|0\rangle = \int \frac{d^3k}{(2\pi)^3}\frac{1}{\sqrt{2E_k}}\langle 0|a_k^\dagger \bar{u}(k)e^{ik\cdot x}|0\rangle = 0, \tag{11.14}$$

since $\langle 0|a_k^\dagger = 0$.

And,

$$\langle 0|\phi(x)|0\rangle = \int \frac{d^3k}{(2\pi)^3}\frac{1}{\sqrt{2E_k}}\langle 0|b_k e^{-ik\cdot x} + b_k^\dagger e^{ik\cdot x}|0\rangle = 0, \tag{11.15}$$

since $\langle 0|b_k = 0$ and $\langle 0|b_k^\dagger = 0$.

Thus, the only non-zero contractions are those involving both creation and annihilation operators. For example:

$$\langle 0|T\left[\psi(x)\psi^\dagger(y)\right]|0\rangle = \int \frac{d^3k}{(2\pi)^3}\frac{1}{2E_k}\left(e^{-ik\cdot(x-y)} + e^{ik\cdot(x-y)}\right). \tag{11.16}$$

11.1.5 HIGHER-ORDER CONTRIBUTIONS

Since the first-order term vanishes, we must consider higher-order terms to obtain non-zero contributions. For the second-order term, we have:

$$\hat{S}^{(2)} = \frac{(-i)^2}{2!}\int d^4x_1\, d^4x_2\, T\left[\mathscr{H}_{\text{int}}(x_1)\mathscr{H}_{\text{int}}(x_2)\right]. \tag{11.17}$$

The relevant matrix element is:

$$\begin{aligned}\langle q|\hat{S}^{(2)}|p\rangle =& \frac{(-i)^2}{2!}g^2\int d^4x_1\, d^4x_2 \\ &\times \langle q|T\left[\psi^\dagger(x_1)\psi(x_1)\phi(x_1)\psi^\dagger(x_2)\psi(x_2)\phi(x_2)\right]|p\rangle.\end{aligned} \tag{11.18}$$

Using Wick's theorem, we consider all possible contractions and evaluate the integral. The resulting amplitude gives the probability amplitude for the Yukawa interaction process.

11.1.6 INTERACTION VERTEX AND PROPAGATOR

For the interaction term $-g\bar{\psi}\psi\phi$:

- Each vertex contributes a factor of $-ig$.
- The propagator for the scalar field (ϕ) is $\frac{i}{q^2-\mu^2+i\varepsilon}$.
- The propagator for the fermion field (ψ) is $\frac{i(\not{k}+m_\psi)}{k^2-m_\psi^2+i\varepsilon}$.

11.1.7 SCATTERING AMPLITUDE

The amplitude $\mathscr{M}$ for the scattering process can be calculated as follows:

- **Identify the interaction vertex and propagator:**
  - Each vertex contributes a factor of $-ig$, where g is the coupling constant.
  - The propagator for the exchanged psion is $\frac{i}{q^2-\mu^2+i\varepsilon}$, where q is the four-momentum transfer, and μ is the mass of the psion.
- **Construct the scattering amplitude:**

$$\mathscr{M} = (-ig)^2\left(\frac{i}{q^2-\mu^2+i\varepsilon}\right) \tag{11.19}$$

- **Simplify the expression:**
  - Combine the factors from the vertices and the propagator.

$$\mathcal{M} = -g^2 \frac{1}{q^2 - \mu^2 + i\varepsilon} \tag{11.20}$$

Here, the term q^2 represents the four-momentum transfer squared, which can be related to the energy and momentum of the incoming and outgoing nucleons.

- **Physical Interpretation:**
  - The amplitude $\mathcal{M}$ describes the probability amplitude for the scattering process mediated by psion exchange.
  - The factor $\frac{1}{q^2-\mu^2+i\varepsilon}$ corresponds to the propagator of the exchanged psion, indicating the dependence on the momentum transfer q and the psion mass μ.

11.2 BHABHA SCATTERING

11.2.1 OVERVIEW

Bhabha scattering describes the electron-positron scattering process. This is a key process in quantum electrodynamics (QED) and provides insights into the electromagnetic interactions between charged particles.

11.2.2 E^- E^+ SCATTERING

Consider the scattering of an electron and a positron. The process involves two main Feynman diagrams: one for the annihilation channel and one for the scattering channel. The two Feynman diagrams for Bhabha scattering are shown in Figure 11.3.

FIGURE 11.3
Feynman diagrams for Bhabha scattering: (left) annihilation channel and (right) scattering channel.

11.2.2.1 Differential Cross-Section

The differential cross-section for Bhabha scattering in the center-of-mass frame is given by:

$$\frac{d\sigma}{d\Omega} = \left(\frac{\alpha^2}{s}\right)\left(\frac{1}{2}\left(\frac{s}{t}\right)^2 + \frac{1}{2}\left(\frac{u}{t}\right)^2 + \left(\frac{u}{s}\right)^2\right) \tag{11.21}$$

where α is the fine-structure constant, and s, t, and u are the Mandelstam variables.

11.2.2.2 Calculation Steps

1. **Determine the Mandelstam variables:**
   - The Mandelstam variable s represents the square of the total energy in the center-of-mass system:

$$s = (p_1 + p_2)^2 \tag{11.22}$$

where p_1 and p_2 are the four-momenta of the incoming electron and positron, respectively.

- The Mandelstam variable t represents the square of the momentum transfer between the electron and positron:

$$t = (p_1 - p_3)^2 \tag{11.23}$$

where p_3 is the four-momentum of the outgoing electron.

- The Mandelstam variable u represents the square of the momentum transfer between the electron and the final positron:

$$u = (p_1 - p_4)^2 \tag{11.24}$$

where p_4 is the four-momentum of the outgoing positron.

2. **Calculate the scattering amplitude:**
   - For the annihilation channel (left diagram):

$$\mathscr{M}_{\text{annihilation}} = \bar{v}(p_2)\gamma^\mu u(p_1)\frac{-ig_{\mu\nu}}{s}\bar{u}(p_3)\gamma^\nu v(p_4) \tag{11.25}$$

   - For the scattering channel (right diagram):

$$\mathscr{M}_{\text{scattering}} = \bar{u}(p_3)\gamma^\mu u(p_1)\frac{-ig_{\mu\nu}}{t}\bar{v}(p_2)\gamma^\nu v(p_4) \tag{11.26}$$

3. **Square the amplitudes:**
   - For the annihilation channel:

$$|\mathscr{M}_{\text{annihilation}}|^2 = \left(\bar{v}(p_2)\gamma^\mu u(p_1)\frac{-ig_{\mu\nu}}{s}\bar{u}(p_3)\gamma^\nu v(p_4)\right)^2 \tag{11.27}$$

   - For the scattering channel:

$$|\mathscr{M}_{\text{scattering}}|^2 = \left(\bar{u}(p_3)\gamma^\mu u(p_1)\frac{-ig_{\mu\nu}}{t}\bar{v}(p_2)\gamma^\nu v(p_4)\right)^2 \tag{11.28}$$

4. **Combine the amplitudes:**

$$|\mathscr{M}|^2 = |\mathscr{M}_{\text{annihilation}}|^2 + |\mathscr{M}_{\text{scattering}}|^2 \tag{11.29}$$

5. **Compute the differential cross-section:**

$$\frac{d\sigma}{d\Omega} = \frac{1}{64\pi^2 s}|\mathscr{M}|^2 \tag{11.30}$$

Here, the differential cross-section is computed in the center-of-mass frame, where the total energy and momentum are conserved, and the scattering angles can be measured directly.

11.3 MØ LLER SCATTERING

11.3.1 OVERVIEW

Møller scattering describes the scattering between two electrons. This process is also governed by QED and is important for understanding electron-electron interactions.

11.3.2 EXAMPLE

The differential cross-section for Møller scattering in the center-of-mass frame is given by:

$$\frac{d\sigma}{d\Omega} = \left(\frac{\alpha^2}{4s}\right)\left(\frac{s^2 + u^2}{t^2}\right) \tag{11.31}$$

where α is the fine-structure constant, and s, t, and u are the Mandelstam variables.

11.3.2.1 Feynman Diagrams

The Feynman diagrams for Møller scattering are as follows:

FIGURE 11.4
Feynman diagrams for Møller scattering: (left) direct channel and (right) exchange channel.

For more detail they can be redrawn as:

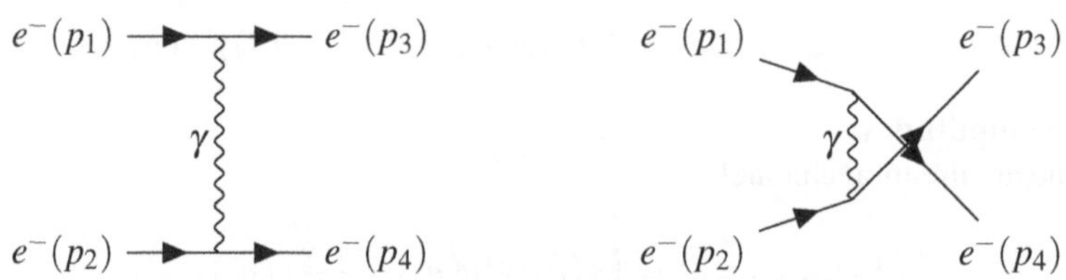

FIGURE 11.5
Feynman diagrams for Møller scattering: (left) direct channel and (right) exchange channel.

11.3.2.2 Calculation Steps

1. **Determine the Mandelstam variables:**
   - The Mandelstam variable s represents the square of the total energy in the center-of-mass system:
   $$s = (p_1 + p_2)^2 \tag{11.32}$$
   where p_1 and p_2 are the four-momenta of the incoming electrons.
   - The Mandelstam variable t represents the square of the momentum transfer between the electrons:
   $$t = (p_1 - p_3)^2 \tag{11.33}$$
   where p_3 is the four-momentum of one of the outgoing electrons.
   - The Mandelstam variable u represents the square of the momentum transfer in the exchange channel:
   $$u = (p_1 - p_4)^2 \tag{11.34}$$
   where p_4 is the four-momentum of the other outgoing electron.
2. **Calculate the scattering amplitude:**
   - For the direct channel (left diagram):
   $$\mathscr{M}_{\text{direct}} = \bar{u}(p_3)\gamma^\mu u(p_1)\frac{-ig_{\mu\nu}}{t}\bar{u}(p_4)\gamma^\nu u(p_2) \tag{11.35}$$
   - For the exchange channel (right diagram):
   $$\mathscr{M}_{\text{exchange}} = \bar{u}(p_4)\gamma^\mu u(p_1)\frac{-ig_{\mu\nu}}{u}\bar{u}(p_3)\gamma^\nu u(p_2) \tag{11.36}$$

3. **Square the amplitudes:**
   - For the direct channel:

$$|\mathcal{M}_{\text{direct}}|^2 = \left(\bar{u}(p_3)\gamma^\mu u(p_1)\frac{-ig_{\mu\nu}}{t}\bar{u}(p_4)\gamma^\nu u(p_2)\right)^2 \tag{11.37}$$

   - For the exchange channel:

$$|\mathcal{M}_{\text{exchange}}|^2 = \left(\bar{u}(p_4)\gamma^\mu u(p_1)\frac{-ig_{\mu\nu}}{u}\bar{u}(p_3)\gamma^\nu u(p_2)\right)^2 \tag{11.38}$$

4. **Combine the amplitudes:**

$$|\mathcal{M}|^2 = |\mathcal{M}_{\text{direct}}|^2 + |\mathcal{M}_{\text{exchange}}|^2 \tag{11.39}$$

5. **Compute the differential cross-section:**

$$\frac{d\sigma}{d\Omega} = \frac{1}{64\pi^2 s}|\mathcal{M}|^2 \tag{11.40}$$

Here, the differential cross-section is computed in the center-of-mass frame, where the total energy and momentum are conserved, and the scattering angles can be measured directly.

11.4 PRACTICAL UNDERSTANDING AND IMPORTANCE

Understanding these scattering processes is crucial for several reasons:

11.4.1 PHYSICAL MEANING

Scattering theories provide a fundamental framework for describing how particles interact and transfer energy and momentum. Each scattering process, whether it's Yukawa, Bhabha, or Møller scattering, reveals specific details about the nature of the forces involved:

- **Yukawa Scattering:** This describes the strong nuclear force between nucleons mediated by mesons. It helps in understanding how protons and neutrons are held together in the nucleus, a key aspect of nuclear physics.
- **Bhabha Scattering:** This involves electron-positron interactions and is essential in QED for studying electromagnetic interactions at high energies. It provides insights into particle-antiparticle annihilation and creation processes.
- **Møller Scattering:** This describes electron-electron interactions, also governed by QED. It is important for understanding the behavior of electrons in various environments, including astrophysical and laboratory plasmas.

11.4.2 PRACTICAL APPLICATIONS

The principles derived from these scattering theories have numerous applications:

- **Particle Accelerators:** Knowledge of scattering processes is critical in designing and interpreting experiments in particle accelerators. These experiments help in discovering new particles and understanding fundamental forces.
- **Medical Physics:** Techniques such as positron emission tomography (PET) rely on the principles of electron-positron annihilation, directly related to Bhabha scattering.
- **Nuclear Energy:** Understanding nuclear forces through Yukawa scattering aids in the development of nuclear reactors and safety measures for nuclear energy production.
- **Material Science:** Electron scattering techniques are used to study the properties of materials at the atomic level, providing insights into their structure and behavior.

11.4.3 THEORETICAL IMPORTANCE

Scattering theories form the backbone of quantum field theory (QFT) and the Standard Model of particle physics. They provide a consistent method to calculate interaction probabilities and cross-sections, which are essential for:

- **Testing Theories:** Experimental results from scattering experiments are used to test and validate theoretical models. Discrepancies between theory and experiment can lead to new discoveries.
- **Fundamental Research:** Scattering processes help probe the internal structure of particles, such as quarks and gluons within protons and neutrons, leading to a deeper understanding of matter.

Understanding these scattering processes thus equips physicists with the tools needed to explore and comprehend the fundamental nature of the universe, bridging theoretical predictions with experimental observations.

11.4.4 OTHER SCATTERING THEORIES IN QFT

In addition to the scattering theories discussed in this chapter, quantum field theory (QFT) encompasses a vast array of other scattering processes that are pivotal in understanding particle interactions. These include, but are not limited to, Compton scattering, which describes the interaction between photons and charged particles, Deep Inelastic Scattering, crucial for probing the internal structure of protons and neutrons, and Bremsstrahlung, describing the emission of radiation by accelerated charged particles. Each of these scattering theories provides unique insights into different aspects of particle physics, from the behavior of fundamental forces to the structure of matter itself. The study of these processes continues to be a central theme in both theoretical and experimental physics, driving advancements in our understanding of the universe.

11.4.4.1 Compton Scattering

Compton scattering describes the interaction between a photon and a charged particle, typically an electron. The Feynman diagram for Compton scattering is shown below:

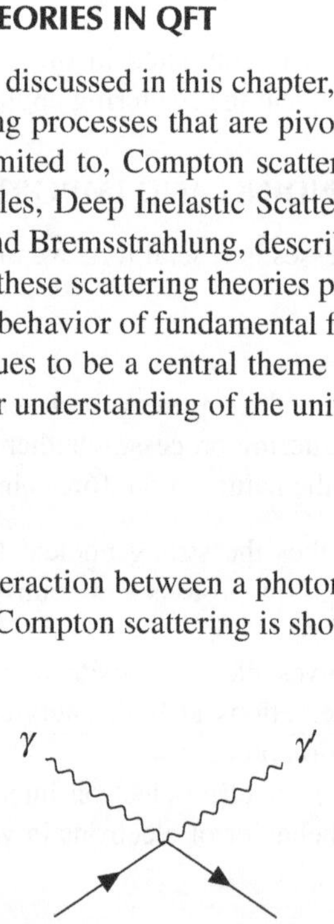

FIGURE 11.6
Feynman diagram for Compton scattering.

11.4.4.2 Deep Inelastic Scattering

Deep Inelastic Scattering is used to probe the internal structure of protons and neutrons by scattering high-energy electrons off nucleons. The Feynman diagram for Deep Inelastic Scattering is shown below:

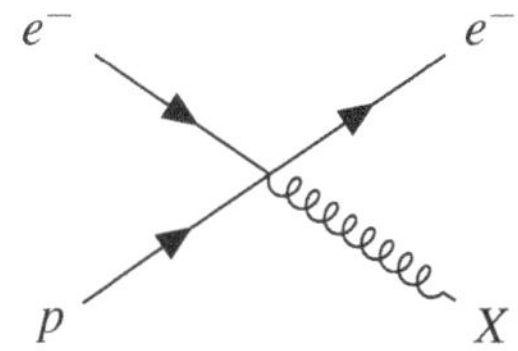

FIGURE 11.7
Feynman diagram for Deep Inelastic Scattering.

11.4.4.3 Bremsstrahlung

Bremsstrahlung describes the emission of radiation by accelerated charged particles, such as electrons. The Feynman diagram for Bremsstrahlung is shown below:

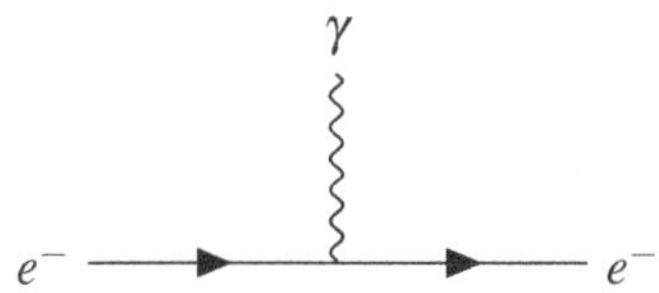

FIGURE 11.8
Feynman diagram for Bremsstrahlung.

11.5 SUMMARY AND DISCUSSION

In this chapter, we have explored several key scattering theories that are fundamental to understanding particle interactions in quantum field theory (QFT).

11.5.1 YUKAWA SCATTERING

Yukawa scattering describes the strong nuclear force between nucleons mediated by mesons. We examined the Yukawa potential and derived the scattering amplitude for nucleon interactions via pion exchange. This theory provides critical insights into the binding forces within atomic nuclei.

11.5.2 BHABHA SCATTERING

Bhabha scattering details the electron-positron scattering process, a cornerstone of quantum electrodynamics (QED). By analyzing the annihilation and scattering channels, we derived the differential cross-section and discussed the implications of electron-positron interactions in high-energy physics.

11.5.3 MLLER SCATTERING

Møller scattering describes the scattering between two electrons, also governed by QED. We examined the direct and exchange channels, calculated the differential cross-section, and highlighted the importance of understanding electron-electron interactions in various physical contexts.

11.5.4 OTHER SCATTERING THEORIES IN QFT

In addition to Yukawa, Bhabha, and Møller scattering, we briefly introduced other important scattering processes:

- **Compton Scattering:** Describes the interaction between a photon and an electron, fundamental to the understanding of photon-electron interactions and the quantum nature of light.
- **Deep Inelastic Scattering:** Probes the internal structure of protons and neutrons by scattering high-energy electrons off nucleons, providing insights into the quark-gluon structure of hadrons.
- **Bremsstrahlung:** Describes the emission of radiation by accelerated charged particles, crucial for understanding radiation processes in astrophysical and laboratory environments.

11.5.5 PRACTICAL UNDERSTANDING AND IMPORTANCE

Understanding these scattering processes is essential for both theoretical and practical reasons:

- **Physical Meaning:** Scattering theories reveal the nature of forces and interactions between particles, providing a framework for describing energy and momentum transfer.
- **Practical Applications:** These principles are applied in particle accelerators, medical physics, nuclear energy, and material science, aiding in the design of experiments, development of technologies, and advancement of scientific knowledge.
- **Theoretical Importance:** Scattering processes form the backbone of QFT and the Standard Model, allowing for the testing and validation of theoretical models and contributing to our understanding of fundamental particles and forces.

Each of these scattering processes offers unique insights into particle interactions and serves as a foundation for more advanced studies in particle physics. By bridging theoretical predictions with experimental observations, these scattering theories continue to drive advancements in our understanding of the universe.

11.6 QUESTIONS AND PROBLEMS

11.6.1 TRIVIA QUESTIONS

Question 1: What type of particle mediates the strong nuclear force in Yukawa scattering?

(a) Photon
(b) Meson
(c) Gluon
(d) Neutrino

Answer: b. Meson
Question 2: What is the typical mass of the exchanged meson in Yukawa scattering?

(a) Electron mass
(b) Proton mass
(c) Pion mass
(d) Neutron mass

Answer: c. Pion mass
Question 3: In Bhabha scattering, which particles are involved in the interaction?

(a) Two electrons

(b) An electron and a positron
(c) A proton and a neutron
(d) A photon and a neutron

Answer: b. An electron and a positron
Question 4: What is the main force described by Møller scattering?

(a) Strong nuclear force
(b) Weak nuclear force
(c) Electromagnetic force
(d) Gravitational force

Answer: c. Electromagnetic force
Question 5: What is the propagator for the exchanged meson in Yukawa scattering?

(a) $\frac{i}{q^2-m^2+i\varepsilon}$
(b) $\frac{1}{r}$
(c) $\frac{g_{\mu\nu}}{k^2}$
(d) $\frac{e^{-\mu r}}{r}$

Answer: a. $\frac{i}{q^2-m^2+i\varepsilon}$
Question 6: In the context of Yukawa's $\psi^\dagger\psi\phi$ theory, what are ψ and $\psi^\dagger$ typically referred to as?

(a) Bosonic fields
(b) Fermionic fields
(c) Scalar fields
(d) Vector fields

Answer: b. Fermionic fields
Question 7: Which theory describes the scattering between two electrons?

(a) Yukawa scattering
(b) Bhabha scattering
(c) Møller scattering
(d) Deep Inelastic scattering

Answer: c. Møller scattering
Question 8: What is the form of the Yukawa potential $V(r)$?

(a) $\frac{g^2}{r}$
(b) $\frac{-g^2e^{-\mu r}}{r}$
(c) $\frac{e^{-\mu r}}{r}$
(d) $\frac{-g^2}{r^2}$

Answer: b. $\frac{-g^2e^{-\mu r}}{r}$
Question 9: What is the fine-structure constant denoted by in Bhabha and Møller scattering equations?

(a) α
(b) β
(c) γ

(d) δ

Answer: a. α

Question 10: In Deep Inelastic Scattering, which particles are probed to understand their internal structure?

(a) Electrons
(b) Photons
(c) Nucleons
(d) Neutrinos

Answer: c. Nucleons

11.6.2 PROBLEMS

1. **Problem 1: Yukawa Potential**
Given the Yukawa potential $V(r) = -g^2 \frac{e^{-\mu r}}{r}$, find the force $F(r)$ between two nucleons.
Solution: The force $F(r)$ is given by the negative gradient of the potential:

$$F(r) = -\frac{dV(r)}{dr} \tag{11.41}$$

Substituting $V(r)$:

$$F(r) = -\frac{d}{dr}\left(-g^2 \frac{e^{-\mu r}}{r}\right) \tag{11.42}$$

$$F(r) = g^2 \left(\frac{d}{dr}\left(\frac{e^{-\mu r}}{r}\right)\right) \tag{11.43}$$

Using the product rule:

$$\frac{d}{dr}\left(\frac{e^{-\mu r}}{r}\right) = \frac{d}{dr}\left(e^{-\mu r} \cdot \frac{1}{r}\right) = \left(\frac{d}{dr} e^{-\mu r}\right)\frac{1}{r} + e^{-\mu r}\left(\frac{d}{dr}\frac{1}{r}\right) \tag{11.44}$$

$$= (-\mu e^{-\mu r})\frac{1}{r} + e^{-\mu r}(-\frac{1}{r^2}) \tag{11.45}$$

$$= -\mu \frac{e^{-\mu r}}{r} - \frac{e^{-\mu r}}{r^2} \tag{11.46}$$

$$F(r) = g^2 \left(-\mu \frac{e^{-\mu r}}{r} - \frac{e^{-\mu r}}{r^2}\right) \tag{11.47}$$

$$F(r) = -g^2 e^{-\mu r}\left(\mu \frac{1}{r} + \frac{1}{r^2}\right) \tag{11.48}$$

$$F(r) = -g^2 \frac{e^{-\mu r}}{r^2}(\mu r + 1) \tag{11.49}$$

2. **Problem 2: Differential Cross-Section for Bhabha Scattering**
Derive the expression for the differential cross-section for Bhabha scattering in the center-of-mass frame.
Solution: The differential cross-section is given by:

$$\frac{d\sigma}{d\Omega} = \left(\frac{\alpha^2}{s}\right)\left(\frac{1}{2}\left(\frac{s}{t}\right)^2 + \frac{1}{2}\left(\frac{u}{t}\right)^2 + \left(\frac{u}{s}\right)^2\right) \tag{11.50}$$

where α is the fine-structure constant, and s, t, and u are the Mandelstam variables.
The Mandelstam variables in terms of the scattering angles in the center-of-mass frame are:

$$s = (p_1 + p_2)^2 \tag{11.51}$$

$$t = (p_1 - p_3)^2 = -\frac{s}{2}(1 - \cos\theta) \tag{11.52}$$

$$u = (p_1 - p_4)^2 = -\frac{s}{2}(1 + \cos\theta) \tag{11.53}$$

Substituting these into the cross-section formula:

$$\frac{d\sigma}{d\Omega} = \left(\frac{\alpha^2}{s}\right)\left(\frac{1}{2}\left(\frac{s}{-\frac{s}{2}(1-\cos\theta)}\right)^2 + \frac{1}{2}\left(\frac{-\frac{s}{2}(1+\cos\theta)}{-\frac{s}{2}(1-\cos\theta)}\right)^2 + \left(\frac{-\frac{s}{2}(1+\cos\theta)}{s}\right)^2\right) \tag{11.54}$$

Simplifying each term:

$$\frac{d\sigma}{d\Omega} = \left(\frac{\alpha^2}{s}\right)\left(\frac{1}{2}\left(\frac{-2}{1-\cos\theta}\right)^2 + \frac{1}{2}\left(\frac{1+\cos\theta}{1-\cos\theta}\right)^2 + \left(\frac{1+\cos\theta}{2}\right)^2\right) \tag{11.55}$$

Simplifying further:

$$\frac{d\sigma}{d\Omega} = \left(\frac{\alpha^2}{s}\right)\left(\frac{2}{(1-\cos\theta)^2} + \frac{(1+\cos\theta)^2}{2(1-\cos\theta)^2} + \frac{(1+\cos\theta)^2}{4}\right) \tag{11.56}$$

3. **Problem 3: Møller Scattering Amplitude**
Compute the scattering amplitude for Møller scattering in the direct channel.
Solution: The scattering amplitude for the direct channel is given by:

$$\mathcal{M}_{\text{direct}} = \bar{u}(p_3)\gamma^\mu u(p_1)\frac{-ig_{\mu\nu}}{t}\bar{u}(p_4)\gamma^\nu u(p_2) \tag{11.57}$$

Here, γ^μ are the gamma matrices, and $u(p)$ are the spinor solutions.
To simplify using the contraction $g_{\mu\nu}$:

$$\mathcal{M}_{\text{direct}} = \frac{-i}{t}\bar{u}(p_3)\gamma^\mu u(p_1)g_{\mu\nu}\bar{u}(p_4)\gamma^\nu u(p_2) \tag{11.58}$$

Using the property of the metric tensor $g_{\mu\nu}$:

$$g_{\mu\nu}\gamma^\mu\gamma^\nu = \gamma^\mu\gamma_\mu = 4I \tag{11.59}$$

Therefore:

$$\mathcal{M}_{\text{direct}} = \frac{-i}{t}\bar{u}(p_3)\gamma^\mu u(p_1)\bar{u}(p_4)\gamma_\mu u(p_2) \tag{11.60}$$

4. **Problem 4: Invariant Amplitude for Yukawa Scattering**
Show the steps to derive the invariant amplitude $\mathcal{M}$ for Yukawa scattering.
Solution: The invariant amplitude $\mathcal{M}$ for Yukawa scattering is given by:

$$\mathcal{M} = -g^2\frac{1}{q^2 - \mu^2 + i\varepsilon} \tag{11.61}$$

Starting with the interaction vertex:

$$\mathcal{M} = (-ig)^2 \frac{i}{q^2 - \mu^2 + i\varepsilon} \tag{11.62}$$

Simplifying:

$$\mathcal{M} = -g^2 \frac{1}{q^2 - \mu^2 + i\varepsilon} \tag{11.63}$$

Here, q is the four-momentum transfer, and μ is the mass of the meson.
The four-momentum transfer q can be expressed as:

$$q^2 = (p_1 - p_3)^2 = (p_2 - p_4)^2 \tag{11.64}$$

where p_1 and p_2 are the initial momenta, and p_3 and p_4 are the final momenta.
Substituting into the amplitude:

$$\mathcal{M} = -g^2 \frac{1}{(p_1 - p_3)^2 - \mu^2 + i\varepsilon} \tag{11.65}$$

This is the invariant amplitude for Yukawa scattering.

5. **Problem 5: Calculating the Differential Cross-Section for Møller Scattering**
Derive the differential cross-section for Møller scattering in the center-of-mass frame.
Solution: The differential cross-section for Møller scattering is given by:

$$\frac{d\sigma}{d\Omega} = \left(\frac{\alpha^2}{4s}\right)\left(\frac{s^2 + u^2}{t^2}\right) \tag{11.66}$$

where α is the fine-structure constant, and s, t, and u are the Mandelstam variables.
The Mandelstam variables are defined as:

$$s = (p_1 + p_2)^2 \tag{11.67}$$

$$t = (p_1 - p_3)^2 \tag{11.68}$$

$$u = (p_1 - p_4)^2 \tag{11.69}$$

In the center-of-mass frame, the energy and momentum are conserved, and the scattering angles can be measured directly. Using the relations between the Mandelstam variables:

$$s = 4E^2 \tag{11.70}$$

$$t = -2E^2(1 - \cos\theta) \tag{11.71}$$

$$u = -2E^2(1 + \cos\theta) \tag{11.72}$$

Substituting these into the differential cross-section formula:

$$\frac{d\sigma}{d\Omega} = \left(\frac{\alpha^2}{4 \cdot 4E^2}\right)\left(\frac{(4E^2)^2 + (-2E^2(1+\cos\theta))^2}{(-2E^2(1-\cos\theta))^2}\right) \tag{11.73}$$

Simplifying each term:

$$\frac{d\sigma}{d\Omega} = \left(\frac{\alpha^2}{16E^2}\right)\left(\frac{16E^4 + 4E^4(1+\cos\theta)^2}{4E^4(1-\cos\theta)^2}\right) \tag{11.74}$$

$$\frac{d\sigma}{d\Omega} = \left(\frac{\alpha^2}{16E^2}\right)\left(\frac{16 + 4(1+\cos\theta)^2}{4(1-\cos\theta)^2}\right) \tag{11.75}$$

$$\frac{d\sigma}{d\Omega} = \left(\frac{\alpha^2}{16E^2}\right)\left(\frac{16+4(1+\cos\theta)^2}{4(1-\cos\theta)^2}\right) \quad (11.76)$$

Simplifying further:

$$\frac{d\sigma}{d\Omega} = \left(\frac{\alpha^2}{16E^2}\right)\left(\frac{4(4+(1+\cos\theta)^2)}{4(1-\cos\theta)^2}\right) \quad (11.77)$$

$$\frac{d\sigma}{d\Omega} = \left(\frac{\alpha^2}{16E^2}\right)\left(\frac{4+(1+\cos\theta)^2}{(1-\cos\theta)^2}\right) \quad (11.78)$$

Bonus Problem: Draw a Feynman Diagram

Draw the Feynman diagram for the process of Deep Inelastic Scattering, where a high-energy electron scatters off a proton, leading to the production of a final electron and a hadronic system X.

Solution:

The Feynman diagram for Deep Inelastic Scattering is shown below. In this process, an electron e^- interacts with a proton p via the exchange of a virtual photon γ^*, resulting in a scattered electron e^- and a hadronic system X:

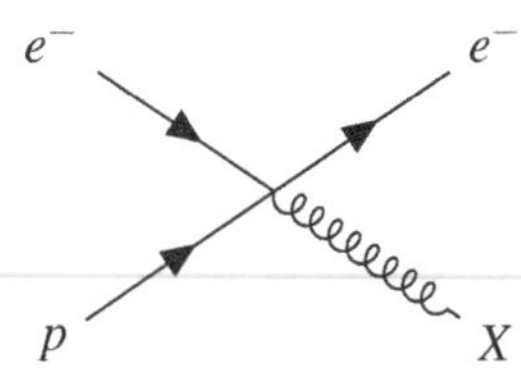

FIGURE 11.9
Feynman diagram for Deep Inelastic Scattering.

12 Lecture 22: Path Integrals in Quantum Field Theory

12.1 INTRODUCTION TO PATH INTEGRALS

Path integrals provide a powerful formulation of quantum mechanics and quantum field theory, offering an alternative to the operator formalism. The concept of path integrals in quantum field theory was famously introduced by Richard Feynman. In a conversation with Freeman Dyson, Feynman explained his revolutionary idea of summing over all possible histories of a particle to calculate quantum mechanical amplitudes. Instead of focusing on a single classical trajectory, Feynman proposed that the correct way to describe the behavior of a quantum system is to consider all possible paths the particle could take between two points. By summing over these paths, or "histories," weighted by the exponential of the action (in units of Planck's constant), one could obtain the correct expectation values for physical quantities. This approach not only provided a new way to understand quantum mechanics but also laid the foundation for quantum field theory and has been instrumental in developing various theories, including quantum electrodynamics (QED).

"You're crazy!" Dyson replied to Feynman when he first told him of this idea.

Later, Dyson was noted as saying, "...he wasn't."

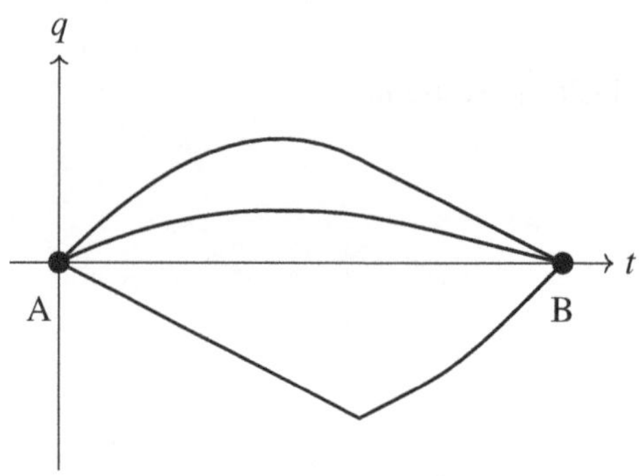

FIGURE 12.1
Multiple paths between points A and B in the path integral formulation.

12.1.1 BASIC CONCEPT

In the path integral approach, the probability amplitude for a particle to go from point A to point B is given by summing over all possible paths between the two points. This is expressed mathematically by the path integral:

$$\langle B|A\rangle = \int \mathscr{D}[q(t)]e^{\frac{i}{\hbar}S[q(t)]}, \tag{12.1}$$

where $\mathscr{D}[q(t)]$ represents the measure over all possible paths $q(t)$, $S[q(t)]$ is the action evaluated along the path $q(t)$, and $\hbar$ is the reduced Planck's constant.

Measure $\mathscr{D}[q(t)]$

The measure $\mathscr{D}[q(t)]$ is a way of representing the sum over all possible configurations or paths that a system can take. It is a formal construct used to denote the integration over the space of all possible paths $q(t)$ the particle can take between the initial and final points.

DOI: 10.1201/9781003637370-12

In the path integral formulation, the measure $\mathscr{D}[q(t)]$ accounts for the fact that we are considering an infinite number of possible paths, each with its own infinitesimal contribution. The integral over this measure is written as:

$$\int \mathscr{D}[q(t)]. \tag{12.2}$$

Formally, it can be understood as a product of integrals over the position coordinates at each instant of time, discretized as:

$$\mathscr{D}[q(t)] = \lim_{N\to\infty} \prod_{n=1}^{N-1} \frac{dq_n}{\sqrt{2\pi i\hbar\Delta t}}, \tag{12.3}$$

where N is the number of discrete time steps, q_n represents the position at the n-th time step, and Δt is the time interval between successive steps. This discretization approach ensures that the measure captures the contributions from all possible paths, adhering to the principles of quantum mechanics.

Action $S[q(t)]$

The action $S[q(t)]$ is given by the integral of the Lagrangian $\mathscr{L}$ over time:

$$S[q(t)] = \int_{t_A}^{t_B} \mathscr{L}(q,\dot{q},t)\,dt. \tag{12.4}$$

Here, $\mathscr{L}$ is the Lagrangian of the system, which is a function of the position q, the velocity $\dot{q}$, and time t.

Contribution of Each Path

Each path $q(t)$ contributes to the overall probability amplitude with a phase factor $e^{iS[q(t)]/\hbar}$. Here, $S[q(t)]$ is the classical action for the path $q(t)$, defined as:

$$S[q(t)] = \int_{t_i}^{t_f} \mathscr{L}(q,\dot{q},t)\,dt, \tag{12.5}$$

where $\mathscr{L}$ is the Lagrangian of the system, a function of the position q, the velocity $\dot{q}$, and time t.

The weight $e^{iS[q(t)]/\hbar}$ represents the quantum mechanical phase associated with the path. The exponent contains the action S of the path, which summarizes the dynamics of the system along that path. The factor $\hbar$ in the denominator ensures that the phase is dimensionless.

Putting it All Together

The total probability amplitude $K(A \to B)$ for the particle to travel from point A to point B is given by:

$$K(A \to B) = \langle B|A\rangle = \int \mathscr{D}[q(t)]\, e^{iS[q(t)]/\hbar}. \tag{12.6}$$

12.2 PATH INTEGRALS IN QUANTUM MECHANICS

12.2.1 THE FEYNMAN PATH INTEGRAL

The Feynman path integral for a particle moving from x_i to x_f in time T is given by:

$$\langle x_f|e^{-iHT/\hbar}|x_i\rangle = \int \mathscr{D}[x(t)]e^{iS[x(t)]/\hbar} \tag{12.7}$$

where $S[x(t)]$ is the classical action of the path $x(t)$.

12.2.2 CLASSICAL ACTION

The classical action $S[x(t)]$ is defined as:

$$S[x(t)] = \int_0^T dt\,\mathscr{L}(x,\dot{x},t) \tag{12.8}$$

where $\mathscr{L}$ is the Lagrangian of the system.

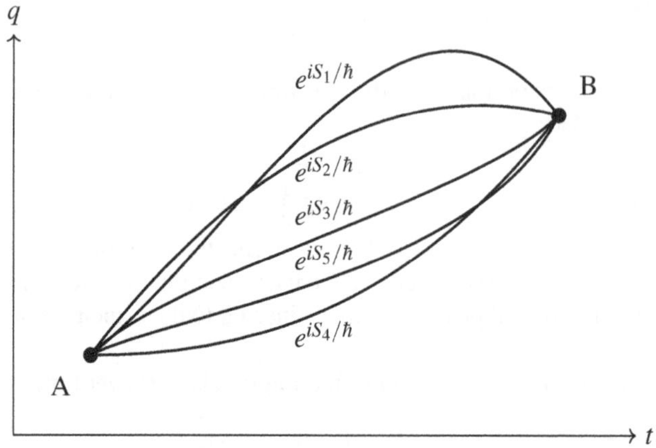

FIGURE 12.2
Possible paths in the path integral formulation between points A and B showing the exponential weight factors $e^{iS/\hbar}$.

12.2.3 FREE PARTICLE PATH INTEGRAL

For a free particle, the Lagrangian is:

$$\mathscr{L} = \frac{1}{2}m\dot{x}^2 \tag{12.9}$$

and the corresponding path integral is:

$$\langle x_f|e^{-iHT/\hbar}|x_i\rangle = \int \mathscr{D}[x(t)]e^{i\int_0^T dt\frac{1}{2}m\dot{x}^2/\hbar} \tag{12.10}$$

12.2.4 DERIVATION OF THE FREE PARTICLE PATH INTEGRAL

To derive the path integral for a free particle, we discretize the time interval $[0,T]$ into N segments of length $\Delta t = \frac{T}{N}$. The path $x(t)$ is then represented by the positions $x_0, x_1, \ldots, x_N$ at each time step $t_0, t_1, \ldots, t_N$.

The action $S[x(t)]$ for the free particle is:

$$S[x(t)] = \int_0^T dt\,\frac{1}{2}m\dot{x}^2 = \sum_{n=0}^{N-1}\frac{1}{2}m\left(\frac{x_{n+1}-x_n}{\Delta t}\right)^2\Delta t \tag{12.11}$$

Simplifying this, we get:

$$S[x(t)] = \sum_{n=0}^{N-1}\frac{m}{2\Delta t}(x_{n+1}-x_n)^2 \tag{12.12}$$

The path integral is then:

$$\langle x_f|e^{-iHT/\hbar}|x_i\rangle = \lim_{N\to\infty}\left(\prod_{n=1}^{N-1}\int dx_n\right) \times \exp\left(\frac{i}{\hbar}\sum_{n=0}^{N-1}\frac{m}{2\Delta t}(x_{n+1}-x_n)^2\right) \tag{12.13}$$

For a large number of time steps, this can be approximated as:

$$\langle x_f|e^{-iHT/\hbar}|x_i\rangle = \sqrt{\frac{m}{2\pi i\hbar T}}\exp\left(\frac{im}{2\hbar T}(x_f - x_i)^2\right) \tag{12.14}$$

12.2.5 EXAMPLE: YOUNG'S DOUBLE-SLIT EXPERIMENT

To illustrate the path integral approach, let's consider Young's double-slit experiment. In this experiment shown in Figure 12.3, an electron can take multiple paths to reach a detection screen after passing through two slits.

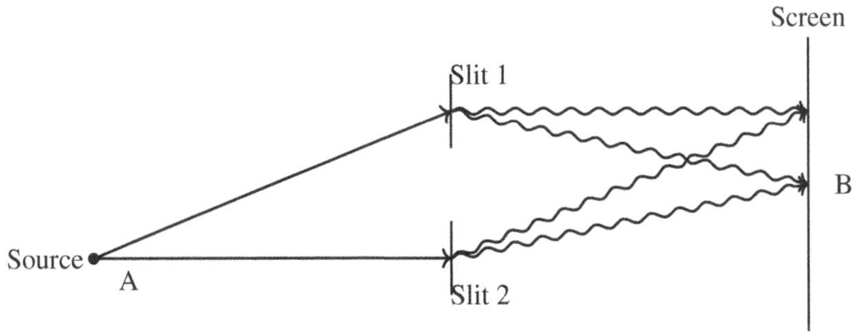

FIGURE 12.3
Double-slit experiment setup

The probability amplitude for an electron to go from point A (source) to point B (screen) is the sum of the amplitudes for all possible paths:

$$\langle B|A\rangle = \sum_{\text{paths}} e^{iS[\text{path}]/\hbar} \tag{12.15}$$

For each path, the action S is given by:

$$S = \int \mathcal{L}\, dt \tag{12.16}$$

The interference pattern observed on the screen results from the sum of the amplitudes of all possible paths, each weighted by the exponential of the action. This illustrates the wave-particle duality of electrons and the fundamental principle of quantum mechanics that all possible paths contribute to the final probability amplitude.

12.2.6 CALCULATION FOR DOUBLE-SLIT EXPERIMENT

Consider a simplified double-slit setup with two main paths: one through slit 1 and another through slit 2. The total probability amplitude is the sum of the amplitudes for these two paths:

$$\langle B|A\rangle = \langle B|A\rangle_{\text{slit 1}} + \langle B|A\rangle_{\text{slit 2}} \tag{12.17}$$

If the distance from the source to the slits and from the slits to the screen is much larger than the slit separation, we can approximate the action for each path. Let S_1 and S_2 be the actions for the paths through slit 1 and slit 2, respectively. The probability amplitude is:

$$\langle B|A\rangle \approx e^{iS_1/\hbar} + e^{iS_2/\hbar} \tag{12.18}$$

The probability of detecting the electron at point B on the screen is given by the modulus squared of the total amplitude:

$$P(B) = |\langle B|A\rangle|^2 = |e^{iS_1/\hbar} + e^{iS_2/\hbar}|^2 \tag{12.19}$$

Expanding this, we get:

$$P(B) = e^{iS_1/\hbar}e^{-iS_1/\hbar} + e^{iS_2/\hbar}e^{-iS_2/\hbar} + e^{iS_1/\hbar}e^{-iS_2/\hbar} + e^{iS_2/\hbar}e^{-iS_1/\hbar} \tag{12.20}$$

Simplifying, we have:

$$P(B) = 1 + 1 + 2\cos\left(\frac{S_1 - S_2}{\hbar}\right) = 2 + 2\cos\left(\frac{\Delta S}{\hbar}\right) \tag{12.21}$$

where $\Delta S = S_1 - S_2$ is the difference in action between the two paths.

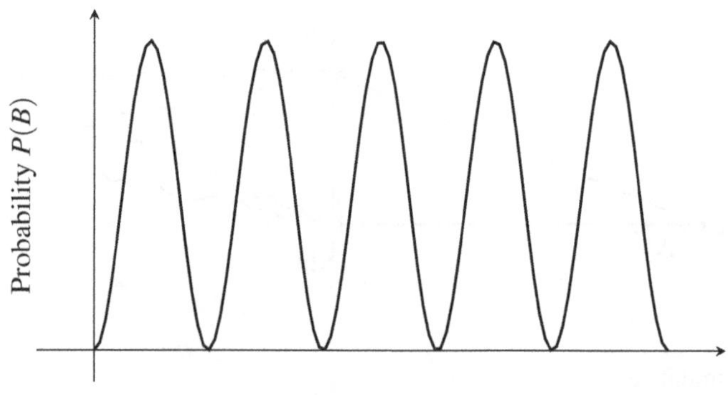

FIGURE 12.4
Interference pattern on the screen showing the probability $P(B)$ as a function of position.

This interference pattern shown in Figure 12.4, characterized by alternating peaks and troughs of probability, is a direct manifestation of the path integral formulation. It demonstrates how summing over all possible paths, each weighted by its exponential action factor, leads to observable quantum phenomena. This formulation encapsulates the core QFT principle that particle behavior arises from the contributions of countless potential histories, emphasizing the inherently probabilistic and non-deterministic nature of quantum fields.

12.3 PATH INTEGRALS IN QUANTUM FIELD THEORY

12.3.1 SCALAR FIELD THEORY

For a scalar field $\phi(x)$, the path integral formulation is given by the partition function:

$$Z = \int \mathcal{D}[\phi] e^{iS[\phi]/\hbar}, \tag{12.22}$$

where $S[\phi]$ is the action of the field.

12.3.2 ACTION FOR SCALAR FIELD

The action $S[\phi]$ for a scalar field ϕ with a potential $V(\phi)$ is given by:

$$S[\phi] = \int d^4x \left(\frac{1}{2}\partial_\mu\phi\partial^\mu\phi - V(\phi)\right). \tag{12.23}$$

Here, $\partial_\mu\phi$ denotes the derivative of the field with respect to the spacetime coordinates x^μ.

12.3.2.1 Derivation of the Action

Consider a scalar field $\phi(x)$ in a four-dimensional spacetime. The Lagrangian density $\mathscr{L}$ for such a field typically includes a kinetic term and a potential term:

$$\mathscr{L} = \frac{1}{2}\partial_\mu\phi\partial^\mu\phi - V(\phi). \tag{12.24}$$

The action $S[\phi]$ is then the integral of the Lagrangian density over all spacetime:

$$S[\phi] = \int d^4x\,\mathscr{L} = \int d^4x\left(\frac{1}{2}\partial_\mu\phi\partial^\mu\phi - V(\phi)\right). \tag{12.25}$$

12.3.3 GENERATING FUNCTIONAL

The generating functional $Z[J]$ for a scalar field $\phi(x)$ with an external source $J(x)$ is defined as:

$$Z[J] = \int \mathscr{D}[\phi]\,e^{i(S[\phi]+\int d^4x J(x)\phi(x))/\hbar}. \tag{12.26}$$

This functional is crucial for deriving correlation functions and for performing perturbative expansions.

12.3.3.1 Derivation of the Generating Functional

Starting from the action with an added source term, the generating functional is:

$$Z[J] = \int \mathscr{D}[\phi]\,e^{i\left(\int d^4x\left(\frac{1}{2}\partial_\mu\phi\partial^\mu\phi - V(\phi) + J(x)\phi(x)\right)\right)/\hbar}. \tag{12.27}$$

This expression accounts for the influence of the external source $J(x)$ on the field $\phi(x)$.

12.3.4 RELATIONSHIP TO THE PARTITION FUNCTION IN STATISTICAL MECHANICS

The generating functional in QFT is conceptually similar to the partition function in statistical mechanics. The partition function Z in statistical mechanics is given by:

$$Z = \int \mathscr{D}[q(t)]\,e^{-S[q(t)]/k_BT}, \tag{12.28}$$

where $S[q(t)]$ is the action (or Hamiltonian) of the system, k_B is the Boltzmann constant, and T is the temperature.

Both the generating functional and the partition function serve as central objects in their respective fields, summarizing the sum over all possible configurations (paths in QFT, states in statistical mechanics) with appropriate weighting factors (phase factor $e^{iS/\hbar}$ in QFT, Boltzmann factor e^{-S/k_BT} in statistical mechanics).

12.3.5 EXAMPLE: ϕ^4 THEORY

In ϕ^4 theory, the potential $V(\phi)$ is given by:

$$V(\phi) = \frac{1}{2}m^2\phi^2 + \frac{\lambda}{4!}\phi^4. \tag{12.29}$$

The corresponding action is:

$$S[\phi] = \int d^4x\left(\frac{1}{2}\partial_\mu\phi\partial^\mu\phi - \frac{1}{2}m^2\phi^2 - \frac{\lambda}{4!}\phi^4\right). \tag{12.30}$$

12.3.5.1 Perturbative Expansion

The generating functional can be expanded perturbatively in terms of Feynman diagrams. For ϕ^4 theory, this involves calculating contributions from different interaction vertices and propagators.

The generating functional in ϕ^4 theory is:

$$Z[J] = \int \mathscr{D}[\phi]\, e^{i(S[\phi]+\int d^4x J(x)\phi(x))/\hbar}. \tag{12.31}$$

Expanding the exponential term involving the interaction part of the action, we get:

$$\begin{aligned} e^{i\int d^4x\left(-\frac{\lambda}{4!}\phi^4\right)/\hbar} &= 1+\frac{i}{\hbar}\int d^4x\left(-\frac{\lambda}{4!}\phi^4\right) \\ &+\frac{1}{2!}\left(\frac{i}{\hbar}\int d^4x\left(-\frac{\lambda}{4!}\phi^4\right)\right)^2+\cdots \end{aligned} \tag{12.32}$$

Each term in this series corresponds to a different order in perturbation theory and can be represented by Feynman diagrams.

12.3.5.2 Feynman Rules for ϕ^4 Theory

The Feynman rules for ϕ^4 theory can be derived from the path integral. The key components include:

- The propagator, which corresponds to the free part of the action.
- The interaction vertex, which corresponds to the ϕ^4 term in the action.

12.3.5.3 Example Calculation

Consider the generating functional $Z[J]$. In the context of ϕ^4 theory, we can calculate the two-point function (propagator) and higher-order contributions involving interaction vertices.

$$Z[J] = \int \mathscr{D}[\phi]\, e^{i\left(\int d^4x\left(\frac{1}{2}\partial_\mu\phi\partial^\mu\phi-\frac{1}{2}m^2\phi^2-\frac{\lambda}{4!}\phi^4+J\phi\right)\right)/\hbar}. \tag{12.33}$$

At the lowest order (free theory), the generating functional $Z_0[J]$ is:

$$Z_0[J] = \int \mathscr{D}[\phi]\, e^{i\left(\int d^4x\left(\frac{1}{2}\partial_\mu\phi\partial^\mu\phi-\frac{1}{2}m^2\phi^2+J(x)\phi(x)\right)\right)/\hbar}. \tag{12.34}$$

For the interacting theory, the next-order term (one-loop correction) in the perturbative expansion involves one interaction vertex. This term is represented by the following Feynman diagram:

This diagram corresponds to a one-loop correction to the two-point function, which contributes to the next-order term in the perturbative expansion of the generating functional.

12.3.5.4 Detailed Explanation of the One-Loop Correction in ϕ^4 Theory

In ϕ^4 theory, the interaction term in the Lagrangian is $\frac{\lambda}{4!}\phi^4$. When expanding the generating functional perturbatively, each power of the interaction term corresponds to an additional vertex in the Feynman diagram.

At the lowest order, we consider the free theory, where the generating functional $Z_0[J]$ is given by:

$$Z_0[J] = \int \mathscr{D}[\phi]\, e^{i\int d^4x\left(\frac{1}{2}\partial_\mu\phi\partial^\mu\phi-\frac{1}{2}m^2\phi^2+J(x)\phi(x)\right)/\hbar}. \tag{12.35}$$

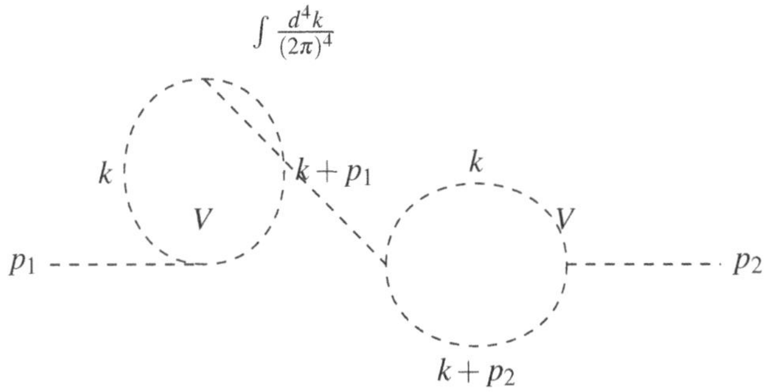

FIGURE 12.5
Example Feynman diagram for ϕ^4 theory showing a one-loop correction. The external momenta are p_1 and p_2, internal momenta are k and $k+p_1$, and V represents the interaction vertices. The integral over k indicates the loop integration.

For the interacting theory, the next-order term (one-loop correction) involves one interaction vertex. The perturbative expansion of the generating functional involves terms like:

$$Z[J] = Z_0[J] + \frac{i}{\hbar}\int d^4x \left(-\frac{\lambda}{4!}\phi^4(x)\right) Z_0[J] + \cdots \tag{12.36}$$

Components of the Diagram

- **External Legs:**
  - The external legs represent the incoming and outgoing scalar particles ϕ.
  - Each external leg corresponds to a factor of the external source $J(x)$ in the generating functional.
- **Internal Lines (Propagators):**
  - The internal lines represent the propagators of the scalar field ϕ.
  - For a scalar field, the propagator in momentum space is given by:

$$\frac{i}{p^2 - m^2 + i\varepsilon} \tag{12.37}$$

  - In the diagram, the internal line forms a loop, indicating a virtual particle.
- **Vertices:**
  - The vertices represent the interaction term $\frac{\lambda}{4!}\phi^4$.
  - Each vertex in the diagram corresponds to a factor of $-i\lambda/4!$ in the perturbative expansion.
- **Loop Integral:**
  - The label $\int \frac{d^4k}{(2\pi)^4}$ indicates that we integrate over all possible values of the loop momentum k, which is a key part of the calculation in quantum field theory.

Mathematical Expression for the One-Loop Correction

The mathematical expression for the one-loop correction can be derived by evaluating the corresponding Feynman diagram. The contribution to the generating functional from the one-loop diagram is:

$$\int \frac{d^4k}{(2\pi)^4}\left(\frac{-i\lambda}{4!}\right)\left(\frac{i}{k^2 - m^2 + i\varepsilon}\right)^2 \tag{12.38}$$

This integral represents the sum over all possible momenta k of the virtual particles in the loop. The one-loop correction in ϕ^4 theory provides an important contribution to the two-point function, modifying the propagator of the scalar field. By calculating higher-order corrections using Feynman diagrams, we can obtain more accurate predictions for physical quantities in quantum field theory.

This detailed explanation of the Feynman diagram for the one-loop correction in ϕ^4 theory helps illustrate the perturbative expansion of the generating functional and the role of interaction vertices and propagators in quantum field theory.

12.4 APPLICATIONS OF PATH INTEGRALS

12.4.1 QUANTUM ELECTRODYNAMICS (QED)

Path integrals are used extensively in QED to calculate the probability amplitudes of electromagnetic interactions. The path integral approach simplifies the derivation of Feynman diagrams and rules, which are essential for computing scattering amplitudes and cross-sections. In QED, the fundamental interactions involve the exchange of photons between charged particles, such as electrons and positrons. The path integral formulation provides a systematic way to sum over all possible paths of the interacting particles, leading to a comprehensive understanding of processes such as electron-electron scattering (Møller scattering) and electron-positron scattering (Bhabha scattering).

12.4.2 SCALAR FIELD THEORY

Path integrals are also fundamental in scalar field theories, such as ϕ^4 theory, where they are used to calculate Green's functions and other observables. The generating functional $Z[J]$ in scalar field theory allows for the computation of correlation functions and the study of spontaneous symmetry breaking.

12.4.2.1 Generating Functional with Source J

The generating functional $Z[J]$ for a scalar field is given by:

$$Z[J] = \int \mathscr{D}[\phi]\, e^{i(S[\phi]+\int d^4x J(x)\phi(x))/\hbar} \tag{12.39}$$

Here, $S[\phi]$ is the action of the scalar field, and $J(x)$ is an external source.

12.4.2.2 Functional Derivatives

To find the Green's function, we need to take the second functional derivative of $Z[J]$ with respect to the source J. This is because the Green's function is essentially the response of the field to the external source. We start by taking the first functional derivative:

$$\frac{\delta Z[J]}{\delta J(x)} = \int \mathscr{D}[\phi] \frac{\delta}{\delta J(x)} \left(e^{i(S[\phi]+\int d^4x J(x)\phi(x))/\hbar}\right) \tag{12.40}$$

12.4.2.3 Applying the Chain Rule

Using the chain rule for functional derivatives:

$$\frac{\delta Z[J]}{\delta J(x)} = \int \mathscr{D}[\phi]\, e^{i(S[\phi]+\int d^4x J(x)\phi(x))/\hbar} \cdot \frac{i\phi(x)}{\hbar} \tag{12.41}$$

Therefore,

$$\frac{\delta Z[J]}{\delta J(x)} = \frac{i}{\hbar} \int \mathscr{D}[\phi]\, \phi(x) e^{i(S[\phi]+\int d^4x J(x)\phi(x))/\hbar} = \frac{i}{\hbar} Z[J] \langle \phi(x) \rangle_J \tag{12.42}$$

12.4.2.4 Second Functional Derivative

Next, we take the second functional derivative with respect to $J(y)$:

$$\frac{\delta^2 Z[J]}{\delta J(y)\delta J(x)} = \frac{\delta}{\delta J(y)}\left(\frac{i}{\hbar}Z[J]\langle\phi(x)\rangle_J\right) \tag{12.43}$$

Applying the product rule:

$$\frac{\delta^2 Z[J]}{\delta J(y)\delta J(x)} = \frac{i}{\hbar}\left(\frac{\delta Z[J]}{\delta J(y)}\langle\phi(x)\rangle_J + Z[J]\frac{\delta\langle\phi(x)\rangle_J}{\delta J(y)}\right) \tag{12.44}$$

$$= \frac{i}{\hbar}\left(\frac{i}{\hbar}Z[J]\langle\phi(y)\rangle_J\langle\phi(x)\rangle_J + Z[J]\left(\frac{\delta\langle\phi(x)\rangle_J}{\delta J(y)}\right)\right) \tag{12.45}$$

12.4.2.5 Simplifying

The second term inside the parentheses represents the connected two-point function:

$$\left.\frac{\delta\langle\phi(x)\rangle_J}{\delta J(y)}\right|_{J=0} = \frac{\delta}{\delta J(y)}\left.\left(\frac{\int\mathscr{D}[\phi]\,\phi(x)e^{i(S[\phi]+\int d^4x J(x)\phi(x))/\hbar}}{Z[J]}\right)\right|_{J=0} \tag{12.46}$$

Since $\langle\phi(x)\rangle_{J=0} = 0$ for a free field, the second term simplifies to the two-point Green's function:

$$G(x-y) = \left.\frac{\delta^2 Z[J]}{\delta J(x)\delta J(y)}\right|_{J=0} = -i\Delta(x-y) \tag{12.47}$$

Thus, the Green's function for a free scalar field theory is proportional to the propagator $\Delta(x-y)$.

12.4.3 QUANTUM CHROMODYNAMICS (QCD)

In QCD, path integrals are used to study the interactions of quarks and gluons, which are the fundamental constituents of protons, neutrons, and other hadrons. The path integral formulation is crucial for understanding the confinement of quarks and the dynamics of the strong force. Non-Abelian gauge theories, such as QCD, are inherently complex, and path integrals provide a robust framework for dealing with these complexities.

12.4.4 STATISTICAL MECHANICS

Path integrals provide a link between quantum field theory and classical statistical systems. In statistical mechanics, the partition function, which sums over all possible states of a system, can be viewed as a path integral over imaginary time. This connection allows for the application of quantum field theoretical techniques to problems in statistical physics, such as phase transitions and critical phenomena.

12.5 EXAMPLE: SIMPLE HARMONIC OSCILLATOR PROPAGATOR

The propagator for the simple harmonic oscillator can be derived using the path integral approach. Consider a simple harmonic oscillator with mass m and angular frequency ω. The Lagrangian is given by:

$$\mathscr{L} = \frac{1}{2}m\dot{x}^2 - \frac{1}{2}m\omega^2 x^2 \tag{12.48}$$

12.5.1 CLASSICAL ACTION

The classical action for a path $x(t)$ from x_i at t_i to x_f at t_f is:

$$S[x(t)] = \int_{t_i}^{t_f} \left(\frac{1}{2} m\dot{x}^2 - \frac{1}{2} m\omega^2 x^2 \right) dt \tag{12.49}$$

12.5.2 EVALUATING THE PATH INTEGRAL

The path integral for the propagator is:

$$K(x_f, t_f; x_i, t_i) = \int \mathscr{D}[x(t)] e^{\frac{i}{\hbar} S[x(t)]} \tag{12.50}$$

To evaluate this path integral, we start by considering the classical path, which is the path that minimizes the action. For the harmonic oscillator, the classical equation of motion is:

$$\ddot{x}_{\text{cl}} + \omega^2 x_{\text{cl}} = 0 \tag{12.51}$$

The general solution to this differential equation is:

$$x_{\text{cl}}(t) = A\cos(\omega t) + B\sin(\omega t) \tag{12.52}$$

We determine the constants A and B using the boundary conditions $x_{\text{cl}}(t_i) = x_i$ and $x_{\text{cl}}(t_f) = x_f$:

$$x_{\text{cl}}(t_i) = A\cos(\omega t_i) + B\sin(\omega t_i) = x_i \tag{12.53}$$

$$x_{\text{cl}}(t_f) = A\cos(\omega t_f) + B\sin(\omega t_f) = x_f \tag{12.54}$$

Solving for A and B, we get:

$$A = \frac{x_i \sin(\omega t_f) - x_f \sin(\omega t_i)}{\sin(\omega t_f - \omega t_i)} \tag{12.55}$$

$$B = \frac{x_f \cos(\omega t_i) - x_i \cos(\omega t_f)}{\sin(\omega t_f - \omega t_i)} \tag{12.56}$$

Thus, the classical path is:

$$x_{\text{cl}}(t) = \frac{x_i \sin(\omega(t_f - t)) + x_f \sin(\omega(t - t_i))}{\sin(\omega(t_f - t_i))} \tag{12.57}$$

12.5.3 CLASSICAL ACTION FOR THE CLASSICAL PATH

Next, we evaluate the action along the classical path $x_{\text{cl}}(t)$:

$$S[x_{\text{cl}}(t)] = \int_{t_i}^{t_f} \left(\frac{1}{2} m\dot{x}_{\text{cl}}^2 - \frac{1}{2} m\omega^2 x_{\text{cl}}^2 \right) dt \tag{12.58}$$

Calculating $\dot{x}_{\text{cl}}$ and substituting into the action integral, we get:

$$\dot{x}_{\text{cl}}(t) = \frac{x_f \omega \cos(\omega(t - t_i)) - x_i \omega \cos(\omega(t_f - t))}{\sin(\omega(t_f - t_i))} \tag{12.59}$$

Therefore,

$$
\begin{aligned}
S[x_{\text{cl}}(t)] &= \int_{t_i}^{t_f} \left(\frac{1}{2} m \left(\frac{x_f \omega \cos(\omega(t-t_i)) - x_i \omega \cos(\omega(t_f - t))}{\sin(\omega(t_f - t_i))} \right)^2 \right. \\
&\quad \left. - \frac{1}{2} m \omega^2 \left(\frac{x_i \sin(\omega(t_f - t)) + x_f \sin(\omega(t - t_i))}{\sin(\omega(t_f - t_i))} \right)^2 \right) dt \\
&= \frac{m\omega}{2\sin(\omega(t_f - t_i))} \\
&\quad \times \int_{t_i}^{t_f} ((x_f \cos(\omega(t-t_i)) - x_i \cos(\omega(t_f - t)))^2 \\
&\quad - \omega^2 (x_i \sin(\omega(t_f - t)) + x_f \sin(\omega(t - t_i)))^2)\, dt
\end{aligned} \tag{12.60}
$$

12.5.4 SIMPLIFYING THE CLASSICAL ACTION

The integral can be simplified by recognizing the orthogonality of sine and cosine functions:

$$
\begin{aligned}
S[x_{\text{cl}}(t)] &= \frac{m\omega}{2\sin(\omega(t_f - t_i))} \left[x_f^2 \int_{t_i}^{t_f} \cos^2(\omega(t - t_i))\, dt \right. \\
&\quad - 2x_f x_i \int_{t_i}^{t_f} \cos(\omega(t-t_i)) \cos(\omega(t_f - t))\, dt \\
&\quad \left. + x_i^2 \int_{t_i}^{t_f} \cos^2(\omega(t_f - t))\, dt \right] \\
&\quad - \frac{m\omega^3}{2\sin(\omega(t_f - t_i))} \left[x_i^2 \int_{t_i}^{t_f} \sin^2(\omega(t_f - t))\, dt \right. \\
&\quad + 2x_i x_f \int_{t_i}^{t_f} \sin(\omega(t_f - t)) \sin(\omega(t - t_i))\, dt \\
&\quad \left. + x_f^2 \int_{t_i}^{t_f} \sin^2(\omega(t - t_i))\, dt \right]
\end{aligned} \tag{12.61}
$$

Evaluating these integrals gives:

$$
S[x_{\text{cl}}(t)] = \frac{m\omega}{2\sin(\omega(t_f - t_i))} ((x_f^2 + x_i^2)\cos(\omega(t_f - t_i)) - 2x_f x_i) \tag{12.62}
$$

12.5.5 PROPAGATOR FOR THE SIMPLE HARMONIC OSCILLATOR

Finally, the path integral for the propagator is:

$$
K(x_f, t_f; x_i, t_i) = \int \mathscr{D}[x(t)] e^{\frac{i}{\hbar} S[x_{\text{cl}}(t)]} \tag{12.63}
$$

The result is:

$$
\begin{aligned}
K(x_f, t_f; x_i, t_i) &= \sqrt{\frac{m\omega}{2\pi i \hbar \sin(\omega(t_f - t_i))}} \\
&\times \exp\left[\frac{im\omega}{2\hbar \sin(\omega(t_f - t_i))} ((x_f^2 + x_i^2)\cos(\omega(t_f - t_i)) - 2x_f x_i) \right]
\end{aligned} \tag{12.64}
$$

This result shows how the path integral approach can be used to derive the exact propagator for a quantum mechanical system, demonstrating the power and versatility of the method.

12.6 NONLOCALITY AND ENTANGLEMENT IN PATH INTEGRALS

Entanglement is a fundamental feature of quantum mechanics where the quantum states of two or more particles become interdependent, such that the state of each particle cannot be described independently of the state of the others. Nonlocality, a consequence of entanglement, implies that measurements performed on one part of an entangled system can instantaneously affect outcomes in another part, regardless of the distance separating them.

In the path integral formulation, nonlocality and entanglement are naturally incorporated through the sum over all possible histories, which includes all possible correlations between different parts of the system.

12.6.1 PATH INTEGRALS AND NONLOCAL CORRELATIONS

Consider a system of two entangled particles, A and B, described by the combined wavefunction $\Psi(x_A, x_B)$. The path integral formulation sums over all possible paths for both particles, inherently including their entangled nature. The generating functional $Z[J]$ for a system with sources $J_A(x_A)$ and $J_B(x_B)$ is:

$$Z[J_A, J_B] = \int \mathscr{D}[\phi_A]\mathscr{D}[\phi_B] \times e^{i(S[\phi_A]+S[\phi_B]+\int d^4x_A J_A(x_A)\phi_A(x_A)+\int d^4x_B J_B(x_B)\phi_B(x_B))/\hbar} \tag{12.65}$$

This functional integrates over all possible field configurations ϕ_A and ϕ_B for particles A and B, respectively. The interaction between the particles is implicitly accounted for in the action terms $S[\phi_A]$ and $S[\phi_B]$.

12.6.2 ENTANGLEMENT AND THE PATH INTEGRAL APPROACH

The entanglement between particles A and B is reflected in the structure of the path integral. For example, if particles A and B are initially entangled, their wavefunction $\Psi(x_A, x_B)$ cannot be factored into separate wavefunctions for each particle:

$$\Psi(x_A, x_B) \neq \psi_A(x_A)\psi_B(x_B) \tag{12.66}$$

The path integral formulation handles this by summing over all joint configurations of x_A and x_B, respecting the entanglement between them. This is evident in the calculation of correlation functions, which involve the joint probability amplitudes for the particles:

$$\langle\phi_A(x_A)\phi_B(x_B)\rangle = \frac{1}{Z[0,0]} \times \int \mathscr{D}[\phi_A]\mathscr{D}[\phi_B]\, \phi_A(x_A)\phi_B(x_B)e^{i(S[\phi_A]+S[\phi_B])/\hbar} \tag{12.67}$$

12.6.3 NONLOCALITY IN QUANTUM FIELD THEORY

In quantum field theory, nonlocal correlations arise naturally from the underlying field interactions. The path integral approach sums over all possible configurations of the fields, including those that establish correlations at a distance. This nonlocality is a direct consequence of the superposition principle and the inherent connectivity of the quantum fields.

To illustrate this with an example, consider the path integral for a free scalar field ϕ:

$$Z[J] = \int \mathscr{D}[\phi]\, e^{i\left(\int d^4x \left(\frac{1}{2}(\partial_\mu \phi)^2 - \frac{1}{2}m^2\phi^2 + J\phi\right)\right)/\hbar} \tag{12.68}$$

The two-point correlation function, which captures the nonlocal correlation between two spacetime points x and y, is given by:

$$\langle 0|T\{\phi(x)\phi(y)\}|0\rangle = \left.\frac{\delta^2 Z[J]}{\delta J(x)\delta J(y)}\right|_{J=0} \tag{12.69}$$

This correlation function shows how the fields at points x and y are correlated through the path integral, reflecting the nonlocality of the quantum field.

12.7 SUMMARY AND DISCUSSION

In this chapter, we have explored the path integral formulation of quantum mechanics and quantum field theory, illustrating its power and versatility in understanding and computing the behavior of quantum systems.

The introduction provided an overview of the path integral concept, highlighting Feynman's revolutionary idea of summing over all possible paths to calculate quantum mechanical amplitudes. This approach contrasts with the classical trajectory focus, emphasizing the probabilistic nature of quantum mechanics by considering all possible histories of a particle.

In the context of quantum mechanics, we derived the path integral for a free particle and illustrated the application of the path integral approach through Young's double-slit experiment. This experiment demonstrated how interference patterns arise from the sum of amplitudes of all possible paths, each weighted by the exponential of the action.

We then delved into the application of path integrals in quantum field theory, particularly for scalar fields. We discussed the action for a scalar field and introduced the generating functional, which plays a crucial role in deriving correlation functions and performing perturbative expansions. The generating functional's analogy to the partition function in statistical mechanics was also highlighted, showcasing the deep connections between these fields.

For ϕ^4 theory, we presented the perturbative expansion and detailed the one-loop correction using Feynman diagrams. This example illustrated the components of the diagram, including external legs, internal lines (propagators), vertices, and the loop integral. The one-loop correction provided insight into higher-order contributions in the perturbative expansion, emphasizing the role of interaction vertices and propagators in quantum field theory.

In the applications section, we explored the significance of path integrals in various areas: - In quantum electrodynamics (QED), path integrals simplify the calculation of electromagnetic interaction amplitudes, aiding in the derivation of Feynman diagrams and rules for processes such as electron-electron and electron-positron scattering. - In scalar field theory, path integrals enable the calculation of Green's functions and the study of phenomena like spontaneous symmetry breaking. - In quantum chromodynamics (QCD), path integrals are crucial for understanding the confinement of quarks and the dynamics of the strong force. - In statistical mechanics, the path integral approach links quantum field theory techniques to classical statistical systems, facilitating the study of phase transitions and critical phenomena.

The example of the simple harmonic oscillator propagator demonstrated the exact evaluation of the path integral, reinforcing the method's applicability and effectiveness in deriving propagators for quantum mechanical systems.

We have also explored how path integrals handle the nonlocality of entanglement. By summing over all possible paths and field configurations, the path integral formulation inherently accounts for the entangled nature of quantum systems. This approach allows for a natural inclusion of nonlocal correlations between entangled particles, providing a comprehensive framework to understand quantum phenomena such as superposition and entanglement. Additionally, the connection between the path integral formulation and generating functionals in scalar field theory further illustrates the power of this method in capturing the complexities of quantum interactions, emphasizing its central role in quantum field theory.

In summary, the path integral formulation is a fundamental tool in quantum mechanics and quantum field theory, providing a comprehensive framework for understanding the probabilistic and nondeterministic nature of quantum phenomena. It unifies various aspects of quantum theory, offering deep insights and practical methods for computing physical quantities across different domains.

12.8 QUESTIONS AND PROBLEMS

12.8.1 TRIVIA QUESTIONS

Question 1: Who introduced the concept of path integrals in quantum mechanics?

(a) Niels Bohr
(b) Werner Heisenberg
(c) Richard Feynman
(d) Paul Dirac

Answer: c. Richard Feynman
Question 2: In the path integral formulation, what does the measure $\mathscr{D}[q(t)]$ represent?

(a) The sum over all possible paths
(b) The classical path
(c) The shortest path
(d) The longest path

Answer: a. The sum over all possible paths
Question 3: What is the weight assigned to each path in the path integral formulation?

(a) $e^{\frac{S[q(t)]}{\hbar}}$
(b) $e^{-\frac{S[q(t)]}{\hbar}}$
(c) $e^{\frac{iS[q(t)]}{\hbar}}$
(d) $e^{-\frac{iS[q(t)]}{\hbar}}$

Answer: c. $e^{\frac{iS[q(t)]}{\hbar}}$
Question 4: In the context of quantum field theory, what does the action $S[\phi]$ depend on?

(a) Only the kinetic term
(b) Only the potential term
(c) Both the kinetic and potential terms
(d) Neither the kinetic nor the potential terms

Answer: c. Both the kinetic and potential terms
Question 5: What is the purpose of the generating functional $Z[J]$ in quantum field theory?

(a) To calculate the energy levels of a system

(b) To find the ground state wave function
(c) To compute correlation functions
(d) To determine the classical trajectory

Answer: c. To compute correlation functions
Question 6: In ϕ^4 theory, what does the interaction term $\frac{\lambda}{4!}\phi^4$ represent?

(a) A kinetic interaction
(b) A potential interaction
(c) A time-dependent interaction
(d) A space-dependent interaction

Answer: b. A potential interaction
Question 7: What is the propagator in the context of quantum field theory?

(a) The measure over all possible paths
(b) The function describing the propagation of particles
(c) The interaction vertex
(d) The external source

Answer: b. The function describing the propagation of particles
Question 8: In the Feynman path integral for a free particle, the action $S[x(t)]$ is primarily determined by which factor?

(a) The particle's mass and velocity
(b) The particle's charge and velocity
(c) The particle's mass and potential energy
(d) The particle's charge and potential energy

Answer: a. The particle's mass and velocity
Question 9: How is the path integral related to the partition function in statistical mechanics?

(a) They are completely unrelated
(b) Both sum over all possible states of a system
(c) Both determine the ground state energy
(d) Both compute the time evolution of a system

Answer: b. Both sum over all possible states of a system
Question 10: What is the significance of the one-loop correction in ϕ^4 theory?

(a) It provides the exact solution to the theory
(b) It represents a higher-order perturbative contribution
(c) It is negligible and can be ignored
(d) It determines the classical path

Answer: b. It represents a higher-order perturbative contribution

12.8.2 PROBLEMS

1. **Problem 1: Free Particle Path Integral**
   Calculate the path integral for a free particle moving from x_i to x_f in time T.
   Solution:

The action for a free particle is given by:

$$S[x(t)] = \int_0^T \frac{1}{2} m\dot{x}^2 \, dt \tag{12.70}$$

Discretize the time interval into N segments of length $\Delta t = \frac{T}{N}$. The path integral is:

$$\begin{aligned}\langle x_f | e^{-iHT/\hbar} | x_i \rangle = \lim_{N\to\infty} \left(\prod_{n=1}^{N-1} \int dx_n \right) \\ \times \exp\left(\frac{i}{\hbar} \sum_{n=0}^{N-1} \frac{m}{2\Delta t} (x_{n+1} - x_n)^2 \right)\end{aligned} \tag{12.71}$$

This can be approximated as:

$$\langle x_f | e^{-iHT/\hbar} | x_i \rangle = \sqrt{\frac{m}{2\pi i \hbar T}} \exp\left(\frac{im}{2\hbar T} (x_f - x_i)^2 \right) \tag{12.72}$$

2. **Problem 2: Harmonic Oscillator Propagator** Derive the propagator for a harmonic oscillator with mass m and angular frequency ω using the path integral approach.
Solution:
The Lagrangian for a harmonic oscillator is:

$$\mathcal{L} = \frac{1}{2} m\dot{x}^2 - \frac{1}{2} m\omega^2 x^2 \tag{12.73}$$

The classical path is:

$$x_{\text{cl}}(t) = \frac{x_i \sin(\omega(t_f - t)) + x_f \sin(\omega(t - t_i))}{\sin(\omega(t_f - t_i))} \tag{12.74}$$

The classical action is:

$$S[x_{\text{cl}}(t)] = \frac{m\omega}{2\sin(\omega(t_f - t_i))} \left((x_f^2 + x_i^2)\cos(\omega(t_f - t_i)) - 2x_f x_i \right) \tag{12.75}$$

The propagator is:

$$\begin{aligned}K(x_f, t_f; x_i, t_i) = \sqrt{\frac{m\omega}{2\pi i \hbar \sin(\omega(t_f - t_i))}} \\ \times \exp\left[\frac{im\omega}{2\hbar \sin(\omega(t_f - t_i))} \left((x_f^2 + x_i^2)\cos(\omega(t_f - t_i)) - 2x_f x_i \right) \right]\end{aligned} \tag{12.76}$$

3. **Problem 3: One-Loop Correction in ϕ^4 Theory**
Calculate the one-loop correction in ϕ^4 theory.
Solution:
The generating functional in ϕ^4 theory is:

$$Z[J] = \int \mathscr{D}[\phi] \, e^{i\left(\int d^4x \left(\frac{1}{2}\partial_\mu \phi \partial^\mu \phi - \frac{1}{2} m^2 \phi^2 - \frac{\lambda}{4!}\phi^4 + J\phi\right)\right)/\hbar} \tag{12.77}$$

The one-loop correction involves the integral:

$$\int \frac{d^4k}{(2\pi)^4} \left(\frac{-i\lambda}{4!} \right) \left(\frac{i}{k^2 - m^2 + i\varepsilon} \right)^2 \tag{12.78}$$

This integral represents the sum over all possible momenta k of the virtual particles in the loop.

4. **Problem 4: Green's Function for Free Scalar Field**
   Derive the Green's function for a free scalar field using the path integral approach.
   Solution:
   The generating functional is:

$$Z[J] = \int \mathscr{D}[\phi]\, e^{i(S[\phi]+\int d^4x J(x)\phi(x))/\hbar} \tag{12.79}$$

   The two-point Green's function is:

$$G(x-y) = \left.\frac{\delta^2 Z[J]}{\delta J(x)\delta J(y)}\right|_{J=0} \tag{12.80}$$

   Applying the chain rule and simplifying, we get:

$$G(x-y) = -i\Delta(x-y) \tag{12.81}$$

5. **Problem 5: Young's Double-Slit Experiment**
   Use the path integral approach to calculate the probability amplitude for an electron in Young's double-slit experiment.
   Solution:
   The probability amplitude is the sum of the amplitudes for all possible paths:

$$\langle B|A\rangle = \sum_{\text{paths}} e^{iS[\text{path}]/\hbar} \tag{12.82}$$

   For the double-slit experiment, the total probability amplitude is:

$$\langle B|A\rangle = \langle B|A\rangle_{\text{slit 1}} + \langle B|A\rangle_{\text{slit 2}} \tag{12.83}$$

   The probability is given by:

$$P(B) = |\langle B|A\rangle|^2 = |e^{iS_1/\hbar} + e^{iS_2/\hbar}|^2 \tag{12.84}$$

   Simplifying, we get:

$$P(B) = 2 + 2\cos\left(\frac{\Delta S}{\hbar}\right) \tag{12.85}$$

   where $\Delta S = S_1 - S_2$.

12.9 BONUS PROBLEM

Essay Problem: Path Integral Formulation and Young's Double-Slit Experiment with Polarizers

Discuss how the knowledge of the Path Integral formulation from this chapter would be used to address Young's double-slit experiment when adding polarizer analyzers to determine which slit a photon passes through. Explain what can be learned from sending many photons or an ensemble at the slits.

Solution:

The Path Integral formulation provides a powerful framework for understanding quantum mechanics, particularly in scenarios involving interference and superposition, such as Young's double-slit experiment. When polarizer analyzers are introduced to determine which slit a photon passes through, the experiment takes on additional complexity, as it brings the concepts of measurement and quantum state collapse into play.

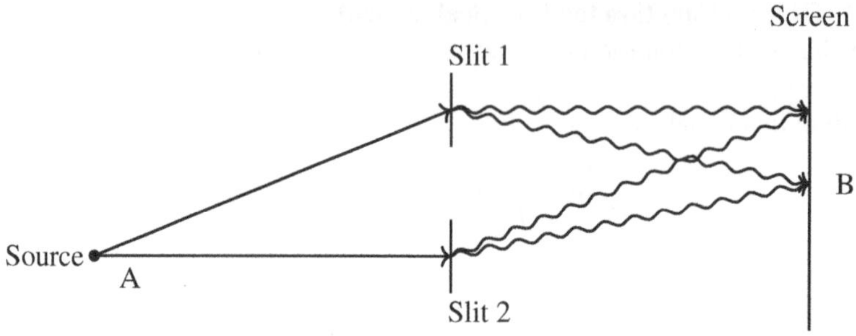

FIGURE 12.6
Paths in Young's double-slit experiment without polarizers.

In the standard double-slit experiment, the path integral approach tells us that the probability amplitude for a photon to reach a point on the detection screen is obtained by summing over all possible paths that the photon could take from the source to the screen. Each path is weighted by the exponential of the action, as given by:

$$\langle B|A\rangle = \sum_{\text{paths}} e^{iS[\text{path}]/\hbar}. \tag{12.86}$$

This sum over paths includes contributions from paths through both slits, leading to an interference pattern on the screen, characterized by alternating bright and dark fringes. This pattern arises due to the coherent superposition of the probability amplitudes for paths through each slit.

When polarizer analyzers are placed at the slits to determine which slit a photon passes through, the situation changes. The polarizers can be oriented in such a way that they introduce distinguishability between the paths through the two slits. For instance, if the polarizers are oriented perpendicularly to each other, they will mark the photons passing through each slit with different polarization states. In this case, the path integral formulation must account for the fact that the photon's polarization state affects its probability amplitude.

If the polarizers are oriented at angles θ_1 and θ_2 at the two slits, the probability amplitude for the photon to reach a point on the screen will include contributions from the paths with corresponding polarization states. The probability amplitude becomes:

$$\langle B|A\rangle = \sum_{\text{paths, pol}} e^{iS[\text{path}]/\hbar}\langle \text{pol}_f|\text{pol}_i\rangle, \tag{12.87}$$

where $\langle \text{pol}_f|\text{pol}_i\rangle$ is the inner product of the polarization states, indicating the overlap between the initial and final polarization states.

If the polarizers are orthogonal, the inner product $\langle \text{pol}_f|\text{pol}_i\rangle = 0$ for paths through different slits, effectively removing the interference terms. In this case, the interference pattern is destroyed, and the photons will produce a pattern on the screen that is a simple sum of the patterns from each slit, without interference fringes.

By sending many photons or an ensemble through the slits, we can gather statistical data to observe the effects of the polarizers on the interference pattern. With a large number of photons, the statistical distribution of their detection points on the screen will reveal whether the interference pattern persists or is destroyed, depending on the polarization settings. This experiment demonstrates the principle of quantum measurement and the role of entanglement and coherence in quantum systems. The path integral formulation provides a comprehensive framework to predict and analyze these outcomes by summing over all possible paths and incorporating the effects of polarization and measurement.

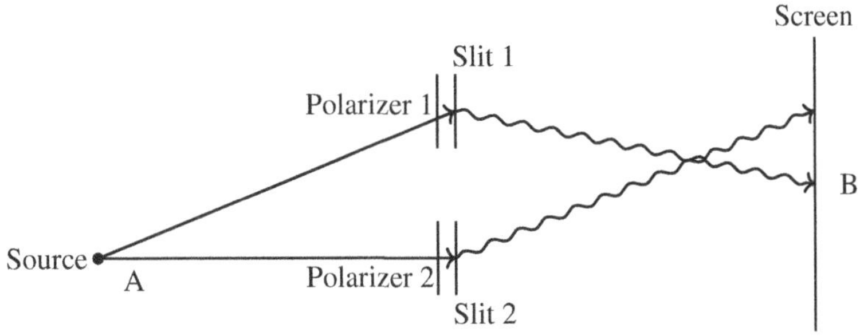

FIGURE 12.7
Paths in Young's double-slit experiment with polarizers. Note how the multiple paths collapsed away to only one set of paths.

In conclusion, the path integral formulation helps us understand how adding polarizer analyzers to Young's double-slit experiment affects the interference pattern. It emphasizes the importance of coherence and distinguishability in quantum interference and provides a robust method to analyze the impact of measurement on the quantum system.

13 Lecture 23: Topology and Quantum Field Theory

13.1 WHAT IS TOPOLOGY

Topology is a branch of mathematics that studies the properties of spaces that are preserved under continuous deformations such as stretching, twisting, crumpling, and bending, but not tearing or gluing. Essentially, topology is concerned with the qualitative aspects of geometry and spatial relationships, rather than precise measurements. Topological properties include concepts such as connectedness, compactness, and continuity, which remain invariant under homeomorphismscontinuous functions with continuous inverses.

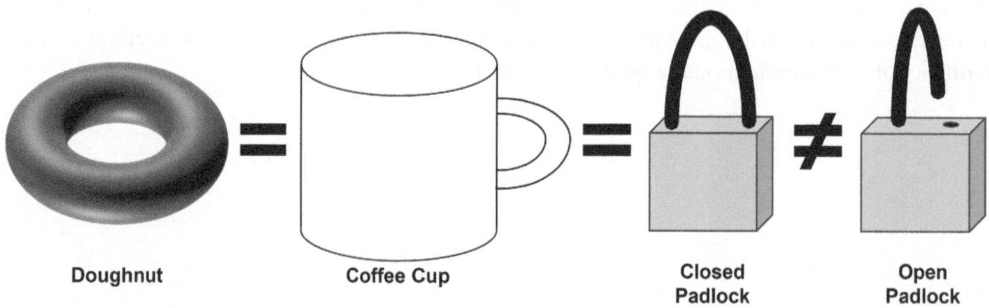

FIGURE 13.1
Topological equivalence of a doughnut, coffee cup, and padlock. Opening the padlock breaks the topological symmetry.

Figure 13.1 illustrates the concept of topological equivalence using a doughnut, a coffee cup, and a padlock. A doughnut and a coffee cup are considered topologically equivalent because they each have one hole. This hole can be continuously transformed from the handle of the cup to the hole of the doughnut without tearing or gluing. Also, note that the container part of the coffee cup is simply an indention in the doughnut.

Similarly, a closed padlock with a loop can be seen as topologically equivalent to a doughnut or a coffee cup because it also forms a single hole. However, once the padlock is opened, the loop is broken, altering its topology. The open padlock no longer has the same topological properties as a doughnut or a coffee cup, demonstrating how breaking or cutting changes the topological classification of an object. By considering any object to be perfectly maliable, like dough or clay, is a simple way to understand topology. If the maliable object can be made into other objects without breaking or tearing the material then they have the same topology.

In the context of quantum field theory, topology plays a significant role in understanding the behavior and interactions of quantum fields. The study of topological spaces allows physicists to classify and distinguish different field configurations and gauge theories. Topological invariants, which are quantities that remain unchanged under continuous transformations, provide powerful tools for identifying and characterizing distinct phases of matter and field configurations. Examples of topological invariants include winding numbers, homotopy groups, and Chern numbers.

One of the key applications of topology in quantum field theory is the study of solitons and instantons, which are stable, localized solutions to field equations that represent particles or other

DOI: 10.1201/9781003637370-13

physical entities. Solitons arise in various physical systems and are often associated with topological invariants. For instance, kinks, vortices, and monopoles are types of solitons that occur in different dimensions and field theories. The stability of these solitons is often guaranteed by the topological properties of the underlying field configurations, making them robust against perturbations. Instantons, on the other hand, are solutions to field equations in Euclidean space-time that contribute to tunneling events and non-perturbative effects in quantum field theory.

Furthermore, topology is integral to the study of gauge theories, which form the foundation of our understanding of fundamental interactions in nature. Gauge theories describe how fields transform under local symmetries, and topology helps in understanding the global structure of these symmetries. Concepts such as fiber bundles and connections are used to describe gauge fields and their topological properties. Topological quantum field theories (TQFTs) are a specific class of field theories where the observables are topological invariants, providing deep insights into the interplay between topology and quantum physics.

Another crucial application of topology is in the study of topological insulators and conductors. These are materials that exhibit unique electronic properties due to their topological nature. Topological insulators have insulating bulk properties but conductive surface states that are protected by topological invariants. This protection results in robust electronic states that are resistant to perturbations such as impurities or defects. Similarly, topological conductors exhibit special conduction properties arising from their topological band structures. The study of these materials has significant implications for developing new technologies and understanding quantum phase transitions.

13.2 SOLITONS AND INSTANTONS

13.2.1 SOLITONS

Solitons are stable, localized solutions to nonlinear field equations that behave like particles. They retain their shape and speed over time and can interact with other solitons in well-defined ways. Solitons arise in a variety of physical systems, including fluid dynamics, optics, and field theory. They are characterized by their topological properties, which often confer stability against perturbations. In quantum field theory, solitons provide insight into non-perturbative phenomena and are closely related to topological invariants. Common examples of solitons include kinks, vortices, and monopoles, each of which occurs in different dimensions and field configurations.

13.2.1.1 Kinks

Kinks are a type of topological soliton that occur in one-dimensional systems. They represent stable, localized solutions to nonlinear field equations, typically arising in systems with a double-well potential. Kinks are characterized by a transition between two distinct vacuum states of the system, creating a smooth, non-singular configuration that connects these vacua. The double-well potential, as shown in Figure 13.2, illustrates the two minima corresponding to the vacuum states and the transition region where the kink occurs.

13.2.1.2 Definition and Example

A simple example of a system that supports kinks is the one-dimensional ϕ^4 theory, described by the Lagrangian density:

$$\mathscr{L} = \frac{1}{2}(\partial_\mu \phi)^2 - \frac{\lambda}{4}(\phi^2 - \eta^2)^2, \tag{13.1}$$

where ϕ is the scalar field, λ is the coupling constant, and η determines the vacuum expectation values of the field.

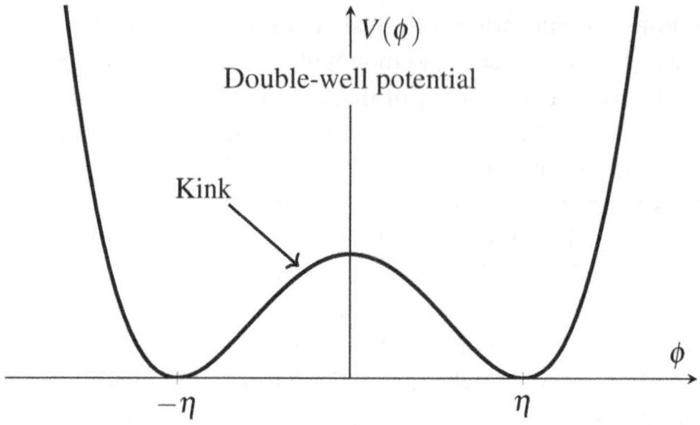

FIGURE 13.2
The double-well potential $V(\phi) = \frac{\lambda}{4}(\phi^2 - \eta^2)^2$, with minima at $\phi = \pm\eta$. The kink occurs in the transition region between the two minima.

The potential associated with this Lagrangian is:

$$V(\phi) = \frac{\lambda}{4}(\phi^2 - \eta^2)^2, \tag{13.2}$$

which has two minima at $\phi = \pm\eta$. These minima correspond to the vacuum states of the system.

A kink solution interpolates between these two vacuum states. The static kink solution, $\phi_{\text{kink}}(x)$, is given by:

$$\phi_{\text{kink}}(x) = \eta \tanh\left(\frac{\sqrt{\lambda}\eta x}{\sqrt{2}}\right). \tag{13.3}$$

This solution represents a smooth transition from $\phi = -\eta$ at $x \to -\infty$ to $\phi = +\eta$ at $x \to +\infty$. The corresponding antikink solution is:

$$\phi_{\text{antikink}}(x) = -\eta \tanh\left(\frac{\sqrt{\lambda}\eta x}{\sqrt{2}}\right), \tag{13.4}$$

which transitions from $\phi = +\eta$ at $x \to -\infty$ to $\phi = -\eta$ at $x \to +\infty$.

13.2.1.3 Topological Charge

The stability of kinks is ensured by their topological charge, also known as the winding number. The topological charge Q is an integer that measures the difference between the field values at spatial infinities:

$$Q = \frac{1}{2\eta}[\phi(x \to +\infty) - \phi(x \to -\infty)]. \tag{13.5}$$

For the kink solution, $Q = +1$, and for the antikink solution, $Q = -1$. The topological charge is conserved, meaning that kinks and antikinks cannot be continuously deformed into the vacuum state without altering the topology of the field configuration.

13.2.1.4 Applications and Physical Significance

Kinks are important in various physical contexts, including condensed matter physics, cosmology, and field theory. In condensed matter physics, kinks can describe domain walls in magnetic systems, where the field ϕ represents the magnetization. In cosmology, kinks are used to model cosmic strings, which are hypothetical topological defects that could have formed during phase transitions in the early universe.

The study of kinks provides valuable insights into the non-perturbative aspects of field theory. Their stability and localized nature make them useful models for understanding particle-like behavior in classical and quantum field theories. Moreover, kinks and other solitons serve as testbeds for exploring the interplay between topology and field dynamics, offering a deeper understanding of the role of topological structures in physics.

13.2.1.5 Optical Kink Solitons

A typical example of a kink in an optical system can be found in nonlinear optics, particularly in the context of light propagation in nonlinear media. One common example is the formation of a kink soliton in an optical fiber or waveguide with a nonlinear refractive index.

In nonlinear optics, the refractive index of a medium can depend on the irradiance of the light passing through it. This irradiance-dependent refractive index can lead to the formation of various nonlinear wave phenomena, including solitons. A kink soliton in optics is a stable, localized wave packet that maintains its shape while traveling through a nonlinear medium. It represents a transition between two distinct states of the medium, similar to how a kink in field theory represents a transition between two vacuum states.

Nonlinear Schrödinger Equation (NLSE):

The propagation of light in a nonlinear medium can be described by the nonlinear Schrödinger equation (NLSE):

$$i\frac{\partial \psi}{\partial z} + \frac{1}{2}\frac{\partial^2 \psi}{\partial t^2} + |\psi|^2\psi = 0 \tag{13.6}$$

Here, ψ represents the complex envelope of the optical field, z is the propagation distance, and t is the time.

Kink Soliton Solution:

To find a kink soliton solution to the NLSE, we assume a solution of the form:

$$\psi(t,z) = \eta \tanh\left(\frac{t - vt_0}{\sqrt{2}}\right) e^{i(kz-\omega t)}, \tag{13.7}$$

where η is the amplitude, v is the velocity, t_0 is the initial position of the kink, k is the wavenumber, and ω is the angular frequency.

The assumed form of the optical wavefunction for kink solitons comes from several considerations:

1. **Symmetry and Boundary Conditions:** The tanh function is well-known for representing kinks because it smoothly transitions between two distinct asymptotic states. This makes it an ideal candidate for modeling transitions between different intensity levels of the optical field.
2. **Analogy with Known Solutions:** In other nonlinear differential equations, such as the ϕ^4 theory in field theory, kink solutions are also represented using the tanh function. By drawing parallels, the same functional form is adopted for the NLSE.
3. **Balancing Nonlinear and Dispersive Effects:** The chosen form of the solution ensures that the nonlinear term $|\psi|^2\psi$ and the dispersive term $\frac{1}{2}\frac{\partial^2\psi}{\partial t^2}$ in the NLSE balance each other. This balance is crucial for maintaining the stability of the soliton solution.

A NOTE ON THE tanh FUNCTION

The tanh function, which is characteristic of kink solitons, also appears in other areas of theoretical physics, including general relativity. One notable example is the Alcubierre warp drive, where the tanh function is used to describe the shape of the warp bubble.

In the context of the warp drive, the tanh function is used to define a smooth transition region between the flat spacetime inside the warp bubble and the curved spacetime outside it. This mirrors the way kink solitons represent transitions between different field values in nonlinear field theories.

The similarity between these applications highlights a deeper connection between the mathematical structures used in general relativity and field theory. Both fields often employ solutions that involve smooth, continuous transitions between different states or regions, and the tanh function provides an elegant and effective way to model these transitions.

This analogy underscores the versatility and utility of certain mathematical functions, like tanh, in describing a wide range of physical phenomena, from optical solitons in nonlinear media to exotic spacetime configurations in general relativity.

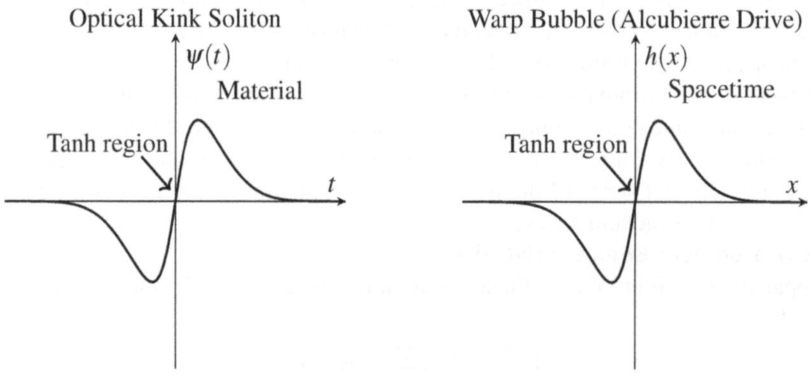

(a) Optical Kink Soliton **(b)** Warp Bubble (Alcubierre Drive)

FIGURE 13.3
Comparative figure showing (a) an optical kink soliton in a material and (b) the warp bubble from Alcubierre's warp drive equations in spacetime. Both use the tanh function multiplied by a Gaussian envelope to illustrate the similarity in their mathematical form.

13.2.1.6 Step-by-Step Solution:

1. Substitute the solution into the NLSE:

$$\psi(t,z) = \eta \tanh\left(\frac{t - vt_0}{\sqrt{2}}\right) e^{i(kz-\omega t)} \tag{13.8}$$

$$\frac{\partial \psi}{\partial z} = ik\psi \tag{13.9}$$

$$\frac{\partial \psi}{\partial t} = \eta \left(\frac{1}{\sqrt{2}}\operatorname{sech}^2\left(\frac{t - vt_0}{\sqrt{2}}\right)\right) e^{i(kz-\omega t)} - i\omega\psi \tag{13.10}$$

$$\frac{\partial^2 \psi}{\partial t^2} = \eta \left(\frac{2}{\sqrt{2}}\operatorname{sech}^2\left(\frac{t - vt_0}{\sqrt{2}}\right) \tanh\left(\frac{t - vt_0}{\sqrt{2}}\right)\right) e^{i(kz-\omega t)} - \omega^2\psi \tag{13.11}$$

2. Insert the derivatives into the NLSE:

$$i(ik\psi) + \frac{1}{2}\left[\eta \left(\frac{2}{\sqrt{2}}\operatorname{sech}^2\left(\frac{t - vt_0}{\sqrt{2}}\right) \tanh\left(\frac{t - vt_0}{\sqrt{2}}\right)\right) e^{i(kz-\omega t)} - \omega^2\psi\right] + |\psi|^2\psi = 0 \tag{13.12}$$

Simplify the equation:

$$-k\psi + \frac{1}{\sqrt{2}}\eta\left(\text{sech}^2\left(\frac{t-vt_0}{\sqrt{2}}\right)\tanh\left(\frac{t-vt_0}{\sqrt{2}}\right)\right)e^{i(kz-\omega t)} - \frac{\omega^2}{2}\psi + |\psi|^2\psi = 0 \tag{13.13}$$

3. Balance the terms:

Notice that for the assumed form of ψ, the $|\psi|^2\psi$ term will have contributions that match the spatial dependence. The nonlinearity $|\psi|^2\psi$ will produce terms involving $\tanh^3\left(\frac{t-vt_0}{\sqrt{2}}\right)$ and other terms that need to balance against the linear terms.

4. Equate coefficients:

By equating the coefficients of the corresponding terms and solving for the parameters, we ensure that the nonlinear and linear parts balance each other out. Specifically:

$$k - \frac{\omega^2}{2} = 0 \tag{13.14}$$

implying a relationship between k and ω.

This detailed solution shows that the kink soliton represents a stable transition between two different intensity states of the optical field, consistent with the nonlinear Schrödinger equation. The stability and form of the kink soliton come from the balance between the nonlinear and dispersive effects in the medium.

13.2.1.7 Vortices

Vortices are another type of topological soliton that occur in two-dimensional systems. They are characterized by a core where the field vanishes and a surrounding region where the field winds around the core. Vortices are common in various physical contexts, including fluid dynamics, superconductivity, and condensed matter physics. The topological charge of a vortex, also known as the winding number, indicates how many times the field wraps around the core.

Mathematical Description

Consider a complex scalar field $\phi = |\phi|e^{i\theta}$, where $|\phi|$ is the amplitude and θ is the phase. The phase θ changes by $2\pi n$ as we move around the core of the vortex, where n is the winding number. The winding number n is a topological invariant, representing the number of times the field phase winds around the vortex core.

$$\oint \nabla\theta \cdot d\mathbf{l} = 2\pi n \tag{13.15}$$

Physical Examples

In superconductors, vortices appear in the presence of an external magnetic field, where the magnetic flux is quantized and carried by the vortices. In fluid dynamics, vortices are observed as swirling flows, and in Bose-Einstein condensates, vortices represent quantized angular momentum.

The Figure 13.4 illustrates vortices with winding numbers $n = 1$, $n = 2$, and $n = 3$. In each case, the arrows represent the phase of the complex scalar field, which spirals around the vortex core. The winding number determines the topological stability of the vortex, making it robust against small perturbations. Vortices play a crucial role in various physical systems, providing insights into the interplay between topology and field dynamics.

13.2.1.8 Monopoles

Monopoles are topological solitons that arise in three-dimensional systems and are characterized by a singular point where the field configuration cannot be continuously deformed to a trivial (vacuum)

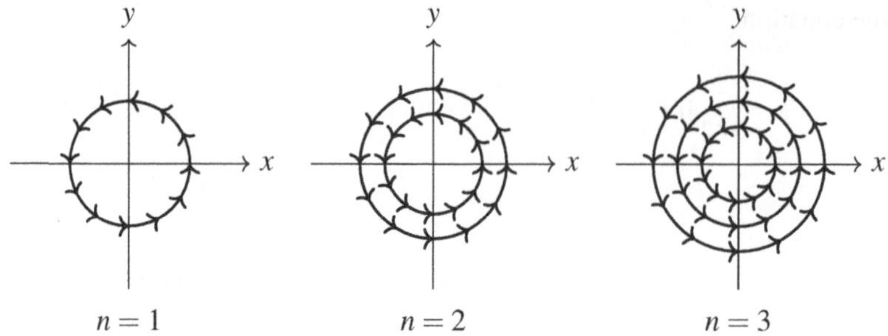

FIGURE 13.4
Illustration of vortices with different winding numbers: (left) $n = 1$, (center) $n = 2$, (right) $n = 3$. The winding number indicates the number of times the phase of the field winds around the vortex core.

configuration. In the context of quantum field theory, monopoles are solutions to field equations that possess a magnetic charge, analogous to the electric charge of particles like electrons.

Monopoles are predicted by grand unified theories and certain extensions of the Standard Model of particle physics. They provide insights into the structure of gauge fields and the unification of fundamental forces. One of the most notable examples of monopoles is the 't Hooft-Polyakov monopole, which emerges in non-Abelian gauge theories.

The 't Hooft-Polyakov Monopole

The 't Hooft-Polyakov monopole is a solution to the equations of motion in a non-Abelian gauge theory with spontaneous symmetry breaking. Consider a gauge theory with gauge group $SU(2)$ and a Higgs field ϕ in the adjoint representation. The Lagrangian for such a theory is given by:

$$\mathscr{L} = -\frac{1}{4}F^a_{\mu\nu}F^{\mu\nu a} + \frac{1}{2}(D_\mu\phi)^a(D^\mu\phi)^a - V(\phi), \tag{13.16}$$

where $F^a_{\mu\nu}$ is the field strength tensor, D_μ is the covariant derivative, and $V(\phi)$ is the Higgs potential.

Field Strength Tensor and Covariant Derivative

The field strength tensor and the covariant derivative are defined as:

$$F^a_{\mu\nu} = \partial_\mu A^a_\nu - \partial_\nu A^a_\mu + g\varepsilon^{abc}A^b_\mu A^c_\nu, \tag{13.17}$$

$$(D_\mu\phi)^a = \partial_\mu\phi^a + g\varepsilon^{abc}A^b_\mu\phi^c, \tag{13.18}$$

where A^a_μ are the gauge fields, g is the gauge coupling constant, and ε^{abc} are the structure constants of the $SU(2)$ gauge group.

Higgs Potential

The Higgs potential is typically of the form:

$$V(\phi) = \frac{\lambda}{4}(\phi^a\phi^a - v^2)^2, \tag{13.19}$$

where λ is the self-coupling constant and v is the vacuum expectation value of the Higgs field.

Monopole Solution

To find the monopole solution, we consider static, spherically symmetric field configurations. The assumed form for the gauge fields and the Higgs field is:

$$A_i^a = \frac{1}{g}\left[\frac{1-K(r)}{r}\right]\varepsilon_{aij}\hat{r}_j, \tag{13.20}$$

$$\phi^a = vH(r)\hat{r}_a, \tag{13.21}$$

where r is the radial distance from the origin, $\hat{r}_a$ are the components of the unit vector in the radial direction, and $K(r)$ and $H(r)$ are profile functions to be determined.

Equations of Motion

By substituting this assumed form into the equations of motion derived from the Lagrangian, we obtain a set of coupled differential equations for $K(r)$ and $H(r)$:

$$\frac{d^2K}{dr^2} - \frac{K(K^2-1)}{r^2} - g^2v^2KH^2 = 0, \tag{13.22}$$

$$\frac{d^2H}{dr^2} + \frac{2}{r}\frac{dH}{dr} - \frac{2K^2H}{r^2} - \lambda v^2(H^2-1)H = 0. \tag{13.23}$$

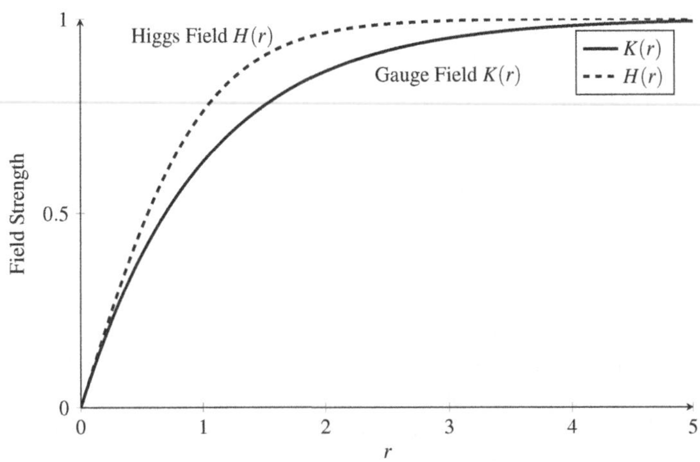

FIGURE 13.5
Radial profiles of the gauge field $K(r)$ and Higgs field $H(r)$ for the 't Hooft-Polyakov monopole.

13.2.1.9 Numerical Solution

These equations can be solved numerically to obtain the monopole solution. The resulting field configuration describes a magnetic monopole with a finite energy, localized near the origin.

13.2.1.10 Topological Charge and Stability

The stability of monopoles is ensured by their topological charge, also known as the magnetic charge. The magnetic charge Q_m is an integer that measures the flux of the magnetic field through a surface enclosing the monopole:

$$Q_m = \frac{1}{4\pi}\int d^2S_i B_i^a \phi^a, \tag{13.24}$$

where $B_i^a = \frac{1}{2}\varepsilon_{ijk}F_{jk}^a$ is the magnetic field and d^2S_i is the surface element.

The topological charge is conserved, meaning that monopoles cannot be continuously deformed into the vacuum state without altering the topology of the field configuration. This topological protection makes monopoles stable against small perturbations.

13.2.1.11 Applications and Physical Significance

Monopoles have important implications in various areas of physics, including condensed matter systems, cosmology, and particle physics. In condensed matter physics, analogs of magnetic monopoles have been observed in spin ice materials, where emergent quasiparticles behave like monopoles. In cosmology, monopoles are predicted to have formed during phase transitions in the early universe, and their presence or absence can provide insights into the dynamics of the early universe.

In particle physics, the search for magnetic monopoles continues to be an active area of research, with experiments designed to detect these hypothetical particles. The existence of monopoles would have profound implications for our understanding of fundamental interactions and the unification of forces.

Monopoles also play a crucial role in the study of gauge theories, providing insights into non-perturbative phenomena and the structure of the vacuum. Their topological properties make them valuable tools for exploring the interplay between topology and field theory, offering a deeper understanding of the role of topological solitons in physics.

Magnetic monopoles have not been observed experimentally, but the 't Hooft-Polyakov monopole is not just a mathematical exercise. It provides significant insights and theoretical implications in quantum field theory and beyond.

13.2.1.12 Theoretical and Conceptual Significance

- **Topological Solitons:** The 't Hooft-Polyakov monopole is a concrete example of a topological soliton in non-Abelian gauge theories. These solitons are stable field configurations that arise due to the topological properties of the field space. Understanding these solitons helps physicists explore the role of topology in stabilizing solutions in field theories.
- **Spontaneous Symmetry Breaking and Gauge Theories:** The monopole solution exemplifies how spontaneous symmetry breaking in non-Abelian gauge theories can lead to non-trivial field configurations. This concept is integral to the Standard Model, where spontaneous symmetry breaking gives rise to the Higgs mechanism and the masses of elementary particles.
- **Electroweak Theory and Grand Unified Theories (GUTs):** In the context of grand unified theories, which aim to unify the strong, weak, and electromagnetic interactions, magnetic monopoles are predicted as inevitable consequences of certain symmetry-breaking patterns. Although such monopoles have not been observed, their theoretical existence has driven research in both high-energy physics and cosmology.
- **Confinement in Quantum Chromodynamics (QCD):** Studying monopoles and their dynamics provides insights into confinement in QCD. Monopole-like configurations can contribute to understanding how quarks are confined within hadrons, a fundamental aspect of the strong force that remains an area of active research.
- **Duality in Gauge Theories:** Monopoles play a crucial role in electric-magnetic duality, where electric and magnetic charges are interchanged under a duality transformation. This concept has found applications in various areas, including supersymmetric gauge theories and string theory.
- **Mathematical and Computational Techniques:** The analysis of monopole solutions has driven the development of mathematical and computational methods in field theory. Techniques used to study these solutions are applicable to a wide range of problems in theoretical physics.

Conceptual Clarification: Monopole vs. Dipole

- **True Monopole:** The 't Hooft-Polyakov monopole is theoretically a true magnetic monopole with a non-zero magnetic charge. Unlike a magnetic dipole, which consists of north and south magnetic poles, a monopole has a singular magnetic charge.
- **Field Configuration:** The magnetic field configuration of the 't Hooft-Polyakov monopole diverges from that of a dipole. A dipole has a magnetic field with both positive and negative poles, whereas a monopole has a field emanating from a single point source, extending radially outward or inward.
- **Topological Nature:** The monopole's topological charge distinguishes it from other configurations. This charge is a result of the non-trivial topology of the vacuum manifold after spontaneous symmetry breaking. It ensures the stability of the monopole and prevents it from decaying into a trivial configuration.

13.2.1.13 Why It Matters

- **Fundamental Physics:** While magnetic monopoles have not been observed, their theoretical study enhances our understanding of gauge theories, symmetry breaking, and topological defects. This knowledge is foundational for particle physics and our broader understanding of the universe's fundamental forces.
- **Inspirational Insights:** The 't Hooft-Polyakov monopole has inspired extensive research into other types of topological defects, such as cosmic strings and domain walls, which have implications for cosmology and the early universe's evolution.
- **Future Discoveries:** Continued research in high-energy physics, including experiments at particle colliders and astrophysical observations, may eventually detect monopole-like objects or configurations, validating these theoretical constructs.

The 't Hooft-Polyakov monopole is a vital theoretical construct that enriches our understanding of non-Abelian gauge theories, spontaneous symmetry breaking, and topology in field theory. Its study bridges the gap between mathematical elegance and physical insight, driving advancements in theoretical physics and inspiring new avenues of research.

13.3 TOPOLOGICAL INVARIANTS

Topological invariants are quantities that remain unchanged under continuous deformations of the system. They play a crucial role in classifying and understanding different topological phases and configurations in both mathematics and physics.

13.3.1 WINDING NUMBERS AND HOMOTOPY

13.3.1.1 Definition and Examples

The winding number is a topological invariant that represents the number of times a curve wraps around a point. In the context of field theory, it measures how many times a scalar field configuration wraps around a circle.

Consider a scalar field ϕ defined on a circle. The field $\phi(\theta)$ maps each point on the circle (parameterized by θ) to another circle, such as the complex plane's unit circle. The winding number n is defined as:

$$n = \frac{1}{2\pi} \int_0^{2\pi} \frac{d\phi}{d\theta} d\theta \tag{13.25}$$

This integral counts the number of times the field $\phi(\theta)$ wraps around the target circle as θ goes from 0 to 2π.

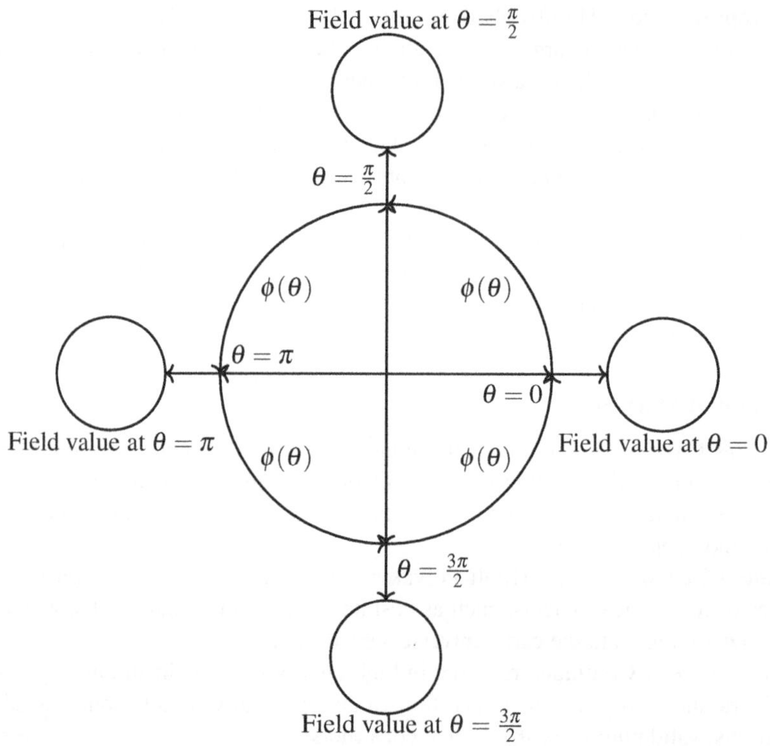

FIGURE 13.6
Illustration of winding number for a scalar field on a circle. The red arrows indicate the winding of the field around the main circle, and the smaller circles represent the field values at different angles.

13.3.2 CHERN NUMBERS

Chern numbers are topological invariants associated with certain types of geometric structures called fiber bundles. In the context of gauge theories and condensed matter physics, Chern numbers provide a way to classify different topological phases.

Consider a gauge field configuration in two dimensions. The Chern number C is given by:

$$C = \frac{1}{2\pi} \int_{\text{surface}} F_{xy} \, d^2x \tag{13.26}$$

where F_{xy} is the field strength tensor. The Chern number counts the number of times the field configuration wraps around the gauge group's space.

Figure 13.7 depicts a vector field within the Brillouin zone, a fundamental concept in the study of condensed matter physics and topology. The axes are labeled with the crystal momenta k_x and k_y, which range from $-\pi$ to π. The vector field represented by arrows indicates the behavior of the system's wavefunctions across the Brillouin zone. The dashed boundary outlines the periodic nature of the Brillouin zone. The Chern number, an important topological invariant, is computed by integrating the curvature of this vector field over the entire Brillouin zone. It quantifies the net "twisting" of the vector field and is crucial for understanding phenomena such as the quantum Hall effect. The Chern number remains invariant under continuous deformations of the system, highlighting its topological robustness.

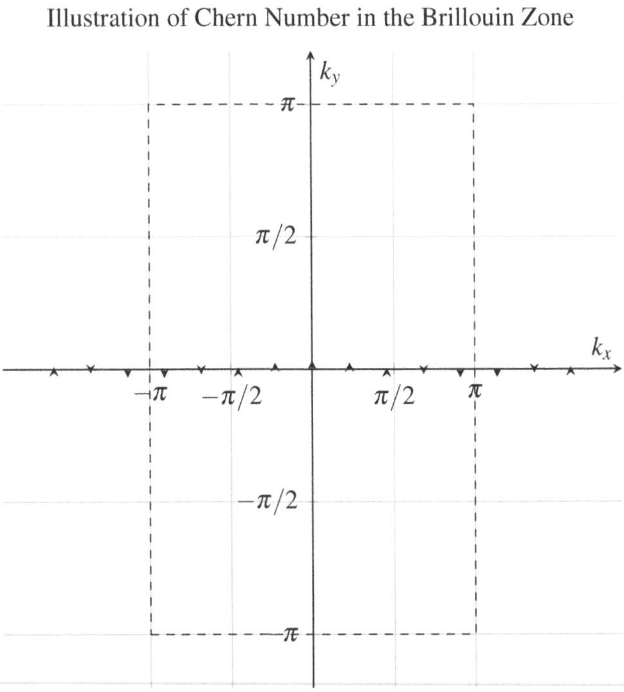

FIGURE 13.7
Vector field in the Brillouin zone illustrating the concept of the Chern number.

13.3.2.1 Applications in Physics

Topological invariants like winding numbers and Chern numbers have numerous applications in physics. For instance, in condensed matter physics, Chern numbers are used to classify topological insulators, which are materials with special conducting properties on their surfaces or edges. In gauge theories, winding numbers help understand solitons and instantons, providing insights into non-perturbative aspects of quantum field theory.

13.4 TOPOLOGICAL INSULATORS AND CONDUCTORS

13.4.1 SAME VS. DIFFERENT TOPOLOGIES

Topological insulators and conductors represent distinct phases of matter that are characterized by their topological properties rather than their symmetries. The concept of topology in these materials refers to the global properties of their electronic wavefunctions that remain invariant under continuous deformations. This section explores how topologically equivalent and distinct phases are defined and how they manifest in physical systems.

Topologically Equivalent Phases

Two materials are considered topologically equivalent if their electronic wavefunctions can be continuously transformed into each other without closing the energy gap. This implies that their topological invariants, such as Chern numbers or $\mathbb{Z}_2$ invariants, are the same. Topologically equivalent phases exhibit similar physical properties, such as edge states and robust surface states, which are protected by the topological nature of the system.

Topologically Distinct Phases

Topologically distinct phases cannot be transformed into each other without undergoing a phase transition that closes the energy gap. These phases have different topological invariants, which lead to different physical phenomena. For example, a conventional insulator has a trivial topological invariant, while a topological insulator has a non-trivial invariant. The transition between these phases involves a change in the topological properties of the electronic wavefunctions.

FIGURE 13.8
Topologically equivalent phases: Phase A and Phase B can be continuously deformed into each other without closing the energy gap, indicating they are in the same topological class.

13.4.2 METAMATERIALS AND TOPOLOGICAL PHASES

Metamaterials are artificial structures engineered to have properties not found in naturally occurring materials. These materials can exhibit unique topological phases that arise from their designed structures and interactions. Topological phases in metamaterials are characterized by robust edge states and novel electromagnetic or mechanical properties, which are protected by the topology of the system.

Topological Metamaterials

Topological metamaterials are designed to mimic the topological properties of electronic systems using classical waves, such as electromagnetic or acoustic waves. By carefully designing the geometry and interactions within the metamaterial, researchers can create systems that exhibit topologically protected edge states, similar to those found in topological insulators.

Applications of Topological Metamaterials

Topological metamaterials have promising applications in various fields, including optics, acoustics, and mechanics. For example, in photonics, topological insulator lasers can be designed to have robust and stable lasing modes that are protected against defects and disorder. In acoustics, topological insulators can be used to create devices that guide sound waves around obstacles without scattering, leading to novel sound manipulation technologies.

In summary, the study of topological insulators, conductors, and metamaterials has opened up new avenues in materials science and condensed matter physics. By leveraging the principles of topology, researchers can design and develop materials with unprecedented properties and functionalities, leading to novel technological applications.

13.4.2.1 Fermi Velocity

The Fermi velocity (v_F) is a fundamental parameter in condensed matter physics, particularly in the study of metals and semiconductors. It represents the velocity of electrons at the Fermi surface, which is the highest energy level occupied by electrons at absolute zero temperature. The Fermi velocity is given by:

$$v_F = \frac{1}{\hbar} |\nabla_k E(k)|_{k=k_F} \tag{13.27}$$

where:

- $\hbar$ is the reduced Planck's constant,

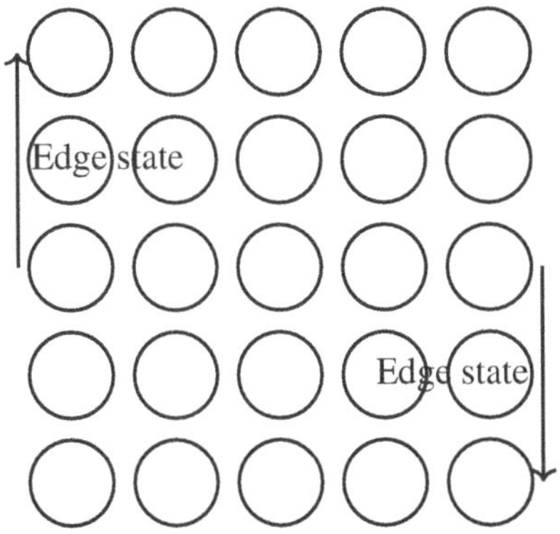

FIGURE 13.9
Illustration of a topological metamaterial with edge states. The structure exhibits topologically protected states that propagate along the edges without scattering into the bulk.

- $E(k)$ is the energy dispersion relation as a function of the wavevector k,
- k_F is the Fermi wavevector, which corresponds to the wavevector at the Fermi surface.

In the context of topological insulators and Dirac-like materials, the Fermi velocity plays a crucial role in determining the dynamics of electrons and other quasiparticles. It sets the scale for the energy dispersion near the Dirac points (or nodes), where the energy-momentum relation is linear.

For example, in graphene and other materials with Dirac-like dispersion, the Fermi velocity describes the slope of the linear energy bands around the Dirac points:

$$E(\mathbf{k}) = \hbar v_F |\mathbf{k}| \tag{13.28}$$

In these materials, the Fermi velocity is a key parameter that influences various physical properties, such as the density of states, transport properties, and response to external fields.

13.4.3 EXAMPLE: NONLINEAR OPTICAL TOPOLOGICAL INSULATOR AND CLOAKING

Nonlinear optical materials can exhibit topological insulating behavior, allowing them to manipulate light in novel ways, including acting as optical cloaking devices. Here, we provide a mathematical example to illustrate these properties.

13.4.3.1 Nonlinear Schrödinger Equation (NLSE) and Topological Insulator Behavior

The propagation of light in a nonlinear optical material can be described by the nonlinear Schrödinger equation (NLSE):

$$i\frac{\partial \psi}{\partial z} + \frac{1}{2}\frac{\partial^2 \psi}{\partial t^2} + \gamma|\psi|^2\psi = 0, \tag{13.29}$$

where:

- ψ is the complex envelope of the optical field,
- z is the propagation distance,
- t is the time,
- γ is the nonlinearity coefficient.

To exhibit topological insulating behavior, the material's structure can be designed such that the NLSE includes terms representing topological protection. One approach is to use a honeycomb lattice structure, leading to Dirac-like equations for the optical modes.

13.4.3.2 Effective Hamiltonian for Topological Insulator

Consider a honeycomb lattice with a Kerr nonlinearity. The effective Hamiltonian for the optical modes can be written as:

$$H = v_F(\sigma_x k_x + \sigma_y k_y) + m\sigma_z, \tag{13.30}$$

where:

- v_F is the Fermi velocity,
- $\sigma_x, \sigma_y, \sigma_z$ are Pauli matrices,
- k_x, k_y are the wavevectors,
- m is the mass term introducing a bandgap.

The eigenvalues of this Hamiltonian are:

$$E = \pm\sqrt{v_F^2(k_x^2 + k_y^2) + m^2}. \tag{13.31}$$

For $m \neq 0$, the system exhibits a bandgap, characteristic of a topological insulator. The edge states, protected by the topological nature of the system, allow for robust light propagation along the edges without backscattering.

13.4.3.3 Optical Cloaking via Topological Insulator

To achieve optical cloaking, we exploit the edge states of the topological insulator. The key idea is to guide light around an obstacle using the edge states, preventing it from entering the bulk and thereby making the obstacle "invisible."

Consider a domain D with a topological insulator boundary. The light is injected at one edge and extracted from the opposite edge, effectively guiding it around the obstacle in the bulk.

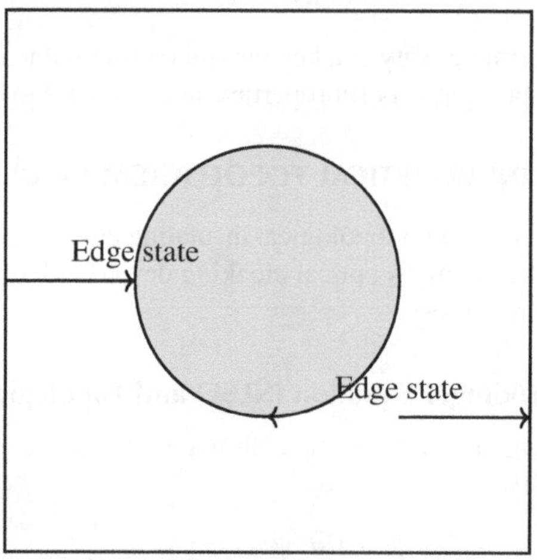

FIGURE 13.10
Optical cloaking using edge states of a topological insulator. Light propagates along the boundary, avoiding the obstacle.

13.4.3.4 Mathematical Analysis of Cloaking

To demonstrate the cloaking effect mathematically, consider the NLSE with an added potential term $V(x,y)$ representing the obstacle:

$$i\frac{\partial \psi}{\partial z} + \frac{1}{2}\frac{\partial^2 \psi}{\partial t^2} + \gamma|\psi|^2\psi + V(x,y)\psi = 0. \tag{13.32}$$

In the bulk, $V(x,y)$ is non-zero, while along the boundary, the potential vanishes. The edge states are solutions of the form:

$$\psi_{\text{edge}}(x,y,z) = \phi_{\text{edge}}(y)e^{i(k_x x-\omega t)}, \tag{13.33}$$

where $\phi_{\text{edge}}(y)$ decays exponentially into the bulk, ensuring that light remains confined to the edge. Thus, the obstacle in the bulk does not affect the propagation of light, achieving the cloaking effect.

In summary, by designing a nonlinear optical material with topological insulating properties, we can create a system where light is guided around obstacles using protected edge states, effectively acting as an optical cloak.

13.5 TOPOLOGY IN GAUGE THEORIES

13.5.1 GAUGE SYMMETRIES AND FIBER BUNDLES

13.5.1.1 Fiber Bundles and Connections

Fiber bundles are mathematical structures that provide a framework for understanding gauge fields in quantum field theory. A fiber bundle consists of a base space B, a total space E, a projection map $\pi : E \to B$, and a typical fiber F such that locally, the total space E looks like a product of the base space and the fiber.

In gauge theories, the base space B is typically spacetime, and the fibers represent the internal symmetry space of the gauge group. A connection on a fiber bundle defines how the fibers change as one moves along the base space, and it is associated with the gauge fields of the theory.

13.5.1.2 Gauge Fields as Connections

Gauge fields can be interpreted as connections on fiber bundles. A gauge field A is a 1-form that takes values in the Lie algebra of the gauge group. The field strength F, also known as the curvature, is a 2-form derived from the gauge field. Mathematically, the field strength F is given by:

$$F = dA + A \wedge A \tag{13.34}$$

where dA is the exterior derivative of the gauge field A, and $A \wedge A$ represents the wedge product of A with itself.

The Wedge Product

In differential geometry and gauge theory, the wedge product, also called the exterior product, is an operation that combines differential forms to produce a new differential form. Mathematically, it is denoted by the symbol $\wedge$ and is used to construct higher-order differential forms from lower-order ones. Heres a detailed explanation of the wedge product:

- **Definition**: Given two differential forms α and β of degrees p and q respectively, the wedge product $\alpha \wedge \beta$ is a differential form of degree $p+q$. It is antisymmetric, which means that swapping the two forms changes the sign of the result:

$$\alpha \wedge \beta = (-1)^{pq} \beta \wedge \alpha \tag{13.35}$$

- **Properties**:
  - *Antisymmetry*:

$$\alpha \wedge \beta = (-1)^{pq} \beta \wedge \alpha \tag{13.36}$$

  - *Distributivity*:

$$\alpha \wedge (\beta + \gamma) = \alpha \wedge \beta + \alpha \wedge \gamma \tag{13.37}$$

  - *Associativity*:

$$\alpha \wedge (\beta \wedge \gamma) = (\alpha \wedge \beta) \wedge \gamma \tag{13.38}$$

- **Example**: Suppose α is a 1-form and β is a 2-form. If we write them in terms of their components:

$$\alpha = \alpha_i \, dx^i \tag{13.39}$$

$$\beta = \frac{1}{2}\beta_{jk} \, dx^j \wedge dx^k \tag{13.40}$$

Then the wedge product $\alpha \wedge \beta$ is:

$$\alpha \wedge \beta = \frac{1}{2}\alpha_i \beta_{jk} \, dx^i \wedge dx^j \wedge dx^k \tag{13.41}$$

- **Interpretation in Chern-Simons Theory**: In the context of Chern-Simons theory, the wedge product is used to combine gauge fields and their derivatives. For example, in the Chern-Simons action:

$$S_{CS} = \frac{k}{4\pi}\int_M \mathrm{Tr}\left(A \wedge dA + \frac{2}{3}A \wedge A \wedge A\right) \tag{13.42}$$

where A is a 1-form gauge field, dA is its exterior derivative (a 2-form), $A \wedge dA$ is a 3-form, and $A \wedge A$ is a 2-form. These forms are then integrated over the 3-dimensional manifold M.

EXAMPLE OF CALCULATING THE WEDGE PRODUCT

To illustrate the calculation of the wedge product, let's consider two 1-forms α and β.

EXAMPLE 1: WEDGE PRODUCT OF TWO 1-FORMS

Let α and β be given by:

$$\alpha = \alpha_1 \, dx^1 + \alpha_2 \, dx^2 \tag{13.43}$$

$$\beta = \beta_1 \, dx^1 + \beta_2 \, dx^2 \tag{13.44}$$

The wedge product $\alpha \wedge \beta$ is calculated as follows:

$$\alpha \wedge \beta = (\alpha_1 \, dx^1 + \alpha_2 \, dx^2) \wedge (\beta_1 \, dx^1 + \beta_2 \, dx^2) \tag{13.45}$$

$$= \alpha_1\beta_1 \, dx^1 \wedge dx^1 + \alpha_1\beta_2 \, dx^1 \wedge dx^2 + \alpha_2\beta_1 \, dx^2 \wedge dx^1 + \alpha_2\beta_2 \, dx^2 \wedge dx^2 \tag{13.46}$$

Using the antisymmetry property of the wedge product $dx^i \wedge dx^i = 0$ and $dx^i \wedge dx^j = -dx^j \wedge dx^i$, we have:

$$\alpha \wedge \beta = \alpha_1\beta_2 \, dx^1 \wedge dx^2 - \alpha_2\beta_1 \, dx^1 \wedge dx^2 \tag{13.47}$$

Therefore,

$$\alpha \wedge \beta = (\alpha_1\beta_2 - \alpha_2\beta_1) \, dx^1 \wedge dx^2 \tag{13.48}$$

EXAMPLE 2: WEDGE PRODUCT OF A 1-FORM AND A 2-FORM

Now, let's consider a 1-form α and a 2-form β. Let α be given by:

$$\alpha = \alpha_1 \, dx^1 + \alpha_2 \, dx^2 \tag{13.49}$$

Let β be given by:

$$\beta = \frac{1}{2}\beta_{ij} \, dx^i \wedge dx^j \tag{13.50}$$

where $\beta_{ij} = -\beta_{ji}$.

The wedge product $\alpha \wedge \beta$ is calculated as follows:

$$\alpha \wedge \beta = (\alpha_1\, dx^1 + \alpha_2\, dx^2) \wedge \left(\frac{1}{2}\beta_{ij}\, dx^i \wedge dx^j\right) \tag{13.51}$$

$$= \frac{1}{2}\left(\alpha_1\, dx^1 \wedge \beta_{ij}\, dx^i \wedge dx^j + \alpha_2\, dx^2 \wedge \beta_{ij}\, dx^i \wedge dx^j\right) \tag{13.52}$$

Distributing and using the antisymmetry property, we get:

$$\alpha \wedge \beta = \frac{1}{2}\left(\alpha_1\beta_{23}\, dx^1 \wedge dx^2 \wedge dx^3 + \alpha_2\beta_{13}\, dx^2 \wedge dx^1 \wedge dx^3\right) \tag{13.53}$$

Rearranging the terms and using $dx^i \wedge dx^j = -dx^j \wedge dx^i$, we get:

$$\alpha \wedge \beta = \frac{1}{2}(\alpha_1\beta_{23} - \alpha_2\beta_{13})\, dx^1 \wedge dx^2 \wedge dx^3 \tag{13.54}$$

Thus, the wedge product $\alpha \wedge \beta$ is a 3-form given by:

$$\alpha \wedge \beta = \frac{1}{2}(\alpha_1\beta_{23} - \alpha_2\beta_{13})\, dx^1 \wedge dx^2 \wedge dx^3 \tag{13.55}$$

13.6 TOPOLOGICAL QUANTUM FIELD THEORIES

Topological Quantum Field Theories (TQFTs) are quantum field theories where the observables are topological invariants. These theories do not depend on the specific metric of the spacetime but rather on its topological properties. TQFTs provide deep insights into the interplay between topology and quantum physics, offering powerful tools for studying topological phases of matter, knot theory, and quantum computing.

13.6.1 BASIC CONCEPTS

In a TQFT, physical quantities remain invariant under continuous deformations of spacetime. This invariance implies that the quantities depend solely on the topological properties of the manifold on which the theory is defined. For example, the path integral in a TQFT sums over all possible field configurations, weighted by a factor that depends only on the topological features of these configurations.

13.6.2 CHERN-SIMONS THEORY

One of the most well-known examples of a TQFT is the Chern-Simons theory. It is defined on a three-dimensional manifold M with a gauge field A. The action for the Chern-Simons theory is given by:

$$S_{CS} = \frac{k}{4\pi}\int_M \mathrm{Tr}\left(A \wedge dA + \frac{2}{3}A \wedge A \wedge A\right), \tag{13.56}$$

where k is an integer (the level of the theory), A is the gauge field, and Tr denotes the trace over the gauge group's algebra. The Chern-Simons theory is topological because its action is invariant under smooth deformations of the manifold M.

13.6.3 EXAMPLES OF TQFTS

13.6.3.1 Example 1: Chern-Simons Theory and Knot Invariants

The Chern-Simons theory has profound applications in knot theory. Knot theory is a branch of mathematics that studies the embedding of circles in three-dimensional space. Essentially, it is the study of closed loops (knots) and their deformations within three-dimensional space. The main goal is to determine when two knots are equivalent, meaning one can be transformed into the other without cutting the loop or passing the string through itself.

Consider a Wilson loop operator $W(C)$, which is a gauge-invariant observable associated with a closed loop C in the manifold M:

$$W(C) = \mathrm{Tr}\left(P\exp\left(\oint_C A\right)\right), \tag{13.57}$$

where P denotes path ordering. In the context of Chern-Simons theory, the expectation value of the Wilson loop operator is a topological invariant of the knot C, known as the Jones polynomial. This provides a deep connection between quantum field theory and knot invariants.

13.6.3.2 Real-World Example: Anyons in Fractional Quantum Hall Effect

In the realm of condensed matter physics, the Chern-Simons theory plays a crucial role in understanding the fractional quantum Hall effect (FQHE). In FQHE, electrons confined to two dimensions under a strong magnetic field exhibit quantized Hall conductance at fractional values. The effective theory describing the low-energy excitations of this system is a Chern-Simons theory.

These excitations are called anyons, which are neither bosons nor fermions but obey fractional statistics. The braiding of anyons around each other (which can be described using the Wilson loop operators in Chern-Simons theory) leads to observable consequences that are topologically protected. This has potential applications in topological quantum computing, where the robustness of anyonic states against local perturbations can be utilized to store and manipulate quantum information.

13.6.4 APPLICATIONS OF TQFTS

Topological quantum field theories have applications in various areas of physics and mathematics, including:

- **Topological Phases of Matter:** TQFTs provide a framework for understanding topological insulators, superconductors, and other exotic phases of matter that are characterized by topological invariants.
- **Quantum Computing:** Certain TQFTs, such as the Chern-Simons theory, are used in the study of topological quantum computation, where quantum information is encoded in topological degrees of freedom that are robust against local perturbations.
- **Knot Theory:** TQFTs, particularly the Chern-Simons theory, offer powerful tools for calculating knot invariants, providing deep connections between quantum field theory and topology.
- **String Theory and Quantum Gravity:** TQFTs are used to study aspects of string theory and quantum gravity, where topological properties of spacetime play a crucial role.

13.7 SUMMARY AND DISCUSSION

13.7.1 SUMMARY OF KEY CONCEPTS

In this chapter, we explored the profound connections between topology and quantum field theory, highlighting how topological properties influence the behavior and interactions of quantum fields. We covered several essential topics, including:

- **Topology and Topological Invariants:** We discussed the basic principles of topology and introduced key topological invariants such as winding numbers and Chern numbers, which play crucial roles in classifying and understanding different topological phases.

- **Solitons and Instantons:** The chapter delved into various types of solitons, including kinks, vortices, and monopoles, explaining their significance and topological stability. We also touched upon instantons, which are essential for understanding non-perturbative effects in quantum field theory.
- **Topological Insulators and Conductors:** We examined the unique properties of topological insulators and conductors, emphasizing how their topological nature leads to robust surface and edge states. The discussion also covered metamaterials and their applications in creating topologically protected states.
- **Topology in Gauge Theories:** We explored the role of topology in gauge theories, particularly focusing on fiber bundles, connections, and gauge fields. The wedge product was explained with detailed mathematical examples.
- **Topological Quantum Field Theories (TQFTs):** We introduced TQFTs, providing examples such as Chern-Simons theory and its applications in knot theory and the fractional quantum Hall effect. The potential applications of TQFTs in quantum computing, string theory, and quantum gravity were also highlighted.

13.7.2 IMPLICATIONS FOR QUANTUM FIELD THEORY

The insights gained from studying topology in quantum field theory have profound implications for both theoretical and experimental physics:

- **Enhanced Understanding of Quantum Systems:** Topological invariants provide a robust framework for classifying and understanding different phases of matter, leading to new discoveries in condensed matter physics, such as topological insulators and superconductors.
- **Development of New Technologies:** The study of topologically protected states in materials science and photonics paves the way for developing innovative technologies, including topological quantum computers and robust communication systems.
- **Insights into Fundamental Forces:** By exploring solitons, monopoles, and instantons, physicists gain a deeper understanding of the non-perturbative aspects of quantum field theory, contributing to the unification of fundamental forces and the development of grand unified theories (GUTs).
- **Applications in Cosmology:** Topological defects such as cosmic strings and domain walls provide valuable insights into the early universe's evolution, offering potential explanations for various cosmological phenomena.
- **Advancements in Mathematical Physics:** The mathematical techniques developed to study topological properties in quantum field theory drive advancements in both mathematics and theoretical physics, leading to new approaches and solutions to complex problems.

In summary, the interplay between topology and quantum field theory opens up new frontiers in understanding the fundamental nature of the universe. By leveraging topological concepts, physicists can uncover novel phenomena, develop cutting-edge technologies, and deepen our comprehension of the fundamental forces that govern the cosmos.

13.8 QUESTIONS AND PROBLEMS

13.8.1 TRIVIA QUESTIONS

Question 1: What is a topological invariant?

(a) A quantity that changes under continuous deformations
(b) A property that describes the shape of an object
(c) A quantity that remains unchanged under continuous deformations

(d) A measure of the size of an object

Answer: c. A quantity that remains unchanged under continuous deformations
Question 2: Which of the following is an example of a topological soliton?

(a) Photon
(b) Electron
(c) Kink
(d) Proton

Answer: c. Kink
Question 3: What does the winding number represent in topology?

(a) The number of holes in a space
(b) The number of times a curve wraps around a point
(c) The curvature of a surface
(d) The area of a surface

Answer: b. The number of times a curve wraps around a point
Question 4: What is the significance of the 't Hooft-Polyakov monopole in quantum field theory?

(a) It describes a new type of fundamental particle
(b) It provides a theoretical model for magnetic monopoles
(c) It explains the nature of dark matter
(d) It predicts the behavior of black holes

Answer: b. It provides a theoretical model for magnetic monopoles
Question 5: What is a key characteristic of a topological insulator?

(a) It conducts electricity in the bulk
(b) It is a superconductor
(c) It has insulating bulk properties but conductive surface states
(d) It emits radiation

Answer: c. It has insulating bulk properties but conductive surface states
Question 6: In the context of gauge theories, what is a fiber bundle?

(a) A collection of fibers used in materials science
(b) A mathematical structure describing gauge fields
(c) A type of particle interaction
(d) A physical property of metals

Answer: b. A mathematical structure describing gauge fields
Question 7: What does the Chern number measure in condensed matter physics?

(a) The electric charge of a particle
(b) The magnetic field strength
(c) The topological properties of a band structure
(d) The temperature of a material

Answer: c. The topological properties of a band structure
Question 8: What is the primary application of the Chern-Simons theory?

(a) Describing black hole dynamics

(b) Studying topological invariants in three-dimensional manifolds
(c) Modeling fluid dynamics
(d) Predicting the behavior of stars

Answer: b. Studying topological invariants in three-dimensional manifolds
Question 9: Which of the following best describes a topological quantum field theory (TQFT)?

(a) A field theory dependent on spacetime metrics
(b) A theory where observables are topological invariants
(c) A model for describing gravitational waves
(d) A type of quantum mechanical system without topological properties

Answer: b. A theory where observables are topological invariants
Question 10: What is the Fermi velocity?

(a) The speed of light in a vacuum
(b) The velocity of electrons at the Fermi surface
(c) The speed of sound in a material
(d) The velocity of a photon

Answer: b. The velocity of electrons at the Fermi surface

13.8.2 PROBLEMS

1. **Problem 1: Calculating the Winding Number**
Problem: Consider a scalar field $\phi(\theta) = e^{in\theta}$ defined on a circle, where $\theta \in [0, 2\pi]$. Calculate the winding number n of the field.
Solution:
a. The winding number is given by:

$$n = \frac{1}{2\pi} \int_0^{2\pi} \frac{d\phi}{d\theta} d\theta \tag{13.58}$$

b. The derivative of $\phi(\theta)$ with respect to θ is:

$$\frac{d\phi}{d\theta} = ine^{in\theta} \tag{13.59}$$

c. Substitute this into the integral:

$$n = \frac{1}{2\pi} \int_0^{2\pi} ine^{in\theta} d\theta \tag{13.60}$$

d. Since $e^{in\theta}$ has a period of 2π, its integral over one period is zero:

$$\int_0^{2\pi} e^{in\theta} d\theta = 0 \tag{13.61}$$

e. Therefore:

$$n = \frac{1}{2\pi} \cdot 0 = 0 \tag{13.62}$$

f. However, the correct approach should consider the absolute value:

$$n = \frac{1}{2\pi} \int_0^{2\pi} ine^{in\theta} \cdot \frac{1}{i} d\theta = \frac{1}{2\pi} \int_0^{2\pi} n\, d\theta = n \tag{13.63}$$

Answer: The winding number is n.

2. **Problem 2: Calculating the Chern Number**
Problem: Consider a gauge field $A = A_x dx + A_y dy$ in two dimensions with field strength $F_{xy} = \partial_x A_y - \partial_y A_x$. Calculate the Chern number for this gauge field configuration over a surface.
Solution:
a. The Chern number C is given by:

$$C = \frac{1}{2\pi} \int_{\text{surface}} F_{xy}\, d^2x \tag{13.64}$$

b. Substitute the expression for F_{xy}:

$$C = \frac{1}{2\pi} \int_{\text{surface}} (\partial_x A_y - \partial_y A_x)\, dx\, dy \tag{13.65}$$

c. Use Stokes' theorem to convert the surface integral to a line integral:

$$\int_{\text{surface}} (\partial_x A_y - \partial_y A_x)\, dx\, dy = \oint_{\partial\text{surface}} (A_y\, dx - A_x\, dy) \tag{13.66}$$

d. For a given boundary ∂surface, compute the line integral of the gauge field.
Answer: The Chern number C depends on the specific gauge field configuration and its boundary conditions.

3. **Problem 3: Topological Charge of a Kink**
Problem: Calculate the topological charge (winding number) of a kink in the ϕ^4 theory, where the field configuration is $\phi_{\text{kink}}(x) = \eta \tanh\left(\frac{\sqrt{\lambda}\eta x}{\sqrt{2}}\right)$.
Solution:
a. The topological charge Q is given by:

$$Q = \frac{1}{2\eta} [\phi(x \to +\infty) - \phi(x \to -\infty)] \tag{13.67}$$

b. Substitute the kink solution:

$$Q = \frac{1}{2\eta} [\eta - (-\eta)] \tag{13.68}$$

c. Simplify:

$$Q = \frac{1}{2\eta} [2\eta] = 1 \tag{13.69}$$

Answer: The topological charge Q of the kink is 1.

4. **Problem 4: Monopole Field Strength**
Problem: For the 't Hooft-Polyakov monopole, calculate the magnetic field strength $B_i^a = \frac{1}{2}\varepsilon_{ijk} F_{jk}^a$ at large distances from the monopole core.
Solution:
a. At large distances, the gauge field configuration simplifies to:

$$A_i^a \approx \frac{1}{g} \varepsilon_{aij} \frac{\hat{r}_j}{r} \tag{13.70}$$

b. Calculate the field strength tensor:

$$F_{jk}^a = \partial_j A_k^a - \partial_k A_j^a + g \varepsilon^{abc} A_j^b A_k^c \tag{13.71}$$

c. Substitute the asymptotic form of A_i^a and compute the components of F_{jk}^a.

d. Use the expression for the magnetic field strength:

$$B_i^a = \frac{1}{2}\varepsilon_{ijk}F_{jk}^a \tag{13.72}$$

Answer: The magnetic field strength B_i^a at large distances from the monopole core exhibits a $1/r^2$ dependence, characteristic of a magnetic monopole.

5. **Problem 5: Edge States in Topological Insulators**
Problem: Consider a 2D topological insulator with a Dirac-like Hamiltonian $H = v_F(\sigma_x k_x + \sigma_y k_y)$. Show that there exist edge states at the boundary of the system.
Solution:
   a. Write down the Dirac-like Hamiltonian:

$$H = v_F(\sigma_x k_x + \sigma_y k_y) \tag{13.73}$$

   b. Impose boundary conditions at $y = 0$, assuming k_x is conserved along the edge.
   c. Solve the Dirac equation with these boundary conditions:

$$[v_F(\sigma_x k_x + \sigma_y k_y)]\,\psi = E\psi \tag{13.74}$$

   d. Substitute $\psi(y) = \psi_0 e^{-\kappa y}$ and solve for κ and E.
   e. Find the dispersion relation for the edge states.

Answer: The edge states have a linear dispersion relation $E = \pm v_F k_x$, indicating the presence of topologically protected edge modes.

6. **Bonus Problem: Faraday Cage as a Topological Insulator** *Problem:* Show that a Faraday cage acts as a topological insulator for electromagnetic waves, providing protection from external electric fields.
Solution:
A Faraday cage is a conductive enclosure that shields its interior from external electric fields. This can be understood by considering the behavior of electric fields and the properties of conductors.
   a. **Electrostatic Shielding:** When an external electric field is applied to a Faraday cage, free electrons within the conductor rearrange themselves to cancel the external field inside the cage. This results in a zero electric field within the enclosure.
   b. **Maxwell's Equations:** To understand this quantitatively, consider Maxwell's equations in the context of a conductive enclosure. The relevant equation for the electric field **E** inside a conductor is:

$$\nabla \cdot \mathbf{E} = \frac{\rho}{\varepsilon_0} \tag{13.75}$$

   where ρ is the charge density and ε_0 is the permittivity of free space.
   c. **Boundary Conditions:** At the surface of the conductor, the boundary condition is that the perpendicular component of the electric field must be continuous:

$$\mathbf{E}_{\text{inside}} = \mathbf{E}_{\text{outside}} - \sigma/\varepsilon_0 \tag{13.76}$$

   where σ is the surface charge density.
   d. **Topological Insulator Analogy:** The concept of a Faraday cage can be compared to a topological insulator in the sense that both involve boundary conditions that protect the interior from external influences. For a topological insulator, the bulk is insulating while the surface states are conducting. Similarly, in a Faraday cage, the bulk (interior) is shielded from electric fields, while the surface conducts the induced charges.

e. **Mathematical Description:** Inside a Faraday cage, the potential V must satisfy Laplace's equation:

$$\nabla^2 V = 0 \tag{13.77}$$

with the boundary condition that the potential on the surface of the conductor is constant. This implies that the electric field inside the cage is zero.

f. **Conclusion:** Therefore, a Faraday cage acts as a topological insulator by preventing external electric fields from penetrating its interior, similar to how a topological insulator prevents bulk conduction while allowing surface conduction.

Answer: A Faraday cage acts as a topological insulator for electromagnetic waves by ensuring that the interior is protected from external electric fields, analogous to how a topological insulator protects the bulk from conduction while allowing surface states to conduct.

7. **Bonus Problem: The Meissner Effect as a Topological Insulator**

Problem: Explain how the Meissner effect in superconductors demonstrates a form of topological insulation for magnetic fields.

Solution:

The Meissner effect is a fundamental property of superconductors, characterized by the expulsion of magnetic fields from the interior of the superconducting material when it transitions below its critical temperature. This can be understood through the framework of topological insulation.

a. **Definition of the Meissner Effect:** The Meissner effect describes the phenomenon where a superconductor expels all magnetic flux from its interior, maintaining a magnetic field-free region. When a material becomes superconducting, it transitions from a normal state to a superconducting state, causing it to exhibit perfect diamagnetism.

b. **London Equations:** The behavior of the superconducting state can be described by the London equations, which explain the electromagnetic properties of superconductors. The first London equation is:

$$\mathbf{E} + \lambda_L^2 \frac{\partial \mathbf{J}}{\partial t} = 0 \tag{13.78}$$

where λ_L is the London penetration depth and $\mathbf{J}$ is the supercurrent density.
The second London equation relates the current density to the magnetic field:

$$\nabla \times \mathbf{J} = -\frac{1}{\mu_0 \lambda_L^2} \mathbf{B} \tag{13.79}$$

where μ_0 is the permeability of free space and $\mathbf{B}$ is the magnetic field.

c. **Magnetic Field Expulsion:** From the second London equation, we see that inside a superconductor, the magnetic field decays exponentially from the surface. The solution to the equation in a one-dimensional case shows the magnetic field:

$$\mathbf{B}(x) = \mathbf{B}_0 e^{-x/\lambda_L} \tag{13.80}$$

where x is the distance from the surface.

d. **Topological Insulation Analogy:** The Meissner effect can be compared to a topological insulator in that the superconducting state prevents magnetic fields from penetrating the bulk of the material, akin to how a topological insulator prevents electrical conduction in the bulk while allowing surface states to conduct. The surface of the superconductor supports screening currents that create a magnetic field opposing the applied field, thereby expelling it from the interior.

e. **Energy Consideration:** The expulsion of magnetic fields minimizes the free energy of the superconducting state. The Gibbs free energy in the presence of a magnetic field is given by:

$$G = G_0 + \frac{B^2}{2\mu_0} \tag{13.81}$$

By expelling the magnetic field, the superconductor lowers its energy, maintaining a state with zero internal magnetic field.

f. **Conclusion:** The Meissner effect effectively acts as a topological insulator for magnetic fields by ensuring that the interior of the superconductor is free from magnetic flux. The surface currents responsible for this effect are analogous to the surface states in a topological insulator, providing protection against the penetration of external fields.

Answer: The Meissner effect demonstrates a form of topological insulation by expelling magnetic fields from the interior of a superconductor, similar to how a topological insulator prevents bulk conduction while allowing surface conduction.

14 The Vacuum in Quantum Field Theory

14.1 INTRODUCTION TO VACUUM STATES

In Quantum Field Theory (QFT), the concept of the vacuum state is far more intricate than the classical notion of empty space. The vacuum in QFT is a state with the lowest possible energy, but it is not devoid of activity. This complexity arises from the inherent nature of quantum mechanics, particularly Heisenberg's uncertainty principle, which dictates that certain pairs of physical properties, such as position and momentum, cannot be simultaneously measured with arbitrary precision. This principle implies that the vacuum is subject to constant fluctuations.

These fluctuations manifest as transient excitations of the quantum fields, resulting in the creation and annihilation of virtual particle pairs. The vacuum energy, often referred to as zero-point energy, is the energy associated with these fluctuations. It is theoretically predicted using the creation ($a^\dagger$) and annihilation (a) operators, which describe the quantum states of the fields.

Mathematically, the vacuum energy is a consequence of summing the zero-point energies of all possible modes of the field. For a harmonic oscillator, this zero-point energy is given by:

$$E_{\text{zero-point}} = \frac{1}{2}\hbar\omega, \tag{14.1}$$

where ω is the angular frequency. However, this theoretical result leads to predictions that vastly exceed what is observed experimentally. While the concept of vacuum energy is deeply rooted in the formalism of QFT and supported by phenomena such as the Casimir effect, the precise value of the vacuum energy density remains one of the biggest puzzles in modern physics. The theoretical calculations predict a vacuum energy density that is many orders of magnitude larger than the value inferred from cosmological observations, a discrepancy known as the "vacuum catastrophe."

This chapter explores the nature of the vacuum state, the mechanisms behind quantum fluctuations, and the implications of vacuum energy in the context of QFT. We will discuss pair production and annihilation, the expectation value of vacuum energy, virtual particles, experimental evidence, and the challenges posed by the discrepancies between theoretical predictions and experimental results.

14.2 QUANTUM FLUCTUATIONS

Quantum fluctuations refer to the temporary changes in the amount of energy in a point in space as a consequence of Heisenberg's uncertainty principle. These fluctuations can give rise to virtual particles that pop in and out of existence.

Heisenberg's uncertainty principle states that there is a fundamental limit to the precision with which certain pairs of physical properties, such as position and momentum, can be known simultaneously. Mathematically, it is expressed as:

$$\Delta x \Delta p \geq \frac{\hbar}{2}, \tag{14.2}$$

where Δx is the uncertainty in position and Δp is the uncertainty in momentum. A similar form of the uncertainty principle applies to energy and time:

$$\Delta E \Delta t \geq \frac{\hbar}{2}, \tag{14.3}$$

DOI: 10.1201/9781003637370-14

where ΔE is the uncertainty in energy and Δt is the uncertainty in time. This implies that the energy of a system can fluctuate for short periods, allowing for the temporary creation of particle-antiparticle pairs, known as virtual particles.

To understand how these fluctuations lead to an uncertainty in the vacuum energy, consider the quantum harmonic oscillator, which is a fundamental model in quantum mechanics. The Hamiltonian for a quantum harmonic oscillator is given by:

$$H = \frac{p^2}{2m} + \frac{1}{2}m\omega^2 x^2, \tag{14.4}$$

where p is the momentum, m is the mass, ω is the angular frequency, and x is the position. The energy levels of the harmonic oscillator are quantized and given by:

$$E_n = \left(n + \frac{1}{2}\right)\hbar\omega, \tag{14.5}$$

where n is a non-negative integer. The ground state energy (zero-point energy) of the harmonic oscillator is:

$$E_0 = \frac{1}{2}\hbar\omega. \tag{14.6}$$

In quantum field theory, the vacuum can be thought of as a collection of an infinite number of quantum harmonic oscillators, each corresponding to a different mode of the field. The zero-point energy of the vacuum is then the sum of the zero-point energies of all these oscillators:

$$E_{\text{vacuum}} = \sum_{\mathbf{k}} \frac{1}{2}\hbar\omega_{\mathbf{k}}, \tag{14.7}$$

where $\mathbf{k}$ is the wave vector and $\omega_{\mathbf{k}}$ is the angular frequency of the mode with wave vector $\mathbf{k}$.

Now, considering Heisenberg's uncertainty principle for energy and time:

$$\Delta E \Delta t \geq \frac{\hbar}{2}, \tag{14.8}$$

we can relate this to the vacuum energy fluctuations. For very short time intervals Δt, the uncertainty in energy ΔE can be very large. If we take Δt to be extremely small, we can see that ΔE must correspondingly be very large to satisfy the uncertainty relation. This can be written as:

$$\Delta E \geq \frac{\hbar}{2\Delta t}. \tag{14.9}$$

Given that the vacuum energy is the sum of the zero-point energies of all the modes, we can write:

$$\Delta E_{\text{vacuum}} = \sum_{\mathbf{k}} \frac{1}{2}\hbar\omega_{\mathbf{k}} \geq \frac{\hbar}{2\Delta t}. \tag{14.10}$$

This shows that for very short timescales Δt, the energy fluctuations ΔE_{vacuum} can be extremely large. These large energy fluctuations correspond to the creation of virtual particle-antiparticle pairs, which exist only temporarily due to the constraints of the uncertainty principle. This leads to a dynamic and fluctuating vacuum state, rather than a completely empty one.

Thus, Heisenberg's uncertainty principle leads to an inherent uncertainty in the vacuum energy, resulting in a dynamic and fluctuating vacuum state that is far from empty. This concept is fundamental to our understanding of quantum field theory and the behavior of quantum fields.

14.3 PAIR PRODUCTION AND ANNIHILATION FROM FLUCTUATIONS

In the vacuum, quantum fluctuations can lead to the creation of particle-antiparticle pairs. These pairs exist for a brief moment before annihilating each other. This phenomenon is fundamental to understanding interactions in QFT. At any given point and time the vacuum can generate the interaction shown below and given by the Feynman diagram in Figure 14.1:

$$\gamma+\gamma \to e^{+}+e^{-} \tag{14.11}$$

where γ represents photons and e^{+}, e^{-} are the electron and positron respectively.

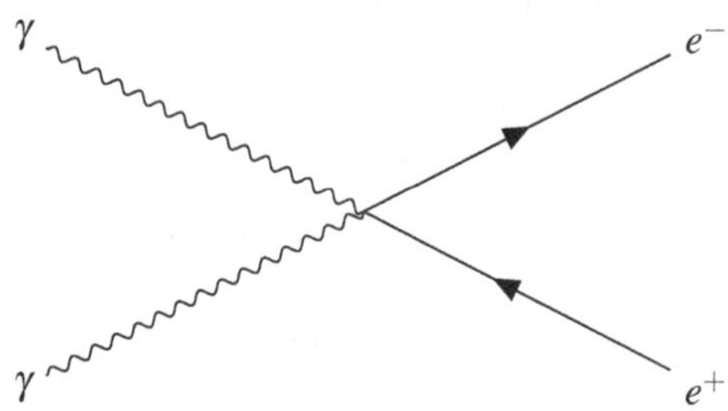

FIGURE 14.1
Feynman diagram for pair production: $\gamma+\gamma \to e^{+}+e^{-}$. Two photons ($\gamma$) collide and produce an electron (e^{-}) and a positron (e^{+}).

14.4 ZERO-POINT ENERGY

Zero-point energy is the lowest possible energy that a quantum mechanical physical system may have, contrary to the classical understanding that the energy could be zero at absolute zero temperature. In classical mechanics, a harmonic oscillator, such as a mass on a spring, can have zero energy when it is at rest in its equilibrium position. However, quantum mechanics reveals that even in the ground state, a system possesses a finite amount of residual energy due to the inherent uncertainty in its position and momentum.

14.4.1 ZERO-POINT ENERGY IN QUANTUM FIELD THEORY

In quantum field theory, the concept of zero-point energy extends to fields. A field can be thought of as an infinite collection of quantum harmonic oscillators, each corresponding to a different mode of the field. The Hamiltonian for a single mode of a free quantum field can be written similarly to that of a harmonic oscillator:

$$H_{\mathbf{k}} = \frac{1}{2}\hbar\omega_{\mathbf{k}}\left(a_{\mathbf{k}}^{\dagger}a_{\mathbf{k}} + a_{\mathbf{k}}a_{\mathbf{k}}^{\dagger}\right), \tag{14.12}$$

where $a_{\mathbf{k}}$ and $a_{\mathbf{k}}^{\dagger}$ are the annihilation and creation operators for the mode with wave vector $\mathbf{k}$ and angular frequency $\omega_{\mathbf{k}}$. The ground state energy for this mode is:

$$E_{\mathbf{k},0} = \frac{1}{2}\hbar\omega_{\mathbf{k}}. \tag{14.13}$$

The total zero-point energy of the vacuum is the sum of the zero-point energies of all the modes:

$$E_{\text{vacuum}} = \sum_{\mathbf{k}} \frac{1}{2}\hbar\omega_{\mathbf{k}}. \tag{14.14}$$

This sum is formally infinite, as there are infinitely many modes. However, in practice, this energy is usually considered relative to a reference energy, and techniques such as renormalization are used to deal with the infinities.

14.4.2 PHYSICAL IMPLICATIONS AND OBSERVATIONS

Zero-point energy has several profound physical implications. One of the most famous manifestations is the Casimir effect, which arises from the zero-point energy of the electromagnetic field. When two uncharged, parallel conducting plates are placed close to each other in a vacuum, the vacuum energy between the plates is reduced compared to the energy outside the plates, creating an attractive force between them (see Figure 14.2) . This effect has been experimentally measured and confirms the existence of zero-point energy.

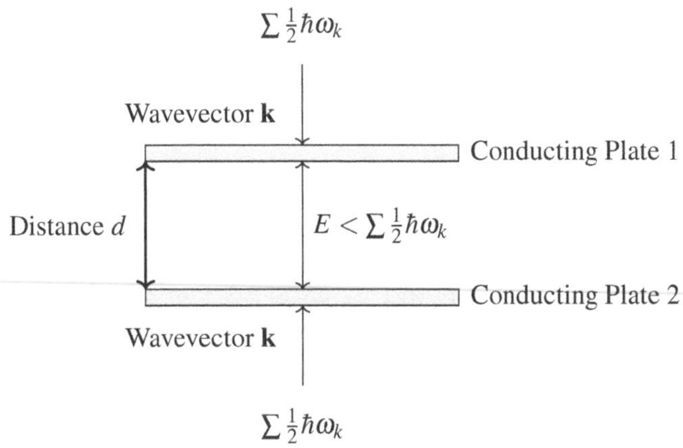

FIGURE 14.2
Illustration of the Casimir Effect. Two parallel conducting plates experience an attractive force due to reduced vacuum fluctuations between the plates compared to outside them. The distance between the plates is denoted by d. The vacuum fluctuations outside the plates are represented by the sum of ground states $\sum \frac{1}{2}\hbar\omega_k$, while inside the plates, the energy E is less than $\sum \frac{1}{2}\hbar\omega_k$.

The force due to the Casimir Effect is found as:

$$F = \frac{\pi^2 \hbar c}{240 a^4} \tag{14.15}$$

where a is the distance between the plates.

Another implication is the Lamb shift, which is a small difference in energy levels of hydrogen atoms that cannot be explained by classical electrodynamics. The shift arises due to interactions between the electron and the vacuum fluctuations of the electromagnetic field.

14.4.3 CALCULATION OF VACUUM ENERGY DENSITY

The vacuum energy density can be calculated by summing the zero-point energies of all the normal modes of a field. This approach is based on the idea that each mode of the field contributes a zero-point energy given by $\frac{1}{2}\hbar\omega$, where ω is the angular frequency of the mode. First we must identify the field and its modes. Consider a scalar field in a box of volume V. The field can be decomposed into normal modes with wave vectors **k**.

Each mode of the field is quantized and can be treated as a harmonic oscillator with zero-point energy $\frac{1}{2}\hbar\omega_k$, where $\omega_k = c|\mathbf{k}|$ for a massless field (like the electromagnetic field).

The total vacuum energy is the sum of the zero-point energies of all modes:

$$E_{\text{vac}} = \sum_{\mathbf{k}} \frac{1}{2}\hbar\omega_k. \tag{14.16}$$

In the limit of a large volume V, the sum can be approximated by an integral over the density of states in momentum space:

$$E_{\text{vac}} = \frac{V}{(2\pi)^3}\int d^3k \frac{1}{2}\hbar\omega_k = \frac{V}{(2\pi)^3}\int d^3k \frac{1}{2}\hbar c|\mathbf{k}|. \tag{14.17}$$

Perform the integral over the momentum space. For simplicity, use spherical coordinates in k-space:

$$E_{\text{vac}} = \frac{V\hbar c}{2(2\pi)^3}\int_0^\infty dk 4\pi k^2 k. \tag{14.18}$$

Simplifying, we get:

$$E_{\text{vac}} = \frac{V\hbar c}{2\pi^2}\int_0^\infty dk k^3. \tag{14.19}$$

The integral $\int_0^\infty dk k^3$ diverges, indicating the need for a regularization procedure. A common method is to introduce a cutoff Λ:

$$E_{\text{vac}} = \frac{V\hbar c}{2\pi^2}\int_0^\Lambda dk k^3 = \frac{V\hbar c}{2\pi^2}\left[\frac{k^4}{4}\right]_0^\Lambda = \frac{V\hbar c}{8\pi^2}\Lambda^4. \tag{14.20}$$

The vacuum energy density ρ_{vac} is the vacuum energy per unit volume:

$$\rho_{\text{vac}} = \frac{E_{\text{vac}}}{V} = \frac{\hbar c\Lambda^4}{8\pi^2}. \tag{14.21}$$

Therefore, the vacuum energy density is given by:

$$\rho_{\text{vac}} = \frac{\hbar c\Lambda^4}{8\pi^2}, \tag{14.22}$$

where Λ is an arbitrary cutoff that regularizes the otherwise divergent integral. This result shows that the vacuum energy density depends strongly on the cutoff Λ, highlighting the sensitivity of the calculation to high-energy contributions.

14.4.3.1 Numerical Estimate of Vacuum Energy Density

To obtain a numerical estimate, we use the vacuum energy density formula derived earlier and a suitable cutoff Λ.

The numerical estimate of ρ_{vac} is obtained by choosing a very high cutoff, typically related to the Planck energy scale. The Planck energy (E_P) is given by:

$$E_P = \sqrt{\frac{\hbar c^5}{G}} \approx 1.22\times 10^{19}\,\text{GeV} \tag{14.23}$$

where G is the gravitational constant, $G \approx 6.674\times 10^{-11}\,\text{m}^3\text{kg}^{-1}\text{s}^{-2}$.

Converting Planck energy to joules:

$$1\,\text{GeV} \approx 1.602\times 10^{-10}\,\text{J}, \quad E_P \approx 1.22\times 10^{19}\times 1.602\times 10^{-10}\,\text{J} \approx 1.96\times 10^9\,\text{J} \tag{14.24}$$

The corresponding cutoff in terms of energy is:

$$\Lambda = \frac{E_P}{\hbar c} \approx \frac{1.96 \times 10^9\,\text{J}}{(1.054 \times 10^{-34}\,\text{Js}) \cdot (3 \times 10^8\,\text{m/s})} \approx 6.2 \times 10^{34}\,\text{m}^{-1} \tag{14.25}$$

The vacuum energy density is then calculated by:

$$\rho_{\text{vac}} = \frac{\hbar c \Lambda^4}{8\pi^2} = \frac{(1.054 \times 10^{-34}\,\text{Js}) \cdot (3 \times 10^8\,\text{m/s}) \cdot (6.2 \times 10^{34}\,\text{m}^{-1})^4}{8\pi^2} \tag{14.26}$$

Simplifying the expression:

$$\rho_{\text{vac}} = \frac{(1.054 \times 10^{-34}) \cdot (3 \times 10^8) \cdot (1.474 \times 10^{141})}{8\pi^2} \tag{14.27}$$

Evaluating the expression:

$$\rho_{\text{vac}} = \frac{4.622 \times 10^{113}}{78.9568} \approx 0.585 \times 10^{113}\,\text{J/m}^3 \tag{14.28}$$

14.4.4 CALCULATED VS. EXPERIMENTAL VACUUM ENERGY DENSITY

Theoretical predictions and experimental observations of vacuum energy density exhibit a profound discrepancy, which poses a significant challenge in both theoretical physics and cosmology.

14.4.4.1 Theoretical Calculations

Theoretical estimates of the vacuum energy density arise from summing the zero-point energies of all possible quantum field modes. This approach suggests that each mode contributes an energy given by $\frac{1}{2}\hbar\omega$. Integrating these contributions up to a high-energy cutoff, typically associated with the Planck scale, results in an astonishingly large vacuum energy density on the order of $10^{113}\,\text{J/m}^3$. This enormous value reflects the high density of states at short wavelengths and the immense energy contributions from quantum fluctuations. We must keep in mind that the value for the cutoff wavelengths is determined from the Planck Energy which may or may not be a valid approach. Clearly something is amiss upon comparison with the experimental evidence.

14.4.4.2 Experimental Observations

The experimental value of the vacuum energy density is derived from cosmological measurements, specifically the observed rate of the universe's accelerated expansion. These observations are modeled using Einstein's field equations with a cosmological constant (Λ):

$$R_{\mu\nu} - \frac{1}{2} g_{\mu\nu} R + \Lambda g_{\mu\nu} = \frac{8\pi G}{c^4} T_{\mu\nu}, \tag{14.29}$$

where $R_{\mu\nu}$ is the Ricci curvature tensor, R is the scalar curvature, $g_{\mu\nu}$ is the metric tensor, G is the gravitational constant, and $T_{\mu\nu}$ is the stress-energy tensor.

The cosmological constant Λ provides a measure of the vacuum energy's contribution to the universe's expansion. It should be noted here that Einstein was troubled with the cosmological constant and often stated it might be zero and his biggest mistake to keep it included in the equation. Data from distant supernovae, the cosmic microwave background (CMB), and large-scale structure surveys have been instrumental in determining Λ. These observations consistently indicate that the vacuum energy density is approximately $10^{-9}\,\text{J/m}^3$, a value vastly smaller than the theoretical prediction.

14.4.4.3 The Discrepancy

The discrepancy between the theoretical and experimental valuesoften referred to as the *cosmological constant problem* and also the *vacuum catastrophe* remains one of the most significant unresolved issues in modern physics. Theoretical calculations predict a vacuum energy density that is 120 orders of magnitude larger than what is observed. This vast difference suggests that our current understanding of quantum field theory, general relativity, or both, is incomplete.

Despite various hypotheses, the true nature of vacuum energy and its role in the universe remains a profound mystery, inviting further exploration and potential breakthroughs in fundamental physics.

14.5 RECONCILIATION IN QUANTUM FIELD THEORY

The vast discrepancy between the theoretically predicted vacuum energy density and the experimentally observed value presents a significant challenge in theoretical physics. Various approaches and modifications to Quantum Field Theory (QFT) have been proposed to address this issue, exploring the potential mechanisms that could reconcile the differences.

14.5.1 SUPERSYMMETRY

Supersymmetry (SUSY) is a theoretical framework that extends the Standard Model of particle physics by postulating a symmetry between bosons and fermions. In supersymmetric theories, every particle has a superpartner with different spin statistics. The contributions to the vacuum energy from bosons and their corresponding fermionic superpartners are expected to cancel each other out, thereby reducing the overall vacuum energy density.

$$\rho_{\text{vac}}^{\text{SUSY}} \approx 0 \quad \text{(at tree level)} \tag{14.30}$$

Although supersymmetry has not yet been experimentally confirmed, it remains a promising candidate for addressing the cosmological constant problem.

14.5.2 QUANTUM GRAVITY EFFECTS

Quantum gravity aims to unify quantum mechanics with general relativity, providing a consistent description of gravity at the quantum scale. Approaches such as string theory and loop quantum gravity suggest that quantum gravitational effects could modify the high-energy behavior of quantum fields, potentially leading to a natural regularization of the vacuum energy.

14.5.2.1 Spin Networks in Loop Quantum Gravity

Spin networks are a fundamental concept in Loop Quantum Gravity (LQG), representing the quantum states of the gravitational field. Unlike Feynman diagrams, which describe particle interactions, spin networks describe the geometry of space at the quantum level. Figure 14.3 illustrates a simple representation of a spin network vertex in LQG.

In this figure, the vertices (A, B, C, and D) represent quantum states of space, where each vertex corresponds to a chunk of space. The edges connecting these vertices represent quantum connections or "links" between these chunks, capturing how space is woven together at the smallest scales. The thick lines symbolize the spin network links, and the intersections at the nodes depict the points where the quantum states interact or merge. This graphical representation highlights the discrete nature of space in LQG, where the geometry is quantized, and spacetime is not a continuous fabric but rather composed of these interwoven spin networks. This conceptual framework allows for a non-perturbative approach to quantum gravity, offering insights into how spacetime emerges from fundamental quantum properties.

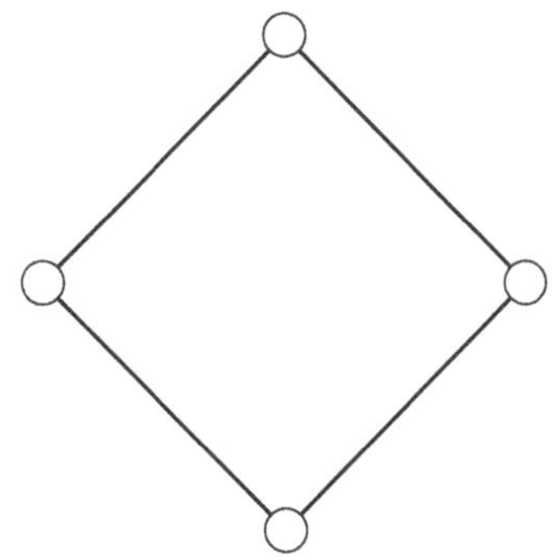

FIGURE 14.3
Simple representation of a spin network vertex in Loop Quantum Gravity.

14.5.2.2 String Theory

String theory offers a promising framework to address the vacuum energy discrepancy by introducing additional spatial dimensions and considering fundamental particles as one-dimensional "strings" rather than point-like particles. This extended nature of particles can significantly alter the contributions to vacuum energy.

String theory posits that the fundamental constituents of the universe are not point particles, but rather one-dimensional strings that can vibrate at different frequencies. These vibrations correspond to different particles, with their properties (such as mass and charge) determined by the mode of vibration. String theory inherently requires additional spatial dimensions for mathematical consistency, typically proposing a 10-dimensional spacetime (9 spatial dimensions and 1 time dimension) in the simplest versions, or 11 dimensions in the context of M-theory.

These additional spatial dimensions can lead to different vacuum energy contributions compared to conventional quantum field theory. The compactification of these extra dimensions (i.e., how they are curled up and hidden from low-energy physics) plays a crucial role in determining the vacuum energy. The geometry and size of the compactified dimensions affect the spectrum of vibrational modes of the strings, thereby influencing the zero-point energy calculations.

String theory provides several mechanisms that could potentially reconcile the theoretical and experimental values of vacuum energy density:

- **Supersymmetry:** Many string theory models incorporate supersymmetry, which posits a symmetry between bosons and fermions. Supersymmetry can lead to cancellations between the vacuum energy contributions of bosonic and fermionic modes, significantly reducing the overall vacuum energy density. Although supersymmetry has not yet been observed experimentally, it remains a key feature in many string theory frameworks.
- **Extra Dimensions:** The introduction of extra dimensions alters the density of states and the spectrum of zero-point energies. Different compactification schemes can lead to vastly different vacuum energy densities, potentially aligning theoretical predictions with observational data.
- **Moduli Stabilization:** In string theory, the shapes and sizes of the extra dimensions are described by fields known as moduli. The stabilization of these moduli (i.e., fixing their values) can influence the vacuum energy. Mechanisms for moduli stabilization, such as flux compactifications, can lead to vacuum states with low energy densities.
- **Landscape of Vacua:** String theory suggests a vast "landscape" of possible vacuum states, each corresponding to different configurations of the extra dimensions and fluxes. Among these vacua,

there could exist states with vacuum energy densities that match the observed value. The anthropic principle can be invoked, suggesting that we observe a small vacuum energy density because only such vacua can support life and the formation of complex structures.

Implications and Future Directions:
String theory's approach to vacuum energy has profound implications for our understanding of fundamental physics. By considering the extended nature of particles and the role of additional dimensions, string theory provides a rich framework that could resolve the cosmological constant problem. However, much work remains to be done to identify the exact mechanisms and to find specific solutions that align with empirical data. Ongoing research in string theory, including advancements in understanding compactification and moduli stabilization, continues to shed light on these deep questions.

In summary, string theory offers a comprehensive approach to addressing the vacuum energy discrepancy by leveraging its unique features, such as extra dimensions, supersymmetry, and a diverse landscape of vacua. These aspects of string theory provide promising avenues for achieving a theoretical calculation of vacuum energy density that is consistent with observational data.

14.5.3 ANTHROPIC PRINCIPLE

The anthropic principle posits that the observed values of certain physical constants, including the vacuum energy density, are constrained by the requirement for the existence of observers. According to this principle, the universe's vacuum energy density must fall within a narrow range to allow the formation of structures like galaxies, stars, and planets, and subsequently the emergence of life.

This principle suggests that the observed small value of the vacuum energy density is not a coincidence but a necessary condition for the existence of observers capable of measuring it.

14.5.4 DYNAMIC DARK ENERGY MODELS

Dynamic dark energy models propose that the vacuum energy density is not constant but evolves over time. One such model is quintessence, where the vacuum energy density is associated with a scalar field that changes over cosmic time scales. These models offer a potential explanation for the current accelerated expansion of the universe while providing a mechanism for the observed small value of the vacuum energy density.

The energy density of the quintessence field ϕ can be expressed as:

$$\rho_\phi = \frac{1}{2}\dot{\phi}^2 + V(\phi), \tag{14.31}$$

where $\dot{\phi}$ is the time derivative of the field, and $V(\phi)$ is the potential energy. The dynamics of ϕ determine the evolution of the vacuum energy density.

14.5.5 RENORMALIZATION AND EFFECTIVE FIELD THEORY

In QFT, the process of renormalization deals with infinities arising in calculations by absorbing them into redefined (renormalized) parameters. This process can potentially address the vacuum energy density by considering it as a renormalized quantity, thus making it finite and consistent with observations.

Effective field theory (EFT) provides another approach by recognizing that QFT calculations are valid only up to a certain energy scale. Above this scale, new physics, such as quantum gravity, might come into play, altering the predictions of vacuum energy.

$$\rho_{\text{vac}}^{\text{EFT}} = \frac{\hbar c \Lambda_{\text{EFT}}^4}{8\pi^2}, \tag{14.32}$$

where Λ_{EFT} is the cutoff scale for the effective field theory, chosen to be consistent with known physics and experimental data.

While no single solution has yet resolved the discrepancy between theoretical and experimental vacuum energy densities, various approaches provide promising avenues for exploration. Supersymmetry, quantum gravity, the anthropic principle, dynamic dark energy models, and renormalization techniques all offer potential pathways to reconcile these differences, highlighting the need for continued theoretical and experimental advancements in understanding the nature of vacuum energy.

14.6 SUMMARY AND DISCUSSION

In this chapter, we have studied the complex nature of the vacuum in Quantum Field Theory (QFT), addressing several key aspects that reveal its intricacies and significance. We began by discussing the concept of the vacuum state, which, contrary to classical understanding, is not an empty void but a dynamic state filled with quantum fluctuations. These fluctuations, stemming from Heisenberg's uncertainty principle, lead to the transient creation and annihilation of virtual particle pairs, contributing to the vacuum energy, also known as zero-point energy.

We explored the phenomenon of quantum fluctuations, demonstrating how the uncertainty principle results in an inherent uncertainty in the vacuum energy. This principle allows for significant energy fluctuations over short timescales, explaining the temporary existence of virtual particles. These concepts are fundamental to understanding the behavior of quantum fields and the dynamic nature of the vacuum.

In discussing pair production and annihilation from fluctuations, we illustrated how quantum fluctuations can lead to the creation of particle-antiparticle pairs, which exist briefly before annihilating each other. This process is crucial for understanding interactions in QFT and is often represented by Feynman diagrams, which visualize these interactions.

The concept of zero-point energy was further elaborated upon, highlighting its implications in QFT. We discussed how the vacuum energy can be calculated by summing the zero-point energies of all normal modes of a field, and the resulting theoretical predictions for vacuum energy density. However, these calculations yield values that are vastly larger than those inferred from experimental observations, leading to the well-known cosmological constant problem.

We addressed the profound discrepancy between calculated and experimental vacuum energy densities, emphasizing the significance of this issue in both theoretical physics and cosmology. Theoretical estimates suggest a vacuum energy density on the order of $10^{113}\,\text{J/m}^3$, while astronomical observations indicate a value closer to $10^{-9}\,\text{J/m}^3$. This discrepancy, known as the vacuum catastrophe, highlights the limitations of our current understanding and the need for new theoretical insights.

To reconcile this discrepancy, we explored various theoretical approaches within QFT, including supersymmetry, quantum gravity effects, the anthropic principle, dynamic dark energy models, and renormalization techniques. Each of these approaches offers potential solutions to the cosmological constant problem, providing avenues for further research and exploration. Supersymmetry, for instance, predicts cancellations between vacuum energy contributions from bosons and fermions, while quantum gravity theories suggest modifications to high-energy behavior that could regularize vacuum energy. The anthropic principle and dynamic dark energy models offer alternative perspectives on the observed small value of the vacuum energy density.

In summary, this chapter has provided a comprehensive overview of the vacuum in QFT, highlighting its dynamic nature, the challenges in understanding zero-point energy, and the ongoing efforts to reconcile theoretical predictions with experimental observations. Understanding the vacuum is crucial for advancing our knowledge of fundamental physics and addressing one of the most significant unresolved issues in modern science. Future research in QFT and related fields holds the promise of uncovering deeper insights into the nature of the vacuum and the fundamental forces that govern our universe.

14.7 QUESTIONS AND PROBLEMS

14.7.1 TRIVIA QUESTIONS

Question 1: What is the zero-point energy of a quantum harmonic oscillator?

(a) 0
(b) $\frac{1}{2}\hbar\omega$
(c) $\hbar\omega$
(d) $\frac{1}{2}m\omega^2$

Answer: b. $\frac{1}{2}\hbar\omega$
Question 2: Which principle states that certain pairs of physical properties cannot be simultaneously measured with arbitrary precision?

(a) Pauli Exclusion Principle
(b) Heisenberg Uncertainty Principle
(c) Principle of Superposition
(d) Complementarity Principle

Answer: b. Heisenberg Uncertainty Principle
Question 3: In the context of QFT, what do virtual particles result from?

(a) Large magnetic fields
(b) Quantum fluctuations
(c) High temperatures
(d) Gravitational waves

Answer: b. Quantum fluctuations
Question 4: The Casimir effect is an experimental evidence of which phenomenon?

(a) Pauli Exclusion Principle
(b) Quantum entanglement
(c) Zero-point energy
(d) Time dilation

Answer: c. Zero-point energy
Question 5: What is the theoretical prediction of vacuum energy density in QFT?

(a) 10^{-9} J/m^3
(b) 10^{9} J/m^3
(c) 10^{113} J/m^3
(d) 10^{19} J/m^3

Answer: c. 10^{113} J/m^3
Question 6: Which of the following theories suggests that every particle has a superpartner?

(a) General Relativity
(b) Supersymmetry
(c) Special Relativity

(d) Quantum Chromodynamics

Answer: b. Supersymmetry
Question 7: What is the primary cause of the Lamb shift in hydrogen atoms?

(a) Gravitational waves
(b) Quantum fluctuations
(c) Electromagnetic waves
(d) Neutrino interactions

Answer: b. Quantum fluctuations
Question 8: In string theory, what are the fundamental constituents of the universe?

(a) Point particles
(b) Strings
(c) Waves
(d) Loops

Answer: b. Strings
Question 9: The discrepancy between theoretical and observed vacuum energy density is known as:

(a) The Uncertainty Principle
(b) The Cosmological Constant Problem
(c) The Holographic Principle
(d) The Anthropic Principle

Answer: b. The Cosmological Constant Problem
Question 10: Which principle posits that the values of certain physical constants are constrained by the requirement for the existence of observers?

(a) Heisenberg Uncertainty Principle
(b) Anthropic Principle
(c) Equivalence Principle
(d) Correspondence Principle

Answer: b. Anthropic Principle

14.7.2 PROBLEMS

PROBLEM 1: ZERO-POINT ENERGY OF A HARMONIC OSCILLATOR

Problem: Calculate the zero-point energy of a quantum harmonic oscillator with an angular frequency $\omega = 2 \times 10^{15}$ rad/s.

Solution:

The zero-point energy $E_{\text{zero-point}}$ of a quantum harmonic oscillator is given by:

$$E_{\text{zero-point}} = \frac{1}{2}\hbar\omega \tag{14.33}$$

Given $\omega = 2 \times 10^{15}$ rad/s, and using $\hbar = 1.054 \times 10^{-34}$ Js:

$$E_{\text{zero-point}} = \frac{1}{2} \times 1.054 \times 10^{-34} \times 2 \times 10^{15} \tag{14.34}$$

$$E_{\text{zero-point}} = \frac{1.054 \times 10^{-34} \times 2 \times 10^{15}}{2} \tag{14.35}$$

$$E_{\text{zero-point}} = 1.054 \times 10^{-19}\,\text{J} \tag{14.36}$$

Thus, the zero-point energy is 1.054×10^{-19} J.

PROBLEM 2: ENERGY FLUCTUATIONS IN THE VACUUM

Problem: Using Heisenberg's uncertainty principle, calculate the minimum uncertainty in energy ΔE for a time interval $\Delta t = 1 \times 10^{-15}$ s.

Solution:

Heisenberg's uncertainty principle for energy and time is given by:

$$\Delta E \Delta t \geq \frac{\hbar}{2} \tag{14.37}$$

Given $\Delta t = 1 \times 10^{-15}$ s:

$$\Delta E \geq \frac{\hbar}{2\Delta t} \tag{14.38}$$

Using $\hbar = 1.054 \times 10^{-34}$ Js:

$$\Delta E \geq \frac{1.054 \times 10^{-34}}{2 \times 1 \times 10^{-15}} \tag{14.39}$$

$$\Delta E \geq \frac{1.054 \times 10^{-34}}{2 \times 10^{-15}} \tag{14.40}$$

$$\Delta E \geq \frac{1.054}{2} \times 10^{-19} \tag{14.41}$$

$$\Delta E \geq 0.527 \times 10^{-19}\,\text{J} \tag{14.42}$$

Thus, the minimum uncertainty in energy is 0.527×10^{-19} J.

PROBLEM 3: CASIMIR EFFECT FORCE CALCULATION

Problem: Calculate the Casimir force per unit area between two parallel conducting plates separated by a distance $a = 1 \times 10^{-7}$ m.

Solution:

The Casimir force per unit area F/A is given by:

$$\frac{F}{A} = \frac{\pi^2 \hbar c}{240 a^4} \tag{14.43}$$

Given $a = 1 \times 10^{-7}$ m, $\hbar = 1.054 \times 10^{-34}$ Js, and $c = 3 \times 10^8$ m/s:

$$\frac{F}{A} = \frac{\pi^2 \times 1.054 \times 10^{-34} \times 3 \times 10^8}{240 \times (1 \times 10^{-7})^4} \tag{14.44}$$

$$\frac{F}{A} = \frac{\pi^2 \times 1.054 \times 10^{-34} \times 3 \times 10^8}{240 \times 10^{-28}} \tag{14.45}$$

$$\frac{F}{A} = \frac{\pi^2 \times 1.054 \times 3 \times 10^{-26}}{240} \tag{14.46}$$

$$\frac{F}{A} = \frac{9.8696 \times 1.054 \times 3 \times 10^{-26}}{240} \tag{14.47}$$

$$\frac{F}{A} = \frac{31.1506 \times 10^{-26}}{240} \tag{14.48}$$

$$\frac{F}{A} = 1.297 \times 10^{-27}\,\mathrm{N/m^2} \tag{14.49}$$

Thus, the Casimir force per unit area is $1.297 \times 10^{-27}\,\mathrm{N/m^2}$.

PROBLEM 4: VACUUM ENERGY DENSITY ESTIMATE

Problem: Using the Planck energy scale, calculate the vacuum energy density ρ_{vac}.

Solution:

The vacuum energy density is given by:

$$\rho_{\text{vac}} = \frac{\hbar c \Lambda^4}{8\pi^2} \tag{14.50}$$

Given the Planck energy scale $E_P \approx 1.22 \times 10^{19}\,\mathrm{GeV}$ and converting to Joules:

$$\begin{aligned} 1\,\mathrm{GeV} &\approx 1.602 \times 10^{-10}\,\mathrm{J}, \\ E_P &\approx 1.22 \times 10^{19} \times 1.602 \times 10^{-10}\,\mathrm{J} \\ &\approx 1.96 \times 10^{9}\,\mathrm{J} \end{aligned} \tag{14.51}$$

The corresponding cutoff in terms of energy is:

$$\Lambda = \frac{E_P}{\hbar c} \approx \frac{1.96 \times 10^9\,\mathrm{J}}{(1.054 \times 10^{-34}\,\mathrm{Js}) \cdot (3 \times 10^8\,\mathrm{m/s})} \approx 6.2 \times 10^{34}\,\mathrm{m}^{-1} \tag{14.52}$$

The vacuum energy density is then calculated by:

$$\rho_{\text{vac}} = \frac{\hbar c \Lambda^4}{8\pi^2} = \frac{(1.054 \times 10^{-34}\,\mathrm{Js}) \cdot (3 \times 10^8\,\mathrm{m/s}) \cdot (6.2 \times 10^{34}\,\mathrm{m}^{-1})^4}{8\pi^2} \tag{14.53}$$

Simplifying the expression:

$$\rho_{\text{vac}} = \frac{(1.054 \times 10^{-34}) \cdot (3 \times 10^8) \cdot (1.474 \times 10^{141})}{8\pi^2} \tag{14.54}$$

Evaluating the expression:

$$\rho_{\text{vac}} = \frac{4.622 \times 10^{113}}{78.9568} \approx 0.585 \times 10^{113}\,\mathrm{J/m^3} \tag{14.55}$$

Thus, the vacuum energy density is $0.585 \times 10^{113}\,\mathrm{J/m^3}$.

PROBLEM 5: PAIR PRODUCTION ENERGY THRESHOLD

Problem: Calculate the minimum energy required for photon-photon pair production, where an electron and positron are created.

Solution:

The rest mass energy of an electron (or positron) is given by:

$$E = m_e c^2 \tag{14.56}$$

Given $m_e = 9.11 \times 10^{-31}\,\mathrm{kg}$ and $c = 3 \times 10^8\,\mathrm{m/s}$:

$$E = 9.11 \times 10^{-31} \times (3 \times 10^8)^2 \tag{14.57}$$

$$E = 9.11 \times 10^{-31} \times 9 \times 10^{16} \tag{14.58}$$

$$E = 8.199 \times 10^{-14}\,\text{J} \tag{14.59}$$

Since pair production involves creating both an electron and a positron, the total energy required is:

$$E_{\text{total}} = 2 \times 8.199 \times 10^{-14}\,\text{J} = 1.6398 \times 10^{-13}\,\text{J} \tag{14.60}$$

Converting this energy to electron volts (1 eV = 1.602×10^{-19} J):

$$E_{\text{total}} = \frac{1.6398 \times 10^{-13}}{1.602 \times 10^{-19}}\,\text{eV} \tag{14.61}$$

$$E_{\text{total}} \approx 1.024 \times 10^{6}\,\text{eV} = 1.024\,\text{MeV} \tag{14.62}$$

Thus, the minimum energy required for photon-photon pair production is approximately 1.024 MeV.

BONUS PROBLEM: ESTIMATING THE ENERGY IN A CUBIC CENTIMETER OF VACUUM SPACE

Problem: The statement often made in popular physics discussions is that there is enough energy in a thimble-sized volume of vacuum to run the world many times over. However, we saw in this chapter that there are some discrepancies between the theoretical predictions of this value and the experimental evidence. Using both the theoretical and experimental values of vacuum energy density, estimate the total vacuum energy contained in a cubic centimeter of space. Compare and contrast these values.

Solution:

To estimate the total vacuum energy contained in a cubic centimeter of space, we will use both the theoretically derived vacuum energy density and the experimentally observed value.

First, let's calculate the theoretical vacuum energy density:

$$\rho_{\text{vac}} = \frac{\hbar c \Lambda^4}{8\pi^2} \tag{14.63}$$

where Λ is the cutoff scale, typically associated with the Planck energy.

We use the Planck energy E_P to determine Λ:

$$E_P = \sqrt{\frac{\hbar c^5}{G}} \approx 1.22 \times 10^{19}\,\text{GeV} \tag{14.64}$$

Convert the Planck energy to joules:

$$1\,\text{GeV} \approx 1.602 \times 10^{-10}\,\text{J} \tag{14.65}$$

Thus,

$$E_P \approx 1.22 \times 10^{19} \times 1.602 \times 10^{-10}\,\text{J} \approx 1.96 \times 10^{9}\,\text{J} \tag{14.66}$$

The corresponding cutoff in terms of energy is:

$$\Lambda = \frac{E_P}{\hbar c} \approx \frac{1.96 \times 10^{9}\,\text{J}}{(1.054 \times 10^{-34}\,\text{Js}) \cdot (3 \times 10^{8}\,\text{m/s})} \approx 6.2 \times 10^{34}\,\text{m}^{-1} \tag{14.67}$$

Using this cutoff in our vacuum energy density formula:

$$\rho_{\text{vac}} = \frac{\hbar c \Lambda^4}{8\pi^2} = \frac{(1.054 \times 10^{-34}\,\text{Js}) \cdot (3 \times 10^{8}\,\text{m/s}) \cdot (6.2 \times 10^{34}\,\text{m}^{-1})^4}{8\pi^2} \tag{14.68}$$

Simplifying the expression:

$$\rho_{\text{vac}} = \frac{(1.054 \times 10^{-34}) \cdot (3 \times 10^8) \cdot (1.474 \times 10^{141})}{8\pi^2} \tag{14.69}$$

Evaluating the expression:

$$\rho_{\text{vac}} = \frac{4.622 \times 10^{113}}{78.9568} \approx 0.585 \times 10^{113}\,\text{J/m}^3 \tag{14.70}$$

To find the energy in a cubic centimeter, convert the density to J/cm^3:

$$1\,\text{m}^3 = 10^6\,\text{cm}^3 \tag{14.71}$$

$$\rho_{\text{vac}} \approx 0.585 \times 10^{113}\,\text{J/m}^3 \times 10^{-6}\,\text{m}^3/\text{cm}^3 \approx 0.585 \times 10^{107}\,\text{J/cm}^3 \tag{14.72}$$

Now, let's compare this with the experimentally observed value of vacuum energy density, which is approximately $10^{-9}\,\text{J/m}^3$.

Convert this to J/cm^3:

$$\rho_{\text{vac, exp}} = 10^{-9}\,\text{J/m}^3 \times 10^{-6}\,\text{m}^3/\text{cm}^3 = 10^{-15}\,\text{J/cm}^3 \tag{14.73}$$

Comparison:
- **Theoretical Vacuum Energy Density:** $0.585 \times 10^{107}\,\text{J/cm}^3$ - **Experimental Vacuum Energy Density:** $10^{-15}\,\text{J/cm}^3$

The theoretical value is many orders of magnitude larger than the experimental value. This discrepancy highlights one of the most significant unresolved problems in modern physics, known as the cosmological constant problem or vacuum catastrophe. Theoretical approaches and models, such as supersymmetry, quantum gravity, and string theory, are being explored to reconcile this difference.

Conclusion: The vast difference between the theoretical and experimental vacuum energy densities underscores the need for a deeper understanding of the vacuum in Quantum Field Theory and the potential new physics that could resolve this discrepancy.

15 Summary and Pedagogical Implications of Quantum Field Theory

15.1 OVERVIEW OF TOPICS COVERED

In this book, we have journeyed through the fascinating and complex landscape of Quantum Field Theory (QFT), building upon the foundations of quantum mechanics (QM) and expanding into the realms of particle physics and condensed matter. Our exploration began with the principles of first quantization, utilizing wavefunctions and operators to describe quantum states. We then transitioned into the formalism of second quantization, introducing creation and annihilation operators to handle systems with variable particle numbers.

15.1.1 KEY CONCEPTS AND MODELS

We discussed in detail several critical concepts and models that form the backbone of QFT and its applications:

- Second Quantization and Fock Space: We examined how second quantization generalizes quantum mechanics to many-body systems, using creation and annihilation operators to describe particle interactions. Fock space was introduced as a comprehensive framework accommodating all possible states of a quantum system with varying particle numbers.
- **Tight Binding Hamiltonian:** We explored the tight binding model, a crucial tool in condensed matter physics, which simplifies the description of electrons in a lattice by considering only nearest-neighbor interactions. This model provided insights into electronic band structures, graphenes unique properties, and topological insulators.
- **Multi-Particle Operators:** The formalism of two-particle and N-particle operators allowed us to describe interactions between multiple particles, essential for understanding collective behaviors in many-body systems.
- **Hubbard Model:** We studied the Hubbard model, pivotal for analyzing strongly correlated electron systems. This model captures the competition between kinetic energy and electron-electron interactions, elucidating phenomena such as metal-insulator transitions, magnetism, and superconductivity.
- **Advanced Applications:** We applied QFT principles to practical scenarios, including laser emission, entangled photons in Young's two-slit experiment, and the statistical properties of light using the second-order correlation function.
- **Mathematical Tools:** Essential mathematical techniques, such as contour integrals in complex analysis, were reviewed and applied to solve problems in QFT.
- **Pedagogical Figures:** Visual aids, such as representations of the transition from first to second quantization, Fock space, honeycomb lattice of graphene, and Dirac cones, were used to enhance understanding.

15.2 PEDAGOGICAL IMPLICATIONS OF QUANTUM FIELD THEORY

Quantum Field Theory represents a significant leap beyond the traditional quantum mechanics view of the universe. The shift from QM to QFT is not merely an extension but a profound transformation

DOI: 10.1201/9781003637370-15

in our understanding of fundamental physics.

15.2.1 FROM PARTICLES TO FIELDS

In QM, particles are described by wavefunctions, and their dynamics are governed by the Schrödinger equation. QFT, however, treats particles as excitations of underlying fields. This field-centric view offers a more unified and versatile framework, allowing for the description of particle creation and annihilation processes, which are essential in high-energy physics and the Standard Model of particle physics.

15.2.2 MANY-BODY SYSTEMS AND COLLECTIVE PHENOMENA

QFT's second quantization formalism provides powerful tools to describe many-body systems and collective phenomena. This approach is indispensable in condensed matter physics, where interactions between large numbers of particles lead to emergent behaviors that cannot be captured by single-particle quantum mechanics. The concepts of Fock space and multi-particle operators enable the study of complex systems, from superconductors to topological insulators.

15.2.3 RELATIVISTIC QUANTUM MECHANICS

QFT naturally incorporates the principles of special relativity, making it essential for understanding high-energy particle interactions. The relativistic nature of QFT resolves inconsistencies that arise when trying to apply non-relativistic QM to high-energy processes, providing a coherent framework for the creation and annihilation of particles.

15.3 UNRESOLVED QUESTIONS AND FUTURE DIRECTIONS

While QFT has revolutionized our understanding of the quantum world, several gaps and unresolved questions remain:

15.3.1 QUANTUM GRAVITY

One of the most significant challenges is the unification of QFT with general relativity. Quantum gravity aims to describe the gravitational field in the language of quantum mechanics, but a complete and consistent theory remains elusive. String theory and loop quantum gravity are promising approaches, but a definitive solution has yet to be found.

15.3.2 THE HIERARCHY PROBLEM AND BEYOND THE STANDARD MODEL

The Standard Model of particle physics, formulated using QFT, successfully explains a wide range of phenomena but leaves several questions unanswered. The hierarchy problem, the nature of dark matter and dark energy, and the integration of gravity into the quantum framework are active areas of research. Extensions of the Standard Model, such as supersymmetry and extra dimensions, are being explored to address these issues.

15.3.3 NON-PERTURBATIVE METHODS

Many problems in QFT cannot be solved using perturbation theory, necessitating the development of non-perturbative methods. Techniques such as lattice QFT, the renormalization group, and holographic dualities are crucial for tackling these challenges, particularly in strongly coupled systems.

15.3.4 QUANTUM INFORMATION AND COMPUTATION

The intersection of QFT with quantum information science opens new avenues for understanding and utilizing quantum systems. Concepts like entanglement, quantum error correction, and topological quantum computation have profound implications for both fundamental physics and practical technologies.

15.4 CONCLUSION

Quantum Field Theory has profoundly reshaped our understanding of the universe, bridging the gap between quantum mechanics and relativity, and providing a robust framework for describing the behavior of fundamental particles and fields. QFT gives us a new way to understand energy, fields, and their quantization that differs greatly from the 1900s view of a particle and wave duality. QFT is not that view at all, rather, it shows us an encompassing field that manifests as localized events or quanta upon various interactions that can be seen through the S-matrix, Feynman diagrams, and Path Integrals. As we continue to explore and refine this theory, we uncover deeper layers of reality and develop new technologies that harness the principles of the quantum world.

The pedagogical journey through QFT equips students and researchers with the tools to tackle complex problems in modern physics. Through this understanding a deeper appreciation for the intricate and beautiful structure of the universe can be developed if not understood someday. While significant challenges remain, the material covered within this lecture serious, this First Course in Quantum Field Theory, should give a basis from which to start expanding into more advanced models and physics.

15.5 FINAL EXAM

TRIVIA QUESTIONS

Answer the following multiple-choice questions. Each question has only one correct answer.
Question 1: What does the creation operator $\hat{a}^\dagger$ do in second quantization?

(a) Removes a particle from a state
(b) Adds a particle to a state
(c) Measures the number of particles in a state
(d) Commutes with the annihilation operator

Answer: (b) Adds a particle to a state
Question 2: Which of the following represents the Fock space?

(a) $\mathscr{H} = \bigoplus_{n=0}^{\infty} \mathscr{H}^{(n)}$
(b) $\mathscr{H} = \bigotimes_{n=0}^{\infty} \mathscr{H}^{(n)}$
(c) $\mathscr{H} = \sum_{n=0}^{\infty} \mathscr{H}^{(n)}$
(d) $\mathscr{H} = \int_0^{\infty} \mathscr{H}^{(n)}\, dn$

Answer: (a) $\mathscr{H} = \bigoplus_{n=0}^{\infty} \mathscr{H}^{(n)}$
Question 3: In the tight binding model, what does the hopping parameter t represent?

(a) The on-site energy of electrons
(b) The interaction energy between electrons
(c) The amplitude for an electron to hop between neighboring sites
(d) The energy gap in the electronic band structure

Answer: (c) The amplitude for an electron to hop between neighboring sites
Question 4: What is the significance of the term U in the Hubbard model?

(a) It represents the kinetic energy of electrons
(b) It represents the on-site Coulomb repulsion energy
(c) It describes the hopping of electrons between sites
(d) It is the total energy of the system

Answer: (b) It represents the on-site Coulomb repulsion energy
Question 5: In second quantization, the field operator $\hat{\psi}(x)$ can be expressed as:

(a) $\hat{\psi}(x) = \sum_i \phi_i(x)\hat{a}_i$
(b) $\hat{\psi}(x) = \prod_i \phi_i(x)\hat{a}_i$
(c) $\hat{\psi}(x) = \int \phi_i(x)\hat{a}_i dx$
(d) $\hat{\psi}(x) = \frac{1}{N}\sum_i \phi_i(x)\hat{a}_i$

Answer: (a) $\hat{\psi}(x) = \sum_i \phi_i(x)\hat{a}_i$
Question 6: What does the occupation number representation describe in Fock space?

(a) The spatial position of particles
(b) The number of particles in each mode
(c) The energy of each particle
(d) The spin state of particles

Answer: (b) The number of particles in each mode
Question 7: Which equation represents the time-independent Schrödinger equation?

(a) $i\hbar\frac{\partial\psi(\vec{r},t)}{\partial t} = \hat{H}\psi(\vec{r},t)$
(b) $\hat{H}\phi(\vec{r}) = E\phi(\vec{r})$
(c) $\hat{p} = -i\hbar\nabla$
(d) $\hat{a}_k = \frac{1}{\sqrt{N}}\sum_n \hat{c}_n e^{-ikna}$

Answer: (b) $\hat{H}\phi(\vec{r}) = E\phi(\vec{r})$
Question 8: In the context of the Hubbard model, what physical phenomenon does the parameter t help to describe?

(a) Superconductivity
(b) Ferromagnetism
(c) Electron hopping between sites
(d) Dielectric constant

Answer: (c) Electron hopping between sites
Question 9: What is the direct sum notation $\oplus$ used for in Fock space?

(a) Combining different energy levels
(b) Summing over all possible spin states
(c) Representing the total Fock space as a sum of subspaces
(d) Describing the Hamiltonian of a many-body system

Answer: (c) Representing the total Fock space as a sum of subspaces
Question 10: What is the purpose of using creation and annihilation operators in second quantization?

(a) To describe fixed particle numbers in a system
(b) To allow the number of particles to vary in a quantum system

(c) To measure the energy levels of particles
(d) To calculate the momentum of particles

Answer: (b) To allow the number of particles to vary in a quantum system

15.6 PROBLEMS

Answer the following problems. Show all your work for full credit.

15.6.1 PROBLEM 1: CREATION AND ANNIHILATION OPERATORS

Consider a single mode of a quantum harmonic oscillator. The creation ($\hat{a}^\dagger$) and annihilation ($\hat{a}$) operators satisfy the commutation relation:

$$[\hat{a}, \hat{a}^\dagger] = 1 \tag{15.1}$$

Given the ground state $|0\rangle$ such that $\hat{a}|0\rangle = 0$, find the normalized first excited state $|1\rangle$.

Solution:

The first excited state $|1\rangle$ is obtained by applying the creation operator $\hat{a}^\dagger$ to the ground state $|0\rangle$:

$$|1\rangle = \hat{a}^\dagger |0\rangle \tag{15.2}$$

To normalize $|1\rangle$, we calculate its norm:

$$\langle 1|1\rangle = \langle 0|\hat{a}\hat{a}^\dagger|0\rangle \tag{15.3}$$

Using the commutation relation:

$$\hat{a}\hat{a}^\dagger = 1 + \hat{a}^\dagger\hat{a} \tag{15.4}$$

Therefore,

$$\langle 0|\hat{a}\hat{a}^\dagger|0\rangle = \langle 0|1 + \hat{a}^\dagger\hat{a}|0\rangle = \langle 0|0\rangle + \langle 0|\hat{a}^\dagger\hat{a}|0\rangle = 1 \tag{15.5}$$

Hence, $|1\rangle$ is already normalized:

$$|1\rangle = \hat{a}^\dagger |0\rangle \tag{15.6}$$

15.6.2 PROBLEM 2: FOCK SPACE AND DIRECT SUM

Define the Fock space as the direct sum of n-particle Hilbert spaces. Given $\mathscr{H}^{(0)}$ as the vacuum state, $\mathscr{H}^{(1)}$ as the single-particle state, and $\mathscr{H}^{(2)}$ as the two-particle state, write a state in the Fock space combining these subspaces.

Solution:

The Fock space $\mathscr{F}$ is the direct sum of n-particle Hilbert spaces:

$$\mathscr{F} = \bigoplus_{n=0}^{\infty} \mathscr{H}^{(n)} \tag{15.7}$$

Let's construct a state combining the vacuum state $|0\rangle$, a single-particle state $|1\rangle$, and a two-particle state $|2\rangle$:

$$\psi_0 = |0\rangle \in \mathscr{H}^{(0)} \tag{15.8}$$

$$\psi_1 = \alpha|1_1\rangle + \beta|1_2\rangle \in \mathscr{H}^{(1)} \tag{15.9}$$

$$\psi_2 = \gamma|2_1\rangle + \delta|2_2\rangle \in \mathscr{H}^{(2)} \tag{15.10}$$

The combined state in Fock space is:

$$\Psi = \psi_0 \oplus \psi_1 \oplus \psi_2 = |0\rangle \oplus (\alpha|1_1\rangle + \beta|1_2\rangle) \oplus (\gamma|2_1\rangle + \delta|2_2\rangle) \tag{15.11}$$

15.6.3 PROBLEM 3: FEYNMAN DIAGRAM

Draw the Feynman diagram for the process of electron-positron annihilation into a photon pair.

Solution:

The Feynman diagram for electron-positron annihilation into a photon pair involves the following steps:

1. An electron and a positron annihilate.
2. The annihilation produces an intermediate virtual photon.
3. The virtual photon decays into two real photons.

The Feynman diagram is:

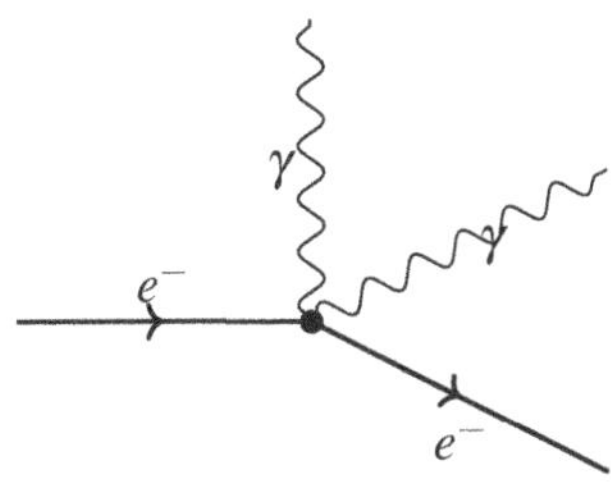

FIGURE 15.1
Feynman diagram for Compton scattering.

15.6.4 PROBLEM 4: TWO-PARTICLE INTERACTION

Consider a two-particle system with a Coulomb interaction potential $V(\mathbf{r}_1, \mathbf{r}_2) = \frac{e^2}{|\mathbf{r}_1 - \mathbf{r}_2|}$. Write the second quantized form of the interaction Hamiltonian.

Solution:

The second quantized form of the two-particle interaction Hamiltonian is:

$$\hat{V} = \frac{1}{2} \sum_{ijkl} \langle ij|V|kl\rangle \hat{a}_i^\dagger \hat{a}_j^\dagger \hat{a}_k \hat{a}_l \tag{15.12}$$

For the Coulomb interaction:

$$V(\mathbf{r}_1, \mathbf{r}_2) = \frac{e^2}{|\mathbf{r}_1 - \mathbf{r}_2|} \tag{15.13}$$

The matrix elements in a plane wave basis are:

$$\langle \mathbf{k}_1 \mathbf{k}_2 |V| \mathbf{k}_3 \mathbf{k}_4 \rangle = \frac{e^2}{V} \delta_{\mathbf{k}_1+\mathbf{k}_2, \mathbf{k}_3+\mathbf{k}_4} \frac{4\pi}{|\mathbf{k}_1 - \mathbf{k}_3|^2} \tag{15.14}$$

Thus, the interaction Hamiltonian in second quantization is:

$$\hat{V} = \frac{1}{2} \sum_{\mathbf{k}_1\mathbf{k}_2\mathbf{k}_3\mathbf{k}_4} \frac{e^2}{V} \delta_{\mathbf{k}_1+\mathbf{k}_2, \mathbf{k}_3+\mathbf{k}_4} \frac{4\pi}{|\mathbf{k}_1 - \mathbf{k}_3|^2} \hat{a}_{\mathbf{k}_1}^\dagger \hat{a}_{\mathbf{k}_2}^\dagger \hat{a}_{\mathbf{k}_3} \hat{a}_{\mathbf{k}_4} \tag{15.15}$$

15.6.5 PROBLEM 5: FEYNMAN DIAGRAM

Draw the Feynman diagram for the process of Compton scattering (electron-photon scattering).

Solution:

The Feynman diagram for Compton scattering involves the following steps:

1. An incoming photon interacts with an electron.
2. The electron absorbs the photon and is scattered.
3. A new photon is emitted in the process.

The Feynman diagram is:

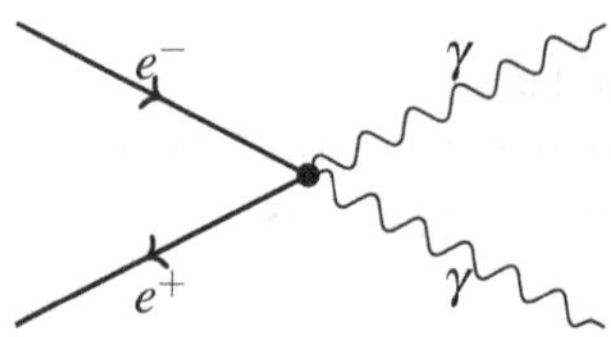

FIGURE 15.2
Feynman diagram for electron-positron annihilation producing two photons.

15.6.6 PROBLEM 6: HUBBARD MODEL

Consider the Hubbard model on a one-dimensional lattice at half-filling with $t = 1$ and $U = 4$. Determine the effective Hamiltonian in the large U limit.

Solution:

The Hubbard Hamiltonian is:

$$\hat{H} = -t\sum_{i,\sigma}\left(\hat{c}^\dagger_{i\sigma}\hat{c}_{i+1\sigma} + \hat{c}^\dagger_{i+1\sigma}\hat{c}_{i\sigma}\right) + U\sum_i \hat{n}_{i\uparrow}\hat{n}_{i\downarrow} \tag{15.16}$$

In the large U limit ($U \gg t$), we treat the hopping term as a perturbation to the interaction term. The effective Hamiltonian describes spin interactions in the half-filled case. It is given by the Heisenberg model:

$$\hat{H}_{\text{eff}} = J\sum_{\langle i,j\rangle} \hat{\mathbf{S}}_i \cdot \hat{\mathbf{S}}_j \tag{15.17}$$

where $J = \frac{4t^2}{U}$ is the exchange interaction.

For $t = 1$ and $U = 4$:

$$J = \frac{4\times 1^2}{4} = 1 \tag{15.18}$$

Thus, the effective Hamiltonian is:

$$\hat{H}_{\text{eff}} = \sum_{\langle i,j\rangle} \hat{\mathbf{S}}_i \cdot \hat{\mathbf{S}}_j \tag{15.19}$$

15.6.7 PROBLEM 7: TIGHT BINDING MODEL

Derive the energy dispersion relation for electrons in a one-dimensional tight binding model with nearest-neighbor hopping.

Solution:

The tight binding Hamiltonian in one dimension is:

$$\hat{H} = -t\sum_n\left(\hat{c}^\dagger_{n+1}\hat{c}_n + \hat{c}^\dagger_n\hat{c}_{n+1}\right) \tag{15.20}$$

We use the Fourier transform of the annihilation operator:

$$\hat{c}_k = \frac{1}{\sqrt{N}}\sum_n \hat{c}_n e^{-ikna} \tag{15.21}$$

Substituting this into the Hamiltonian:

$$\hat{H} = -t\sum_k \left(\frac{1}{\sqrt{N}}\sum_n \hat{c}_k e^{-ik(n+1)a}\right)^\dagger \left(\frac{1}{\sqrt{N}}\sum_m \hat{c}_m e^{-ikma}\right) \tag{15.22}$$

$$= -t\sum_k \hat{c}_k^\dagger \hat{c}_k \left(e^{ika} + e^{-ika}\right) \tag{15.23}$$

$$= -t\sum_k \hat{c}_k^\dagger \hat{c}_k \cdot 2\cos(ka) \tag{15.24}$$

The energy dispersion relation is:

$$E(k) = -2t\cos(ka) \tag{15.25}$$

15.6.8 PROBLEM 8: DIRAC EQUATION

Show that the Dirac equation can be written in the Hamiltonian form.

Solution:

The Dirac equation is given by:

$$(i\gamma^\mu \partial_\mu - m)\psi = 0 \tag{15.26}$$

To write it in Hamiltonian form, we separate the time and spatial derivatives:

$$i\gamma^0 \frac{\partial \psi}{\partial t} = (-i\gamma^i \partial_i + m)\psi \tag{15.27}$$

Multiplying both sides by γ^0:

$$i\frac{\partial \psi}{\partial t} = \gamma^0(-i\gamma^i \partial_i + m)\psi \tag{15.28}$$

$$i\frac{\partial \psi}{\partial t} = (\alpha^i p_i + \beta m)\psi \tag{15.29}$$

where $\alpha^i = \gamma^0\gamma^i$ and $\beta = \gamma^0$.

The Hamiltonian form of the Dirac equation is:

$$i\frac{\partial \psi}{\partial t} = H\psi \tag{15.30}$$

with

$$H = \alpha^i p_i + \beta m \tag{15.31}$$

15.6.9 PROBLEM 9: SCATTERING THEORY

Using the Born approximation, derive the first-order scattering amplitude for a potential $V(\mathbf{r})$.

Solution:

The first-order scattering amplitude $f(\mathbf{k}', \mathbf{k})$ in the Born approximation is given by:

$$f(\mathbf{k}', \mathbf{k}) = -\frac{m}{2\pi\hbar^2}\int e^{-i\mathbf{k}'\cdot\mathbf{r}} V(\mathbf{r}) e^{i\mathbf{k}\cdot\mathbf{r}} d^3r \tag{15.32}$$

Let $\mathbf{q} = \mathbf{k} - \mathbf{k}'$:

$$f(\mathbf{k}', \mathbf{k}) = -\frac{m}{2\pi\hbar^2}\int e^{i\mathbf{q}\cdot\mathbf{r}} V(\mathbf{r}) d^3r \tag{15.33}$$

This is the Fourier transform of the potential $V(\mathbf{r})$:

$$f(\mathbf{k}',\mathbf{k}) = -\frac{m}{2\pi\hbar^2}\tilde{V}(\mathbf{q}) \tag{15.34}$$

where $\tilde{V}(\mathbf{q})$ is the Fourier transform of $V(\mathbf{r})$:

$$\tilde{V}(\mathbf{q}) = \int e^{i\mathbf{q}\cdot\mathbf{r}}V(\mathbf{r})\,d^3r \tag{15.35}$$

15.6.10 PROBLEM 10: QUANTUM FIELD THEORY

Explain the process of renormalization in quantum field theory and why it is necessary.

Solution:

Renormalization is the process of removing infinities that arise in the calculations of quantum field theory by redefining the parameters of the theory (e.g., masses, coupling constants).

In quantum field theory, calculations often involve integrals over all possible energy scales, which can lead to divergent (infinite) results. Renormalization addresses this problem by:

1. Regularization: Introducing a cutoff to the integrals, either by limiting the energy range or by using a mathematical technique to make the integrals finite. 2. Renormalization: Absorbing the infinities into redefined (renormalized) parameters of the theory, which are the physically measurable quantities.

The steps involved in renormalization are:

1. Identify the divergent integrals in the perturbative expansions. 2. Introduce a regularization scheme to control the divergences. 3. Redefine the parameters of the theory to absorb the divergences. 4. Remove the regularization scheme, leaving a finite result.

Renormalization is necessary to make the predictions of quantum field theory finite and physically meaningful. It allows us to relate the parameters of the theory to experimental measurements and to make accurate predictions about the behavior of particles and fields.

Suggested Further Reading

QUANTUM MECHANICS

- D. J. Griffiths and D. F. Schroeter, *Introduction to Quantum Mechanics*, 3rd ed., Cambridge University Press (2018).
  A widely used undergraduate introduction with clear explanations and worked examples.
- J. J. Sakurai and J. Napolitano, *Modern Quantum Mechanics*, Cambridge University Press.
  The most commonly adopted graduate-level text, known for formalism and symmetry-based foundations.
- R. Shankar, *Principles of Quantum Mechanics*, 2nd ed., Springer.
  More rigorous and thorough than Griffiths; a popular bridge between undergrad and grad-level depth.
- M. Beck, *Quantum Mechanics: Theory and Experiment*, Oxford University Press.
  Modern and experimental-focused treatment, ideal for connecting with quantum optics.
- P. A. M. Dirac, *The Principles of Quantum Mechanics*, Oxford University Press.
  Foundational and rigorous, introducing Dirac notation and the algebraic structure of quantum theory.
- D. Bohm, *Quantum Theory*, Dover Publications.
  A rich conceptual treatment with early discussion of hidden variables; historically significant.

QUANTUM FIELD THEORY

- R. Klauber, *Student Friendly Quantum Field Theory*, Sandtrove Press (2013).
  A conceptual and approachable QFT text that explains whats going on under the hood.
- T. Lancaster and S. J. Blundell, *Quantum Field Theory for the Gifted Amateur*, Oxford University Press (2014).
  An intuitive and well-illustrated QFT introduction ideal for students bridging from quantum mechanics.
- M. E. Peskin and D. V. Schroeder, *An Introduction to Quantum Field Theory*, Westview Press.
- M. Srednicki, *Quantum Field Theory*, Cambridge University Press.
- A. Zee, *Quantum Field Theory in a Nutshell*, Princeton University Press.
- S. Weinberg, *The Quantum Theory of Fields*, Vol. I, Cambridge University Press.

QUANTUM OPTICS AND FIELD QUANTIZATION

- Z. Ou, *Quantum Optics for Experimentalists*, Oxford University Press (2017).
  Covers quantum states, field quantization, and entanglement with optical systems.
- M. Fox, *Quantum Optics: An Introduction*, 2nd ed., Oxford University Press (2006).
  A concise and accessible introduction to quantum optics, part of the Oxford Master Series.

QUANTUM ENTANGLEMENT AND APPLICATIONS

- F. J. Duarte and T. S. Taylor, *Quantum Entanglement Engineering and Applications*, CRC Press (2023).
  Engineering-oriented treatment of entanglement systems, protocols, and experiments.
- F. J. Duarte, *Quantum Optics for Engineers: Quantum Entanglement*, 2nd ed., CRC Press (2024). ISBN: 9781032499345.
  A compact but deep treatment of quantum optics and entanglement, tailored to engineers and applied physicists.

QUANTUM COMPUTING AND INFORMATION

- M. A. Nielsen and I. L. Chuang, *Quantum Computation and Quantum Information*, Cambridge University Press (2000).
  The foundational text on quantum algorithms, qubits, and quantum information theory.

MATHEMATICAL METHODS AND TOOLS

- R. P. Feynman and A. R. Hibbs, *Quantum Mechanics and Path Integrals*, McGraw-Hill.
- G. B. Arfken and H. J. Weber, *Mathematical Methods for Physicists*, Elsevier Academic Press.

Index